Genetic Engineering for Crop Improvement

Genetic Engineering for Crop Improvement

Editor: Lawrence Davenport

R CALLISTO REFERENCE

www.callistoreference.com

Callisto Reference,
118-35 Queens Blvd., Suite 400,
Forest Hills, NY 11375, USA

Visit us on the World Wide Web at:
www.callistoreference.com

ISBN: 978-1-64116-196-1 (Hardback)

Cataloging-in-Publication Data

Genetic engineering for crop improvement / edited by Lawrence Davenport.
    p. cm.
Includes bibliographical references and index.
ISBN 978-1-64116-196-1
1. Crop improvement. 2. Genetic engineering. 3. Agricultural biotechnology.
4. Plant biotechnology. I. Davenport, Lawrence.
SB106.I47 G46 2019
630--dc23

Table of Contents

Preface

Every book is initially just a concept; it takes months of research and hard work to give it the final shape in which the readers receive it. In its early stages, this book also went through rigorous reviewing. The notable contributions made by experts from across the globe were first molded into patterned chapters and then arranged in a sensibly sequential manner to bring out the best results.

Genetic engineering is the science that is concerned with the direct manipulation of an organism's genes using the techniques of biotechnology. It has vast applications in the field of agriculture. The development of genetically modified crops for the production of genetically modified food is a modern example of crop engineering. Such utilization of genetic engineering tools to augment agricultural productivity is in the domain of agricultural biotechnology. It involves the use of molecular diagnostics and markers, vaccines and tissue culture to develop modifications in crops. Other techniques of crop modification are induced mutagenesis and polyploidy, crossbreeding, genome editing, etc. The traits in crops that are of economic interest are herbicide tolerance, disease resistance, insect resistance and temperature tolerance. This book unravels the recent studies in the field of crop improvement. Also included in this book is a detailed explanation of the various concepts and techniques of genetic engineering for crop production. Students, researchers, experts and all associated with this field will benefit alike from this book.

It has been my immense pleasure to be a part of this project and to contribute my years of learning in such a meaningful form. I would like to take this opportunity to thank all the people who have been associated with the completion of this book at any step.

Editor

Genome-wide dissection of AP2/ERF and HSP90 gene families in five legumes and expression profiles in chickpea and pigeonpea

Gaurav Agarwal[1], Vanika Garg[1], Himabindu Kudapa[1], Dadakhalandar Doddamani[1], Lekha T. Pazhamala[1], Aamir W. Khan[1], Mahendar Thudi[1], Suk-Ha Lee[2,3] and Rajeev K. Varshney[1,4,*]

[1]*International Crops Research Institute for the Semi-Arid Tropics (ICRISAT), Hyderabad, India*

[2]*Department of Plant Science, Research Institute for Agriculture and Life Sciences, Seoul National University, Seoul, Korea*

[3]*Plant Genomics and Breeding Institute, Seoul National University, Seoul, Korea*

[4]*School of Plant Biology, Institute of Agriculture, The University of Western Australia, Crawley, WA, Australia*

*Correspondence

email r.k.varshney@cgiar.org

Keywords: AP2/ERF, HSP90, chickpea, pigeonpea.

Summary

APETALA2/ethylene response factor (AP2/ERF) and heat-shock protein 90 (HSP90) are two significant classes of transcription factor and molecular chaperone proteins which are known to be implicated under abiotic and biotic stresses. Comprehensive survey identified a total of 147 AP2/ERF genes in chickpea, 176 in pigeonpea, 131 in Medicago, 179 in common bean and 140 in Lotus, whereas the number of HSP90 genes ranged from 5 to 7 in five legumes. Sequence alignment and phylogenetic analyses distinguished AP2, ERF, DREB, RAV and soloist proteins, while HSP90 proteins segregated on the basis of their cellular localization. Deeper insights into the gene structure allowed ERF proteins to be classified into AP2s based on DNA-binding domains, intron arrangements and phylogenetic grouping. RNA-seq and quantitative real-time PCR (qRT-PCR) analyses in heat-stressed chickpea as well as Fusarium wilt (FW)- and sterility mosaic disease (SMD)-stressed pigeonpea provided insights into the modus operandi of AP2/ERF and HSP90 genes. This study identified potential candidate genes in response to heat stress in chickpea while for FW and SMD stresses in pigeonpea. For instance, two DREB genes (Ca_02170 and Ca_16631) and three HSP90 genes (Ca_23016, Ca_09743 and Ca_25602) in chickpea can be targeted as potential candidate genes. Similarly, in pigeonpea, a HSP90 gene, C.cajan_27949, was highly responsive to SMD in the resistant genotype ICPL 20096, can be recommended for further functional validation. Also, two DREB genes, C.cajan_41905 and C.cajan_41951, were identified as leads for further investigation in response to FW stress in pigeonpea.

Introduction

Tropical food legumes like chickpea (*Cicer arietinum*), pigeonpea (*Cajanus cajan*) and common bean (*Phaseolus vulgaris*) play an important role in reducing poverty, improving human health and nutrition, besides leading to ecosystem resilience. Globally 71.7 million tons of pulses (chickpea, pigeonpea and beans) were produced during 2013 and consumed in various forms (http://www.fao.org/docrep/019/i3751e/i3751e.pdf). Temperate legume species like Medicago (*Medicago truncatula*) and Lotus (*Lotus japonicus*) are considered as model legumes for genomics and physiological studies. Moreover, their syntenic relationship with other related legume crops could be helpful in better understanding the gene families relevant to both biotic and abiotic stress tolerances (Young and Udvardi, 2009). Legume crops of economic importance such as chickpea, common bean and pigeonpea have not witnessed expected increase in production and productivity in recent past (Varshney *et al.*, 2010). Abiotic stresses such as drought, heat, cold, high salinity and biotic stresses such as Fusarium wilt (FW), Ascochyta blight and sterility mosaic disease (SMD) have been reported to reduce the average yield drastically in these crops.

In addition to conventional breeding strategies, several omics technologies are being deployed for improvement of these crops. However, mining and characterization of stress-responsive genes will facilitate their use in crop improvement programmes. Previously, characterization of various stress-responsive genes or gene families has mostly been limited to model crops such as Arabidopsis (*Arabidopsis thaliana*), rice (*Oryza sativa*) and maize (*Zea mays*). However, in the case of chickpea, comprehensive resources for gene discovery were developed in response to drought and salinity stresses (Varshney *et al.*, 2009). In addition, comprehensive transcriptome assemblies were also developed (Hiremath *et al.*, 2011; Kudapa *et al.*, 2014). Similarly, in the case of pigeonpea, candidate genes associated with FW and SMD were mined (Raju *et al.*, 2010) and transcriptome assembly was developed using Sanger sequencing as well as next-generation sequencing (NGS) technologies (Kudapa *et al.*, 2012).

Among stress-responsive gene families, APETALA2/ethylene response factor (AP2/ERF) superfamily and heat-shock protein 90 (HSP90) family are important, as they not only regulate responses against various biotic and abiotic stresses in plants, but also play an important role in various developmental processes (Mizoi *et al.*, 2012; Wessler, 2005). Until recently, lack of information on legume genomes restricted the genome-wide survey of genes implicated in biotic and abiotic stresses. In recent years, genome sequences of chickpea (Varshney et al., 2013), pigeonpea (Varshney et al., 2012), common bean (Schmutz et al., 2014), Medicago (Young et al., 2011) and Lotus (Sato et al., 2008) have become available.

Heat stress in chickpea and FW and SMD in pigeonpea are major yield reducers in these crops. AP2/ERF and HSP90 genes are known to be implicated in both biotic and abiotic stresses and AP2/ERF family TFs were found involved both in developmental regulation and stress response in plants. The induction of HSP expression against high temperatures is one of the best-characterized responses. HSP90 chaperones are constitutively expressed in most organisms under normal conditions, while their expression increases significantly under stress. HSP90s play a vital role in plant development, stress response and disease resistance (Lindquist and Jarosz, 2010; Sangster and Queitsch, 2005; Takahashi et al., 2003; Xu et al., 2012). Recent studies have also shown relation between heat stress-induced gene expression and DREB2A gene (Sato et al., 2014).

In view of above, we identified AP2/ERF and HSP90 gene family in chickpea, pigeonpea, common bean, Medicago and Lotus in the present study. We also conducted phylogenetic, syntenic, evolutionary studies, apart from their gene and protein structure analysis. In addition, expression profiling of these genes using RNA-seq data of heat stress in chickpea and FW and SMD stress in pigeonpea has also been performed. Furthermore, quantitative real-time PCR (qRT-PCR) validation of selected AP2/ERF and all HSP90 genes in different tissues of contrasting genotypes for heat stress in chickpea and pathogen stress in pigeonpea was examined to confirm the expression patterns of the selected genes.

Results and discussion

Identification of AP2/ERF transcription factor superfamily genes

To identify AP2/ERF transcription factor (TF) superfamily genes, BLASTP and HMM searches were performed against reference

genomes of chickpea, pigeonpea, Medicago, common bean and Lotus. To identify the members of AP2/ERF subfamilies, the sequences were checked for the presence of AP2 and B3 domains. Sequences with single AP2 domain were classified as ERFs and the ones with two AP2 domains were categorized as AP2, while sequences sharing AP2 and B3 domains were classified under RAV (related to ABI3/VP1) subfamily. Sequences having low homology with ERF members were termed as 'soloist'. ERF proteins were further subclassified into ERF and DREB proteins based on the variation in the amino acid sequences. ERFs were most conspicuously distributed followed by DREBs and AP2s in genomes of legumes (Xu et al., 2011). Different AP2/ERF family members which include AP2, ERF, DREB and soloist were identified, and their chromosomal distribution in five legume crops was determined. As a result, a total of 147, 176, 131, 179 and 140 AP2/ERF family members were identified in chickpea, pigeonpea, Medicago, common bean and Lotus, respectively (Figure 1a; Table 1). The chromosomal distribution of this AP2/ERF family of TFs revealed their localization on the pseudo-molecules and scaffolds (Figures 1b and S1–S5). In a separate study, 16 AP2 and 120 putative ERF TFs were identified in chickpea (Deokar et al., 2015). In that study, the AP2s were identified and characterized strictly based on the presence of two AP2 domains whereas in our present study, despite the presence of one AP2 domain, three ERFs in chickpea, two in pigeonpea and one in common bean were clustered with AP2 sequences. Similar observations were also made in the case of Arabidopsis, potato (Solanum tuberosum) and rubber (Hevea brasiliensis), where four, five and seven sequences with single AP2 domain, respectively, were classified as AP2 (Duan et al., 2013). One possible reason could be the presence of larger number of introns compared to other ERFs, which is a peculiar feature of AP2 sequences. In the case of soybean, 98 unigenes with full-length AP2/ERF domains were identified in an earlier study (Zhang et al., 2008). Variations in biochemical attributes like isoelectric point, protein length and molecular weight of the members of same family indicate the presence of putative novel variants (Tables S1–S5) and are in accordance with the findings in foxtail millet (Lata et al., 2014).

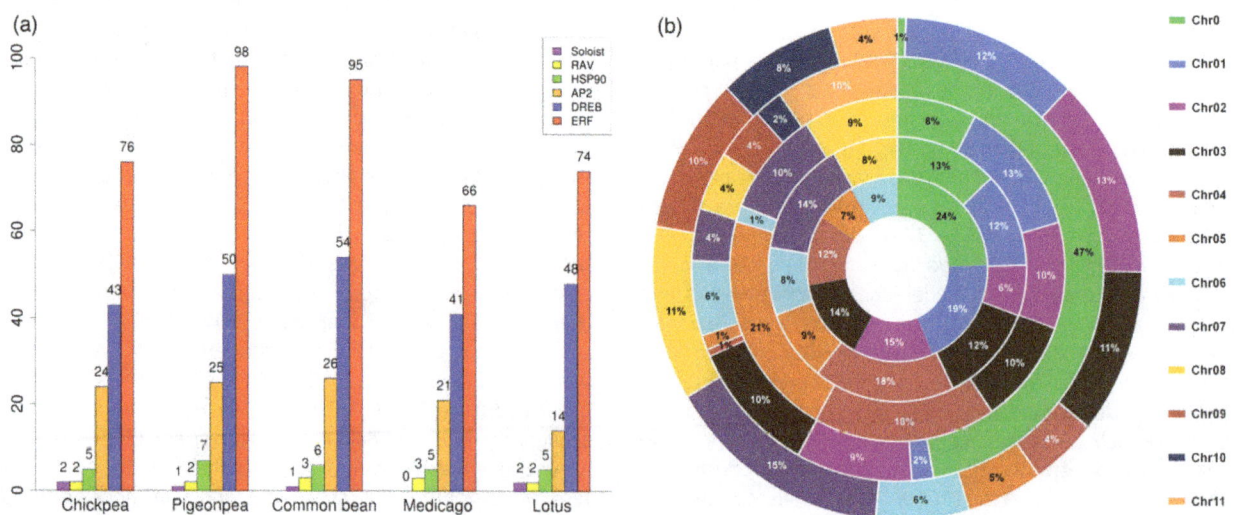

Figure 1 Distribution of genes encoding transcription factors of AP2/ERF family and HSP90 across five legumes. Graphical representation of number of AP2, ERF, DREB, RAV, soloist and HSP90 genes in chickpea, pigeonpea, common bean, Medicago and Lotus (a). Chromosomal distribution and percentage share of AP2/ERF genes in five different legumes. The innermost ring represent chromosomes of Lotus, followed by chickpea, Medicago, pigeonpea and the outer most represents common bean (b).

Table 1 Summary of the structure of AP2/ERF transcription factor superfamily in five legumes

Subfamily	Subgroup	Chickpea	Pigeonpea	Common bean	Medicago	Lotus
DREB	A1	6	5	8	4	7
	A2	5	9	8	7	4
	A3	1	1	1	1	1
	A4	14	18	19	14	18
	A5	10	10	10	11	11
	A6	7	7	8	4	7
	Total	43	50	54	41	48
ERF	B1	12	16	17	17	12
	B2	5	5	4	6	4
	B3	23	39	33	16	26
	B4	14	8	9	6	7
	B5	8	7	8	5	6
	B6	14	23	24	16	19
	Total	119	148	149	107	122
AP2		14	16	16	14	11
AINTEGUMENTA		10	9	10	7	3
RAV		2	2	3	3	2
Soloist		2	1	1	0	2
Total AP2/ERF family genes		147	176	179	131	140
Total genes in genome		28 269	48 680	31 638	45 888	37 971
AP2/ERF transcription factor genes (%)		0.52	0.36	0.57	0.29	0.37
Genome size (Mb)		738	833	521	257.60	472
Average number of AP2/ERF TFs per Mb		0.20	0.21	0.34	0.51	0.30

We observed that the genome sizes and the number of gene family members of AP2/ERF were not directly correlated in these legumes. For instance, even though Medicago and chickpea have large variations in their genome size, did not show much variation in the number of AP2/ERF genes. Similarly, the number of AP2/ERF genes in common bean (521 Mb) did not differ significantly with the number of AP2/ERF genes in pigeonpea (833 Mb), although their genome sizes varied significantly. Nevertheless, in general, the cool season legumes (Lotus and Medicago) possessed low number of AP2/ERF members when compared to warm season legumes (pigeonpea and common bean). In spite of the considerable difference in genome size of respective legumes, little variation in the number of AP2/ERF transcription factors indicated that this family remained conserved during the evolution of legumes. RAV is considered to be one of the most conserved subfamilies among dicot species and is generally known to have six members (Licausi et al., 2010). The number of RAV genes (two to three) identified in this study was similar to what were found in dicots like tomato (Solanum lycopersicum) and potato. However, in Chinese cabbage (Brassica pekinensis L.), as many as 14 RAV genes out of a total of 291 AP2/ERF genes were identified (Song et al., 2013). Additional information on the isoelectric points, molecular weight and variation in the amino acid sequences are provided as Supplementary information.

Identification of HSP90 family genes

To identify HSP90 family in five legume species, protein sequences were scanned for the presence of histidine kinase-like ATPases (HATPase_c) and HSP90 motifs. As a result, five HSP90 genes in chickpea, seven in case of pigeonpea, six in common bean, and

five each in Medicago and Lotus were identified (Figure 1a). The proteins encoded by HSP90 genes ranged from 648 to 818 amino acids in length with isoelectric points ranging from 4.79 to 5.45 (Table S6), suggesting the conserved nature of HSP90 proteins across the five legumes. The number of amino acids in soybean HSP90s ranged from 699 to 847 (Xu et al., 2012). Interestingly, all HSP90 genes in common bean, Medicago and Lotus were found on the pseudomolecules, whereas three of the HSP90 genes identified in each chickpea and pigeonpea were found on pseudomolecules, while two and four HSP90 genes were identified on scaffolds, respectively (Figures S1–S5).

Classification of ERF and DREB members

In the present study, ERF and DREB members of ERF subfamily were distinguished based on the sequence alignment. The sequences with alanine and aspartic acid conserved at 14th and 19th position, respectively, were classified as ERF, while those with valine and glutamic acid conserved at 14th and 19th position were classified under DREB. In addition, the amino acids were also found to be conserved in the tertiary structure of these proteins (Figure 2a,b). The domains with conserved 14V, irrespective of a residue at 19th position were also classified as DREBs because of the importance of 14V over 19E in determining the DNA-binding specificity of DREB transcription factor to the DRE cis-element (Sakuma et al., 2002).

The conserved amino acids V14 and E19 in the ERF/AP2 domains of DREB proteins play a quintessential role in DNA binding and substitution at these amino acids with alanine (A) and aspartic acid (D), hallmark of ERF proteins leads to reduced DNA-binding activity and specificity (Sakuma et al., 2002). Further, 16 conserved amino acids specific to ERF and DREB proteins were also

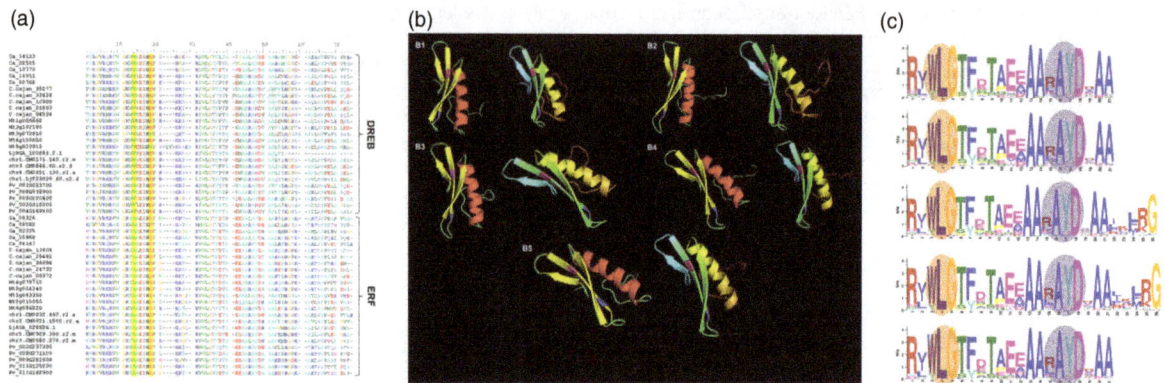

Figure 2 Sequence alignment, prediction of structure and conserved motifs of ERF and DREB proteins across five legumes. (a) Multiple sequence alignment (MSA) representative sequences of amino acid sequences of ERF and DREB subfamily proteins of chickpea, pigeonpea, Medicago, common bean and Lotus using ClustalW. (b) Conserved amino acids in ERF and DREB sequences across the five legumes chickpea (B1), pigeonpea (B2), Medicago (B3), common bean (B4) and Lotus (B5) predicted by I-TASSER. Structures with red alpha helix represent DREB and yellow represents ERF. Pink and blue residues represent the conserved valine and glutamic acid on beta sheets of DREB; alanine and aspartic acid on beta sheets of ERF. (c) Highly conserved WLG and RAYD elements found in motif 1 of AP2/ERF domain across chickpea (C1), pigeonpea (C2), Medicago (C3), common bean (C4) and Lotus (C5).

identified in more than 90% proteins of the five legumes. Earlier studies in *Hevea brasiliensis* (Duan *et al.*, 2013) identified ten such signature amino acids, 14 each in Arabidopsis, cotton (*Gossypium hisutum* L.) and rice (Champion *et al.*, 2009) and were recognized as group markers of the ERF family. The sequence alignment also revealed two additional elements, WLG and RAYD which were conserved among the legumes, studied for most of the AP2/ERF family members (Figure 2c). More details are provided under the section on motif prediction.

Phylogeny of AP2/ERF and HSP90 proteins

Phylogeny of AP2/ERF proteins

Phylogenetic analysis based on conserved domains in DREB, ERF, AP2, and RAV subfamilies, grouped the AP2/ERF proteins of the five legumes into 11–15 groups. In the case of chickpea, AP2/ERF proteins were grouped into 12 major groups (Groups I–XII). Among these groups, Groups I–III comprised of DREB subfamily,

Groups IV–X possessed both ERF and RAV subfamilies and Group XI consists of AP2 subfamily, while two soloists (Ca_11707 and Ca_17230) were placed in Group XII (Figure 3a). The AP2 family was further classified into two groups including ten AINTEGU-MENTA (ANT) and 11 AP2 members. Three ERF sequences with single AP2 domain that clustered with AP2 sequences were considered as AP2s instead of ERFs. Eight members of the Group I were identified with a consensus core sequence ATDS [SD], a representative feature of cytokinin response factor (CRF) proteins (Liu *et al.*, 2013; Table 2). To date, 21 BrCRFs (Liu *et al.*, 2013), 12 AtCRFs and 11 SlCRFs (Shi *et al.*, 2012) have been identified and characterized in detail. In general, the proportion of CRFs in AP2/ERF protein family is expected to be in the range of 5%–10%; for instance, rice and poplar (*Populus trichophora*) were reported to have 6.5% in each (Nakano *et al.*, 2006; Zhuang *et al.*, 2008). In our study, it was found to be 5.36% in chickpea. In case of chickpea, 37.5% CRFs were found to contain a C-terminal SP[T/V]SVL motif, which functions as a putative MAP

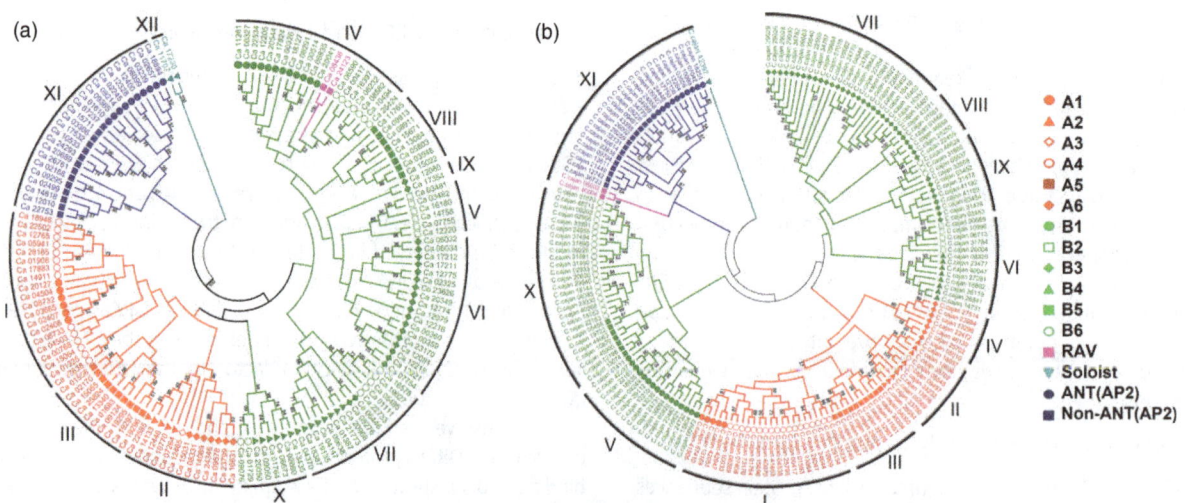

Figure 3 Phylogenetic analysis of AP2/ERF genes in chickpea and pigeonpea. Conserved domains of AP2/ERF genes were used to construct phylogenetic tree for chickpea (a) and pigeonpea (b). Legends on the right represent the respective subfamily members. Bootstrap values greater than 50% support are indicated.

Table 2 Summary of motifs identified specific to each subfamily in each legume crop

| Subfamily | Subgroup | Motif number in legumes | | | | |
		Chickpea	Pigeonpea	Common bean	Medicago	Lotus
DREB	A1	7	8, 17	8, 10	7, 8, 15	7, 11, 15
	A2	14	14	13	–	–
	A3	–	–	–	–	–
	A4	7	8, 17	8, 10	7	7
	A5	7	8, 23	8	7, 23	7, 19
	A6	–	–	–	–	–
ERF	B1	–	–	–	13	12
	B2	–	–	–	–	–
	B3	13	9, 24	9	14, 17, 18, 19	17, 18
	B4	10	–	–	–	–
	B5	21	20	14	–	–
	B6	15	11, 16, 18	9, 4, 12, 15, 18, 20	8, 21	9, 13, 18, 20, 24
CRF		11	20	–	–	13
AP2		6, 8, 9	4, 5, 12, 15	4, 5, 6	4, 5, 9, 10, 11, 16	4, 6, 10, 14
AINTEGUMENTA		12	19	19	12	–
RAV		–	–	17	24	–

–, no motif identified.

kinase and/or casein kinase 1 phosphorylation site speculated to be involved in cytokinin signalling pathway (Xu et al., 2008) along with CRF domain [ATDxSS] motif. In pigeonpea, all six CRFs were marked only by the presence of CRF domain, while the putative MAPK phosphorylation site was not present. In case of Medicago, common bean and Lotus the AP2/ERF sequences were not marked by the presence of either CRF or the putative MAPK phosphorylation domains. However, this doesn't rule out the possibility that ERFs without CRF domain will not respond to cytokinins, as reported in rice, where up-regulation in AP2/ERF expression in response to cytokinins was reported despite the absence of CRF domains (Hirose et al., 2007).

DREBs were grouped as A1–A6 and ERFs were grouped as B1–B6 in all the legumes based on their phylogeny (Figure S6). DREBs (A1–A6) were classified mainly into Groups I to III in the case of chickpea, whereas these were grouped into Groups I to IV in the case of pigeonpea and Medicago and into Groups I to V in the case of common bean and Lotus. However, in case of peanut, two subgroups (A1 and A3) were found to be absent (Wan et al., 2014). A total of 11 major groups were identified (Groups I–XI) in pigeonpea. Groups I–IV consisted of DREBs, Groups V–X contained ERFs, Group XI contained AP2 and RAV, and one soloist remained ungrouped (C.cajan_42397). Two ERFs that clustered with AP2 group were considered as AP2. Eleven ANTs were found to cluster together in AP2 family. Six ERF members of Group I were identified with CRF domain (Figure 3b). Similar studies were also conducted in Medicago (Figure S7), common bean (Figure S8) and Lotus (Figure S9) for which details are provided as Supplementary information.

Phylogeny of HSP90 proteins

HSP90 proteins from all five legumes were analysed *in silico* for their location in cellular milieu using ProtComp v9.0 of Softberry (http://www.softberry.com/berry.phtml). In chickpea and pigeonpea, 3/5 and 4/7 proteins were predicted to be localized in cytoplasm, and the others were either localized on chloroplast (1/5 and 2/7) or endoplasmic reticulum (1/5 and 1/7). However, in Medicago, common bean and Lotus, 3/5, 3/6, and 2/5 proteins,

respectively, were predicted to be cytoplasmically localized and the others on chloroplast, mitochondria and endoplasmic reticulum. None of the identified HSP90 proteins in chickpea and pigeonpea were predicted to be localized in mitochondria (Table S6). Based on subcellular organization, HSP90 proteins were grouped into two major groups (Groups I and II) using neighbour-joining method. As shown in Figure S10a, Group I consisted of cytosolic and Group II consisted of organellar HSP90 proteins.

Gene structure and motif prediction of AP2/ERF and HSP90 families

Gene structure of AP2/ERF

In chickpea, 53/79 ERFs were found to be intronless, and 23 had one intron. However, three ERF genes clustered with AP2 showed higher number of introns (seven to eight), as seen in AP2 sequences. 41/43 DREBs were intronless, and the rest two had one intron. In case of AP2 genes, introns ranged from 6 to 11 and none of them were intronless, while the two soloists contained three and five introns each (Table S1). In pigeonpea, 61/100 ERFs were intronless, 34 had one intron and three had two introns, and the two ERFs clustered with AP2 contained four and eight introns each. AP2 genes contained 5–12 introns, 38/50 DREB genes were intronless, 11 contained one intron each and the remaining one had two introns. Soloist contained only one intron (Table S2).

In Medicago, 31/67 ERF genes were without introns, 30 genes had one, and six genes contained two. None of the AP2 genes were intronless and introns ranged from 5 to 14, 31/41 DREBs were intronless, four had one intron and the other six genes had two introns (Table S3). In common bean, 68/94 ERFs were intronless, 21 ERFs contained one intron each, three had two and two had three introns. Forty-seven of 55 DREB genes were intronless and the other eight had one intron. All 28 AP2 genes contained introns ranging from 6 to 18 (Table S4). The only soloist had 11 introns. Lotus had all 48 DREBs without introns, 58/74 ERFs were without introns, 15 had one intron and one gene was with two introns. All 17 AP2 genes contained two to nine introns (Table S5). The two soloists had one and five introns each. None of the RAV sequences in chickpea, Lotus and common bean contained introns. However, in pigeonpea, one to two introns were present in the identified two RAVs, and in Medicago, two of three RAVs were intronless, the other one contained only one intron.

In general, the ERF and DREB subfamily members outnumbered the AP2 and RAV subfamilies, which have more complex gene structure with two AP2 domains, more number of introns and a RAV-specific B3 domain. The lesser number could be attributed to speculation of early addition of introns, or perhaps the second DNA-binding structure resulting in impaired duplication of ancestral HNH endonuclease during early evolution of this family of genes. Otherwise, the transposition of longer DNA segment might have prevented the duplication, thus resulting in lesser number of AP2s and RAVs (Magnani et al., 2004).

Gene structure of HSP90

In the case of HSP90, the exon–intron boundaries of Group I consisted of lesser introns compared to Group II. Members of Group I had two to four introns, whereas Group II contained genes with 14–19 introns. Similarly, in soybean three such groups with genes having two to three introns in one Group, 14–16 in second Group and ≥18 in the third Group have been reported in an earlier study (Xu et al., 2012). The splicing phases were designated as: phase 0, splicing happened after third nucleotide of the codon; phase 1, splicing after first nucleotide of the codon; and phase 2, splicing after the second nucleotide. The splicing phases were conserved within Group I and Group II members, but showed stark differences among the other two groups (Figure S10b). The exon–intron organization of paralogous pairs of HSP90 present on pseudomolecules was also examined to identify traceable intron loss/gain within these genes. Intron loss/gain within one pair of paralogous genes was observed in chickpea (Ca_17680/Ca_09743). Two pairs of paralogous genes, each in pigeonpea (Cc_15978/Cc_07342, Cc_15978/Cc_05971) and common bean (Pv008G281300/Pv004G107700, Pv008G281400/Pv004G107700), showed intron loss/gain. One paralogous gene pair, each in pigeonpea (Cc_05971/Cc_07342), common bean (Pv008G281300/Pv008G281400) and Medicago (Mt5 g096430/Mt5 g096460), showed conserved exon/intron structures in terms of number of introns. Interestingly, Mt1 g099840 contained an additional C-terminal exon, in comparison with other Group I members. No paralogous genes/ duplications events were seen in Lotus (Table S7).

Motif prediction

A total of 25 motifs were screened for each legume using MEME (default parameters). Among them, two motifs (motif 1 and 2) were seen in almost all AP2/ERF members in chickpea, pigeonpea, Medicago, common bean and Lotus. A total of 146/147, 176/ 176, 120/131, 179/179 and 135/140 AP2/ERF members in chickpea, pigeonpea, Medicago, common bean and Lotus, respectively, contained motif 1 (Figures S11–S15). The other motif was observed in 137/147, 169/176, 177/179, 131/131 and 119/140, genes in respective legumes as specified above. Some WLG elements (motif 1) (Figure 2c) were found to be converted into YLG elements in AP2 subfamily. Further, the conserved RAYD (motif 1) element in AP2/ERF superfamily was converted to RAHD in a very few sequences, contrary to complete conversion to RAHD in two subgroups of DREB in Chinese cabbage (Song et al.,

2013). Motifs 5, 6, 8 and 9 were chickpea AP2 subfamily-specific, shared by 22, 18, 16 and 8 members, respectively. Similarly, motifs 4, 5, 12 and 15 were present in 18, 20, 5, 11 and 22 pigeonpea AP2 sequences; in Medicago, motifs 4 and 5 were shared by 16 and 12 members; common bean shared 4th and 6th motif by 17 and 18 proteins of this subfamily; and in Lotus, motif 4 and 6 were shared by 13 and 6 AP2 proteins. Of 25 motifs predicted, 7th motif was DREB-specific in chickpea and observed in 27 proteins and was also shared by 25 members in Lotus. Motif 8 was shared by 26, 29 and 21 DREB members in pigeonpea, common bean and Medicago, respectively. More elaborative motif distribution in each of the five legumes is listed in Table 2. CRF-specific N-terminus [ATDxSS] domain was found in four legumes except Medicago. However, the TEH motif at the start of N-terminus was missing, which is in accordance with earlier findings (Rashotte and Goertzen, 2010; Zwack et al., 2012). The ERF associated amphiphilic repression (EAR) motif known to repress the transcription (Ohta et al., 2001), like DEAR1, a DREB sequence containing EAR motif mediates crosstalk between signalling pathways for stress responses (Tsutsui et al., 2009). Similar results have been reported in rubber (Duan et al., 2013), tomato (Sharma et al., 2010) and Arabidopsis (Licausi et al., 2010).

The RAV subfamily is known to regulate gene expression in response to ethylene (Alonso et al., 2003), other biotic (Sohn et al., 2006) and abiotic stresses (Li et al., 2011), by binding to a bipartite recognition sequence with the B3 and AP2 recognizing the sequences, CACCTG and CAACA. The motif [YEAHLWD] specific to AP2 subfamily in chickpea, pigeonpea, Medicago, common bean and Lotus are known to form a long linker between the two β-sheets like the linker residues in AINTEGU-MENTA (ANT) protein, expected to be involved in activating the function of TFs in AP2 subfamily. Another motif, [IHEYQAKS] LNFP was found specific to ERF subfamily among the five concerned legumes. The motif is found to be characterized by three blocks of conserved amino acid residues: LPRP, D [IV] QAA/ DIR [RA] specific to ERF and [IHEYQAKS] LNFP specific to DREB. These residues are known to interact with CBL-interacting serine/ threonine proteins kinase-12 (Albrecht et al., 2001) and ethylene-responsive factor, ERF037 (Qu and Zhu, 2006). Xu et al. (2013) also identified similar motifs in castor bean (*Ricinus communis* L.). Details of motif prediction in case of HSP90 genes are provided in Supplementary information. Gene ontology analysis indicated that large number of genes were annotated for biological processes in all five legumes (Figure S16; Tables S18–S22).

Chromosomal distribution, duplication and orthologs of AP2/ERF and HSP90 genes

Of 147 AP2/ERF and five HSP90 genes identified in chickpea, 128 and three were found on eight chickpea pseudomolecules (Ca1– Ca8). Similarly, 93/176, 3/7 in pigeonpea, 122/13, 4/6 in Medicago, 179/179, 6/6 in common bean and 105/140, 5/5 in Lotus, AP2/ERF and HSP90 genes could be found on the pseudomolecules. In chickpea, maximum number (27) of genes with 16 ERFs, eight DREBs and one each of RAV, AP2 and HSP90 were located on Ca4, followed by Ca7 (21 genes) and Ca3 (18 genes). Other two HSP90s were identified on Ca2 and Ca5. Six AP2/ERF genes including three ERFs, two DREBs and one AP2 (Ca_09050, Ca_09076, Ca_09124, Ca_09214, Ca_14911 and Ca_18387) were identified in the '*QTL-hotspot*' region (35.8– 46.7 Mb) on chromosome 4 (Table S1) of chickpea for drought tolerance (Kale et al., 2015). Chromosomal distribution of genes in pigeonpea was maximum on CcLG2 and CcLG3, with 18 genes each including, six ERFs, seven DREBs, two each of AP2s and HSP90s and one RAV on CcLG2 and 11 ERFs, four AP2s and three DREBs on CcLG3, followed by 17 genes on CcLG11. The other HSP90 gene was found on CcLG8. Gene distribution in Medicago was found to be maximum on Mt5 (*Medicago truncatula* chromosome 5) with 31 including, eight ERFs, 16 DREBs, three AP2s, one RAV and three HSP90 genes followed by Mt4 (23 genes) and Mt1 (19 genes including one HSP90 genes). In case of common bean, maximum number of genes (27) were identified on Pv7 (*Phaseolus vulgaris* chromosome 7) with 19 ERFs, five DREBs, two RAVs and one AP2 followed by Pv2 and Pv8 (24 and 23 respectively). The HSP90 genes were found on Pv1, Pv2 and Pv4. Lotus shared the maximum number of genes (27) on Lj1 (*Lotus japonicus* chromosome 1) with 18 ERFs, nine DREBs, four AP2s, and one soloist followed by Lj2 (24 including four HSP90 genes) and Lj3 with 20 genes.

To identify the contribution of segmental and tandem gene duplications in genome-wide expansion of AP2/ERF family in the considered five legumes, genes which were found within the 5-Mb regions with 80% and higher similarity with e-value threshold of 1e-10 were considered as tandemly duplicated genes, and the ones separated by >5 Mb distance were identified as segmentally duplicated genes (Figures S1–S5). We found a total of 13 duplication events (paralogous genes) in chickpea, 18 in pigeonpea, 14 in Medicago, 13 in common bean and 17 in Lotus (Table S7). Of these duplication events, two groups of tandemly duplicated ERF genes in chickpea, one ERF in pigeonpea, three in Medicago (two of ERF and one of HSP90), one group of HSP90 genes in common bean and two in Lotus, and one group each of ERF and DREB genes were identified. It was observed that most of the groups were formed by ERF subfamily. However, in case of Medicago and common bean, one group of HSP90 genes was also identified, apart from just one tandemly duplicated group of DREB genes identified in Lotus among the five legumes (Figures S1–S5). HSP90 genes were segmentally duplicated compared to tandem duplications. It is obvious from these findings that segmental duplications outnumbered the tandem duplications, thus signifying a major role of segmental duplications in expansion of this gene family. Orthologs of chickpea in pigeonpea, Medicago, common bean, and Lotus were found using the best bidirectional BLAST approach with an e-value threshold of 1e-10. We identified, 100, 103, 84 and 80 such orthologs of chickpea AP2/ERF in pigeonpea, Medicago, common bean, and Lotus (Figure 4a–d; Table S8). Similarly, six orthologs of chickpea HSP90 were found in Medicago, pigeonpea and common bean and three in Lotus (Figure 4e; Table S9). All the legumes considered in this study were identified with segmental duplications to play a key role in the expansion of AP2/ERF family. Similar results have been reported in rice (Sharoni et al., 2011), Arabidopsis (Nakano et al., 2006) and *Brassica rapa* ssp. pekinensis (Liu et al., 2013), indicating that mechanisms underlying AP2/ERF family expansion vary from species to species.

Gene expression patterns in chickpea

To gain insights into the expression pattern of AP2/ERF and HSP90 genes in chickpea under heat stress, RNA-seq data generated from leaf, root and flower tissues at vegetative and reproductive stages from three tolerant (ICCV 92944, ICC 1356, ICC 15614) and three sensitive (ICC 5912, ICC 4567, ICC 10685) genotypes was used. Expression patterns were compared

between respective controls and (i) heat-stressed leaf tissue before flowering, (ii) heat-stressed root tissue before flowering, (iii) heat-stressed root tissue after flowering, (iv) heat-stressed leaf tissue after flowering and (v) vegetative (leaf and root) and reproductive (flower) tissues in two heat-tolerant and one heat-sensitive chickpea genotype (Figure 5a–e; Tables S10–S14). Unique set of 58/147 AP2/ERF genes were expressed in different tissues in chickpea with 39, 43, 50, 30 and 37 genes in each of the five comparisons as mentioned above. Hierarchical clustering for each comparison broadly classified them into clusters representing gene expression levels.

Validation using qRT-PCR (Primers sequence provided in Table S15) demonstrated differential, temporal, spatial and genotype-specific expression of genes (Figure 6). In vegetative leaf tissue, Ca_01566 and Ca_14133 were down-regulated among all the six heat-tolerant and heat-sensitive genotypes, whereas Ca_14133 was significantly up-regulated in one of the three sensitive genotypes, and Ca_02170 was up-regulated in one tolerant and two sensitive genotypes (Figure 6a). In vegetative root tissue, Ca_02170 and Ca_16631 were up-regulated in tolerant and almost negligible expression in sensitive genotypes when compared to its control. However, Ca_09578 and Ca_14133 were up-regulated in all tolerant and one sensitive genotype and significantly up-regulated in two sensitive genotypes, respectively, while Ca_22585 and Ca_23799, on the other hand, were down- and up-regulated in sensitive and tolerant genotypes, respectively (Figure 6b). Ca_15031 was found to be significantly up-regulated in one tolerant genotype with almost more than 10-fold expression (Figure S6c). In case of reproductive leaf tissue, Ca_02170 was up-regulated in both tolerant and sensitive genotypes except one sensitive genotype where the expression was almost negligible, and Ca_08436 and Ca_23799 were up-regulated in tolerant and down-regulated in sensitive genotypes (Figure 6d). In flower tissues, Ca_02170 was almost unexpressed and Ca_00673 was up-regulated in tolerant and with almost zero expression in sensitive genotypes, whereas Ca_08436 and Ca_15031 were up-regulated and Ca_22585 was insignificantly down-regulated across all genotypes (Figure 6e). Expression results prompt the identification of probable tissue and stage-specific candidate genes which can counteract the given stress condition. Similar studies in peanut against heat stress resulted in stress tolerant *AhERF019* transgenic Arabidopsis plants (Wan et al., 2014). In soybean, expression analysis and transgenic tobacco plants developed using GmERF057 and GmERF089 revealed enhanced tolerance to salt and drought stress, but not to pathogen stress under GmERF089 overexpression. However, GmERF057 overexpression resulted in enhanced tolerance to salt and pathogen stress (Zhang et al., 2008), conferring different roles of ERFs under different stress conditions. Another study in soybean showed transactivation of DREB2A;2 under drought, heat and low temperature (Mizoi et al., 2013).

HSP90 genes in flower tissues of heat- tolerant and -sensitive chickpea genotypes were observed to be up-regulated in tolerant compared to the sensitive genotypes, except Ca_17680 (Figure 7). Its expression was found to be induced even in the sensitive genotype compared to its control. However, expression of the same HSP90 genes in the vegetative leaf and root tissues were found to be up-regulated, compared to control except for Ca_09743 (Figure 7a,b). It was observed to be up-regulated in root and down-regulated in leaf tissue. In reproductive leaf and root tissues, Ca_25602 and Ca_23016 were found to be consistently almost negligibly expressed or

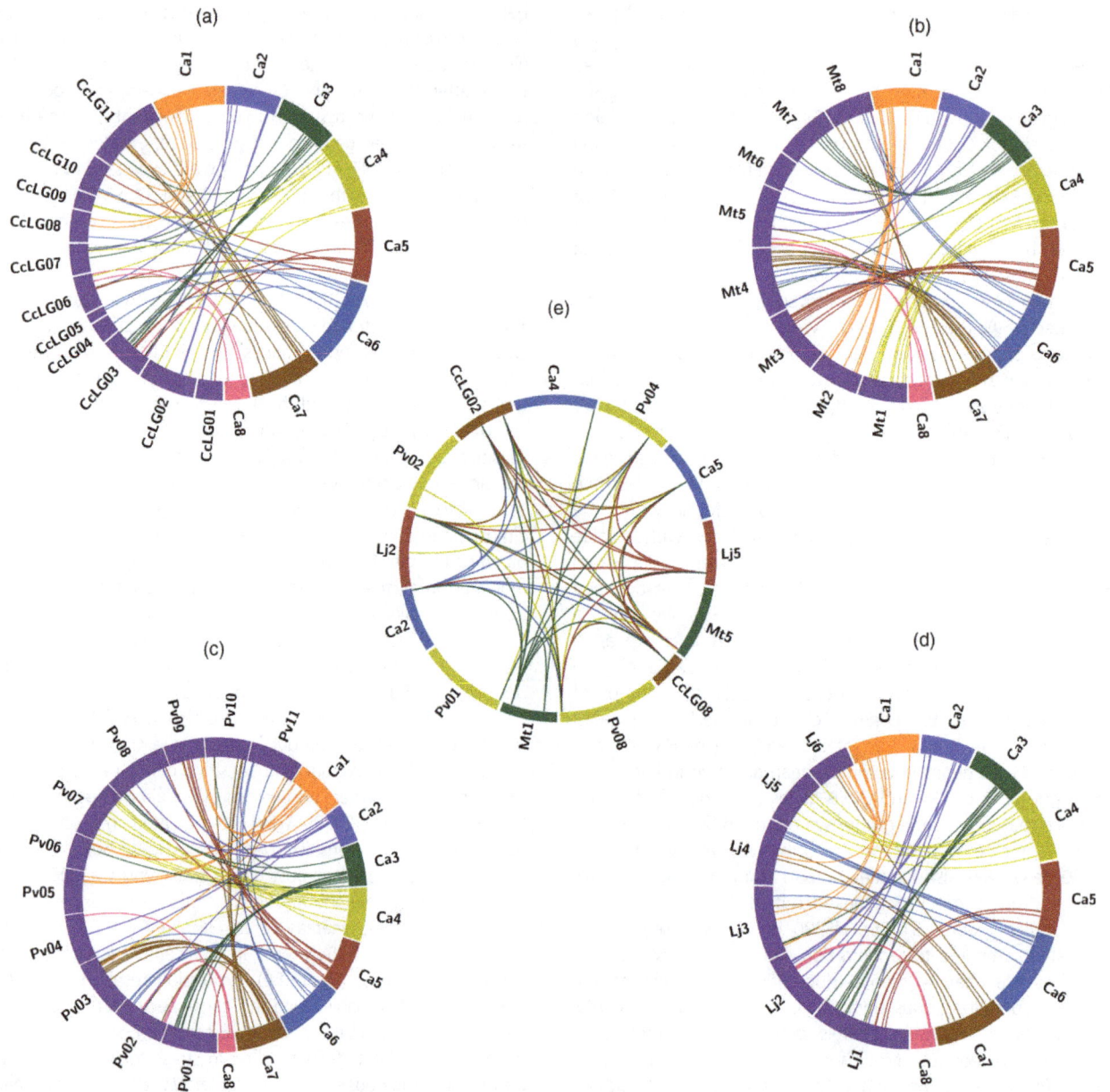

Figure 4 Comparative analysis of orthologous relationship among five legumes for AP2/ERF and HSP90 genes. The Circos plot represents orthology of chickpea genes with (a) pigeonpea, (b) Medicago, (c) common bean and (d) Lotus. Eight pseudomolecules of chickpea are represented with different colour and chromosomes of four species in blue. The strokes originating with the same colour from chickpea pseudomolecules landing on a different species represent an orthology of a given gene between the two species. (e) Orthologous relationships among the five legumes for HSP90 genes.

down-regulated across the tolerant and sensitive genotypes (Figure 7c,d). Ca_25602 was up-regulated in the two tolerant genotypes, whereas Ca_23016 was up-regulated in a sensitive genotype (Figure 7e). Overall, the HSP90 genes were found to be up-regulated across all genotypes and tissues. We observed that there was no single gene which was consistently up- or down-regulated throughout the tissues and genotypes suggesting that HSP90 genes have temporal, spatial and genotype-specific gene expression in chickpea. Ca_25602 in vegetative leaf, Ca_09743 in vegetative root, Ca_17680 in flower, Ca_23016 in reproductive leaf and Ca_17680 in reproductive root were found to be up-regulated in heat-stressed tissues compared to their control.

Gene expression patterns in pigeonpea

A total of 76 AP2/ERF genes were quantified and depicted through heatmaps based on their FPKM values in ten pathogen-stressed pigeonpea genotypes (Figure 5f; Table S16), which included five parental combinations (see materials and methods). Hierarchical clustering formed clusters of genes based on their FPKM values. Among the contrasting parents, ICPL 20096 and ICPL 332 showed the most contrasting expression followed by ICPB 2049 and ICPL 99050.

Among the resistant and susceptible genotypes, genes were mostly found to be up-regulated in susceptible ones (Table S16). Validation using qRT-PCR of 16 ERF, nine DREB

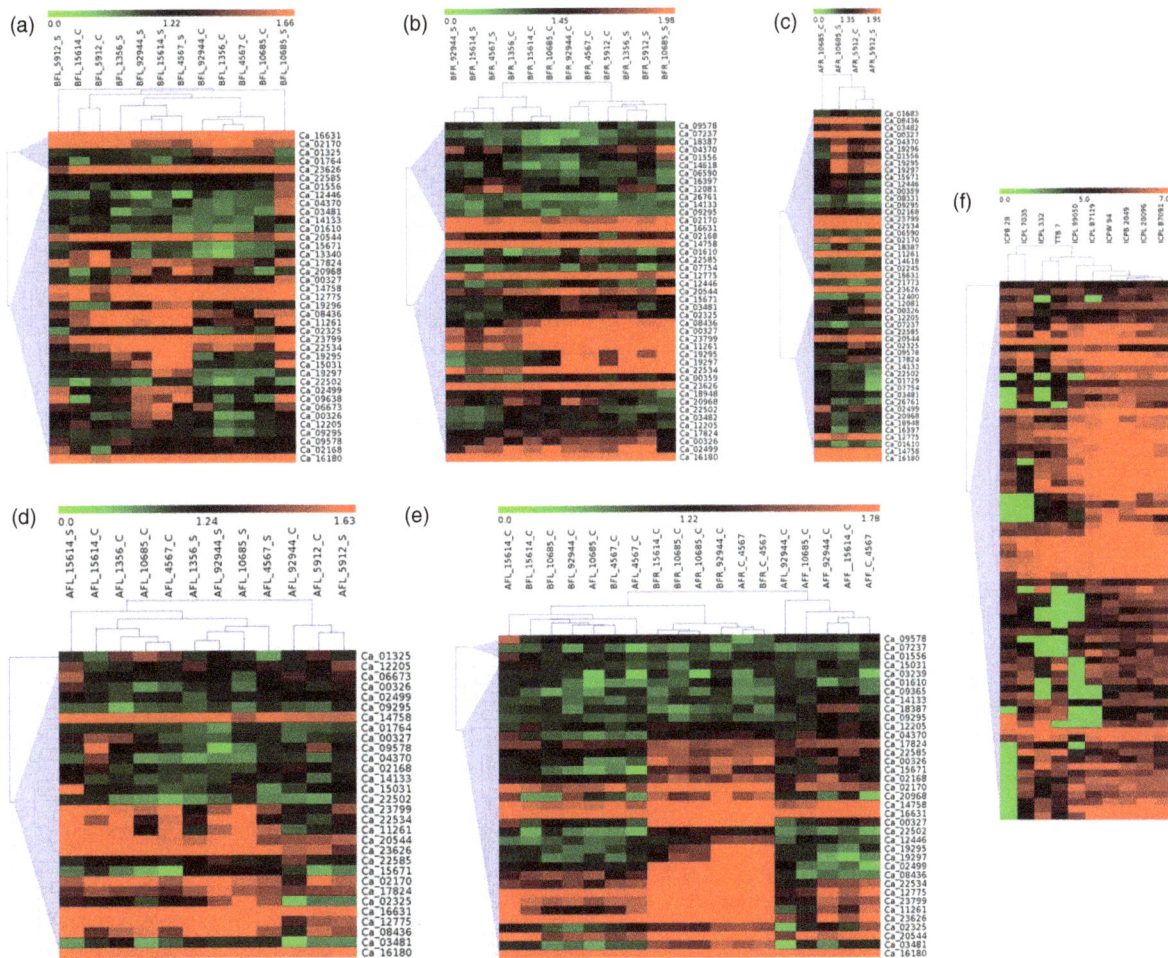

Figure 5 Expression profiles of AP2/ERF and HSP90 genes in chickpea and pigeonpea. Hierarchical clustering of AP2/ERF and HSP90 genes in chickpea (a–e) and pigeonpea (f) using log$_{10}$-transformed FPKM values. BFL, before flowering in leaf; BFR, before flowering in roots; AFL, after flowering in leaf; AFR, after flowering in roots; C, control; S, stressed. (a) Two genes, namely Ca_09638 and Ca_02499, were found to be specifically expressed in the stressed leaf tissues, while repressed in the control. (b) In root tissues, Ca_04370 was found to be specifically expressed under stress. (c) Three genes, Ca_19295, Ca_19296 and Ca_19297, were found to be highly expressed in all the stressed root tissues. (d) Ca_02325 showed specific expression, while Ca_08436 was highly expressed in comparison with control leaf tissues. (e) A gene cluster was identified which is specifically expressed in root tissues unlike genes Ca_16180, Ca_16631 and Ca_14758, which were constitutively expressed in all the tissues under controlled conditions irrespective of the developmental stages. (f) C.cajan_24047 showed high expression in all the Fusarium wilt-resistant genotypes, while repressed in the susceptible ones. Two genes, C.cajan_25793 and C.cajan_24044, were highly expressed in the SMD-resistant genotypes when compared to the susceptible genotypes.

and six HSP90 genes in ICPL 20096 and ICPL 332 was performed (Primers sequence provided in Table S17; Figure 8a–c). FW and SMD stress imposed root and leaf tissues from the two genotypes were analysed for the expression analysis. HSP90s also showed their role in disease resistance in several plant species. High throughput virus-induced gene silencing in plants is implicated through HSP90 in disease resistance, like in case of barley for powdery mildew resistance (Hein et al., 2005; Lu et al., 2003). Similarly, in wheat, cytosolic HSP90 genes are known to be involved in seedling growth and disease resistance (Wang et al., 2011). The TF and chaperone genes in general, were observed to be down-regulated in both the genotypes under biotic stress conditions instead of an expected up-regulation. Five genes (C.cajan_06713, C.cajan_27281, C.cajan_28250, C.cajan_36094 and C.cajan_25702) in particular, showed a profound dip in expression in the FW-stressed leaf tissues of ICPL 22096 genotype. Of these DNA-binding TF genes, the one with the maximum down-regulation (C.ca-

jan_36094) is known to negatively regulate the transcription, and ethylene-mediated signalling is also known to play role in defence response during respiratory burst and induced systemic resistance. Interestingly, the gene was slightly up-regulated in leaf tissues in both the genotypes against viral infection, however, was down-regulated in root tissues against fungal invasion with deep repression in ICPL 20096 and a minor repression in ICPL 332. Meanwhile, C.cajan_27281 was found to be significantly down-regulated in root tissues in the resistant genotype (ICPL 22096) but was found to be slightly up-regulated in the susceptible genotype (ICPL 332). An obvious trend of down-regulation of AP2/ERF and HSP90 genes in the stressed tissues of these genotypes prompt towards a more complex mechanism of resistance and susceptibility against biotic stresses like FW and SMD mediated by this class of TF and chaperones in pigeonpea. Another gene (C.cajan_27949) with 50-fold down-regulation observed in ICPL 22096 is a HSP90 gene which is known to interact with a

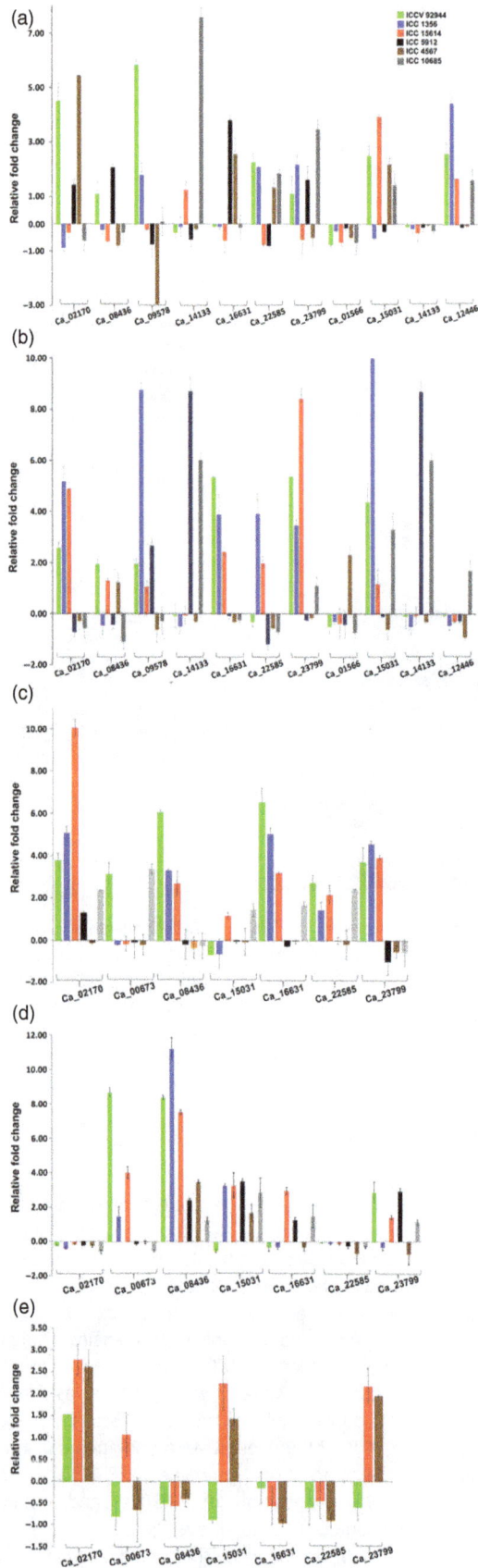

Figure 6 Expression profiling of AP2/ERF genes in chickpea. Quantitative real-time PCR validation of differential expression of genes in vegetative leaf (a), vegetative root (b), reproductive leaf (c), reproductive root (d) and flower (e) tissues in heat- tolerant and -sensitive chickpea varieties. Expression in flower tissues was performed in two tolerant and one sensitive genotype. The other genotypes could not withstand the heat stress till flowering stage.

Experimental procedures

Identification of AP2/ERF and HSP90 proteins from five legume proteomes

Two different approaches were used to mine the AP2/ERF and HSP90 domain containing sequences in chickpea, pigeonpea, common bean, Medicago and Lotus. (i) BLASTP search (e-value 1e-5) using Arabidopsis and rice AP2/ERF and HSP90 sequences. (ii) Hidden Markov model (HMM) scan using AP2 (PF00847) and HSP90 (PF00183) Pfam profiles. A final unique set of the protein sequences identified using above approaches were further scanned for the presence of AP2/ERF, HSP90 and HATPase_c domains. Only the sequences containing these domains were retained.

Classification of AP2/ERF and HSP90 genes

Phylogenetic and evolutionary analyses were carried out by using MEGA6.0 (http://www.megasoftware.net/). Neighbour-joining method with pairwise deletion option was used for construction of phylogenetic trees for all five legumes using domain peptide sequences of AP2/ERF. Reliability of the constructed trees was assessed by using boot strapping with 5000 replicates. The conserved motifs of AP2/ERF and HSP90 were predicted using standalone version of motif based sequence analysis tool (MEME) (version 4.9.0) (Bailey *et al.*, 2009) with default parameters, number of motifs set at 25, optimum width of 10–200 amino acids and any number of repetitions of a motif. Gene structure prediction was made using online server, Gene Structure Display Server based on full-length mRNA alignments with corresponding genomic sequences. Protein structures of AP2/ERF were predicted using I-TASSER (Zhang, 2008) and viewed using PyMOL version 1.5.0.4 (www.pymol.org).

Identification of orthologous AP2/ERF and HSP90 genes, their distribution and duplication

Generic feature format (GFF) files for the genomes of five legumes were used to mark the location of each gene on their physical maps. The distribution of AP2/ERF and HSP90 genes was visualized using MapChart (Voorrips, 2002). Orthologous genes with respect to chickpea were predicted using best bidirectional hit (BBH) approach with e-value threshold of 1e-10. The chickpea AP2/ERF genes were used as query against the database of pigeonpea, Medicago, common bean and Lotus AP2/ERF genes. The predicted orthologous genes were depicted using Circos program (Krzywinski *et al.*, 2009). The AP2/ERF and HSP90 genes in each legume species were searched for duplication events at an e-value of ≤1e-10 and sequence identity of ≥80%.

Gene expression studies

Gene expression patterns of AP2/ERF and HSP90 genes in chickpea were studied using RNA-seq data generated from leaf, root and flower tissues at vegetative and reproductive stages, while in case of pigeonpea RNA-seq data downloaded from

NBS-LRR protein, RPM1 in Arabidopsis and a mis-sense mutation in HSP90 resulted in diminished levels of RPM1 (Hubert *et al.*, 2003).

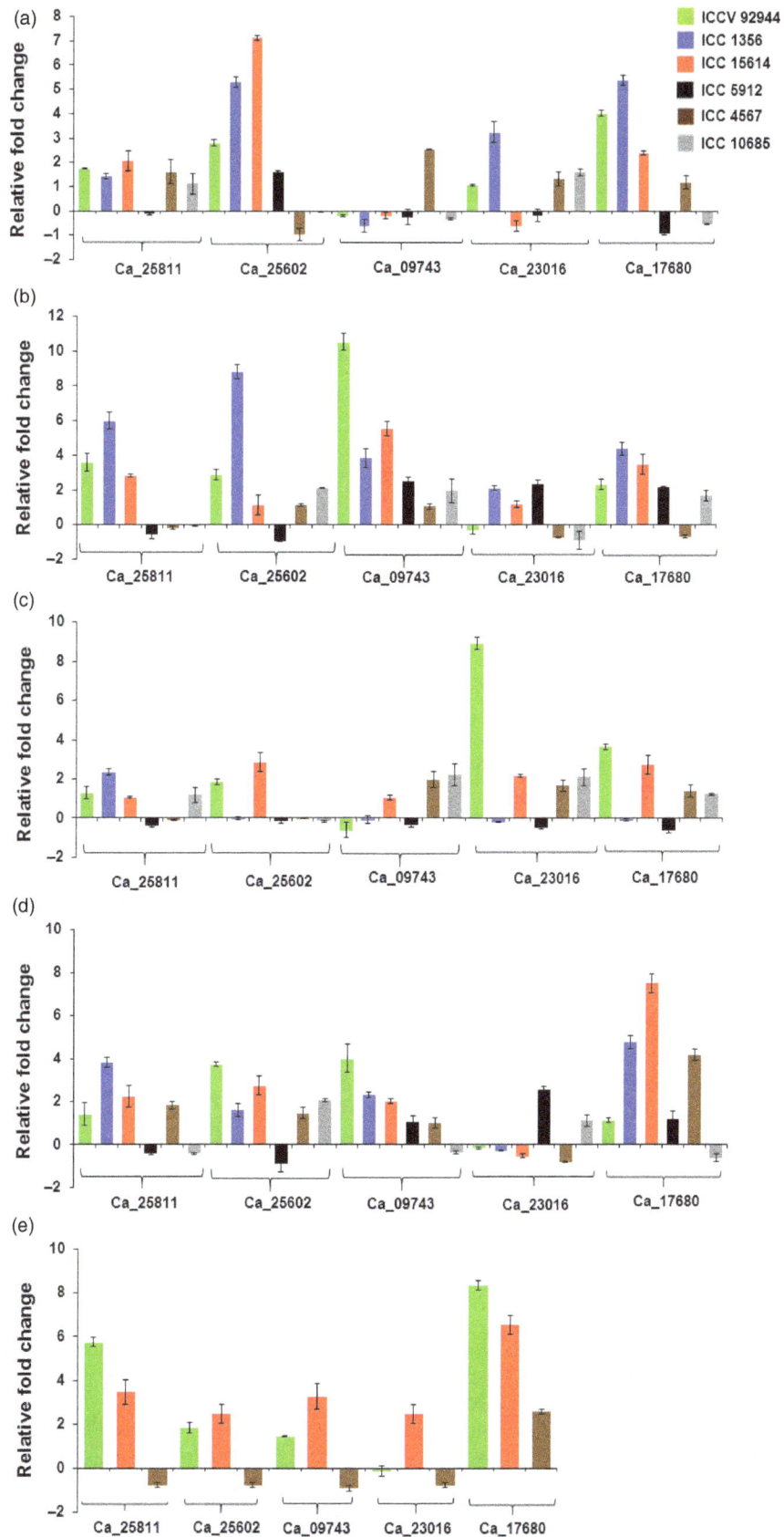

Figure 7 Expression profiling of HSP90 genes in chickpea. Quantitative real-time PCR validation of differential expression of genes in vegetative leaf (a), vegetative root (b), reproductive leaf (c), reproductive root (d) and flower (e) tissues in heat-tolerant and heat-sensitive chickpea varieties. Expression in flower tissues was performed in two tolerant and one sensitive genotype. The other genotypes could not withstand the heat stress till flowering stage.

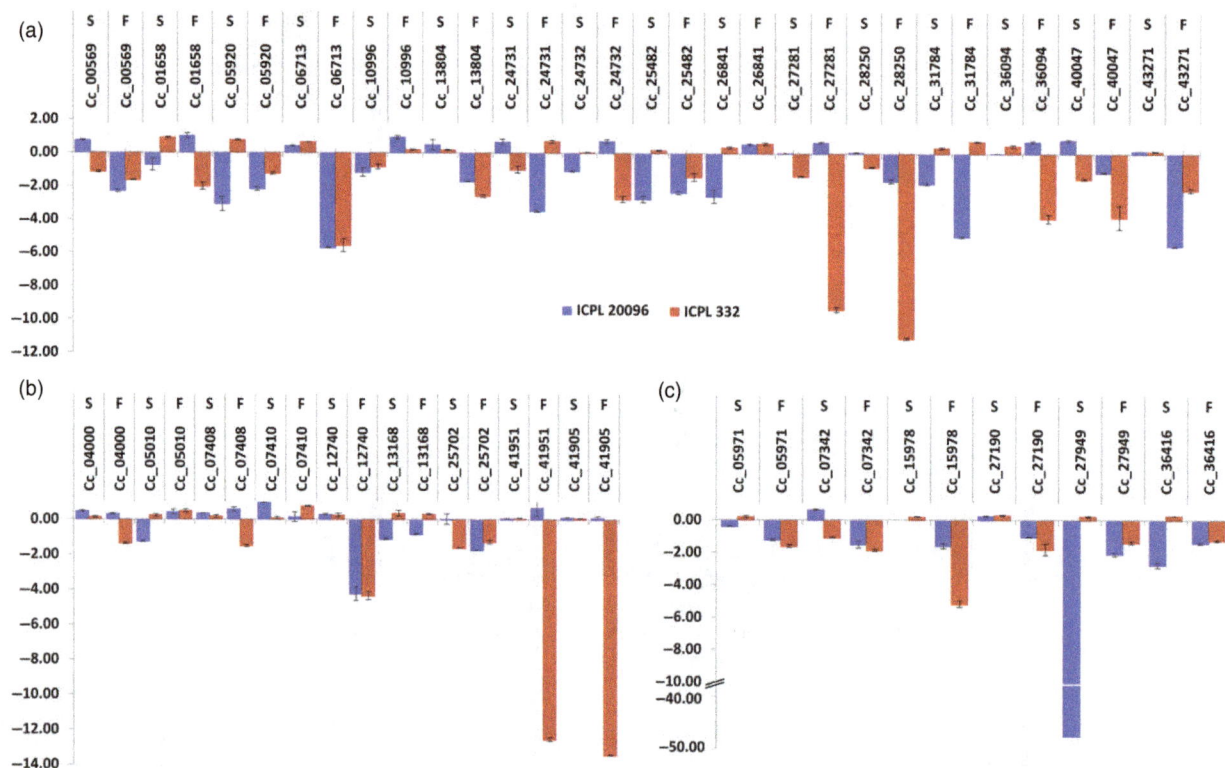

Figure 8 Expression profiling of AP2/ERF and HSP90 genes in pigeonpea. Quantitative real-time PCR validation of differential expression of ERF genes (a), DREB genes (b) and HSP90 genes (c). F and S denote FW and SMD.

sequence read archive (http://www.ncbi.nlm.nig.gov/sra) (SRA030523.1 to SRP005971.1) were used. RNA was isolated in three biological replicates from each sample. The quality filtered reads were mapped to respective chickpea and pigeonpea genomes with spliced read mapper, TopHat (Trapnell et al., 2009). Cufflinks followed by cuffcompare (Trapnell et al., 2010) was used to estimate the abundance of reads mapped to genes by calculating FPKM values (fragments per kilobase of transcript per million). Genes with class code '=' were considered for expression studies. The heatmaps showing expression profiles were generated by using log_{10}-transformed FPKM values by MultiExperiment Viewer (MeV 4.8.1; Saeed et al., 2003).

Validation of gene expression profiles using qRT-PCR

Six chickpea genotypes, three tolerant (ICCV 92944, ICC 1356, ICC 15614) and three sensitive (ICC 5912, ICC 4567, ICC 10685) to heat as mentioned above, were used to validate the expression profiles of select candidate genes. Plants of each genotype were grown in five replications each for vegetative and reproductive stages. Three seeds of each genotype were sown in a pot (2.4 L volume) containing a mixture of black Vertisol soil, sand and vermicompost (4 : 2 : 1 by volume) and the plants were grown at 27/16 °C in a greenhouse for 20 days and then transferred to a growth room to expose them to high temperatures at vegetative stage. The control plants were continued to grow in the glasshouse at 27/16 °C. The temperature in the growth room was increased daily by 1 °C, for example 28–40 °C during the day and 16–25 °C during night. Therefore, the plants were exposed to a gradual increase in temperature for stress imposition. Leaf and root tissues at vegetative stage were harvested 15 days after heat stress imposition. Similarly, for reproductive stage, the plants were

grown in greenhouse conditions until the first appearance of flowers. Then the plants were subjected to heat stress in growth room for 15 days as described above. Leaf, root and flower tissues were harvested at reproductive stage. At least three biological replicates of each tissue sample were harvested and stored at −80 °C until RNA extraction. In case of pigeonpea, two genotypes, one resistant and one susceptible, namely ICPL 20096 and ICPL 332, respectively, were used to validate the putative candidate genes identified in the present study. FW and SMD stresses were imposed on 10-day-old seedlings of ICPL 20096 and ICPL 332 grown separately for each stress. Root dip inoculation (FW) and leaf staple techniques (SMD) were followed for stress imposition under glasshouse conditions, and tissues (roots—FW; and leaves—SMD) were harvested after 7 days of stress. The seedlings were stressed with Fusarium udum Butler (6×10^6 conidia/ml) for FW inoculation (Sharma et al., 2012) and with viruliferous mites (Aceria cajani) for SMD infection, at the two-leaf stage by stapling the primary leaves with SMD-infected pigeonpea leaves containing at least five live mites (Nene and Reddy, 1976). The RNA-seq data for the selected genes were validated through qRT-PCR using Applied Biosystems 7500 Real-Time PCR System with the SYBR green chemistry (Applied Biosystems, Foster City, CA). The gene-specific primers were designed using Primer3 software (http://bioinfo.ut.ee/primer3-0.4.0/primer3/) applying the default parameters with slight modification which includes product size 80–150 bp and primer size 18–25 bp (Tables S21 and S22). The qRT-PCRs were performed using SYBR Green Master Mix in 96-well plates with two technical replicates and three biological replicates using GAPDH (chickpea) and actin (pigeonpea) as endogenous controls. The PCR conditions used are as follows: 2 min at 50 °C, 10 min at 95 °C, and 40 cycles of 15 s at 95 °C and 1 min at 60 °C. The relative transcriptional level in terms of fold change was calculated using the $2^{-\Delta\Delta C_T}$ method (Livak and Schmittgen, 2001).

Acknowledgements

This work has been undertaken as part of the CGIAR Research Program on Grain Legumes. ICRISAT is a member of CGIAR Consortium. Suk-Ha Lee acknowledges a grant from the Next-Generation BioGreen 21 Program (Code No. PJ00811701), Rural Development Administration, Republic of Korea.

References

Albrecht, V., Ritz, O., Linder, S., Harter, K. and Kudla, J. (2001) The NAF domain defines a novel protein-protein interaction module conserved in Ca2+-regulated kinases. *EMBO J.* **20**, 1051–1063.

Alonso, J.M., Stepanova, A.N., Leisse, T.J., Kim, C.J., Chen, H., Shinn, P., Stevenson, D.K. *et al.* (2003) Genome-wide insertional mutagenesis of *Arabidopsis thaliana. Science*, **301**, 653–657.

Bailey, T.L., Boden, M., Buske, F.A., Frith, M., Grant, C.E., Clementi, L., Ren, J. *et al.* (2009) MEME SUITE: tools for motif discovery and searching. *Nucleic Acids Res.* **37**, W202–W208.

Champion, A., Hebrard, E., Parra, B., Bournaud, C., Marmey, P., Tranchant, C. and Nicole, M. (2009) Molecular diversity and gene expression of cotton ERF transcription factors reveal that group IXa members are responsive to jasmonate, ethylene and Xanthomonas. *Mol. Plant Pathol.* **10**, 471–485.

Deokar, A.A., Kondawar, V., Kohli, D., Aslam, M., Jain, P.K., Karuppayil, S.M., Varshney, R.K. *et al.* (2015) The CarERF genes in chickpea (*Cicer arietinum* L.) and the identification of CarERF116 as abiotic stress responsive transcription factor. *Funct. Integr. Genomics*, **15**, 27–46.

Duan, C., Argout, X., Gebelin, V., Summo, M., Dufayard, J.F., Leclercq, J., Piyatrakul, P. *et al.* (2013) Identification of the *Hevea brasiliensis* AP2/ERF superfamily by RNA sequencing. *BMC Genom.* **14**, 30.

Hein, I., Barciszewska-Pacak, M., Hrubikova, K., Williamson, S., Dinesen, M., Soenderby, I.E., Sundar, S. *et al.* (2005) Virus-induced gene silencing-based functional characterization of genes associated with powdery mildew resistance in barley. *Plant Physiol.* **138**, 2155–2164.

Hiremath, P.J., Farmer, A., Cannon, S.B., Woodward, J., Kudapa, H., Tuteja, R., Kumar, A. *et al.* (2011) Large-scale transcriptome analysis in chickpea (*Cicer arietinum* L.), an orphan legume crop of the semi-arid tropics of Asia and Africa. *Plant Biotechnol. J.* **9**, 922–931.

Hirose, N., Makita, N., Kojima, M., Kamada-Nobusada, T. and Sakakibara, H. (2007) Overexpression of a type-A response regulator alters rice morphology and cytokinin metabolism. *Plant Cell Physiol.* **48**, 523–539.

Hubert, D.A., Tornero, P., Belkhadir, Y., Krishna, P., Takahashi, A., Shirasu, K. and Dangl, J.L. (2003) Cytosolic HSP90 associates with and modulates the Arabidopsis RPM1 disease resistance protein. *EMBO J.* **22**, 5679–5689.

Kale, S.M., Jaganathan, D., Ruperao, P., Chen, C., Ramu, P., Kudapa, H., Thudi, M. *et al.* (2015) Prioritization of candidate genes in "QTL-hotspot" region for drought tolerance in chickpea (*Cicer arietinum* L.). *Sci. Rep.* **5**, 1–14.

Krzywinski, M., Schein, J., Birol, I., Connors, J., Gascoyne, R., Horsman, D., Jones, S.J. *et al.* (2009) Circos: an information aesthetic for comparative genomics. *Genome Res.* **19**, 1639–1645.

Kudapa, H., Bharti, A.K., Cannon, S.B., Farmer, A.D., Mulaosmanovic, B., Kramer, R., Bohra, A. *et al.* (2012) A comprehensive transcriptome assembly of pigeonpea (*Cajanuscajan* L.) using Sanger and second-generation sequencing platforms. *Mol. Plant*, **5**, 1020–1028.

Kudapa, H., Azam, S., Sharpe, A.G., Taran, B., Li, R., Deonovic, B., Cameron, C. *et al.* (2014) Comprehensive transcriptome assembly of chickpea (*Cicer arietinum* L.) using Sanger and next generation sequencing platforms: development and applications. *PLoS ONE*, **9**, e86039.

Lata, C., Mishra, A.K., Muthamilarasan, M., Bonthala, V.S., Khan, Y. and Prasad, M. (2014) Genome-wide investigation and expression profiling of AP2/ERF transcription factor superfamily in foxtail millet (*Setaria italica* L.). *PLoS ONE*, **9**, e113092.

Li, C.W., Su, R.C., Cheng, C.P., You, S.J., Hsieh, T.H., Chao, T.C. and Chan, M.T. (2011) Tomato RAV transcription factor is a pivotal modulator involved in the AP2/EREBP-mediated defense pathway. *Plant Physiol.*, **156**, 213–227.

Licausi, F., Giorgi, F.M., Zenoni, S., Osti, F., Pezzotti, M. and Perata, P. (2010) Genomic and transcriptomic analysis of the AP2/ERF superfamily in *Vitis vinifera. BMC Genom.* **11**, 719.

Lindquist, S. and Jarosz, D.F. (2010) Hsp90 and environmental stress transform the adaptive value of natural genetic variation. *Science*, **330**, 1820–1824.

Liu, Z., Kong, L., Zhang, M., Lv, Y., Liu, Y., Zou, M., Lu, G. *et al.* (2013) Genome-wide identification, phylogeny, evolution and expression patterns of AP2/ERF genes and cytokinin response factors in *Brassica rapa* ssp. pekinensis. *PLoS ONE*, **8**, e83444.

Livak, K.J. and Schmittgen, T.D. (2001) Analysis of relative gene expression data using real-time quantitative PCR and the $2^{-\Delta\Delta C_T}$ method. *Methods*, **25**, 402–408.

Lu, R., Malcuit, I., Moffett, P., Ruiz, M.T., Peart, J., Wu, A.J., Rathjen, J.P. *et al.* (2003) High throughput virus-induced gene silencing implicates heat shock protein 90 in plant disease resistance. *EMBO J.* **22**, 5690–5699.

Magnani, E., Sjolander, K. and Hake, S. (2004) From endonucleases to transcription factors: evolution of the AP2 DNA binding domain in plants. *Plant Cell*, **16**, 2265–2277.

Mizoi, J., Shinozaki, K. and Yamaguchi-Shinozaki, K. (2012) AP2/ERF family transcription factors in plant abiotic stress responses. *Biochim. Biophys. Acta*, **1819**, 86–96.

Mizoi, J., Ohori, T., Moriwaki, T., Kidokoro, S., Todaka, D., Maruyama, K., Kusakabe, K. *et al.* (2013) GmDREB2A;2, a canonical DEHYDRATION-RESPONSIVE ELEMENT-BINDING PROTEIN2-type transcription factor in soybean, is posttranslationally regulated and mediates dehydration-responsive element-dependent gene expression. *Plant Physiol.* **161**, 346–361.

Nakano, T., Suzuki, K., Fujimura, T. and Shinshi, H. (2006) Genome-wide analysis of the ERF gene family in Arabidopsis and rice. *Plant Physiol.* **140**, 411–432.

Nene, Y.L. and Reddy, M.V. (1976) A new technique to screen pigeonpea for resistance to sterility mosaic. *Trop. Grain Legume Bull.* **5**, 23.

Ohta, M., Matsui, K., Hiratsu, K., Shinshi, H. and Ohme-Takagi, M. (2001) Repression domains of class II ERF transcriptional repressors share an essential motif for active repression. *Plant Cell*, **13**, 1959–1968.

Qu, L.J. and Zhu, Y.X. (2006) Transcription factor families in Arabidopsis: major progress and outstanding issues for future research. *Curr. Opin. Plant Biol.* **9**, 544–549.

Raju, N.L., Gnanesh, B.N., Lekha, P., Jayashree, B., Pande, S., Hiremath, P.J., Byregowda, M. *et al.* (2010) The first set of EST resource for gene discovery and marker development in pigeonpea (*Cajanuscajan* L.). *BMC Plant Biol.* **10**, 45.

Rashotte, A.M. and Goertzen, L.R. (2010) The CRF domain defines cytokinin response factor proteins in plants. *BMC Plant Biol.* **10**, 74.

Saeed, A.I., Sharov, V., White, J., Li, J., Liang, W., Bhagabati, N., Braisted, J., *et al.* (2003) TM4: a free, open-source system for microarray data management and analysis. *Biotechniques*, **34**, 374–378.

Sakuma, Y., Liu, Q., Dubouzet, J.G., Abe, H., Shinozaki, K. and Yamaguchi-Shinozaki, K. (2002) DNA-binding specificity of the ERF/AP2 domain of Arabidopsis DREBs, transcription factors involved in dehydration- and cold-inducible gene expression. *Biochem. Biophys. Res. Commun.* **290**, 998–1009.

Sangster, T.A. and Queitsch, C. (2005) The HSP90 chaperone complex, an emerging force in plant development and phenotypic plasticity. *Curr. Opin. Plant Biol.* **8**, 86–92.

Sato, S., Nakamura, Y., Kaneko, T., Asamizu, E., Kato, T., Nakao, M., Sasamoto, S. *et al.* (2008) Genome structure of the legume, *Lotus japonicus. DNA Res.* **15**, 227–239.

Sato, H., Mizoi, J., Tanaka, H., Maruyama, K., Qin, F., Osakabe, Y., Morimoto, K. *et al.* (2014) Arabidopsis DPB3-1, a DREB2A interactor, specifically enhances heat stress-induced gene expression by forming a heat stress-specific transcriptional complex with NF-Y subunits. *Plant Cell*, **26**, 4954–4973.

Schmutz, J., McClean, P.E., Mamidi, S., Wu, G.A., Cannon, S.B., Grimwood, J., Jenkins, J. et al. (2014) A reference genome for common bean and genome-wide analysis of dual domestications. Nat. Genet. **46**, 707–713.

Sharma, M.K., Kumar, R., Solanke, A.U., Sharma, R., Tyagi, A.K. and Sharma, A.K. (2010) Identification, phylogeny, and transcript profiling of ERF family genes during development and abiotic stress treatments in tomato. Mol. Genet. Genom. **284**, 455–475.

Sharma, M., Rathore, A., Mangala, U.N., Ghosh, R., Sharma, S., Upadhyay, H. and Pande, S. (2012) New sources of resistance to Fusarium wilt and sterility mosaic disease in a mini-core collection of pigeonpea germplasm. Eur. J. Plant Pathol. **133**, 707–714.

Sharoni, A.M., Nuruzzaman, M., Satoh, K., Shimizu, T., Kondoh, H., Sasaya, T., Choi, I.-R. et al. (2011) Gene structures, classification and expression models of the AP2/EREBP transcription factor family in rice. Plant Cell Physiol. **52**, 344–360.

Shi, X., Gupta, S. and Rashotte, A.M. (2012) Solanum lycopersicum cytokinin response factor (SlCRF) genes: characterization of CRF domain-containing ERF genes in tomato. J. Exp. Bot. **63**, 973–982.

Sohn, K.H., Lee, S.C., Jung, H.W., Hong, J.K. and Hwang, B.K. (2006) Expression and functional roles of the pepper pathogen-induced transcription factor RAV1 in bacterial disease resistance, and drought and salt stress tolerance. Plant Mol. Biol. **61**, 897–915.

Song, X., Li, Y. and Hou, X. (2013) Genome-wide analysis of the AP2/ERF transcription factor superfamily in Chinese cabbage (Brassica rapa ssp. pekinensis). BMC Genom. **14**, 573.

Takahashi, A., Casais, C., Ichimura, K. and Shirasu, K. (2003) HSP90 interacts with RAR1 and SGT1, and is essential for RPS2-mediated disease resistance in Arabidopsis. Proc. Natl Acad. Sci. USA, **100**, 11777–11782.

Trapnell, C., Pachter, L. and Salzberg, S.L. (2009) TopHat: discovering splice junctions with RNA-Seq. Bioinformatics, **25**, 1105–1111.

Trapnell, C., Williams, B.A., Pertea, G., Mortazavi, A., Kwan, G., van Baren, M.J., Salzberg, S.L. et al. (2010) Transcript assembly and quantification by RNA-Seq reveals unannotated transcripts and isoform switching during cell differentiation. Nat. Biotechnol. **28**, 511–515.

Tsutsui, T., Kato, W., Asada, Y., Sako, K., Sato, T., Sonoda, Y., Kidokoro, S. et al. (2009) DEAR1, a transcriptional repressor of DREB protein that mediates plant defense and freezing stress responses in Arabidopsis. J. Plant. Res. **122**, 633–643.

Varshney, R.K., Hiremath, P.J., Lekha, P., Kashiwagi, J., Balaji, J., Deokar, A.A., Vadez, V. et al. (2009) A comprehensive resource of drought- and salinity-responsive ESTs for gene discovery and marker development in chickpea (Cicer arietinum L.). BMC Genom. **10**, 523.

Varshney, R.K., Thudi, M., May, G.D. and Jackson, S.A. (2010) Legume genomics and breeding. Plant Breed. Rev. **33**, 257–304.

Varshney, R.K., Chen, W., Li, Y., Bharti, A.K., Saxena, R.K., Schlueter, J.A., Donoghue, M.T. et al. (2012) Draft genome sequence of pigeonpea (Cajanuscajan), an orphan legume crop of resource-poor farmers. Nat. Biotechnol. **30**, 83–89.

Varshney, R.K., Song, C., Saxena, R.K., Azam, S., Yu, S., Sharpe, A.G., Cannon, S. et al. (2013) Draft genome sequence of chickpea (Cicer arietinum) provides a resource for trait improvement. Nat. Biotechnol. **31**, 240–246.

Voorrips, R.E. (2002) MapChart: software for the graphical presentation of linkage maps and QTLs. J. Hered. **93**, 77–78.

Wan, L., Wu, Y., Huang, J., Dai, X., Lei, Y., Yan, L., Jiang, H. et al. (2014) Identification of ERF genes in peanuts and functional analysis of AhERF008 and AhERF019 in abiotic stress response. Funct. Integr. Genomics, **14**, 467–477.

Wang, G.F., Wei, X., Fan, R., Zhou, H., Wang, X., Yu, C., Dong, L. et al. (2011) Molecular analysis of common wheat genes encoding three types of cytosolic heat shock protein 90 (Hsp90): functional involvement of cytosolic Hsp90s in the control of wheat seedling growth and disease resistance. New Phytol. **191**, 418–431.

Wessler, S.R. (2005) Homing into the origin of the AP2 DNA binding domain. Trends Plant Sci. **10**, 54–56.

Xu, Z.S., Chen, M., Li, L.C. and Ma, Y.Z. (2008) Functions of the ERF transcription factor family in plants. Botany, **86**, 969–977.

Xu, Z.S., Chen, M., Li, L.C. and Ma, Y.Z. (2011) Functions and application of the AP2/ERF transcription factor family in crop improvement. J. Integr. Plant Biol. **53**, 570–585.

Xu, Z.S., Li, Z.Y., Chen, Y., Chen, M., Li, L.C. and Ma, Y.Z. (2012) Heat shock protein 90 in plants: molecular mechanisms and roles in stress responses. Int. J. Mol. Sci. **13**, 15706–15723.

Xu, J., Xue, C., Xue, D., Zhao, J., Gai, J., Guo, N. and Xing, H. (2013) Overexpression of GmHsp90s, a heat shock protein 90 (Hsp90) gene family cloning from soybean, decrease damage of abiotic stresses in Arabidopsis thaliana. PLoS ONE, **8**, e69810.

Young, N.D. and Udvardi, M. (2009) Translating Medicago truncatula genomics to crop legumes. Curr. Opin. Plant Biol. **12**, 193–201.

Young, N.D., Debellé, F., Oldroyd, G.E., Geurts, R., Cannon, S.B., Udvardi, M.K., Benedito, V.A. et al. (2011) The Medicago genome provides insight into the evolution of rhizobial symbioses. Nature, **480**, 520–524.

Zhang, Y. (2008) I-TASSER server for protein 3D structure prediction. BMC Bioinformatics, **9**, 40.

Zhang, G., Chen, M., Chen, X., Xu, Z., Guan, S., Li, L.C., Li., A. et al. (2008) Phylogeny, gene structures, and expression patterns of the ERF gene family in soybean (Glycine max L.). J. Exp. Bot., **59**, 4095–4107.

Zhuang, J., Cai, B., Peng, R.H., Zhu, B., Jin, X.F., Xue, Y., Gao, F. et al. (2008) Genome-wide analysis of the AP2/ERF gene family in Populus trichocarpa. Biochem. Biophys. Res. Commun. **371**, 468–474.

Zwack, P.J., Shi, X., Robinson, B.R., Gupta, S., Compton, M.A., Gerken, D.M., Goertzen, L.R. et al. (2012) Vascular expression and C-terminal sequence divergence of cytokinin response factors in flowering plants. Plant Cell Physiol. **53**, 1683–1695.

Small RNA and degradome deep sequencing reveals drought-and tissue-specific micrornas and their important roles in drought-sensitive and drought-tolerant tomato genotypes

Bilgin Candar-Cakir[1,2], Ercan Arican[3]* and Baohong Zhang[2]*

[1]Programme of Molecular Biology and Genetics, Institute of Science, Istanbul University, Vezneciler, Istanbul, Turkey
[2]Department of Biology, East Carolina University, Greenville, NC, USA
[3]Department of Molecular Biology and Genetics, Faculty of Science, Istanbul University, Vezneciler, Istanbul, Turkey

*Correspondence (e-mail:earican@istanbul.edu.tr)
email zhangb@ecu.edu

Summary

Drought stress has adverse impacts on plant production and productivity. MicroRNAs (miRNAs) are one class of noncoding RNAs regulating gene expression post-transcriptionally. In this study, we employed small RNA and degradome sequencing to systematically investigate the tissue-specific miRNAs responsible to drought stress, which are understudied in tomato. For this purpose, root and upground tissues of two different drought-responsive tomato genotypes (*Lycopersicon esculentum* as sensitive and *L. esculentum* var. *cerasiforme* as tolerant) were subjected to stress with 5% polyethylene glycol for 7 days. A total of 699 conserved miRNAs belonging to 578 families were determined and 688 miRNAs were significantly differentially expressed between different treatments, tissues and genotypes. Using degradome sequencing, 44 target genes were identified associated with 36 miRNA families. Drought-related miRNAs and their targets were enriched functionally by Gene Ontology (GO) and Kyoto Encyclopedia of Genes and Genomes (KEGG) pathway analyses. Totally, 53 miRNAs targeted 23 key drought stress- and tissue development-related genes, including DRP (dehydration-responsive protein), GTs (glycosyltransferases), ERF (ethylene responsive factor), PSII (photosystem II) protein, HD-ZIP (homeodomain-leucine zipper), MYB and NAC-domain transcription factors. miR160, miR165, miR166, miR171, miR398, miR408, miR827, miR9472, miR9476 and miR9552 were the key miRNAs functioning in regulation of these genes and involving in tomato response to drought stress. Additionally, plant hormone signal transduction pathway genes were differentially regulated by miR169, miR172, miR393, miR5641, miR5658 and miR7997 in both tissues of both sensitive and tolerant genotypes. These results provide new insight into the regulatory role of miRNAs in drought response with plant hormone signal transduction and drought-tolerant tomato breeding.

Keywords: deep sequencing, degradome, drought, microRNA, signal transduction, tomato.

Introduction

Drought is one of the destructive environmental stress condition that restricts crop production and reproduction in plants (Ding *et al.*, 2013). Plants cope with drought stress by recruiting drought avoidance and/or drought tolerance mechanisms at the morphological, physiological, biochemical, cellular and molecular levels such as decreased stomatal conductance, photosynthesis and respiration alterations, production of antioxidant and scavenger compounds, osmotic re-adjustment and hormonal metabolism changes (Bartels and Sunkar, 2005; Bhargava and Sawant, 2013; Ding *et al.*, 2013; Fang and Xiong, 2015). All of these strategies cause gene expression induction and accumulation of some enzymes and drought-related proteins (Ding *et al.*, 2013; Ramachandra Reddy *et al.*, 2004; Shinozaki and Yamaguchi-Shinozaki, 2007). Control of gene expression is one of the regulatory mechanisms on plant response to drought (Golldack *et al.*, 2011). Epigenetic regulations such as methylation, histone modifications and post-transcriptional alterations are stress-inducible mechanisms and have important roles in stress tolerance (Bhargava and Sawant, 2013).

In plants, small RNA-mediated regulatory mechanisms play crucial roles on biological processes including growth, development, maturation, transposon silencing, response to abiotic stress and pathogen defence (Xie *et al.*, 2012; Li and Zhang, 2016). Small RNAs, especially microRNAs (miRNAs), regulate abiotic stress-related gene expression post-transcriptionally, down-regulating their target genes, while their expression changes conversely (Carrington and Ambros, 2003; Ding *et al.*, 2013; Sunkar, 2010; Zhang, 2015; Zhang and Wang, 2015). Many miRNAs response to drought stress via signal transduction pathways such as auxin signalling, ABA-mediated regulation, osmoprotectant biosynthesis and scavenging of antioxidants (Ding *et al.*, 2013). Drought-related miRNAs and their targets have been identified in different plants such as *Arabidopsis* (Liu *et al.*, 2008; Sunkar and Zhu, 2004), cotton (Xie *et al.*, 2015), switchgrass (Xie *et al.*, 2014) rice (Zhao *et al.*, 2007; Zhou *et al.*,

2010) and soya bean (Kulcheski *et al.*, 2011). However, no systematical study has been performed on tomato.

Tomato (*Lycopersicon esculentum*) is an economically important crop due to demand for fresh vegetable market and processed food industry worldwide (Klee and Giovannoni, 2011). Tomato contains strong antioxidant enzymes, high level of lycopene, rich iron content, and vitamins A and C (Rai *et al.*, 2013). Tomato is also one of the most favoured models for fleshly fruit ripening and epigenetic researches (Gonzalez *et al.*, 2013; Klee and Giovannoni, 2011). Tomato growth, productivity and nutritional quality are generally affected by environmental stress factors such as drought, salt, flooding and pathogen infection (Rai *et al.*, 2013). Cultivated tomatoes (*L. esculentum*) are known as sensitive to drought at all stages of plant development, while seed germination and early seedling stages are mostly affected (Foolad *et al.*, 2003). However, their wild types such as *L. pimpinellifolium*, *L. pennellii*,*L. chilense* and *L. esculentum* var. *cerasiforme* are drought-tolerant species (Sadashiva *et al.*, 2013). Researches are in progress to clarify molecular pathways on drought response of plants (Sadashiva *et al.*, 2013), and miRNAs can be alternatives to identify drought metabolism, as they play crucial role on abiotic stress response regulating the stress-related mRNAs (Kumar, 2014). Determination and functional characterization of stress-related miRNAs and target identification are important in breeding programmes and can contribute to developing new strategies for improving stress tolerance (Barrera-Figueroa *et al.*, 2013). Although several studies have been performed on tomato miRNAs (Cao *et al.*, 2014; Feng *et al.*, 2014; Karlova *et al.*, 2013; Korir *et al.*, 2013; Moxon *et al.*, 2008; Pilcher *et al.*, 2007), there is no study on genomewide drought-responsive miRNA identification of tomato in tissue-specific manner.

In this study, we aimed to identify miRNAs and determine their expressions in different tissues under drought stress. For this purpose, we sequenced both root and upground tissues of drought-sensitive (*L. esculentum*) and drought-tolerant (*L. esculentum* var. *cerasiforme*) tomato genotypes. Also we carried out degradome sequencing for identifying the targets of drought-related miRNAs. From here, we identified a total of 699 conserved miRNAs belonging to 578 families, in which 688 miRNAs were differentially expressed between different treatments, tissues and genotypes. We also identified 44 target genes associated with 36 miRNA families. These miRNAs and their targets play an important role in tomato response to drought stress.

Results

Data mining of small RNA sequencing

To reveal tissue-specific and tolerance-related miRNAs under drought conditions, eight small RNA libraries were constructed and sequenced. These samples included four drought-treated (D) and four untreated (C) samples, in which each contained two roots (R) and two upgrounds (U) tissues belonging to drought-sensitive *L. esculentum* (S) and drought-tolerant *L. esculentum* var. *cerasiforme* (T) genotypes. After sequencing, a total of 194 625 986 raw and 192 387 328 clean reads were obtained with the average of 24 328 248 and 24 048 416 (98.85%) reads in each library, respectively (Table S1). The clean tags were used for analysing the distribution of 16- to 28-nt-length small RNAs and 94.78% of them were determined among 20–24 nt, while the most abundant sRNAs were 21 and 24 nt lengths with the percentage of 21.75% and 35.60%, respectively (Figure 1). However, the small RNA distributions of two libraries (Sensitive C-R and Sensitive D-R) were a little different from common results, in which 20-nt small RNAs are also dominant except 21-nt and 24-nt small RNAs. This has not been reported in other plant species. However, the reason for this is unclear; 20-nt small RNAs may have some role in this situation. About 70.29% and 69.10% unique and 83.72% and 83.01% redundant reads were mapped to tomato genome database (ITAG Release 2.4) in root and upground libraries, respectively (Tables 1 and 2). Using the blastn and blastall alignments against Genbank and Rfam databases, small RNAs were annotated to root and upground tissues of tomato genotypes which have different response to drought stress. In root libraries, most abundant RNA class was rRNA for unique reads with the mean value of 2.32%, followed by tRNA (0.23%), snRNA (0.13%) and snoRNA (0.12%). As for redundant reads, tRNAs were most abundant (21.41%) and other RNAs had quantity with the average of 10.47% (rRNA), 0.23% (snRNA) and 0.17% (snoRNA) (Table 1). For upground libraries, the annotation results were similar and rRNA proportion was highest (1.26%) in unique reads followed by tRNA (0.13%), snRNA (0.06%) and snoRNA (0.04%). In redundant reads, tRNA amount was

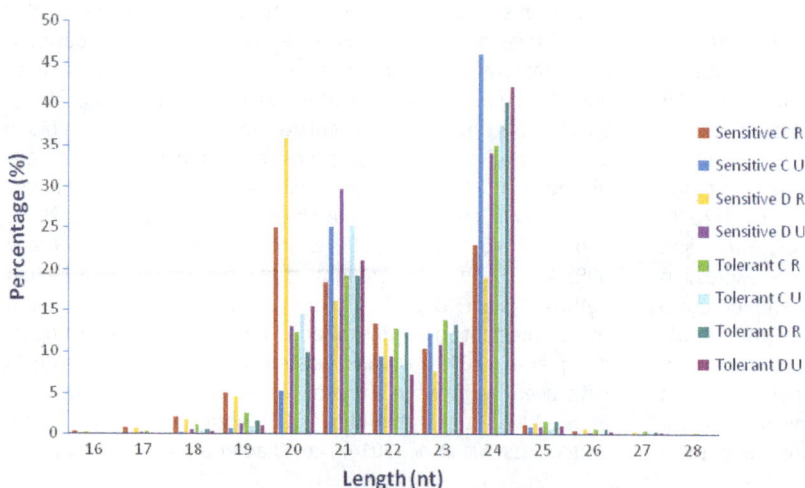

Figure 1 Length distribution of unique small RNAs in drought-sensitive and drought-tolerant tomato roots and upgrounds. C, control; D, drought; R, root; U, upground.

Table 1 Small RNA categorization in tomato roots

	Unique (Chavez Montes, #305)	(%)	Redundant (Chavez Montes, #305)	(%)	Unique (S-D)	(%)	Redundant (S-D)	(%)	Unique (T-C)	(%)	Redundant (T-C)	(%)	Unique (T-D)	(%)	Redundant (T-D)	(%)
Mapped	3 494 084	64.49	18 673 483	80.77	2 851 287	64.46	18 884 586	82.60	5 121 311	76.84	19 065 768	85.72	5 941 378	75.35	23 827 315	85.80
rRNA	124 649	2.30	2 780 848	12.03	139 283	3.15	2 446 445	10.70	139 546	2.09	2 269 466	10.20	138 216	1.75	2 485 796	8.95
tRNA	11 811	0.22	5 888 558	25.47	13 324	0.30	8 635 509	37.77	14 586	0.22	2 757 133	12.40	15 134	0.19	2 774 291	9.99
snRNA	7938	0.15	62 136	0.27	7213	0.16	50 819	0.22	7336	0.11	57 384	0.26	7788	0.10	45 713	0.16
snoRNA	7281	0.13	45 602	0.20	6740	0.15	44 012	0.19	6359	0.10	37 937	0.17	6244	0.08	34 406	0.12
miRNA	38 135	0.70	1 922 795	8.32	32 852	0.74	1 688 075	7.38	49 802	0.75	2 213 510	9.95	55 393	0.70	2 803 937	10.10
Unann	5 228 253	96.50	12 418 033	53.72	4 223 685	95.49	9 996 894	43.73	6 446 901	96.73	14 906 564	67.02	7 662 721	97.17	19 628 169	70.68
Total	5 418 067		23 117 972		4 423 097		22 861 754		6 664 530		22 241 994		7 885 496		27 772 312	

Table 2 Small RNA categorization in tomato upgrounds

	Unique (Chavez Montes, #305)	(%)	Redundant (Chavez Montes, #305)	(%)	Unique (S-D)	(%)	Redundant (S-D)	(%)	Unique (T-C)	(%)	Redundant (T-C)	(%)	Unique (T-D)	(%)	Redundant (T-D)	(%)
Mapped	4 140 172	62.59	17 577 661	76.50	3830 562	62.49	18 565 085	79.66	4 962 483	76.19	21 966 600	88.26	5 101 529	75.13	22 104 275	87.64
rRNA	79 061	1.20	1 093 011	4.76	92 586	1.51	867 283	3.72	70 614	1.08	642 359	2.58	85 932	1.27	896 898	3.56
tRNA	7116	0.11	827 402	3.60	10 453	0.17	2 691 444	11.55	7321	0.11	3 271 601	13.14	8825	0.13	3 740 846	14.83
snRNA	3817	0.06	9709	0.04	4754	0.08	17 003	0.07	3634	0.06	9779	0.04	4130	0.06	15 089	0.06
snoRNA	1862	0.03	4032	0.02	3490	0.06	10405	0.04	2202	0.03	6337	0.03	2709	0.04	7594	0.03
miRNA	46 961	0.71	3 974 298	17.30	42 070	0.69	5 638 271	24.19	44 071	0.68	4 795 634	19.27	46 018	0.68	4 031 156	15.98
Unann	6 475 993	97.90	17 067 997	74.28	5 976 291	97.50	14 080 950	60.42	6 385 673	98.04	16 163 653	64.94	6 642 704	97.83	16 530 545	65.54
Total	6 614 810		22 976 449		6 129 644		23 305 356		6 513 515		24 889 363		6 790 318		25 222 128	

S, sensitive; T, tolerant; C, control; D, drought; Mapped, mapped to tomato genome (ITAG Release 2.4); Unann, unannotated tags.

10.78% with the highest value, followed by rRNA (3.65%), snRNA (0.05%) and snoRNA (0.03%) (Table 2).

Identification of miRNAs from deep sequencing

To identify miRNAs, the clean sequence tags were aligned to all plant miRNA mature sequences deposited in miRBase database (Kozomara and Griffiths-Jones, 2014). The average of 44 046 (0.72%) and 44 780 (0.69%) unique, and 2 157 079 (8.94%) and 4 609 840 (19.19%) redundant reads were matched to the currently known miRNA sequences for root and upground libraries, respectively (Tables 1 and 2). Totally, 699 miRNAs were obtained from sequencing in eight libraries which belong to 578 families (Table S2). Among 578 families, sly-miR171 was represented with seven members, as the largest one, followed by sly-miR166 and sly-miR319 with five members (Table S2). Eleven

miRNA families (sly-miR156, sly-miR157, sly-miR164, sly-miR166, sly-miR167, sly-miR168, sly-miR4414, sly-miR6022, sly-miR6027, sly-miR7822 and sly-miR9471) were represented with the top read abundance above 10 000 at all libraries (Table S2).

We evaluated the miRNA distribution among libraries and determined that 197 miRNAs were common among control and drought-treated samples of both sensitive and tolerant genotypes in root libraries (Figure 2a). When we compared the genotypes separately, 35 miRNAs were common between control and drought samples of sensitive genotype and 32 miRNAs such as sly-miR166k, sly-miR408-3p and sly-miR9552b-3p were specific to control plants, whereas 25 miRNAs such as sly-miR1101-3p, sly-miR2628 and sly-miR3932b-3p were expressed only in drought-treated roots of sensitive genotype (Table S2). In tolerant genotype, 32 miRNAs such as sly-miR165b-5p, sly-miR2867-3p

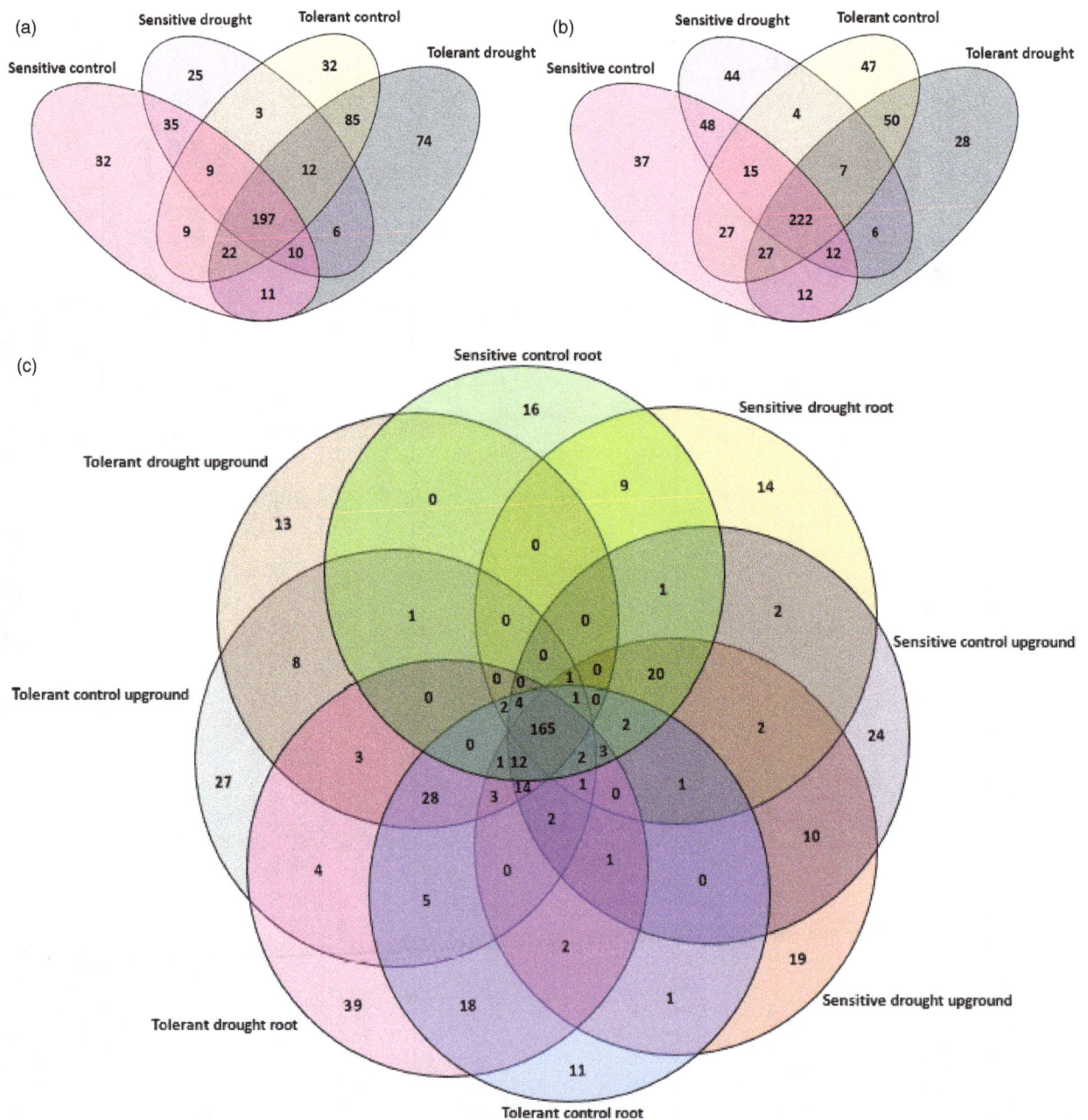

Figure 2 Distribution of tomato miRNAs (a) in root tissues, (b) in upground tissues, (c) in all samples.

and sly-miR7520 were expressed only in control conditions, while 74 miRNAs such as sly-miR171f-5p, sly-mir838-5p and sly-miR8046-3p were expressed specifically after drought exposure (Table S2). Additionally, 85 miRNAs were commonly expressed between control and drought treatments in tolerant genotype (Figure 2a). In upground samples, 222 miRNAs were common among four samples. Additionally, in control and drought-treated samples, 37 miRNAs such as sly-miR171a-3p, sly-miR6024-3p and sly-miR7997a were expressed specifically in control conditions, while 44 miRNAs such as sly-miR845a, sly-miR5797 and sly-miR8762d were specific to drought-treated upground tissues in sensitive genotype (Table S2, Figure 2b). In terms of tolerant genotype, 47 miRNAs such as sly-miR166d-5p, sly-miR4392 and sly-miR6288a showed specific expression in control sample, whereas 28 miRNAs such as sly-miR904a, sly-miR5171a and sly-miR6485 were expressed specifically in response to dehydration stress (Table S2, Figure 2b). Furthermore, sensitive control and drought libraries shared 48 miRNAs, while tolerant control and drought libraries had 50 common miRNAs (Figure 2b). When we compared eight libraries consisting of 699 miRNAs, we determined 165 miRNAs to be expressed commonly in all samples, while 63 and 90 miRNAs were expressed only in sensitive (C-R, 16; D-R, 14; C-U, 24; D-U, 19) and tolerant genotypes (C-R, 11; D-R, 39; C-U, 27; D-U, 13), respectively (Figure 2c).

Expression analyses of miRNAs

A total of 688 of 699 (98.4%) miRNAs belonging to eight libraries were expressed significantly based on fold change (≥1 or ≤−1) and P/q-value (<0.05) criteria (Table S2) and at least 130 miRNAs expressed approximately in all tomato libraries (Figure 3). Some miRNAs expressed differentially in root and upground tissues of sensitive and tolerant genotypes under control and drought conditions. Generally, the majority of miRNAs were down-regulated in sensitive genotype (mostly in upgrounds), while up-regulated in tolerant genotype (mostly in roots) by drought stress treatment (Table S2). A total of 11 miRNAs showed significant expression in all tissues of two genotypes in response to drought stress (Figure 4a, Table S2). For example, the expression of sly-miR169a-5p was decreased in all tissues, even the decrease was higher in root tissues. However, the expression of sly-miR6261 was decreased in root and upground tissues in sensitive genotype, whereas increased in tolerant genotype. In root, specific expression changes were observed. Some miRNAs were down-regulated in sensitive genotype, while up-regulated in tolerant by drought exposure; these miRNAs include sly-miR403-3p and sly-miR845a-3p. Contrary to this, sly-miR5512a and sly-miR9559-5p were up-regulated in sensitive genotype, but down-regulated in tolerant genotype. Additionally, the expression patterns were similar for certain miRNAs in both sensitive and tolerant genotypes. For instance, sly-miR399a-5p and sly-miR5282 were down-regulated with drought, while sly-miR4346 and sly-miR6269 were up-regulated in root tissues of two genotypes (Figure 4b, Table S2). Similarly, same expression alteration patterns were observed in upground tissues. For example, the expressions of sly-miR7494b and sly-miR7997c were increased, whereas sly-miR5029 were decreased in two genotypes. In addition, some adverse alterations were observed in two genotypes. Sly-miR479 and sly-miR837-3p were down-regulated, but up-regulated in sensitive and tolerant genotypes, respectively; sly-miR3954 and sly-miR9471a-5p were up-regulated, but down-regulated in sensitive and tolerant genotypes, respectively (Figure 4c, Table S2).

Some miRNAs showed tissue- and genotype-specific expression patterns. For example, several miRNAs were expressed only in root tissues in sensitive genotype and repressed/downregulated by drought exposure such as sly-miR166k, sly-miR408-3p and sly-miR9552b-3p (Table S2). Similarly, some miRNAs showed expression only in upground tissues of sensitive genotype and the expression level was decreased or repressed after drought stress such as sly-miR171a-3p, sly-miR1426, sly-miR5239, sly-miR6024-3p and sly-miR7997a (Table S2). In tolerant genotype, the expression of some miRNAs (sly-miR2867-3p, sly-miR3514-5p, sly-miR5251, sly-miR5763, sly-miR7520, sly-miR7730-5p, sly-mi8751b and sly-miR9493) was also specific to root tissues and suppressed/decreased by dehydration (Table S2). Likewise, some miRNAs were expressed specifically in drought-tolerant upground tissues and down-regulated or repressed with drought treatment such as sly-miR166d-5p, sly-miR408a-3p, sly-miR1507c-5p, sly-miR1857-5p, sly-miR4392, sly-miR6288a and sly-miR9722 (Table S2). Certain miRNAs were induced/upregulated by drought treatment in all tissues and genotypes. Among them, sly-miR319a, sly-miR1101a-3p, sly-miR2628 and sly-miR3932b-5p were specific to sensitive root tissues; sly-miR845a, sly-miR1511-3p, sly-miR5259, sly-miR5797 and sly-miR6440b to sensitive upground tissues; sly-miR171f-5p, sly-miR838-5p, sly-miR946a-5p, sly-miR1846a-5p, sly-miR3637-3p, sly-miR5035-3p, sly-miR5760 and sly-miR6172 to tolerant root; and sly-miR904a, sly-miR5171a, sly-miR6173, sly-miR6225-3p and sly-miR6485 to tolerant upground tissues (Table S2). When we evaluated the drought-sensitive and tolerant genotypes, we determined that a total of 20 miRNAs were expressed in all four libraries belonging to sensitive genotype such as sly-miR529 g and sly-miR5817 (Figure 5a, Table S2). Similarly, 28 miRNAs showed expression only in all tolerant-related libraries such as sly-miR415, sly-miR1520 g and sly-miR2111 (Figure 5b, Table S2).

Target prediction and degradome analyses

psRNATarget server (http://plantgrn.noble.org/psRNATarget/) was used to predict miRNA target transcripts; a total of 41 300 miRNA target pairs were obtained which contain 688 significantly expressed miRNAs (Table S3). These targets were mainly transcription factors, stress- and dehydration-/drought-related proteins, resistance-associated proteins and significant cellular enzymes like kinases, transferases and phosphatases (Table S3). Among these, 360 miRNAs (such as sly-miR319-3p, sly-miR2089-3p and sly-miR5671a) targeted plant development and stress response-related transcription factors, such as MYB, WRKY, GRAS, TCP, NAC-domain, ARF (auxin response factor), SBP (squamosa promoter-binding-protein-like), LEA (late embryogenesis abundant) protein and ERF (ethylene responsive factor) families (Table S3). Additionally, stress-related proteins known as stress-responsive protein, stress-enhanced protein, universal stress protein and stress-induced protein were potentially targeted by 49 miRNAs, such as sly-miR164a, sly-miR1074, sly-miR1873, sly-miR2628 and sly-miR5029. Also, especially some miRNA target genes were associated with dehydration/drought stress directly. These transcripts included dehydration-responsive family protein, DRP (dehydration-responsive protein), ERD (early responsive to dehydration-like) protein, DREB (dehydration-responsive element binding) and Di19 protein (dehydration-/drought-induced 19 protein) and potentially regulated by 38 tomato miRNAs; these miRNAs included sly-miR160a-3p, sly-miR170-3p, sly-miR1074, sly-miR3948, sly-miR5081, sly-miR5758, sly-miR8001b-5p and sly-miR9748.

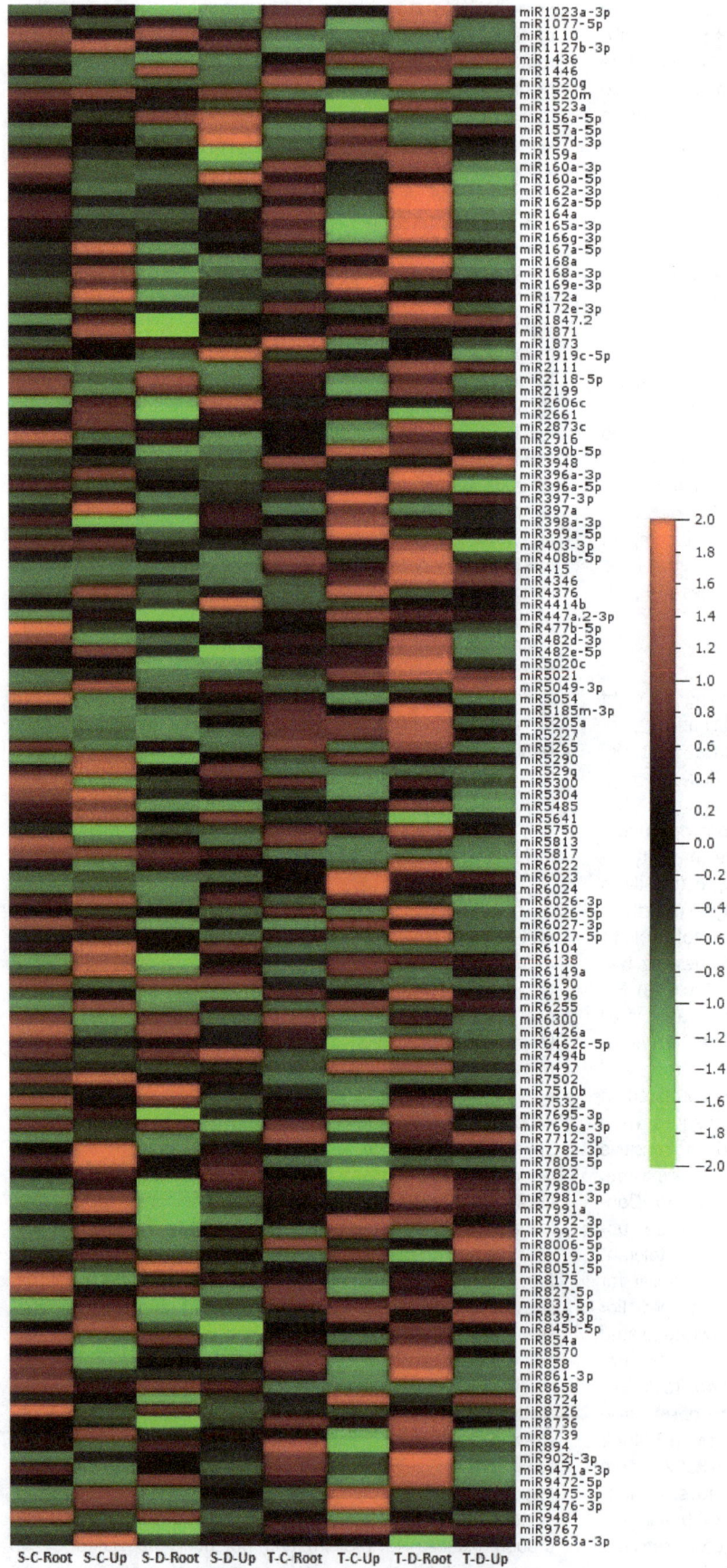

Figure 3 Differential expression of 130 most abundant conserved miRNAs in drought-sensitive and drought-tolerant tomato roots and upgrounds. The miRNA expressions were shown as Z-score. S, sensitive; T, tolerant; C, control; D, drought; Up, upground.

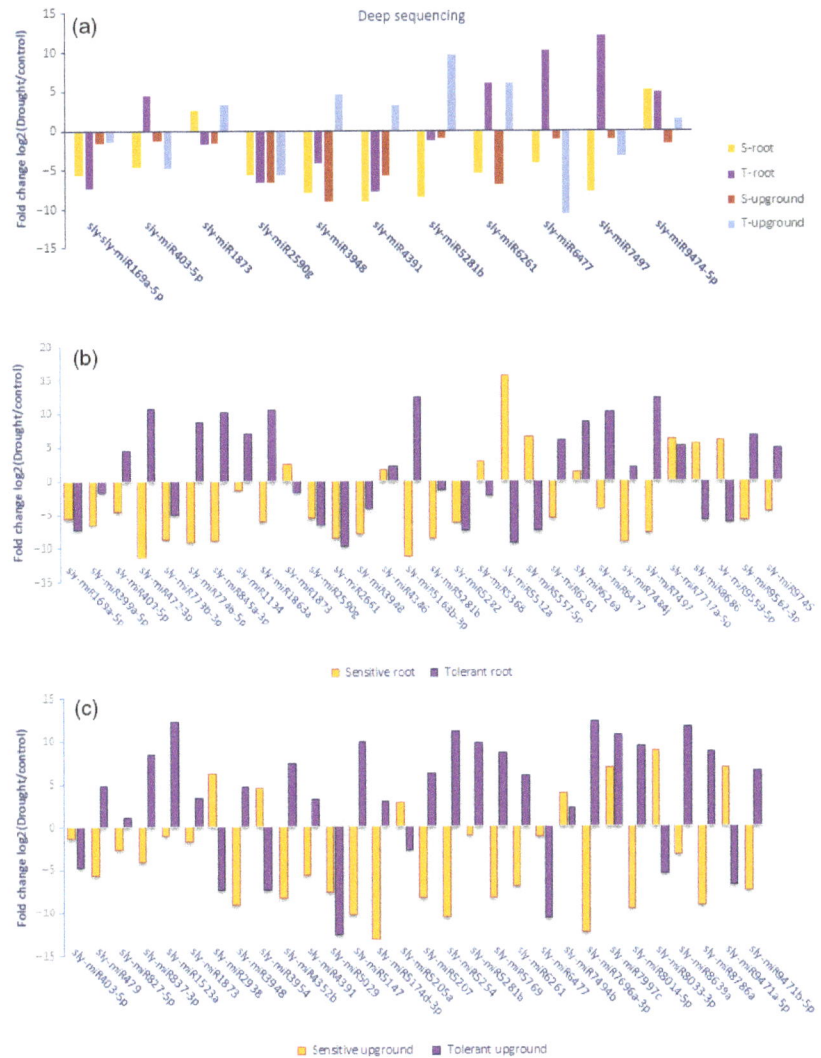

Figure 4 Comparisons of conserved miRNA expression changes after drought exposure according to the deep sequencing results. (a) Common miRNAs among root and upground tissues of sensitive and tolerant genotypes; (b) Common miRNAs only in root tissues of sensitive and tolerant genotypes; (c) Common miRNAs only in upground tissues of sensitive and tolerant genotypes.

We employed degradome sequencing to identify miRNA targets. After sequencing, a total of 10 819 148 raw and 10 799 028 clean tags (99.81%) were obtained. Then, the clean tags were mapped to the reference genome database of tomato ITAG 2.4 Release (ftp://ftp.solgenomics.net/genomes/Solanum_lycopersicum/annotation/ITAG2.4_release/ITAG2.4 genomic.fasta) by SOAP2.20 (Li et al., 2009) and 7 422 369 matched tags (68.73%) were determined. With the classification of these tags, cDNA_sense tags were selected and the identified 1 249 158 (46.86%) unique 5′ cDNA tags were used for the prediction of cleavage sites of tomato miRNAs. After prediction using CleaveLand v3.0.1 pipeline (Addo-Quaye et al., 2009), a total of 59 cleavage sites were determined associated with 62 miRNAs and 44 target genes and 115 specific miRNA–mRNA pairs were predicted at cleavage sites with P-value <0.05 (Table S4). For the identification of cleavage sites, degradome peaks are classified into five categories according to the peak height at mRNA position and the targets with category 0 or 1 were evaluated as the most significant (Karlova et al., 2013). In tomato degradome results, 15 target genes were related to stimulus response such as ARF and disease resistance proteins were identified in category 0 cleaved by sly-miR160, sly-miR168, sly-miR172, sly-miR396, sly-miR482, sly-miR6023 and sly-miR6024 families (Figure 6, Table S4). In category 1, only two genes cleaved by miR156 and miR162 families were

obtained (Table S4). In another significant class of peaks, category 2, mostly stimulus- and cellular component organization-associated 47 mRNAs like PSII (photosystem II) protein, NAD(H) kinase, phosphorus transporter, ATP-sulfurylase and SCL (scarecrow-like) protein were determined. These targets were cleaved by sly-miR156, sly-miR164, sly-miR166, sly-miR169, sly-miR171, sly-miR395 and sly-mir9477 (Table S4). The rest of 51 target genes were belonged to less significant categories 3 and 4. One of the stimulus response-associated target PSII degraded by 16 miRNA families such as sly-miR167, sly-miR319, sly-miR390, sly-miR482, sly-mir1919, sly-miR5302 and sly-miR9479. Other two stress-related genes, SBP and ARF, were cleaved by sly-miR156 and sly-miR160 families, respectively. Sly-miR171, sly-miR403 and sly-miR6027 families targeted histone-arginine methyltransferase involved in cellular and developmental process, and sly-miR395 and sly-miR9477 degraded acyltransferase gene, the cellular and metabolic process-associated target (Table S4).

GO enrichment and KEGG pathway analyses

Gene ontology (GO)-based enrichment analysis was performed to further investigate the potential role of miRNAs in tomato response to drought stress. A total of 9810 potential miRNA targets were classified into 665 biological processes, 45 molecular functions and 72 cellular components (Table S5). Some of the significant

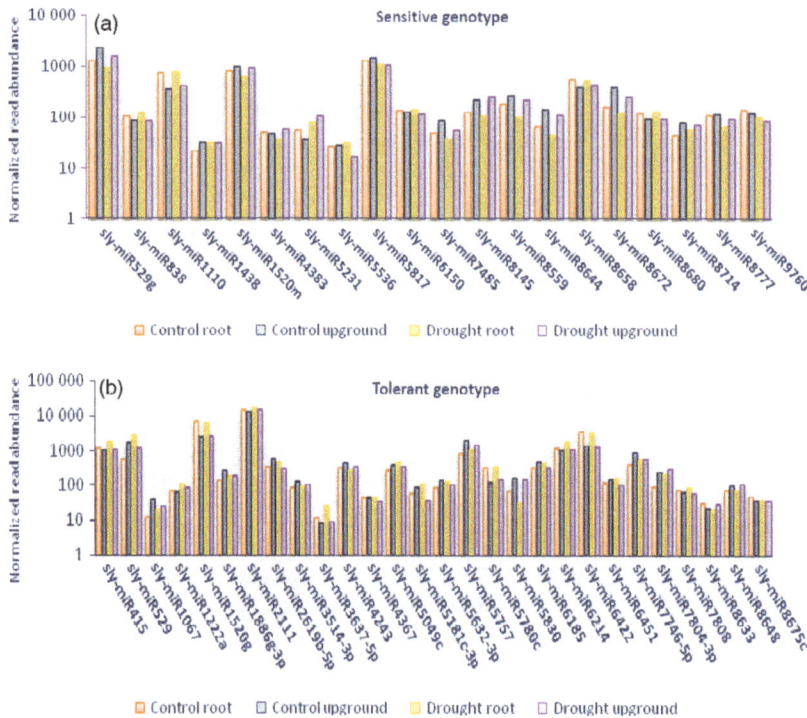

Figure 5 Normalized read abundance comparisons of conserved miRNAs belong to control and drought-treated root and upground tissues specific to (a) sensitive genotype, (b) tolerant genotype.

biological processes with the highest target numbers were cellular process (GO:0009987), cellular metabolic process (GO:0044237), response to stimulus (GO:0050896), response to abiotic stimulus (GO:0009628), response to stress (GO:0006950) and response to water deprivation (GO:0009414) (Table S5). Binding (GO:0005488), catalytic activity (GO:0003824), nucleotide binding (0000166) and hydrolase activity (0016787) were among the most abundant classes in molecular function category (Table S5). The common cellular component terms were cell part (GO:0044464), intracellular part (GO:0044424), organelle part (GO:0043226) and cytoplasmic part (GO:0044444) (Table S5). We employed agriGO web-based tool to visualize the enriched biological process, molecular function and cellular component categories and draw hierarchical graphs of significantly enriched GO terms (Du et al., 2010). In biological process, stress-associated terms such as response to hormone stimulus (GO:0009725) like abscisic acid (GO:0009737), jasmonic acid (GO:0009753), auxin (GO:0009733), ethylene (GO:0009723) and salicylic acid (GO:0009751), response to reactive oxygen species (GO:0000302), response to water deprivation (GO:0009414), response to salt stress (GO:0009651) and signal transduction (GO:0007165) were related to each other (Figures 7a and 8) and development-related ontologies like fruit development (GO:0010154), shoot development (GO:0048367), seed development (GO:004831), root development (GO:0048364) and leaf development (GO:0048366) were determined (Figure 8). Pyrophosphatase activity (GO:0016452), nucleoside-triphosphatase activity (GO:0017111), ATPase activity (GO: 0016887), purine ribonucleotide binding (GO: 003255) and ATP binding (GO:0005524) were some of the significantly enriched molecular function terms (Figure 7b, Table S5). In cellular component category, organelle subcompartment (GO:0031984), cell junction (GO:0030054), vacuole (GO:0031984) and chloroplast thylakoid (GO:0009534) were categorized significantly (Figure 7c, Table S5).

KEGG (The Kyoto Encyclopedia of Gene and Genome) annotation was carried out for pathway analysis, and 289 KEGG pathways were obtained. Thirty-six pathways including 2414

genes were detected as closely associated with drought stress including plant–pathogen interaction, biosynthesis of plant hormones, plant hormone signal transduction, oxidative phosphorylation, calcium signalling pathway, carotenoid metabolism, alpha-Linolenic acid metabolism, brassinosteroid biosynthesis and photosynthesis (Figure 9). Plant hormone signal transduction pathway was represented by 134 genes and contained auxin-, cytokinine-, gibberellin-, abscisic acid-, ethylene-, brassinosteroid-, jasmonic acid- and salicylic acid-associated signalling genes inducible via abiotic and biotic stresses (Figure 10). For example, carotenoid biosynthesis was involved in the pathway, contributes to abscisic acid synthesis, contained enriched phosphatase 2C (PP2C), plant-specific serine/threonine kinase (SnRK2) and abscisic acid-responsive element binding factor (ABF). These genes were regulated by sly-miR172a, sly-miR172e-3p, sly-miR393a, sly-miR2628, sly-miR5265, sly-miR5641, sly-miR6020a-5p and sly-miR7696a-3p (Figure 10). Similarly, jasmonic acid–amido synthetase JAR1, jasmonate ZIM domain (JAZ) and MYC2 genes were enriched to alpha-Linolenic acid metabolism associated with jasmonic acid synthesis, and these genes were targeted by sly-miR162a-3p, sly-miR169a-5p, sly-miR172a, sly-miR827-5p, sly-miR5083, sly-miR5298a, sly-miR5658, sly-miR6476a and sly-miR8576 (Figure 10).

Cytoscape platform was further employed to build the network between the drought-responsive miRNAs and their targets. Totally, 53 miRNAs (such as sly-miR164, sly-miR166 and sly-miR408) targeted 23 genes associated with drought response and tissue development, which included NAC-domain, HD-ZIP and BCP (blue copper protein). Sly-miR5641 targeted 10 drought- and development-related genes such as DRP, LEA, ERF and SBP (Figure 11a). Different gene families were targeted by different numbers of miRNAs. For example, DRP was targeted by 20 miRNAs, followed by GTs (glycosyltransferases) with 16 miRNAs; MYB with 11 miRNAs; NAC with 8 miRNAs; HD-ZIP with 7 miRNAs; and PSII and ERF with 6 miRNAs (Figure 11a). In plant hormone signal transduction pathway, 23 miRNAs targeted 19

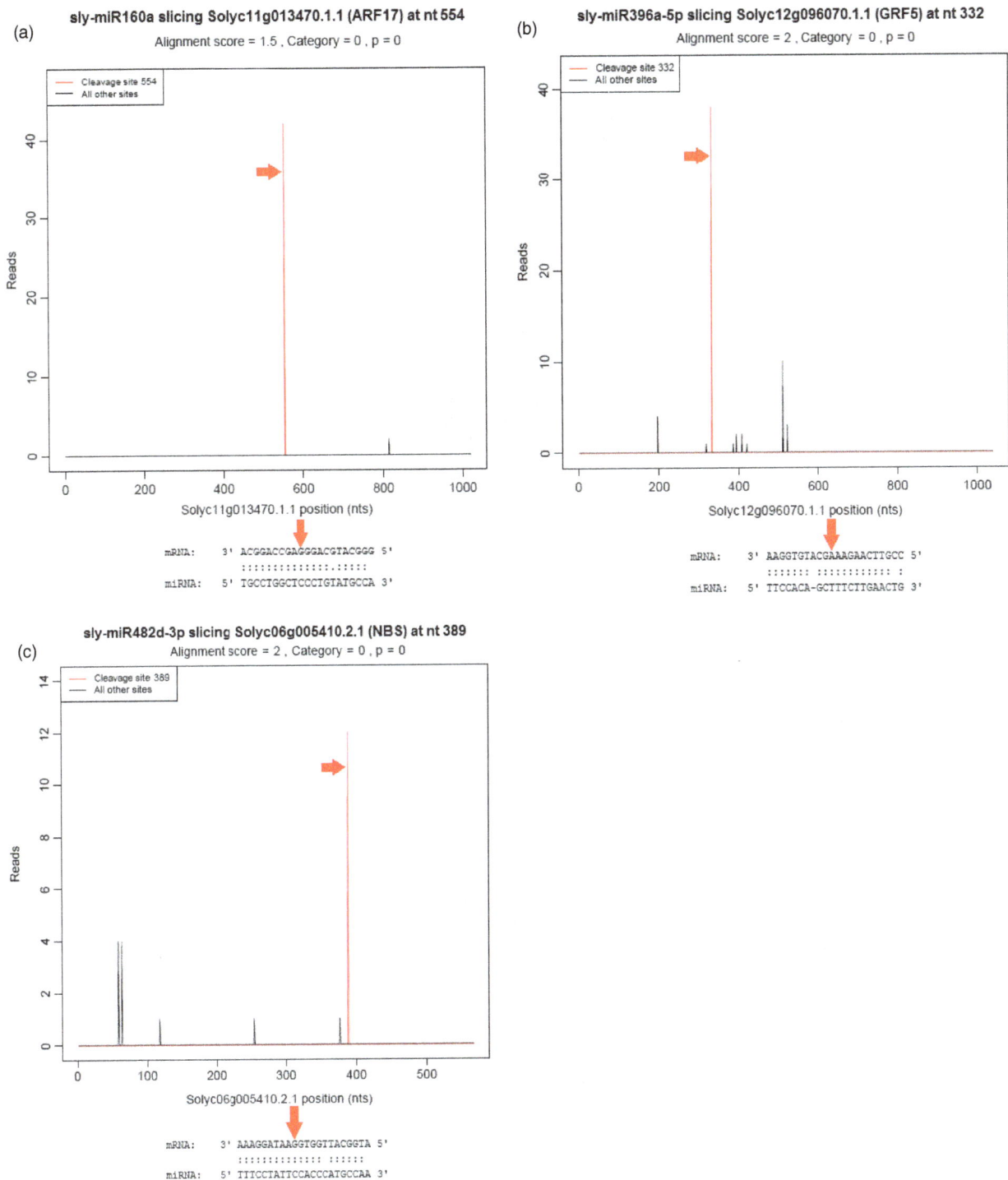

Figure 6 Target (T) plots and miRNA–mRNA alignments of tomato. (a) sly-miR160a cleaves ARF17 gene; (b) sly-miR396a-5p cleaves GRF5 gene; (c) sly-miR482d-3p cleaves NBS protein. The red arrows represent the cleavage nucleotide positions on the target genes.

genes that play roles in cell division, plant growth, stomatal closure and stress response. For example, sly-miR5641 potentially targeted seven genes involved in this pathway. Additionally, sly-miR393/miR6476 and sly-miR172/miR5658 potentially targeted 5 and 4 related genes, respectively (Figure 11b).

qRT-PCR validation of miRNA expressions

We performed qRT-PCR to validate the deep sequencing results with randomly selected eight miRNAs (sly-miR156a-5p,

sly-miR169e-3p, sly-miR172a, sly-miR393a-5p, sly-miR399a-5p, sly-miR408b-5p, sly-miR482d-3p and sly-miR9472-5p). For this purpose, we used root tissues of tolerant genotype under control and drought conditions and we figured out the expression levels between drought-treated versus control samples using log2 fold change ($2^{-\Delta\Delta C_t}$) values with three technical and three biological replicates. The expression results of root tissues in tolerant genotype exposed to drought were similar to the deep sequencing data (Figure 12). Three miRNAs (sly-miR172a, sly-miR482d-3p

(a)

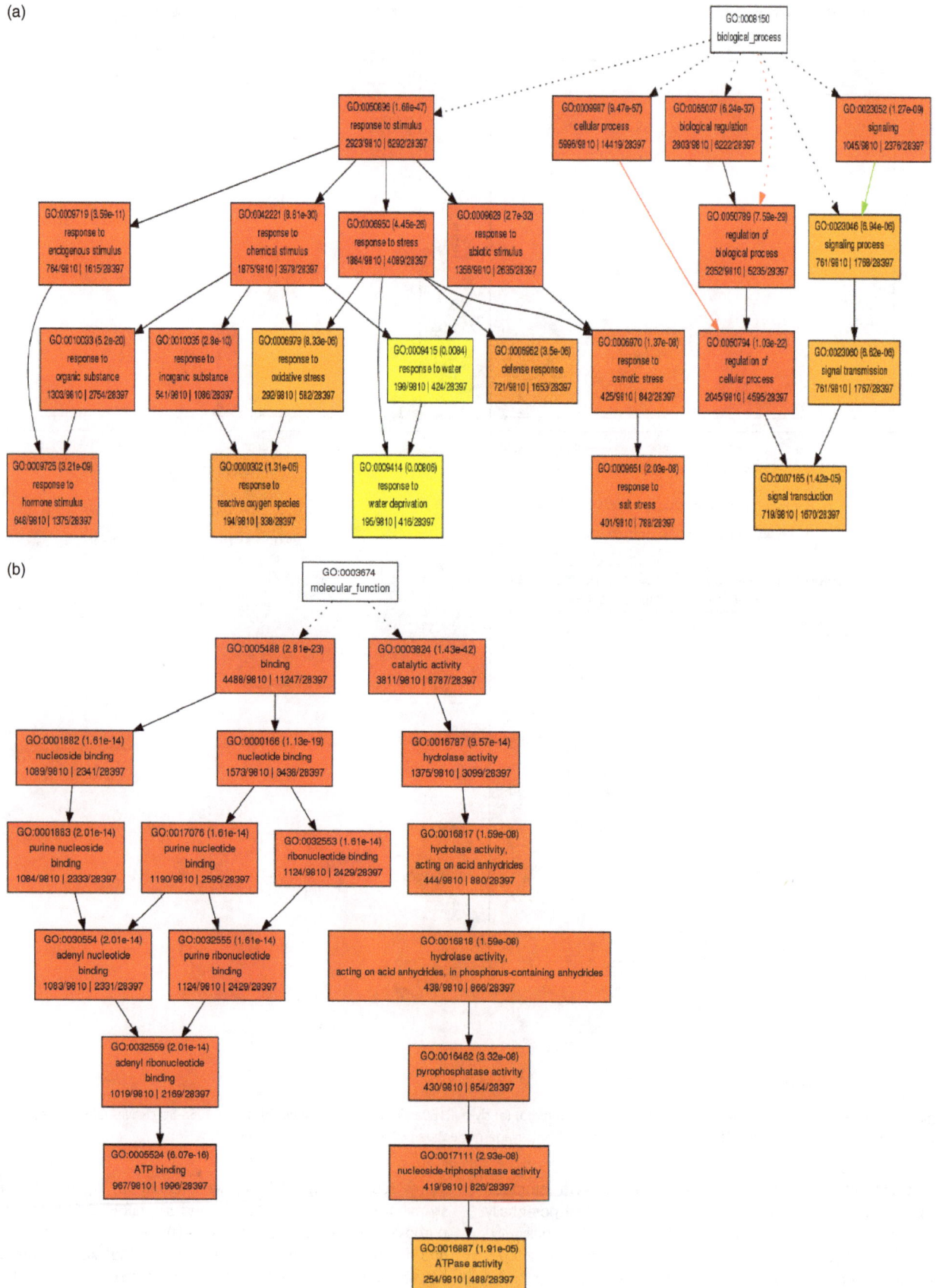

(b)

Figure 7 Diagrams of enriched GO terms of drought-responsive miRNA-associated tomato targets constructed by AGRIGO. (a) biological process, (b) molecular function and (c) cellular component. Red to yellow colours represent decreasing significance levels (Red is most, yellow is least significant). Red and green arrows mean positive and negative regulation of terms.

(c)

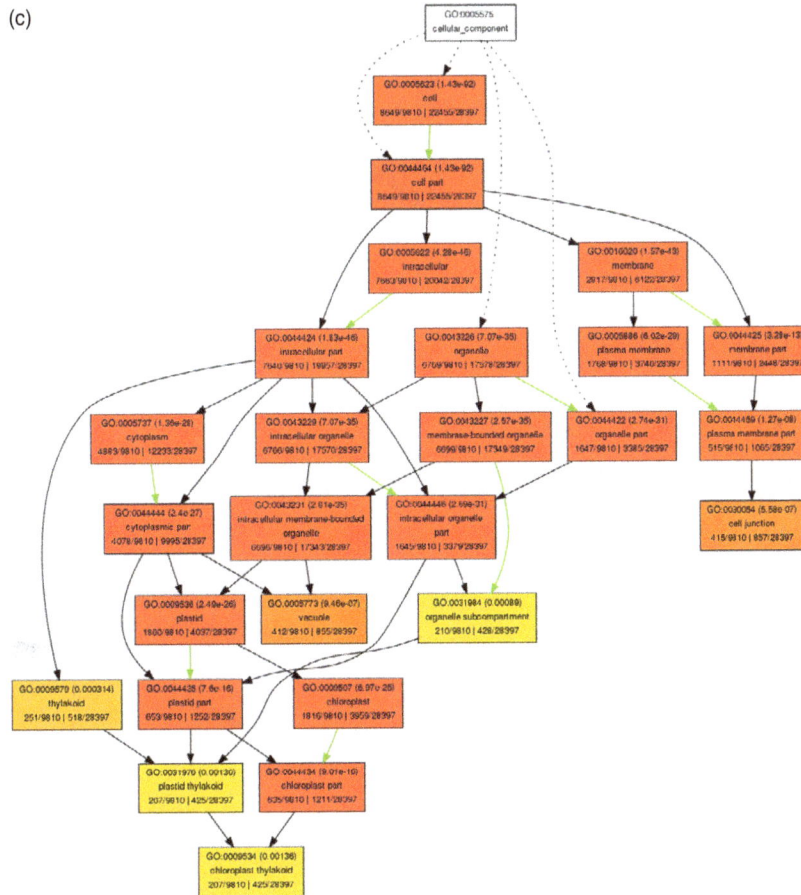

Figure 7 Continued.

and sly-miR9472-5p) were up-regulated in qRT-PCR analysis showing a positive correlation with deep sequencing results. Similarly, the other miRNAs were down-regulated in both qRT-PCR and high-throughput sequencing results (Figure 12).

Discussion

MicroRNAs, the post-transcriptional gene regulators, are associated with drought stress evidenced by that miRNAs target a wide range of drought-related genes, such as ARF, MYB, TCP and GRAS family transcription factors, dehydrins, glutathione S-transferase (GST) and abscisic acid-related genes (Ferdous et al., 2015). Using microarray and deep sequencing approaches, several drought-responsive miRNAs were identified in wheat (Kantar et al., 2011), sorghum (Pasini et al., 2014), sugarcane (Thiebaut et al., 2014), tobacco (Yin et al., 2014), potato (Zhang et al., 2014) and barley (Hackenberg et al., 2015). However, there is no study on systematical identification and expression analysis of miRNAs in tomato response to drought stress using microarray or deep sequencing approaches. In this study, we employed deep sequencing technology to systematically study the effect of drought exposure on miRNA expression in two different tomato genotypes with different drought sensitivity in a tissue-specific manner. Based on our study, a total of 775 miRNAs were differentially affected by drought; among them, 438 miRNAs were down-regulated in sensitive genotype (Table S2). We also performed degradome sequencing to identify miRNA

targets. Our results were different from that in tomato fruit development (Karlova et al., 2013). In our study, we found 44 target genes, but only the targets of 4 miRNAs (sly-miR156, sly-miR160, sly-miR166 and sly-miR482) were matched with their results (Table S4). We identified not only the common targets as reported previously but also new targets in tomato. For example, miR156 targets SBP transcription factors, but sly-miR156d-3p targeted NAD(H)-like kinase protein, while sly-miR156e-5p targeted four genes including SBP6 and PSII protein. Except the previous report that miR160 was found to target ARF10, ARF16 and ARF17 (Karlova et al., 2013), we also found that miR160 targeted ARF18-like and PSII protein (Table S4). Several other drought-related miRNAs targeted several new genes.

Despite the similarity of each member in a same miRNA family (Zhang et al., 2006), they may response differently to drought stress with a tissue- and genotype-dependent manner (Table S2). For example, miR165/166 families were known to target HD-ZIP III transcription factor which is crucial for leaf polarity, lateral root development and vascular patterning (Elhiti and Stasolla, 2009; Williams et al., 2005; Zhong and Ye, 2007). The expressions of miR165/166 family were generally decreased by drought treatment in other plant species such as rice (Zhou et al., 2010), cotton (Xie et al., 2015), wheat (Kantar et al., 2011) and peach (Eldem et al., 2012). However, the expression of miR166 was increased in Medicago truncatula by drought stress and the expression level was higher in roots in comparison with upper parts (Trindade et al., 2010). Moreover, overexpressions of

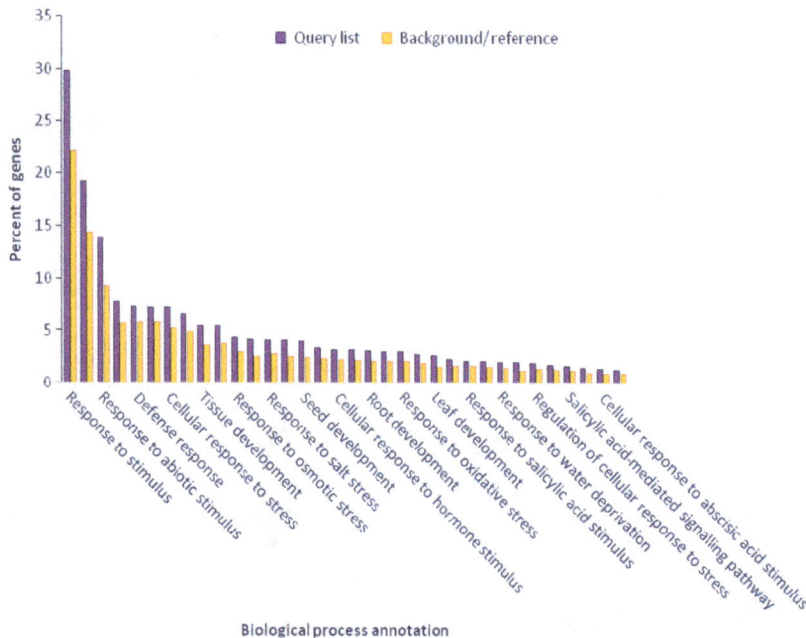

Figure 8 Functional annotation of GO biological process related with drought stress in tomato by AGRIGO.

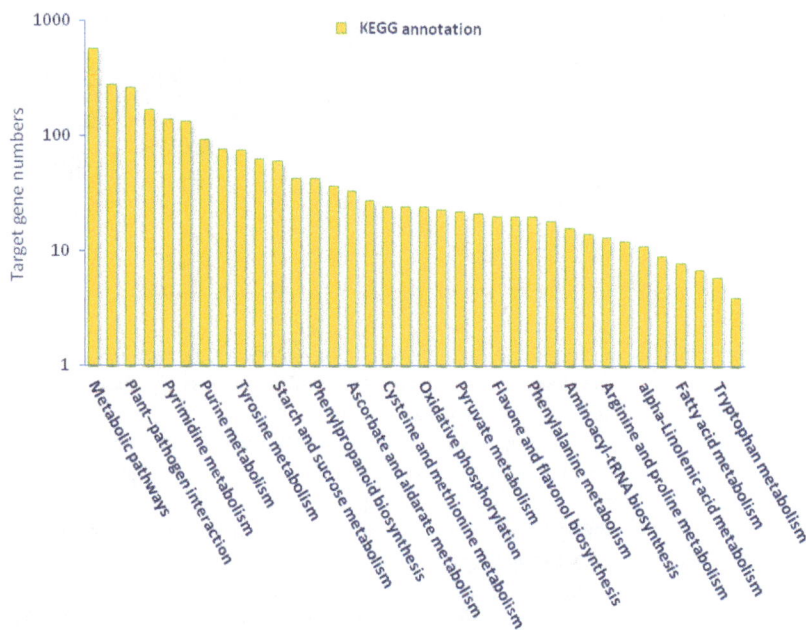

Figure 9 Functional annotation of KEGG pathways in tomato by KEGG database.

miR165/166 caused formations of vascular bundles and decreased lateral roots, respectively (Boualem et al., 2008; Zhong and Ye, 2007). These results suggest that miR165/166 families regulate root development and drought response. In our study, drought majorly induced the expression of miR165, but inhibited the expression of miR166 in roots. Additionally, sly-miR165a-5p/miR166k and sly-miR165b-5p were root specific in sensitive and tolerant genotypes, respectively, whereas sly-miR166d-5p was specific to upground tissues of tolerant genotype. Drought did not affect the sensitive genotype significantly in miR165 family. However, in root tissues of tolerant genotype, the expression of miR165b-5p was decreased by −8.25-fold after drought treatment, whereas miR165a-3p was up-regulated (3.10-fold) in upground tissues. Sly-miR166a and sly-miR166 g-3p were up-regulated by 19.50- and 2.34-fold, respectively, in upground

tissues of tolerant genotype with dehydration, while miR166d-5p expression was decreased sharply (−10.98-fold). In sensitive root samples, sly-miR166 g-5p was up-regulated (11.64-fold) by drought stress, whereas miR166k expression was decreased at the same time by −18.71-fold (Table S2). These results suggest that miR165 and miR166 regulated concurrently the drought-responsive gene expression as positively or negatively.

One of the main targets of miR398 is copper/zinc superoxide dismutase (Cu/Zn-SOD, CSD), a scavenger enzyme of ROS (reactive oxygen species) detoxifying superoxide radicals (Sunkar et al., 2006). miR398 was down-regulated by drought in tomato (Luan et al., 2014), M. truncatula (Wang et al., 2011) and cotton (Xie et al., 2015), whereas up-regulated in another M. truncatula species (Trindade et al., 2010) and wheat (Kantar et al., 2011). Down-regulation of this miRNA results in increase in CSD

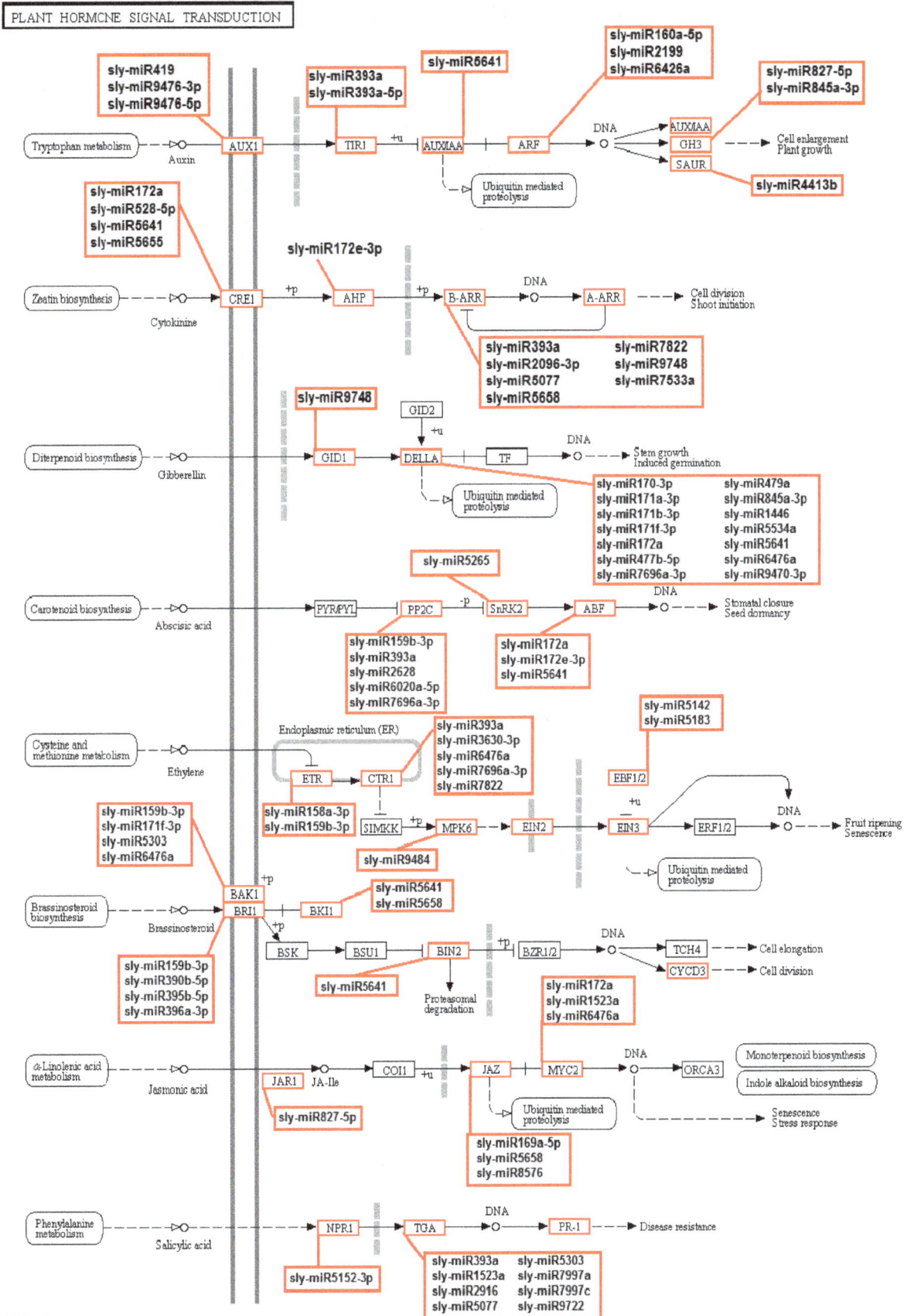

Figure 10 Plant hormone signal transduction pathway and related miRNAs.

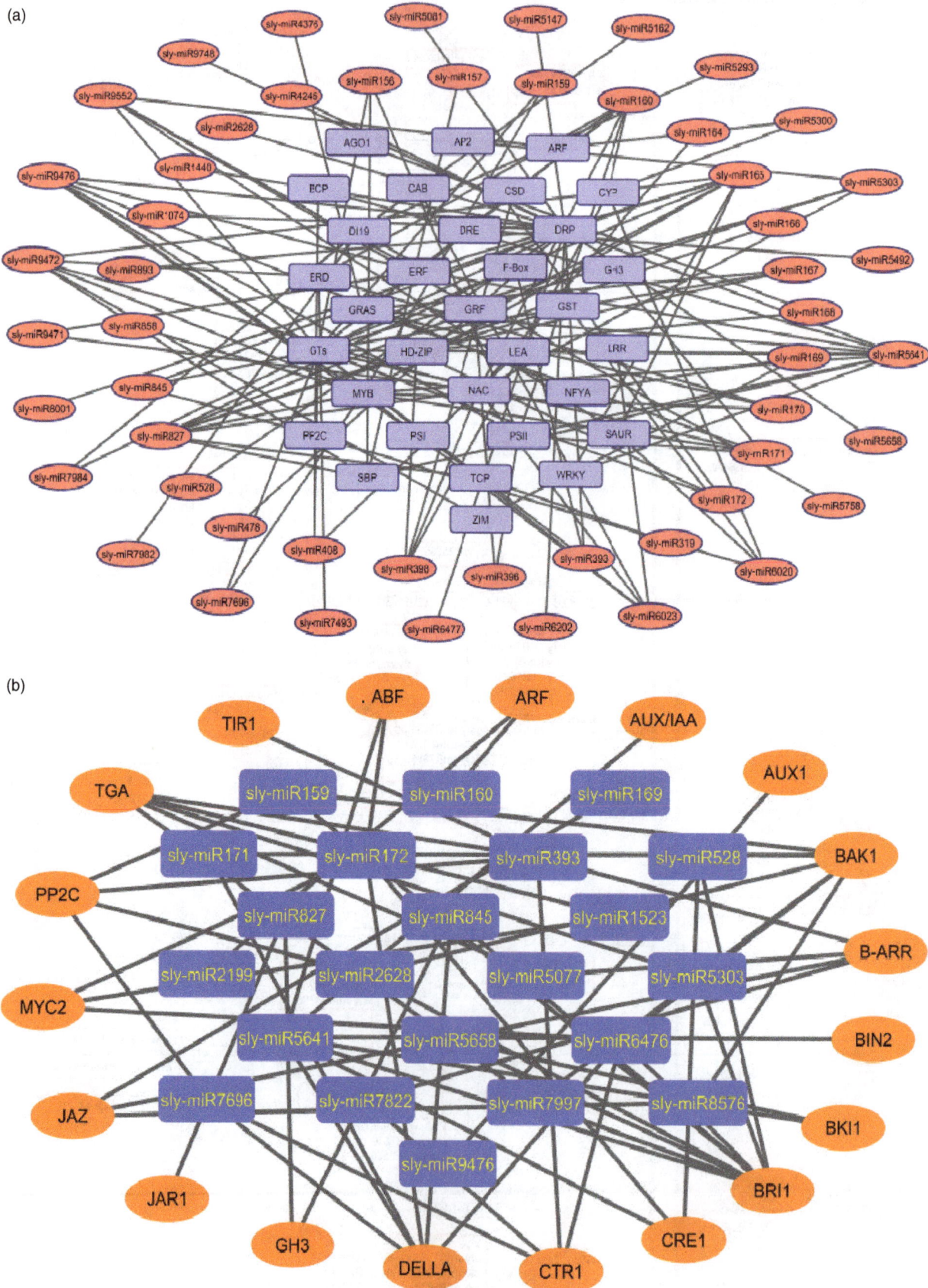

Figure 11 Relationships between miRNAs and their targets associated with (a) drought response, (b) plant hormone signal transduction pathway.

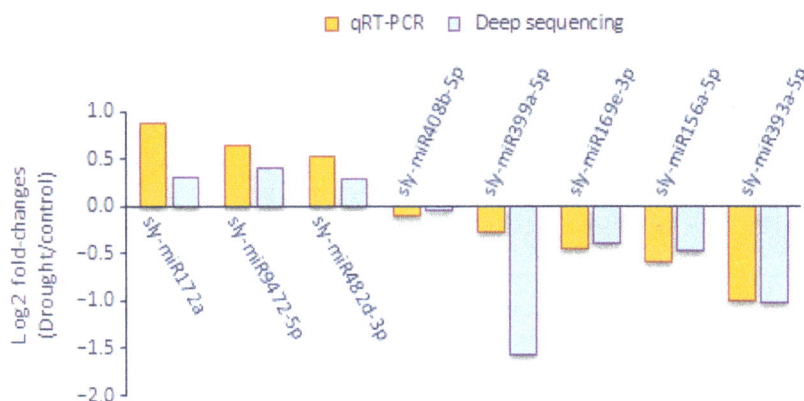

Figure 12 qRT-PCR validation of randomly selected drought-responsive eight miRNAs in tolerant root tissues.

expression and tolerance to oxidative stress (Ding et al., 2013). Consistently, the overexpression of miR398 led to down-regulation of CSD1 and 2 enzymes and caused sensitivity to drought stress in rice (Lu et al., 2010). Our target identification results were consistent with the literature and sly-miR398a-3p targeted CSD enzyme also (Table S3). However, sly-miR398a-3p expression did not change significantly by drought (Figure 3, Table S2). Sly-miR398a-5p was down-regulated in sensitive upground and tolerant root tissues by 2.65- and 1.17-fold, respectively, with drought treatment (Table S2). This miRNA targeted the development and drought-related NAC-domain, HD-ZIP III and auxin response proteins according to the target prediction results (Table S3). Sly-miR398a-5p expressed higher in roots of sensitive genotype in control (1.29-fold) and drought (4.09-fold) conditions, but generally miR398 family expression was decreased in upground tissues excluding sly-miR398a-3p in sensitive genotype. This up-regulation (2.03-fold) may be responsible for the sensitivity of L. esculentum decreasing CSD activity. The other copper proteins plantacyanin (basic blue) and laccase were the predicted targets of miR408 family (Abdel-Ghany and Pilon, 2008), and the expression of this family usually decreased by dehydration in plant species such as rice (Zhou et al., 2010), M. truncatula (Trindade et al., 2010), Populus (Ren et al., 2012), peach (Eldem et al., 2012) and cotton (Xie et al., 2015). miR408 is necessary for the adjustment of copper levels in cells as the copper deficiency causes production of ROS and oxidative stress (Abdel-Ghany and Pilon, 2008). The overexpression of miR408 in chickpea led to the inhibition of plantacyanin expression and accumulation of copper and also induction of DREB expression by drought (Hajyzadeh et al., 2015) and overexpression of DREB transcription factors increased drought tolerance in rice (Chen et al., 2008) and Arabidopsis (Xu et al., 2009). According to our results, miR408a-3p targeted laccase protein as well as other drought-related genes GTs and LEA proteins (Table S3). The expression of miR408 was decreased after drought treatment. Especially, miR408-3p was down-regulated by −10.09-fold in root tissues of sensitive genotype and this miRNA was suggested as root specific to sensitive genotype as it was expressed only in root tissues of L. esculentum. Besides, miR408a-3p expression was decreased by −1.02-fold in upground tissues of tolerant genotypes and was suggested as upground specific in L. esculentum var. cerasiforme. High-level of decrease in miR408-3p expression in L. esculentum may be the reason of drought sensitivity by comparison with miR408a-3p expression change in tolerant genotype.

The expression of miR9552 showed drought-, tissue- and genotype-specific pattern. Sly-miR9552a-3p was only expressed

in the roots of sensitive genotype and induced (12.38-fold) by drought treatment, whereas repressed (−14.16-fold and −13.64-fold) in upground tissues of sensitive and tolerant genotypes (Table S2). miR9552 targeted SAUR protein whose expression is regulated by auxin (Abel and Theologis, 1996). Overexpression of SAUR39 gene caused the formation of shorter plants with less leaves in rice indicating negative correlation with auxin biosynthesis (Kant et al., 2009). In contrast, overexpression of SAUR19-24 genes led to large leaves and hypocotyls implying cell enlargement and plant growth function of SAUR proteins induced by auxin (Spartz et al., 2012). Biotic and abiotic stress induced the differential expression of SAUR genes. For instance, 11 randomly selected SAUR genes were expressed in different tomato tissues and mostly down-regulated by drought treatment (Wu et al., 2012). However, in our results, SAUR expression was found to be down-regulated in root tissues, while up-regulated in upground tissues contrary to miRNA expression profiles indicating tissue-specific function of drought signalling in tomato. Sly-miR9552b-3p was expressed only in root tissues of sensitive genotype under control conditions and suppressed in response to stress treatment, so this miRNA might be suggested as root specific in L. esculentum. One of the predicted targets of this miRNA was UDP-glucosyltransferase (UGT) (Table S3). Glycosyltransferase enzymes of plants (GTs, EC 2.4) function in secondary metabolism and hormone modification catalysing sugar addition to acceptor molecules such as auxin, ABA and flavonoids (Bowles et al., 2005; Tognetti et al., 2010) and play a role in biotic and abiotic stress tolerance (Vogt and Jones, 2000). One of the common glycosyltransferase UGT71C5 was investigated for the elucidation of ABA impact on drought adaptation in Arabidopsis plants, and when the gene was down-regulated, drought tolerance increased but decreased after overexpression of UGT71C5 (Liu et al., 2015). These results suggest that the suppression of sly-miR9552b-3p expression with drought may increase UGT level in roots and decrease drought tolerance in sensitive genotype.

Plant hormone signal transduction pathway

Plant hormones play a key role in signalling networks involving in plant development and stress response (Golldack et al., 2014). Different miRNAs regulate the expression of plant hormone-associated genes in response to different environmental stresses. For example, a stress-responsive gene ARF which is related with auxin signalling was targeted by sly-miR160, sly-miR2199 and sly-miR6426 in response to drought stress in our study. However, ARF was targeted by miR167 after selenium treatment in Astragalus (Cakir et al., 2015). One of the important phytohormones, ABA, functions centrally in drought and salinity tolerance regulating main transcriptional processes (Cutler et al., 2010). The carotenoid biosynthesis of plant signal transduction pathway is regulated by ABA signals and these signals finally stimulate ABA-responsive genes regulating the activation/inactivation of type 2C protein phosphatases (PP2Cs), SNF1-related protein kinases (SnRK2s) and ABA-responsive promoter elements binding factors (ABFs) (Golldack et al., 2014; Ma et al., 2009; Vlad et al., 2009). Our results found that PP2C was targeted by 5 miRNAs such as sly-miR393a and sly-miR7696a-3p (Figures 10 and 11b). miR393 was up-regulated in rice (Zhou et al., 2010), Arabidopsis (Liu et al., 2008) and wheat (Kantar et al., 2011), while down-regulated in cotton (Xie et al., 2015) and peach (Eldem et al., 2012). Overexpression of miR393 led to decrease in drought tolerance affecting growth in rice (Xia et al., 2012). However, in this study sly-miR393 was down-regulated in root tissues of sensitive genotype, whereas up-regulated in upground tissues of drought-responsive tomato (Table S2). miR7696 also targeted ABA signalling pathway, whose expression was significantly altered in upground tissues of sensitive and tolerant genotypes and down-regulated sharply by −12.33-fold in sensitive genotype, whereas up-regulated by 12.25-fold in tolerant genotype after drought exposure. However, this miRNA was expressed higher in root tissues under control conditions (Figure 3, Table S2). These results indicate differential regulation of PP2C by several miRNAs in root and upground tissues of drought-sensitive and tolerant genotypes. Similarly, ABF was targeted by sly-miR172a/miR172e-3p and sly-miR5641 (Figures 10 and 11b). When the tomato plants were exposed to drought stress, miR172 family expressed significantly only in upground tissues of sensitive genotype and sly-miR172a and sly-miR172e-3p were down-regulated in response to drought by −2.01- and −1.07-fold, respectively (Figure 3, Table S2). Sly-miR172 expression was decreased after drought treatment in rice (Zhou et al., 2010), barley (Hackenberg et al., 2015) and cotton (Xie et al., 2015), whereas up-regulated in Arabidopsis (Jones-Rhoades and Bartel, 2004), wheat (Kantar et al., 2011) and Populus (Ren et al., 2012). Sly-miR5641 was also down-regulated by −4.31-fold in root tissues of tolerant genotype (Figure 3, Table S2). The results show that miR172 is different in response to drought among plant species and ABF gene is regulated by different miRNAs in different tissues and genotypes under drought stress.

ABA usually interacts with gibberellic acid (GA) and jasmonate (JA) during plant development and response to drought stress (Golldack et al., 2014). GA signalling is controlled by GID1 (GIBBERELLIN INSENSITIVE DWARF1) receptors and DELLA proteins, the subgroup of GA-responsive GRAS family transcription factors (Griffiths et al., 2006; Tyler et al., 2004). DELLA protein was targeted by 12 tomato miRNAs containing sly-miR172, sly-miR845, sly-miR5641 and sly-miR7696 (Figures 10 and 11b). miR845 was expressed differentially in two tissues and genotypes. Sly-miR845a-3p was down-regulated (−8.86-fold) in sensitive root tissues, whereas up-regulated (10.09-fold) in tolerant roots (Figures 3 and 4b, Table S2). In upground tissues, sly-miR845a-3p and miR845b-5p were down-regulated (~ −11.00-fold), while miR845a was up-regulated by 8.84-fold (Figures 3 and 4c, Table S2). The results not only indicate the different regulatory roles of unique miRNA members in different tissues and genotypes, but also show the function of the miRNAs in more than one signalling way. In JA signalling, there are three key receptor proteins known as Jasmonate Resistant 1 (JAR1), Jasmonate ZIM Domain (JAZ) and Jasmonate Insensitive 1 (JIN1, also known as

MYC2) (Kazan and Manners, 2008). JAR1 was targeted by sly-miR827-5p (Figures 10 and 11b) whose expression was decreased (−2.50-fold) in sensitive upground tissues, while increased (1.02-fold) in tolerant upground samples in response to dehydration stress (Figures 3 and 4c, Table S2). The root tissues were not affected by drought, but generally miR827-5p expression was higher in roots (Figure 3, Table S2). In same signalling cascade, JAZ receptor was targeted by sly-miR169a-5p (Figures 10 and 11b). In our results, sly-miR169a-5p expression was decreased in all tissues with drought (Figure 4a, b, Table S2). miR169 targets Jasmonate ZIM Domain (JAZ) and nuclear transcription factor Y subunit A-3 (NFYA-3) in tomato fruits (Karlova et al., 2013) and this is further validated in our study (Table S3). Additionally, sly-miR169 expression was increased in tomato by drought treatment, while three SlNFYA (1/2/3) genes were down-regulated (Zhang et al., 2011). Moreover, overexpression of sly-miR169c caused significantly down-regulation of tomato target genes and induced the increased drought tolerance of tomato (Zhang et al., 2011), whereas overexpression of miR169a led to drought sensitivity in Arabidopsis plants (Li et al., 2008b). Our results were similar with Arabidopsis result indicating NFYA up-regulation by drought in an ABA-dependent manner (Li et al., 2008b). ABA-dependent signalling of drought results in stomatal closure (Figure 10), and stomatal closure is the first response of plants to drought stress (Schroeder et al., 2001). This response is controlled by not only ABA, but also interactions of ABA with the other phytohormones JA, ethylene, auxin and cytokinine (Nemhauser et al., 2006). ABA and JA positively regulate the stoma closure, while auxin and cytokinine regulate negatively and ethylene response depends on tissues and stresses (Daszkowska-Golec and Szarejko, 2013; Huang et al., 2008; Nemhauser et al., 2006). Excitingly, 48 miRNAs target plant hormone signal transduction pathway (Figures 10 and 11b).

In conclusion, we identified 699 known miRNAs and the majority of them were expressed significantly in response to drought stress in a tissue- and genotype-specific manner. According to the GO and KEGG analyses, the majority of these miRNAs involved in response to hormone stimulus/reactive oxygen species/water deprivation/salt stress, signal transduction, fruit, shoot, seed and root development (Figures 7a and 8) and plant–pathogen interaction, biosynthesis of plant hormones and plant hormone signal transduction pathways (Figures 9 and 10). Drought-responsive miRNAs (such as sly-miR160, sly-miR165, sly-miR166, sly-miR171, sly-miR398, sly-miR408, sly-miR827, sly-miR9472, sly-miR9476 and sly-miR9552) regulated drought and development-associated genes like DRP, HD-ZIP, MYB, NAC and PSII in root and upground tissues (Figure 11a). Likewise, sly-miR169, sly-miR172, sly-miR393, sly-miR5641, sly-miR5658 and sly-miR7997 function in plant hormone signal transduction pathway and related proteins (Figures 10 and 11b). These results reveal drought-responsive miRNA profiles of drought-sensitive and drought-tolerant tomato genotypes in tissue-specific pattern and contribute to the development of drought-tolerant tomato plants.

Materials and methods

Plant material and drought treatment

The seeds of drought-sensitive (CGN24169: Lycopersicon esculentum, L.M.I-56) and drought-tolerant (CGN18399: L. esculentum var. cerasiforme, Tomatillo; PI 187002 selection 1) tomato genotypes were obtained from the Centre for Genetic Resources,

The Netherlands (CGN), Wageningen University and Research Centre, The Netherlands. The seeds were surface-sterilized with 75% (v/v) ethanol for 15 s, followed by 20% bleach (v/v) for 15 min and washed with sterilized distilled water for at least three times. Sterilized seeds were germinated on MS (Murashige and Skoog, 1962) medium (pH 5.8), containing 3% sucrose and 0.8% agar in a growth chamber with fluorescent light (~1400/mol^2/ms) under 16-h light /8-h dark cycle at 25 ± 2 °C for 14 days. For drought treatment, 14-day-old seedlings were exposed to 5% polyethylene glycol for 7 days. For control and drought treatments, four seeds were germinated in Magenta boxes and the experiments were replicated six times in six individual vessels. After stress treatment for 7 days, the root and upground tissues were collected from seedlings and immediately frozen in liquid nitrogen. The samples were stored at −80 °C till RNA extraction.

Small RNA library construction and sequencing

Total RNAs were extracted from root and upground tissues of drought-sensitive and tolerant tomato plants using the mirVana™ miRNA Isolation Kit (Ambion, Austin, TX) according to the manufacturer's instructions. The quality and quantity of RNAs were measured with a NanoDrop ND-2000 spectrophotometer (Thermo Scientific, Wilmington, DE). RNA isolation was carried out individually for each sample with four biological replicates, then the RNAs were sent to BGI (Shenzen, China) for small RNA library construction and high-throughput sequencing using Illumina HiSeq2000 platform.

Identification of tomato miRNAs using deep sequencing

The raw reads were first cleaned up, including removing adapter sequences and eliminating low-quality reads. Then, the length distribution of clean reads was categorized to analyse the composition of small RNA data, and 16- to 28-nt-length small RNAs were used for further analysis. High-quality clean small RNA tags were mapped to tomato genome (ftp://ftp.solge-nomics.net/tomato_genome/annotation/ITAG2.4_release/ITAG2.4_genomic.fasta) by SOAP (short oligonucleotide alignment program) to find out their expression and distribution on the genome (Li et al., 2008a). Then, the matched tags were aligned to NCBI GenBank (http://www.ncbi.nlm.nih.gov/genbank/) (Benson et al., 2015) and Rfam 11.0 (http://rfam.xfam.org/) (Burge et al., 2013) databases using BLASTall and BLASTn to determine rRNA, tRNA, snRNA and snoRNAs. Following this search, repeat-containing RNAs, sense and antisense exon and intron sequences were detected, and fully matched all RNA types excluding miRNAs were gotten rid of. Then, for annotation of remaining sequences, conserved miRNAs were mapped to miRBase Release (v21 on June 26th, 2014) database (http://www.mirbase.org/ftp.shtml) (Kozomara and Griffiths-Jones, 2014) and researched for L. esculentum miRNAs. The expression of miRNAs was calculated simultaneously by summing the count of tags overlapping at least 16 nt and/or two mismatches with aligned known miRNAs in database. DEGseq package was used for differential expression analysis of miRNAs, after normalizing raw read numbers with trimmed mean of M-values (TMM) normalization method (Robinson and Oshlack, 2010; Wang et al., 2010). For normalization, firstly the normalization factors were calculated and then normalized read numbers were determined using following formula: [raw read counts/(library size × normalization factor) × 10^6] (Bai et al., 2014). Fold changes were calculated with the formula [log$_2$(normalized read numbers of

group2/group1)]. Then, for the identification of significantly expressed miRNAs, the criteria were used as if (fold change ≥ 1 or ≤ -1) and (P or q-value <0.05) (Storey, 2003). To show differential expression profile among drought-sensitive and drought-tolerant tomato root and upgrounds, heatmap was constructed for the most abundant 130 conserved using Qlucore Omics Explorer 3.0 (Qlucore AB) (http://www.qlucore.com/).

Target prediction, GO enrichment and KEGG pathway analysis

Significantly expressed miRNAs were used for target prediction using psRNATarget server (http://plantgrn.noble.org/psRNATarget/) with default parameters, including maximum expectation score (3.0), length for complementarity (17 bp) and range of central mismatch (10–11 nt) (Dai and Zhao, 2011). Solanum lycopersicum (tomato) cDNA library, version 2.4 (ftp://ftp.solgenomics.net/tomato_genome/annotation/ITAG2.4_release/ITAG2.4_cdna.fasta) was used to predict target genes, and for functional annotation and enrichment analysis of target genes, agriGO (GO Analysis Toolkit and Database for Agricultural Community) web-based tool was used (Du et al., 2010). Firstly, the protein sequences of target genes (ftp://ftp.solgenomics.net/tomato_genome/annotation/ITAG2.4_release/ITAG2.4_proteins.fasta) were aligned against Arabidopsis protein sequences (ftp://ftp.arabidopsis.org/home/tair/Sequences/blast_datasets/TAIR10_blastsets/TAIR10_pep_20101214_updated) to find out the homologues. Then, the matched gene list was submitted to agriGO query list as TAIR10 locus ID, and GO classification was performed. The enriched GO terms of biological process, molecular function and cellular component categories were visualized with DAGs (directed acyclic hierarchical graph) and bar charts and pathway analyses were performed using KEGG (The Kyoto Encyclopedia of Gene and Genome) database (http://www.genome.jp/kegg/kegg1.html) using target gene IDs as queries (Kanehisa et al., 2014). The relationship between drought-responsive miRNAs and their putative targets was visualized using Cytoscape network platform (Saito et al., 2012).

Validation of miRNA expressions with qRT-PCR analysis

Total RNAs of tolerant root tissues belonging to drought-treated and control samples isolated for deep sequencing were used to validate miRNA expression results. Firstly, stem-loop reverse transcription (RT) was carried out using TaqMan® MicroRNA Reverse Transcription Kit (Applied Biosystems, Foster City, CA). A total of 15 µL RT reaction contained 1 mM dNTPs with dTTP, 1× RT buffer, 50U MultiScribe™ Reverse Transcriptase, 3.8U RNase inhibitor and 500 ng total RNAs and nuclease-free water. Also 1.3 µM miRNA-specific stem-loop RT primers were used to generate single-stranded cDNA for miRNAs (Table S6). The following temperature program was used to perform RT reaction as 30 min at 16 °C, 30 min at 42 °C, 5 min at 85 °C, and then holding at 4 °C. Before quantitative real-time PCR (qRT-PCR), the cDNAs were diluted in 100 µL DNase/RNase-free water. Eight miRNAs (sly-miR156a-5p, sly-miR169e-3p, sly-miR172a, sly-miR393a-5p, sly-miR399a-5p, sly-miR408b-5p, sly-miR482d-3p and sly-miR9472-5p) were randomly selected, and specific forward primers and universal reverse primer were designed to amplify the miRNAs (Table S6). qRT-PCRs were carried out using 2× SensiFAST SYBR® Hi-ROX mix (Bioline, Taunton, MA) on a Applied Biosystems 7300 Real-Time PCR System. Briefly, each 20 µL reactions contained 10 µL SensiFAST mix, 6 µL nuclease-free water, 2 µL cDNA product and 2 µL primer mix. The reactions were performed with the following temperature program: 10 min at 95 °C for enzyme activation, 40 cycles of 15 s at 95 °C for denaturation and 60 s at 60 °C for annealing/extension, followed by a dissociation step for 1 cycle of 95 °C for 15 s, 60 °C for 60 s, 95 °C for 15 s and 60 °C for 15 s. Three technical replicates for each biological reactions and three biological replicates were performed for root tissues of tolerant genotype. Actin-7 gene was used for normalization of qRT-PCR data. The fold changes were calculated using $2^{-(\Delta\Delta Ct)}$ values, and relative expressions were shown as $\log_2$ fold changes.

Degradome library construction, sequencing and data analysis

For degradome sequencing, the RNAs were first pooled from all samples at a same amount. Firstly, polyadenylated RNAs were isolated with oligo-d(T) bead extraction. Then, MmeI recognition site carrying 5′-RNA adapter was ligated to 5′-end that has mRNA fragments of miRNA-induced cleavage. Afterwards, the fragments were converted to cDNA by reverse transcription and amplified by PCR (German et al., 2009). The PCR products of degradome library were sequenced on Illumina HiSeq2000 sequencing system, and the adapter sequences, low-quality reads and N-containing fragments were filtered from the raw reads. After preprocessing, sRNAs were eliminated by Genbank and Rfam 11.0 databases, and then KEGG (Kanehisa et al., 2014) and NR (nonredundant) (ftp://ftp.ncbi.nih.gov/blast/db/FASTA/nr.gz) databases were used for the annotation of cleaved target genes. Then, clean tags were matched to the tomato genome (ITAG2.4 Release cDNA library) by SOAP2.20 (Li et al., 2009) with allowing only two mismatches, and with the classification of clean tags, the sense strand of tomato cDNA library were used to predict miRNA cleavage sites using CleaveLand v3.0.1 (August 26, 2011) pipeline (Addo-Quaye et al., 2009). The potential targets of miRNAs were analysed by PAREsnip software with P-value <0.05 (Folkes et al., 2012), and T-plot figures were drawn.

Acknowledgements

We appreciate Dr. Ozgur Cakir for bioinformatic analysis. This work was kindly supported by Scientific Research Projects Coordination Unit of Istanbul University with project numbers 42638 (PhD Thesis) and 45084 (IRP). B. Candar-Cakir was supported partially by TUBITAK (The Scientific and Technological Research Council of Turkey)-BIDEB 2211-C National Scholarship Programme for Phd Students and 2214-A International Doctoral Research Fellowship Programme for this study.

References

Abdel-Ghany, S.E. and Pilon, M. (2008) MicroRNA-mediated systemic down-regulation of copper protein expression in response to low copper availability in Arabidopsis. J. Biol. Chem. **283**, 15932–15945.

Abel, S. and Theologis, A. (1996) Early genes and auxin action. Plant Physiol. **111**, 9–17.

Addo-Quaye, C., Miller, W. and Axtell, M.J. (2009) CleaveLand: a pipeline for using degradome data to find cleaved small RNA targets. Bioinformatics, **25**, 130–131.

Bai, Y., Huang, J.M., Liu, G., Zhang, J.B., Wang, J.Y., Liu, C.K. and Fang, M.Y. (2014) A comprehensive microRNA expression profile of the backfat tissue from castrated and intact full-sib pair male pigs. BMC Genom. **15**, 47.

Barrera-Figueroa, B.E., Wu, Z. and Liu, R. (2013) Abiotic stress-associated microRNAs in plants: discovery, expression analysis, and evolution. Front. Biol. **8**, 189–197.

Bartels, D. and Sunkar, R. (2005) Drought and salt tolerance in plants. Crit. Rev. Plant Sci. **24**, 23–58.

Benson, D.A., Clark, K., Karsch-Mizrachi, I., Lipman, D.J., Ostell, J. and Sayers, E.W. (2015) GenBank. Nucleic Acids Res. **43**, D30–D35.

Bhargava, S. and Sawant, K. (2013) Drought stress adaptation: metabolic adjustment and regulation of gene expression. Plant Breeding, **132**, 21–32.

Boualem, A., Laporte, P., Jovanovic, M., Laffont, C., Plet, J., Combier, J.P., Niebel, A. et al. (2008) MicroRNA166 controls root and nodule development in Medicago truncatula. Plant J.: Cell Mol. Biol. **54**, 876–887.

Bowles, D., Isayenkova, J., Lim, E.K. and Poppenberger, B. (2005) Glycosyltransferases: managers of small molecules. Curr. Opin. Plant Biol. **8**, 254–263.

Burge, S.W., Daub, J., Eberhardt, R., Tate, J., Barquist, L., Nawrocki, E.P., Eddy, S.R. et al. (2013) Rfam 11.0: 10 years of RNA families. Nucleic Acids Res. **41**, D226–D232.

Cakir, O., Candar-Cakir, B. and Zhang, B. (2015) Small RNA and degradome sequencing reveals important microRNA function in Astragalus chrysochlorus response to selenium stimuli. Plant Biotechnol. J. **14**, 543–556.

Cao, X., Wu, Z., Jiang, F., Zhou, R. and Yang, Z. (2014) Identification of chilling stress-responsive tomato microRNAs and their target genes by high-throughput sequencing and degradome analysis. BMC Genom. **15**, 1130.

Carrington, J.C. and Ambros, V. (2003) Role of microRNAs in plant and animal development. Science, **301**, 336–338.

Chen, J.Q., Meng, X.P., Zhang, Y., Xia, M. and Wang, X.P. (2008) Over-expression of OsDREB genes lead to enhanced drought tolerance in rice. Biotechnol. Lett. **30**, 2191–2198.

Cutler, S.R., Rodriguez, P.L., Finkelstein, R.R. and Abrams, S.R. (2010) Abscisic acid: emergence of a core signaling network. Annu. Rev. Plant Biol. **61**, 651–679.

Dai, X. and Zhao, P.X. (2011) psRNATarget: a plant small RNA target analysis server. Nucleic Acids Res. **39**, W155–W159.

Daszkowska-Golec, A. and Szarejko, I. (2013) Open or close the gate – stomata action under the control of phytohormones in drought stress conditions. Front. Plant Sci. **4**, 138.

Ding, Y., Tao, Y. and Zhu, C. (2013) Emerging roles of microRNAs in the mediation of drought stress response in plants. J. Exp. Bot. **64**, 3077–3086.

Du, Z., Zhou, X., Ling, Y., Zhang, Z. and Su, Z. (2010) agriGO: a GO analysis toolkit for the agricultural community. Nucleic Acids Res. **38**, W64–W70.

Eldem, V., Celikkol Akcay, U., Ozhuner, E., Bakir, Y., Uranbey, S. and Unver, T. (2012) Genome-wide identification of miRNAs responsive to drought in peach (Prunus persica) by high-throughput deep sequencing. PLoS ONE, **7**, e50298.

Elhiti, M. and Stasolla, C. (2009) Structure and function of homodomain-leucine zipper (HD-Zip) proteins. Plant Signal. Behav. **4**, 86–88.

Fang, Y. and Xiong, L. (2015) General mechanisms of drought response and their application in drought resistance improvement in plants. Cell. Mol. Life Sci. **72**, 673–689.

Feng, J., Liu, S., Wang, M., Lang, Q. and Jin, C. (2014) Identification of microRNAs and their targets in tomato infected with Cucumber mosaic virus based on deep sequencing. Planta, **240**, 1335–1352.

Ferdous, J., Hussain, S.S. and Shi, B.J. (2015) Role of microRNAs in plant drought tolerance. Plant Biotechnol. J. **13**, 293–305.

Folkes, L., Moxon, S., Woolfenden, H.C., Stocks, M.B., Szittya, G., Dalmay, T. and Moulton, V. (2012) PAREsnip: a tool for rapid genome-wide discovery of small RNA/target interactions evidenced through degradome sequencing. Nucleic Acids Res. **40**, e103.

Foolad, M.R., Zhang, L.P. and Subbiah, P. (2003) Genetics of drought tolerance during seed germination in tomato: inheritance and QTL mapping. Genome, **46**, 536–545.

German, M.A., Luo, S., Schroth, G., Meyers, B.C. and Green, P.J. (2009) Construction of parallel analysis of RNA ends (PARE) libraries for the study of cleaved miRNA targets and the RNA degradome. Nat. Protoc. **4**, 356–362.

Golldack, D., Luking, I. and Yang, O. (2011) Plant tolerance to drought and salinity: stress regulating transcription factors and their functional significance in the cellular transcriptional network. Plant Cell Rep. **30**, 1383–1391.

Golldack, D., Li, C., Mohan, H. and Probst, N. (2014) Tolerance to drought and salt stress in plants: unraveling the signaling networks. *Front. Plant Sci.* **5**, 151.

Gonzalez, R.M., Ricardi, M.M. and Iusem, N.D. (2013) Epigenetic marks in an adaptive water stress-responsive gene in tomato roots under normal and drought conditions. *Epigenetics*, **8**, 864–872.

Griffiths, J., Murase, K., Rieu, I., Zentella, R., Zhang, Z.L., Powers, S.J., Gong, F. *et al.* (2006) Genetic characterization and functional analysis of the GID1 gibberellin receptors in Arabidopsis. *Plant Cell*, **18**, 3399–3414.

Hackenberg, M., Gustafson, P., Langridge, P. and Shi, B.J. (2015) Differential expression of microRNAs and other small RNAs in barley between water and drought conditions. *Plant Biotechnol. J.* **13**, 2–13.

Hajyzadeh, M., Turktas, M., Khawar, K.M. and Unver, T. (2015) miR408 overexpression causes increased drought tolerance in chickpea. *Gene*, **555**, 186–193.

Huang, D., Wu, W., Abrams, S.R. and Cutler, A.J. (2008) The relationship of drought-related gene expression in *Arabidopsis thaliana* to hormonal and environmental factors. *J. Exp. Bot.* **59**, 2991–3007.

Jones-Rhoades, M.W. and Bartel, D.P. (2004) Computational identification of plant microRNAs and their targets, including a stress-induced miRNA. *Mol. Cell*, **14**, 787–799.

Kanehisa, M., Goto, S., Sato, Y., Kawashima, M., Furumichi, M. and Tanabe, M. (2014) Data, information, knowledge and principle: back to metabolism in KEGG. *Nucleic Acids Res.* **42**, D199–D205.

Kant, S., Bi, Y.M., Zhu, T. and Rothstein, S.J. (2009) SAUR39, a small auxin-up RNA gene, acts as a negative regulator of auxin synthesis and transport in rice. *Plant Physiol.* **151**, 691–701.

Kantar, M., Lucas, S.J. and Budak, H. (2011) miRNA expression patterns of *Triticum dicoccoides* in response to shock drought stress. *Planta*, **233**, 471–484.

Karlova, R., van Haarst, J.C., Maliepaard, C., van de Geest, H., Bovy, A.G., Lammers, M., Angenent, G.C. *et al.* (2013) Identification of microRNA targets in tomato fruit development using high-throughput sequencing and degradome analysis. *J. Exp. Bot.* **64**, 1863–1878.

Kazan, K. and Manners, J.M. (2008) Jasmonate signaling: toward an integrated view. *Plant Physiol.* **146**, 1459–1468.

Klee, H.J. and Giovannoni, J.J. (2011) Genetics and control of tomato fruit ripening and quality attributes. *Annu. Rev. Genet.* **45**, 41–59.

Korir, N.K., Li, X., Xin, S., Wang, C., Changnian, S., Kayesh, E. and Fang, J. (2013) Characterization and expression profiling of selected microRNAs in tomato (*Solanum lycopersicon*) 'Jiangshu14'. *Mol. Biol. Rep.* **40**, 3503–3521.

Kozomara, A. and Griffiths-Jones, S. (2014) miRBase: annotating high confidence microRNAs using deep sequencing data. *Nucleic Acids Res.* **42**, D68–D73.

Kulcheski, F.R., de Oliveira, L.F., Molina, L.G., Almerao, M.P., Rodrigues, F.A., Marcolino, J., Barbosa, J.F. *et al.* (2011) Identification of novel soybean microRNAs involved in abiotic and biotic stresses. *BMC Genom.* **12**, 307.

Kumar, R. (2014) Role of microRNAs in biotic and abiotic stress responses in crop plants. *Appl. Biochem. Biotechnol.* **174**, 93–115.

Li, C. and Zhang, B.H. (2016) MicroRNAs in control of plant development. *J. Cell Physiol.* **231**, 303–313.

Li, R., Li, Y., Kristiansen, K. and Wang, J. (2008a) SOAP: short oligonucleotide alignment program. *Bioinformatics*, **24**, 713–714.

Li, W.X., Oono, Y., Zhu, J., He, X.J., Wu, J.M., Iida, K., Lu, X.Y. *et al.* (2008b) The Arabidopsis NFYA5 transcription factor is regulated transcriptionally and posttranscriptionally to promote drought resistance. *Plant Cell*, **20**, 2238–2251.

Li, R., Yu, C., Li, Y., Lam, T.W., Yiu, S.M., Kristiansen, K. and Wang, J. (2009) SOAP2: an improved ultrafast tool for short read alignment. *Bioinformatics*, **25**, 1966–1967.

Liu, H.H., Tian, X., Li, Y.J., Wu, C.A. and Zheng, C.C. (2008) Microarray-based analysis of stress-regulated microRNAs in *Arabidopsis thaliana*. *RNA*, **14**, 836–843.

Liu, Z., Yan, J.P., Li, D.K., Luo, Q., Yan, Q., Liu, Z.B., Ye, L.M. *et al.* (2015) UDP-glucosyltransferase71c5, a major glucosyltransferase, mediates abscisic acid homeostasis in Arabidopsis. *Plant Physiol.* **167**, 1659–1670.

Lu, Y., Feng, Z., Bian, L., Xie, H. and Liang, J. (2010) miR398 regulation in rice of the responses to abiotic and biotic stresses depends on CSD1 and CSD2 expression. *Funct. Plant Biol.* **38**, 44–53.

Luan, Y., Wang, W. and Liu, P. (2014) Identification and functional analysis of novel and conserved microRNAs in tomato. *Mol. Biol. Rep.* **41**, 5385–5394.

Ma, Y., Szostkiewicz, I., Korte, A., Moes, D., Yang, Y., Christmann, A. and Grill, E. (2009) Regulators of PP2C phosphatase activity function as abscisic acid sensors. *Science*, **324**, 1064–1068.

Moxon, S., Jing, R., Szittya, G., Schwach, F., Rusholme Pilcher, R.L., Moulton, V. and Dalmay, T. (2008) Deep sequencing of tomato short RNAs identifies microRNAs targeting genes involved in fruit ripening. *Genome Res.* **18**, 1602–1609.

Murashige, T. and Skoog, F. (1962) A revised medium for rapid growth and bioassays with tobacco tissue cultures. *Physiol. Plant.* **15**, 473–497.

Nemhauser, J.L., Hong, F. and Chory, J. (2006) Different plant hormones regulate similar processes through largely nonoverlapping transcriptional responses. *Cell*, **126**, 467–475.

Pasini, L., Bergonti, M., Fracasso, A., Marocco, A. and Amaducci, S. (2014) Microarray analysis of differentially expressed mRNAs and miRNAs in young leaves of sorghum under dry-down conditions. *J. Plant Physiol.* **171**, 537–548.

Pilcher, R.L., Moxon, S., Pakseresht, N., Moulton, V., Manning, K., Seymour, G. and Dalmay, T. (2007) Identification of novel small RNAs in tomato (*Solanum lycopersicum*). *Planta*, **226**, 709–717.

Rai, A.C., Singh, M. and Shah, K. (2013) Engineering drought tolerant tomato plants over-expressing BcZAT12 gene encoding a C(2)H(2) zinc finger transcription factor. *Phytochemistry*, **85**, 44–50.

Ramachandra Reddy, A., Chaitanya, K.V. and Vivekanandan, M. (2004) Drought-induced responses of photosynthesis and antioxidant metabolism in higher plants. *J. Plant Physiol.* **161**, 1189–1202.

Ren, Y., Chen, L., Zhang, Y., Kang, X., Zhang, Z. and Wang, Y. (2012) Identification of novel and conserved Populus tomentosa microRNA as components of a response to water stress. *Funct. Integr. Genomics*, **12**, 327–339.

Robinson, M.D. and Oshlack, A. (2010) A scaling normalization method for differential expression analysis of RNA-seq data. *Genome Biol.* **11**, R25.

Sadashiva, A.T., Christopher, M.G. and Krithika, T.K. (2013) Genetic Enhancement of Tomato Crop for Abiotic Stress Tolerance. In *Climate-Resilient Horticulture: Adaptation and Mitigation Strategies* (Singh, H.P., ed.), pp. 113–124. India: Springer.

Saito, R., Smoot, M.E., Ono, K., Ruscheinski, J., Wang, P.L., Lotia, S., Pico, A.R. *et al.* (2012) A travel guide to Cytoscape plugins. *Nat. Methods*, **9**, 1069–1076.

Schroeder, J.I., Allen, G.J., Hugouvieux, V., Kwak, J.M. and Waner, D. (2001) Guard cell signal transduction. *Annu. Rev. Plant Physiol. Plant Mol. Biol.* **52**, 627–658.

Shinozaki, K. and Yamaguchi-Shinozaki, K. (2007) Gene networks involved in drought stress response and tolerance. *J. Exp. Bot.* **58**, 221–227.

Spartz, A.K., Lee, S.H., Wenger, J.P., Gonzalez, N., Itoh, H., Inze, D., Peer, W.A. *et al.* (2012) The SAUR19 subfamily of SMALL AUXIN UP RNA genes promote cell expansion. *Plant J.: Cell Mol. Biol.* **70**, 978–990.

Storey, J.D. (2003) The positive false discovery rate: a Bayesian interpretation and the q-value. *Ann. Stat.* **31**, 2013–2035.

Sunkar, R. (2010) MicroRNAs with macro-effects on plant stress responses. *Semin. Cell Dev. Biol.* **21**, 805–811.

Sunkar, R. and Zhu, J.K. (2004) Novel and stress-regulated microRNAs and other small RNAs from Arabidopsis. *Plant Cell*, **16**, 2001–2019.

Sunkar, R., Kapoor, A. and Zhu, J.K. (2006) Posttranscriptional induction of two Cu/Zn superoxide dismutase genes in Arabidopsis is mediated by downregulation of miR398 and important for oxidative stress tolerance. *Plant Cell*, **18**, 2051–2065.

Thiebaut, F., Grativol, C., Tanurdzic, M., Carnavale-Bottino, M., Vieira, T., Motta, M.R., Rojas, C. *et al.* (2014) Differential sRNA regulation in leaves and roots of sugarcane under water depletion. *PLoS ONE*, **9**, e93822.

Tognetti, V.B., Van Aken, O., Morreel, K., Vandenbroucke, K., van de Cotte, B., De Clercq, I., Chiwocha, S. *et al.* (2010) Perturbation of indole-3-butyric acid homeostasis by the UDP-glucosyltransferase UGT74E2 modulates

Arabidopsis architecture and water stress tolerance. *Plant Cell*, **22**, 2660–2679.

Trindade, I., Capitao, C., Dalmay, T., Fevereiro, M.P. and Santos, D.M. (2010) miR398 and miR408 are up-regulated in response to water deficit in Medicago truncatula. *Planta*, **231**, 705–716.

Tyler, L., Thomas, S.G., Hu, J., Dill, A., Alonso, J.M., Ecker, J.R. and Sun, T.P. (2004) Della proteins and gibberellin-regulated seed germination and floral development in Arabidopsis. *Plant Physiol.* **135**, 1008–1019.

Vlad, F., Rubio, S., Rodrigues, A., Sirichandra, C., Belin, C., Robert, N., Leung, J. *et al.* (2009) Protein phosphatases 2C regulate the activation of the Snf1-related kinase OST1 by abscisic acid in Arabidopsis. *Plant Cell*, **21**, 3170–3184.

Vogt, T. and Jones, P. (2000) Glycosyltransferases in plant natural product synthesis: characterization of a supergene family. *Trends Plant Sci.* **5**, 380–386.

Wang, L., Feng, Z., Wang, X., Wang, X. and Zhang, X. (2010) DEGseq: an R package for identifying differentially expressed genes from RNA-seq data. *Bioinformatics*, **26**, 136–138.

Wang, T., Chen, L., Zhao, M., Tian, Q. and Zhang, W.H. (2011) Identification of drought-responsive microRNAs in *Medicago truncatula* by genome-wide high-throughput sequencing. *BMC Genom.* **12**, 367.

Williams, L., Grigg, S.P., Xie, M., Christensen, S. and Fletcher, J.C. (2005) Regulation of Arabidopsis shoot apical meristem and lateral organ formation by microRNA miR166 g and its AtHD-ZIP target genes. *Development*, **132**, 3657–3668.

Wu, J., Liu, S., He, Y., Guan, X., Zhu, X., Cheng, L., Wang, J. *et al.* (2012) Genome-wide analysis of SAUR gene family in Solanaceae species. *Gene*, **509**, 38–50.

Xia, K., Wang, R., Ou, X., Fang, Z., Tian, C., Duan, J., Wang, Y. *et al.* (2012) OsTIR1 and OsAFB2 downregulation via OsmiR393 overexpression leads to more tillers, early flowering and less tolerance to salt and drought in rice. *PLoS ONE*, **7**, e30039.

Xie, Z., Jia, G. and Ghosh, A. (2012) Small RNAs in Plants. In *MicroRNAs in Plant Development and Stress Responses* (Sunkar, R., ed.), pp. 1–28. Heidelberg: Springer.

Xie, F., Stewart Jr, C.N., Taki, F.A., He, Q., Liu, H. and Zhang, B. (2014) High-throughput deep sequencing shows that microRNAs play important roles in switchgrass responses to drought and salinity stress. *Plant Biotechnol. J.* **12**, 354–366.

Xie, F., Wang, Q., Sun, R. and Zhang, B. (2015) Deep sequencing reveals important roles of microRNAs in response to drought and salinity stress in cotton. *J. Exp. Bot.* **66**, 789–804.

Xu, Z.S., Ni, Z.Y., Li, Z.Y., Li, L.C., Chen, M., Gao, D.Y., Yu, X.D. *et al.* (2009) Isolation and functional characterization of HvDREB1-a gene encoding a

dehydration-responsive element binding protein in *Hordeum vulgare*. *J. Plant. Res.* **122**, 121–130.

Yin, F., Gao, J., Liu, M., Qin, C., Zhang, W., Yang, A., Xia, M. *et al.* (2014) Genome-wide analysis of water-stress-responsive microRNA expression profile in tobacco roots. *Funct. Integr. Genomics*, **14**, 319–332.

Zhang, B.H. (2015) MicroRNA: a new target for improving plant tolerance to abiotic stress. *J. Exp. Bot.* **66**, 1749–1761.

Zhang, B. and Wang, Q. (2015) MicroRNA-based biotechnology for plant improvement. *J. Cell. Physiol.* **230**, 1–15.

Zhang, B.H., Pan, X.P., Cannon, C.H., Cobb, G.P. and Anderson, T.A. (2006) Conservation and divergence of plant microRNA genes. *Plant J.* **46**, 243–259.

Zhang, X., Zou, Z., Gong, P., Zhang, J., Ziaf, K., Li, H., Xiao, F. *et al.* (2011) Over-expression of microRNA169 confers enhanced drought tolerance to tomato. *Biotech. Lett.* **33**, 403–409.

Zhang, N., Yang, J., Wang, Z., Wen, Y., Wang, J., He, W., Liu, B. *et al.* (2014) Identification of novel and conserved microRNAs related to drought stress in potato by deep sequencing. *PLoS ONE*, **9**, e95489.

Zhao, B., Liang, R., Ge, L., Li, W., Xiao, H., Lin, H., Ruan, K. *et al.* (2007) Identification of drought-induced microRNAs in rice. *Biochem. Biophy. Res. Commun.* **354**, 585–590.

Zhong, R. and Ye, Z.H. (2007) Regulation of HD-ZIP III genes by MicroRNA 165. *Plant Signal. Behav.* **2**, 351–353.

Zhou, L., Liu, Y., Liu, Z., Kong, D., Duan, M. and Luo, L. (2010) Genome-wide identification and analysis of drought-responsive microRNAs in Oryza sativa. *J. Exp. Bot.* **61**, 4157–4168.

3

Functional analyses of cellulose synthase genes in flax (*Linum usitatissimum*) by virus-induced gene silencing

Maxime Chantreau[1], Brigitte Chabbert[2,3], Sylvain Billiard[4], Simon Hawkins[1] and Godfrey Neutelings[1,*]

[1]*UMR INRA 1281 Stress Abiotiques et Différenciation des Végétaux Cultivés, Université Lille Nord de France Lille 1, Villeneuve d'Ascq, France*
[2]*INRA, UMR 614 Fractionnement des AgroRessources et Environnement, Reims, France*
[3]*UMR 614 Fractionnement des AgroRessources et Environnement, Université de Reims Champagne-Ardenne, Reims, France*
[4]*UMR CNRS 8198 Laboratoire de Génétique & Evolution des Populations Végétales, Université Lille Nord de France Lille 1, Villeneuve d'Ascq, France*

*Correspondence
email godfrey.neutelings@univ-lille1.fr

Keywords: cellulose, fibres, flax, gene expression, silencing, virus-induced gene silencing.

Summary

Flax (*Linum usitatissimum*) bast fibres are located in the stem cortex where they play an important role in mechanical support. They contain high amounts of cellulose and so are used for linen textiles and in the composite industry. In this study, we screened the annotated flax genome and identified 14 distinct cellulose synthase (*CESA*) genes using orthologous sequences previously identified. Transcriptomics of 'primary cell wall' and 'secondary cell wall' flax *CESA* genes showed that some were preferentially expressed in different organs and stem tissues providing clues as to their biological role(s) *in planta*. The development for the first time in flax of a virus-induced gene silencing (VIGS) approach was used to functionally evaluate the biological role of different *CESA* genes in stem tissues. Quantification of transcript accumulation showed that in many cases, silencing not only affected targeted *CESA* clades, but also had an impact on other *CESA* genes. Whatever the targeted clade, inactivation by VIGS affected plant growth. In contrast, only clade 1- and clade 6-targeted plants showed modifications in outer-stem tissue organization and secondary cell wall formation. In these plants, bast fibre number and structure were severely impacted, suggesting that the targeted genes may play an important role in the establishment of the fibre cell wall. Our results provide new fundamental information about cellulose biosynthesis in flax that should facilitate future plant improvement/engineering.

Introduction

The *Linum* genus contains about 180 different species spread across the six continents (Sveinsson *et al.*, 2014). Some of these species have been exploited by man since at least the upper Palaeolithic period when wild flax fibres were used by hunter-gatherers to make cords (Kvavadze *et al.*, 2009). Common flax (*Linum usitatissimum*) was then domesticated during the Neolithic era and became one of the earliest cultivated plants used for weaving (Zohary and Hopf, 2004) but also as an oil source (Oomah, 2001). Since then, commercial flax has been progressively selected to optimize stem content in cellulose-rich phloem fibres (bast fibres) and/or seed oil naturally rich in omega-3 alpha-linolenic acid (ALA).

Bast fibres are very long cells located beneath the epidermis in the stem cortex of some plants such as flax, hemp, ramie and jute (Huis *et al.*, 2012). They are considered among the longest cells in terrestrial plants where they can reach up to 55 cm in ramie (*Boehmeria nivea*) (Aldaba, 1927). These fibres are characterized by a very thick secondary cell wall (CWII) generally containing high amounts of crystalline cellulose, significant amounts of hemicellulose and low lignin contents in contrast to the much more heavily lignified cell walls typically found in the xylem tissue of most plants (Day *et al.*, 2005b; Gorshkova and Morvan, 2006; Neutelings, 2011; Chantreau *et al.*, 2014). In addition to their particular chemical composition, the bast fibre CWII is also characterized by parallel cellulose microfibrils aligned with respect to the fibre axis thereby conferring tensile strength (Gorshkova *et al.*, 2012). This special composition and organization of cell wall polymers explains the particular physical properties of these fibres that are used in the textile (linen) industry and more recently in composite materials for the car and aeronautic industries Day et al., 2013.

In plants, cellulose microfibrils are believed to contain 36 glucan chains, each one synthesized by an individual cellulose synthase (CESA) subunit organized in a heteromeric rosette complex (CSC) in which the exact number of units is still uncertain (Somerville, 2006). CESAs are part of the CAZy glycosyltransferase family 2 (GT2) and catalyse the polymerization of ß-(1,4) glucans using UDP-glucose as a substrate. They are large proteins bound to the plasma membrane by 8 transmembrane helices with a cytosolic region containing the D_D_D_QxxRW motif involved in catalytic activity. They belong to a large protein superfamily including nine additional subfamilies named cellulose synthase-like (CSL) defined by sequence similarity to CESAs. Different *CESA* genes are generally believed to be more specifically associated with either primary or secondary cell wall cellulose biosynthesis. In fact while cellulose is present in both the primary and secondary plant cell wall, the two wall types show important differences in the degree of polymerization and crystallinity (Lei *et al.*, 2012). In *Arabidopsis*, there are 10 *CESA* genes; three different CESAs (CESA4, 7 and 8) are needed to synthesize the cellulose microfibrils in CWII, while in the primary cell wall, the CSCs contain CESA1 and 3 and CESA6-like isoforms (CESA2, 5, 6 and 9). The role of AtCESA10 is less clear. When sequences from other plant species are included for phylogenetic analyses, six clades can be highlighted (Nairn and Haselkorn, 2005). Three of these clades are suspected to include the CWII-associated CESAs with each clade containing at least one *CESA* gene from each species. Among the 3 remaining 'primary cell wall clades', one contains the family three sequences, the second contains AtCESA1 and AtCESA10, and the last clade contains AtCESA2, 5, 6 and 9.

Given the cellulose-rich nature of the flax bast fibre cell wall and the importance of this polymer in determining fibre performance and quality, there are surprisingly only few reports on *CESA* genes in this species (Galinousky *et al.*, 2014; Mokshina *et al.*, 2014). The practical interest of such an investigation for flax is underlined by a recent study demonstrating that a functional relationship exists between CESA structures, cellulose crystallinity and saccharification efficiency in *Arabidopsis* (Harris *et al.*, 2012). While the information available from the recently sequenced genome of *Linum usitatissimum* (Wang *et al.*, 2012) has already been used to describe some cell wall-related gene families in this species (Barvkar *et al.*, 2012; Babu *et al.*, 2013; Hobson and Deyholos, 2013; Mokshina *et al.*, 2014), no studies have yet undertaken a functional approach on members of the *CESA* gene family.

Currently, functional genomics of a high number of genes in flax are difficult to carry out using a standard transgenic approach. Although stable transgene integration has been successfully performed in this species, the plant transformation rate remains low and obtaining regenerated plants is a time-consuming process (Caillot *et al.*, 2009). For this reason, a gene inactivation approach by VIGS (virus-induced gene silencing) (Baulcombe, 2006) is an interesting alternative because this technique is relatively easy to implement on many different plant species (Senthil-Kumar and Mysore, 2011). VIGS utilizes the naturally occurring phenomenon of post-transcriptional gene silencing (PTGS) that is believed to derive from an ancient mechanism involved in the defence of host cells against foreign nucleic acids, including viruses and active transposable elements (Voinnet, 2009). In the VIGS approach, a fragment of the targeted gene is cloned into a modified viral vector inserted in a plasmid. The viral DNA is then delivered to the plant by *Agrobacterium tumefaciens* infection. The most popular and efficient vector is the bi-partite TRV-VIGS vector (Liu *et al.*, 2002; a) derived from the tobacco rattle virus, a member of the genus *Tobravirus*. The TRV vectors are the most commonly used and have been tested successfully on at least a dozen of different species (Tian *et al.*, 2014). As VIGS vectors derive from plant viruses, the success of the gene silencing depends on the compatibility between the virus and the host plant. *Linum usitatissimum* has been reported to be susceptible to TRV (Brunt *et al.*, 1996), and in this paper, we report the use of this vector to induce silencing of flax *CESA* genes.

Results

Characterization and expression of flax *CESA* genes

Flax *CESA* genes were identified by BLAST screening the annotated genome (Wang *et al.*, 2012) and published (Day *et al.*, 2005a; Fenart *et al.*, 2010; Venglat *et al.*, 2011) or unpublished (http://www.ncbi.nlm.nih.gov) EST sequences followed by manual inspection. Altogether, we identified 14 potential *LusCESA* genes based on their homology to orthologous characterized sequences (Figure 1). The *in silico* characterization of the genomic sequences showed that a mistake occurred in the annotation of *Lus10008225* and *Lus10008226* which were considered as two short genes separated by a 361 nt fragment.

This was probably due to an inappropriate splicing design resulting from a stretch of unknown nucleotides present on the scaffold_157. Both sequences were associated and compared to the sequenced ESTs to remodel the splice junctions. We obtained the *Lus10008225-6* genomic sequence of 4193 bp resulting in a predicted protein of 1057 amino acids (Figure 1).

These results confirm the data recently published by Mokshina *et al.* (2014) during the revision of this manuscript. These authors identified 14 *CESA* genes based on the published flax genome, as well as two additional genes (*CESA7A*, *CESA7B*) on the scaffolds 57 and 464, thereby bringing the potential total number of flax *CESA* genes to 16. However, inspection of the CESA7A protein structure showed that it was missing 2 transmembrane domains, and we therefore re-aligned flax ESTs (Fenart *et al.*, 2010) and genomic data from Phytozome to generate a complete flax CESA7 protein (Figure 1). Similarly, the proposed CESA7B protein lacks an N-terminal Zn domain and is most likely a CSL protein, rather than a true CESA. These observations seem to show that the flax genome only contains a single *CESA7* gene as indicated in another recent paper (Galinousky *et al.*, 2014). Taken together, our results suggest that the flax genome contains 15 *CESA* genes (Figure 1).

The predicted *LusCESA* genes (Figure 1) have a size ranging from 3867 to 5836 bp and contain 11 to 14 exons which form coding regions between 2721 and 3294 bp (906 to 1097 amino acids). All the resulting proteins show characteristic features of plant CESAs (Delmer, 1999; Saxena *et al.*, 2001) including a Zn-binding motif and two transmembrane domains in the N-Terminal region, a central domain with a class-specific region and a GT2 motif (D_D_D_QxxRW) and finally six C-terminal transmembrane domains. A phylogenetic tree containing the protein sequences from *Arabidopsis thaliana*, *Populus trichocarpa* and *Linum usitatissimum* (Figure 2) shows that the 15 flax CESA genes are separated into 6 different clades in agreement with recently published data (Mokshina *et al.*, 2014).

Previous studies indicate that the genes involved in the formation of either the primary or secondary cell wall rosette structure are coexpressed in *Arabidopsis* (Persson *et al.*, 2005) and in rice (Wang *et al.*, 2010). To get a first overview of the potential functions of the flax *CESA* genes identified in this study, we determined the expression profiles of three genes from 'primary cell wall' clades (*LusCESA1-A*, *LusCESA3-C* and *LusCESA6-E*) and two genes from 'secondary cell wall' clades (*LusCESA4* and *LusCESA8-B*). Samples were collected at four different heights in the flax stem representing a developmental gradient from young (sample 1, upper) to older (sample 4, lower) stem tissues (Figure 3a). Stem samples from positions 2, 3 and 4 were separated into outer tissues (rich in bast fibres) and inner tissues (mainly xylem). *CESA* gene expression was also evaluated in the flowers, leaves and roots (Figure 3b). *LusCESA1-A* and *LusCESA6-D* showed very similar expression profiles. For both genes, no significant differences could be detected between inner- and outer-stem tissues regardless of stem height. Both genes were expressed in flowers, leaves and roots but at lower levels than in stem tissues. On the other hand, both *LusCESA4* and *LusCESA8-B* were more highly expressed in the inner stem tissues compared to the outer-stem tissues. Both genes were only very weakly expressed in flowers, leaves and roots. Interestingly, *LusCESA3-C* showed an opposite expression profile to *LusCESA4* and *LusCESA8-B* with significantly higher expression in outer-stem tissues even though this difference was reduced at the bottom of the stem. In

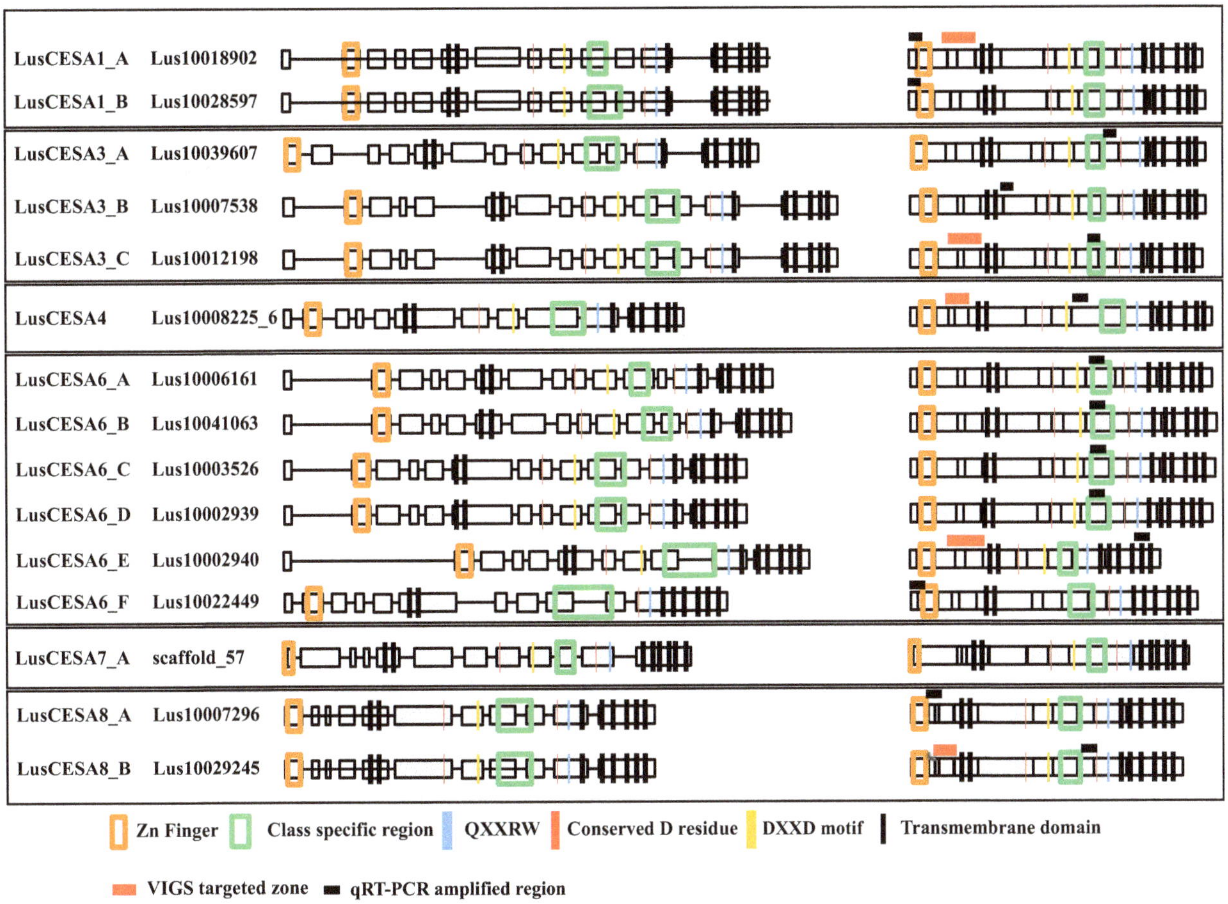

Figure 1 Gene structure of the 15 flax *CESA* genes. The genomic sequence (left) and predicted coding sequence (right) structures are shown with the CESA characteristics. Exons are represented by boxes and introns by lines.

flowers, leaves and roots, *LusCESA3-C* showed a similar expression to *LusCESA1-A* and *LusCESA6-D*.

Induction and optimization of TRV-based silencing in flax

To optimize VIGS in flax, different infection procedures (Table 1) were tested using a 450-bp cDNA fragment of *LusPDS* (Phytozome: Lus10021967) cloned into TRV2 and introduced in *Agrobacterium*. The *PDS* gene encodes a phytoene desaturase required for the biosynthesis of chlorophyll-protecting carotenoids (Kumagai *et al.*, 1995). Silencing of *PDS* genes leads to an easily detectable photobleaching phenotype (Ruiz *et al.*, 1998), and they are therefore often used as positive controls in VIGS. Of the different tested protocols, agrodrench, carborundum abrasion and infection with preamplified virus particles in tobacco only gave very low efficiency values. The number of photobleached plants increased when they were soaked in the bacterial suspension under vacuum, but the best results were obtained by syringe infection. Developmental stages and density of *Agrobacterium* inoculum are known to impact the efficiency of VIGS in some species (Velasquez *et al.*, 2009), and so we also evaluated these parameters. Plants were infected at 12, 15 and 20 days after germination with suspensions containing increasing bacterial densities (OD_{600} = 0.5; 1; 1.5 and 2), but no significant differences in silencing phenotypes were observed (data not shown). Slight virus infection symptoms appeared 13 days

postinoculation (dpi), and the first photobleached leaves and stems were visible 24 dpi (Figure 4a,d). When plants were infected with nonrecombinant TRV vectors, leaves exhibited only very mild wrinkling and occasional small pale zones (Figure 4c). No impact was observed on flax stems.

In other species, weak phenotypes have been associated with uneven virus distribution between the main stem and lateral branches (e.g. peanut stripe potyvirus, PStV) (Jain *et al.*, 2000). In tobacco, topping of plants resulted in a more efficient spread of the virus in the leaves of lateral shoots produced from the activated axillary buds (Wijdeveld *et al.*, 1992). We therefore removed the shoot apical meristem just after *Agrobacterium* inoculation in an attempt to improve the systemic spread of the virus. The plants then developed two secondary basal shoots 3 days after apex removal. These conditions led to a much higher silencing efficiency (up to 95%, see Table 1) and a more rapid (8 dpi versus 13 dpi) and generalized appearance of photobleached zones. This observation would suggest that the latent period associated with apex removal allows improved viral multiplication before lateral shoot regrowth.

Silencing of flax cellulose synthase genes

To explore the role of the flax *CESA* genes using a VIGS approach, we constructed independent recombinant TRV2 vectors containing sequences specific to the 3 'primary' (clades 1, 3 and 6) and 2 'secondary' (clades 4 and 8) cell wall *CESA* genes analysed in this

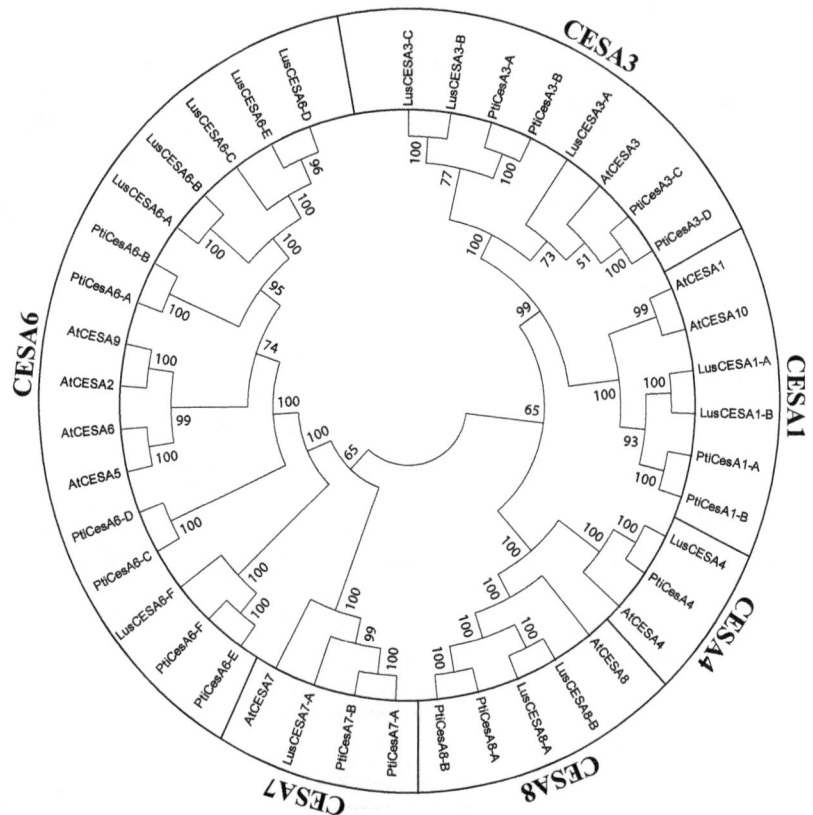

Figure 2 Neighbour-joining phylogenetic tree of CESA proteins. *Arabidopsis thaliana* (At), *Populus trichocarpa* (Pti) and *Linum usitatissimum* (Lus) sequences of the CESA superfamily were aligned with ClustalW (Thompson *et al.*, 1994) and the tree created with MEGA5 (Tamura *et al.*, 2011). Bootstrap values over 50% of 500 replications are indicated in the nodes.

study. These sequences (Figure 1) corresponded to the structurally heterogeneous N-terminal region located between the zinc finger and the first transmembrane domain of the CESA protein (Somerville, 2006) and allowed us to theoretically distinguish between the different clades. At the same time, the similarity between different CESA genes in the same clade also allowed us to use a single vector to target all genes within a given clade (Table S1). The 5 vectors were then used to infect flax plants that were named V_CS1, V_CS3, V_CS4, V_CS6 and V_CS8, respectively.

Expression of 14 *LusCESA* genes was evaluated by qRT-PCR in individual axillary stems of infected plants harvested at 35 dpi (Figure 5) and compared to the empty vector controls. The melting curves of all final products were carefully inspected to check that only one fragment was amplified with each primer couple (Table S2). Of the 14 tested *LusCESA* genes, all were expressed at detectable levels except for *LusCESA6-A* and *LusCESA6-F*. When an appropriate vector targeted a specific clade, expression levels of the corresponding genes were always significantly lower when compared to the control (100%), except for *LusCESA3-A* and *LusCESA3-C* (V_CS3 plants) and *LusCESA4* (V_CS4 plants). In the case of the V_CS3 silenced plants, this was not due to the time of sampling as the expression of the 2 genes was not significantly lower at 10, 21 and 31 days after the infection (Figure S1). Specific transcript level decreases varied from 24% (*LusCESA3-B*) to 74% (*LusCESA6-C*), respectively, and were >47% for clades 1, 6 and 8. Interestingly, when a specific *CESA* clade was targeted, at least one other gene outside this clade was also affected except in the case of V_CS3 plants. For example, when clade 1 or clade 8 CESAs were targeted, the mRNA levels of all *LusCESA* genes were significantly reduced compared with control plants. Similarly, the targeting of clade 6 *CESAs* affected all clades except for the clade 1. On the contrary,

the only effect observed when the clade 4 was targeted was a 35% reduction in *LusCESA6-E* transcripts. In some cases, the reduction in transcript accumulation for a given gene was lower with the clade-specific vector than with other nonspecific vectors. For example, the reduction in the expression of the *LusCESA8-A* gene was greater in V_CS1 (58%) and V_CS6 (74%) plants when compared to V_CS8 plants (47%).

Phenotypic analyses of *CESA*-silenced plants

At 35 dpi, all silenced plants were smaller than the controls (Figure 6a). A curly leaf phenotype was also visible in V_CS1, V_CS3 and V_CS6 plants (Figure 6b). To see whether stem anatomy and cell wall formation were affected in silenced plants, we examined thin transverse stem sections stained with toluidine-blue-O (TBO). In agreement with the observed reduction in plant size, all stem sections contained fewer secondary xylem cell layers and a complete pith region. More modifications were observed in V_CS1, V_CS3 and V_CS6 plants (Figure 7) than in V_CS4 and V_CS8 plants (data not shown). *CESA* gene silencing had a marked effect on outer-stem tissues in both V_CS1 and V_CS6 plants. This effect was particularly strong in V_CS1 plants (Figure 7c,d) where the outline of cortical parenchyma cells was highly irregular and epidermal cells appeared more rounded than in control plants (Figure 7a). Bast fibre number was greatly reduced, and cell shape was irregular. Vascular cambial cells and young differentiating xylem cells, but not more mature lignified xylem cells, were also characterized by an irregular cell contour. Dense purple-coloured deposits were observed between pith cells and probably represent cell wall material of crushed cells. In V_CS3 plants (Figure 7e,f), the overall cell shape was much less affected and outer-stem tissues appeared normal with no marked modifications to bast fibre number and shape. In contrast, an important effect was observed at the xylem–pith junction where

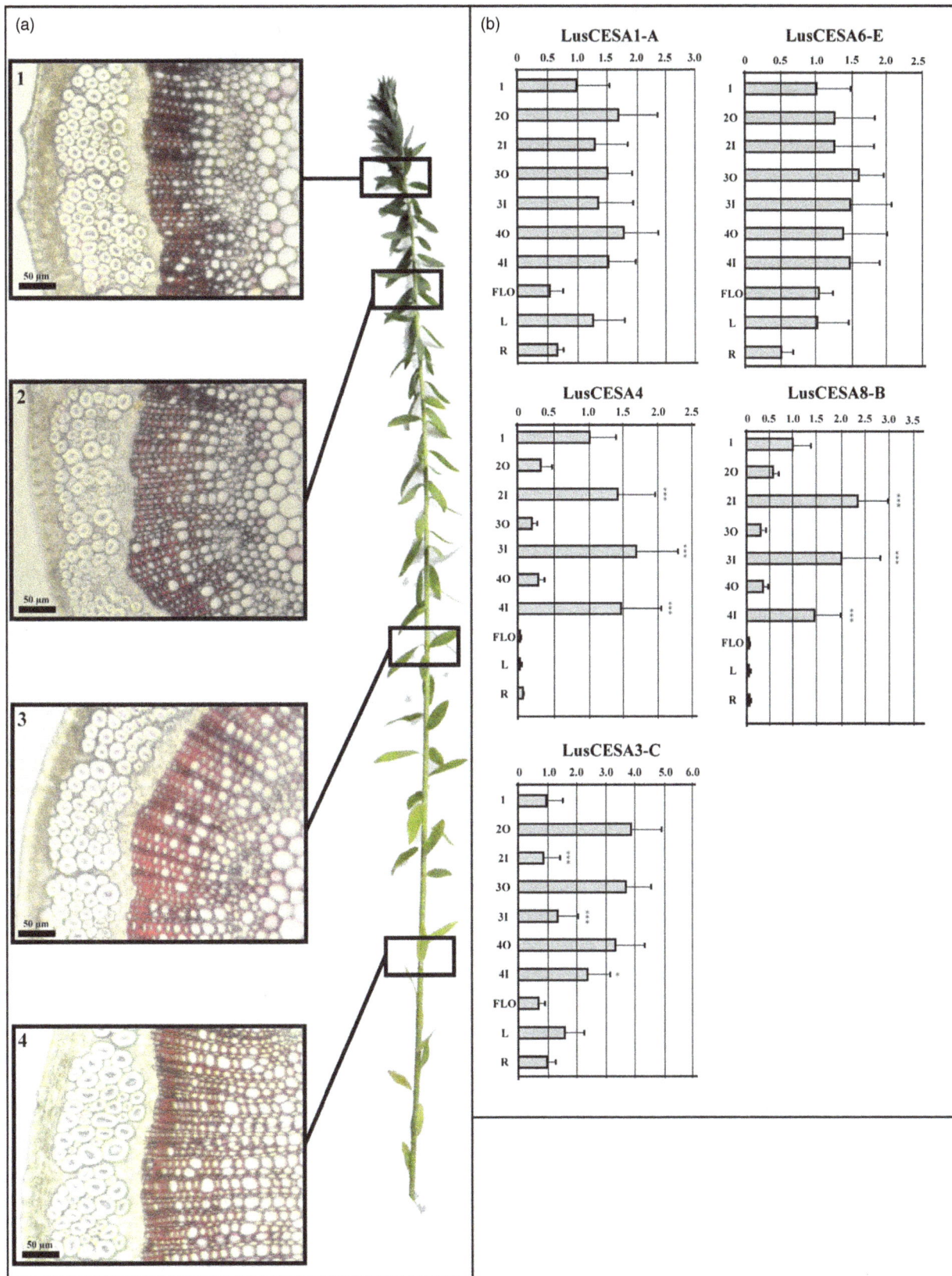

Figure 3 Clade-specific gene expression in flax stems at different heights. (a): Four flax wild-type stem fragments were collected, and transversal sections were stained with phloroglucinol-HCl. The lignified cell walls are coloured in red. (b): Relative quantification of a member from 5 CESA clades by qRT-PCR. The value of the top whole fragment (1) was set to 1. The numbers 1, 2, 3 and 4 indicate the position of each fragment (4, 10, 25 and 40 cm from the shoot apex, respectively) as shown in (a). I: Inner tissues; O: outer tissues; FLO: flower buds; L: leaves; R: roots. Significant differences between the outer and inner tissues were shown at $P < 0.001$ (***), $P < 0.01$ (**) and $P < 0.05$ (*). Mean expression values and standard deviation were presented.

Table 1 Virus-induced gene silencing-induced PDS silencing with previous published protocols

Infection method	Number of tested plants	Infection efficiency (% infected flax plants)	Protocol reference
Agrodrench	63	0%	(Ryu et al., 2004)
Abrasion	56	2%	(Ruiz et al., 1998)
Sap inoculation	15	20%	(Brigneti et al., 2004)
	55	0%	(Valentine et al., 2004)
Vacuum	39	72%	(Hileman et al., 2005)
Syringe	39	85%	(Fu et al., 2005)
Syringe (removed apical meristem)	95	95%	This paper

numerous lignified cells appeared to be crushed and/or showed irregular outlines. The impact of *CESA* silencing was most extreme in V_CS6 plants where an important disorganization was observed (Figure 7g–i). The secondary xylem ring was incomplete although small poles of primary xylem containing 2–3 lignified cells could be observed in certain zones. In outer tissues, the epidermis was unaffected, but the cortical parenchyma was highly disorganized with very irregular blue-coloured deposits of cell wall material. Similar deposits could also be observed at the xylem–pith interface. As for V_CS1 plants, bast fibre number was greatly reduced.

Figure 4 Phenotype of the TRV2-*PDS*-transformed plants. (a): Leaves and a stem fragment of a photobleached plant. (b): The same organs from a wild-type plant and (c): from an empty vector control-transformed plant. (d): Young lateral stem emerging from a TRV2-*PDS*-transformed plant.

An analysis of the sugar content was also performed to identify potential differences in the cell wall composition between the control and silenced plants (Figure 8). *CESA* silencing produced significant modifications in sugar composition. Slight but significant reductions in the relative proportion of glucose were determined in V_CS1 (8.6%), V_CS3 (10.5%) and V_CS6 (14.4%) plants compared to the control group. Modifications in sugar content associated with other cell wall polymers were also observed. The most important impact was observed in V_CS6 plants where the rhamnose content increased by 200% and the galactose and galacturonic acid content increased by 160% ($P < 0.001$). Significant increases for these sugars were also observed in V_CS1 and V_CS3 plants. In contrast, xylose content remained constant in all VIGS infiltrated plants.

Discussion

In flax, the role of the bark is not only to protect the plant against outside attacks but also to confer rigidity to the stem. This latter property is fundamental to prevent lodging because the ratio between the length and the width of the stem can reach values close to 400 in cultivated *Linum* species. The stem rigidity is reinforced by the outlying bast fibres organized in bundles, each one containing tens of elementary units cemented together by pectins (Charlet et al., 2010). At the end of their development process, their CWII are very well developed and are usually larger than the cell lumen itself. The presence of high amounts of cellulose is an extremely important feature of these cell walls, and engineering/breeding of flax plants for better fibre quality requires an improved understanding of the mechanisms controlling its synthesis.

Based on the flax published genome (Wang et al., 2012) and re-evaluation of recent data (Mokshina et al., 2014), we identified 15 gene models that meet the criteria described in previously characterized CESA sequences, that is the presence of specific motifs such as an N-terminal zinc finger, 8 transmembrane domains and a class-specific region (Richmond and Somerville, 2000). The 15 flax *CESA* genes could be assigned to 6 different clades as observed in other plant species (Carroll and Specht, 2011; Yoo and Wendel, 2014).

Analyses of flax *CESA* gene expression from 'primary' and 'secondary' cell wall clades revealed a number of organ-specific differences. Of particular interest is the observation that *LusCESA3-C* was much more highly expressed in outer-stem tissues containing bast fibres than in inner stem tissues. A similar, but less important, difference in the expression of this gene was also recently observed in these tissues (Mokshina et al., 2014). While it is tempting to speculate that this gene (and/or other clade 3 genes) may play a specific role in the synthesis of the bast fibre cell wall, further characterization is necessary. Organ-/tissue-specific differential expression of *CESA* genes is also supposed to be related to cell wall formation in cellulose-rich cotton fibres. In this species, a clade 1 gene, *GhCESA6,* is more highly expressed in the cotton bolls as compared to the stem (Li et al., 2013) and may therefore have a more important role in the construction of the cell wall of cotton fibres (single cell outgrowths of seed epidermal cells) as compared to other CESA genes involved in xylem cell wall formation. In contrast to the higher expression of the *LusCESA3-C* gene in outer-stem tissues, 2 other genes (*LusCESA4* and *LusCESA8-B*) were more highly expressed in inner stem tissues in agreement with recently published results (Mokshina et al., 2014). This study also reported a similar expression pattern for the flax *CESA7* gene clade, and it is

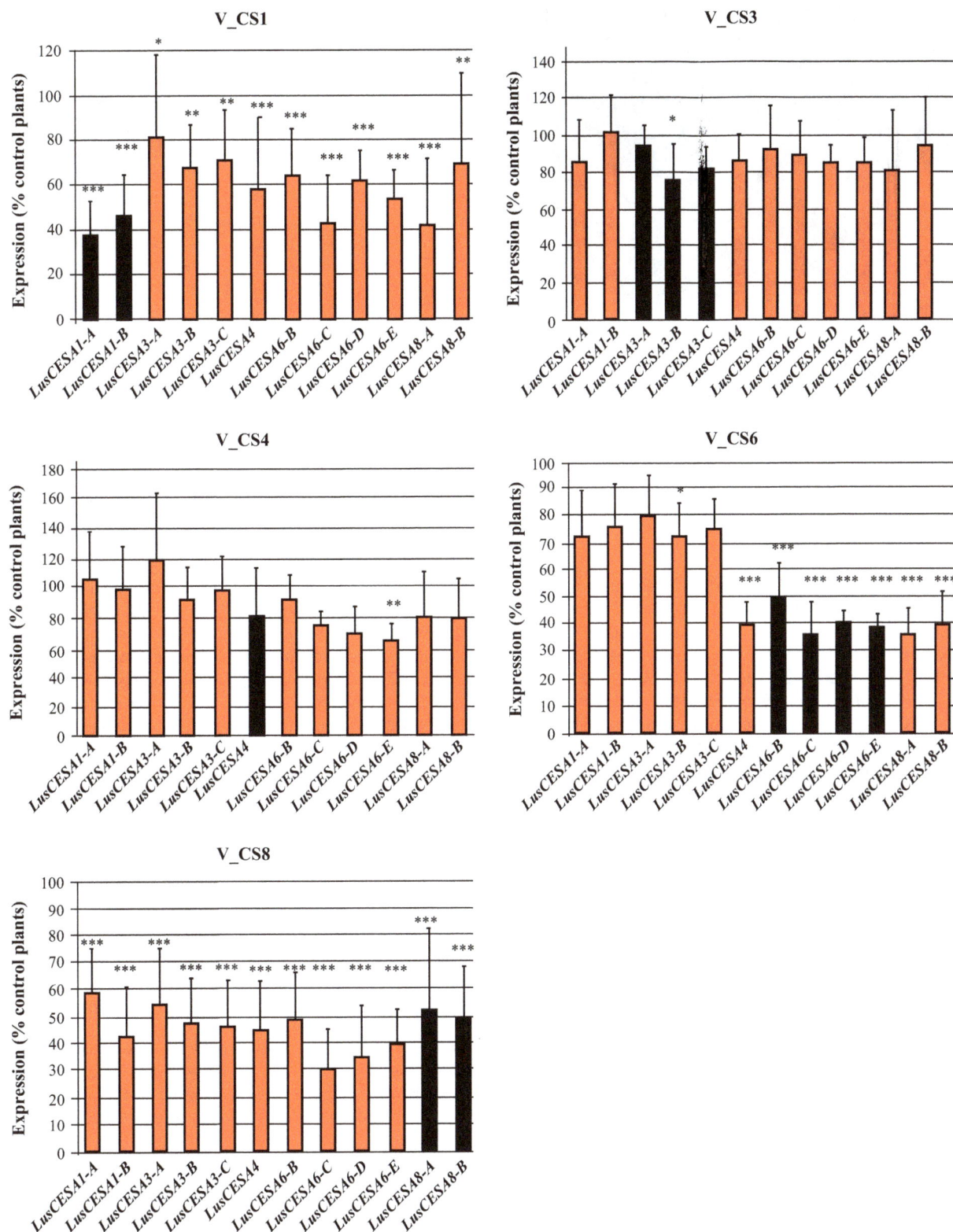

Figure 5 qRT-PCR analysis of *LusCESA* genes in TRV2-*CESA*-transformed plants. The number associated with the V_CS nomenclature indicates the corresponding clade-specific fragment. The bars indicating the genes targeted by VIGS are in black. Significant differences between the transformed and the control plants were shown at $P < 0.001$ (***), $P < 0.01$ (**) and $P < 0.05$ (*). Mean expression values and standard deviation were presented.

likely that these 3 flax clades (*LusCESA4*, *LusCESA8* and *LusCESA7*) are functional orthologs of the CWII-specific *AtCESA4*, *AtCESA7* and *AtCESA8* genes in *Arabidopsis*.

The role of different flax *CESA* genes in cell wall metabolism was then functionally characterized by a reverse genetics approach. In flax, functional genomics is hampered because

Figure 6 Phenotype of the TRV2-*CESA*-transformed plants. (a): Plant height; (b): leaf form/size. The number associated with the V_CS nomenclature indicates the corresponding clade-specific fragment. CONT: empty vector control.

stable transformation is time-consuming. We therefore decided to see whether we could develop a VIGS protocol for this species. To our knowledge, this is the first report of successful gene knock-down in the *Linaceae* family by this method. Previous published data reported an interaction between the oat blue dwarf virus and *Linum usitatisimum* (Banttari and Zeyen, 1972), but no VIGS vector has been constructed from this genome. In other plant species, the TRV vector system (Liu *et al.*, 2002) has been widely used because of its large host spectrum. Flax was reported to be susceptible to TRV in the Virus Identification Data Exchange project (Brunt *et al.*, 1996), and we therefore used this vector to silence selected *LusCESA* genes. We first used a flax *PDS* gene as a positive control because silencing causes a photobleaching phenotype due to the inhibition of carotenoid biosynthesis (Kumagai *et al.*, 1995). Initial testing of different infection protocols revealed that syringe infiltration of young leaves with an equal mix of TRV1 and recombinant TRV2 was the most efficient. In our first attempts, we noticed that bleached leaves were distant from the origin of infection and that silencing only spreads to a limited number of leaves and only poorly to the stem. However, when the apex of the plant was removed, the two secondary shoots that developed at the bottom of the stem immediately gave rise to organs with a strong bleached phenotype. This protocol enabled us to obtain plants showing modified gene expression at only 3 weeks after germination as compared to the average time (>3 months) necessary to select stably transformed transgenic lines. The observed photobleaching of stem cortical chlorenchyma confirmed that *PDS* silencing was also effective in this organ, indicating that VIGS can be considered as a quick and simple method for screening several cell wall genes within stem bast fibres. This approach could also be used to functionally characterize other key genes in flax. For example, flax seeds are considered as major sources of polyunsaturated fatty acids used as feed and food products and also contain high amounts of the biologically active lignan secoisolariciresinol diglucoside (SDG), and it would therefore be interesting to use the VIGS approach to investigate these metabolic pathways. VIGS

has indeed already been successfully achieved on seeds or grains of several crop species such as *Triticum* and *Physalis* (Ma *et al.*, 2012; Zhang *et al.*, 2014).

For the VIGS silencing of flax *CESA* genes, we constructed 5 vectors that targeted 3 'primary' and 2 'secondary' cell wall clades. In the light of the recent identification (Mokshina *et al.*, 2014) of the flax *CESA7* gene, it would also be interesting to target this gene in future studies. Modifications in the expression of 14 identified flax *CESA* genes were then monitored by qRT-PCR. Our results indicated that targeted silencing was successful in all plants except for the clade 3 (V_CS3 plants), in which only *LusCESA3-B* was significantly down-regulated, and clade 4. The reduction in transcript abundance compared to the control was variable depending on the clade and reached 74% for *LusCESA6-C*. The efficiency of VIGS can depend on the position of the insert with respect to the cDNA (Liu and Page, 2008) and may also vary between different members of the same multigenic family even when the fragment is designed to target the same area. Our results also indicated that there was a strong cross-regulation effect between the members of the different CESA clades. This was most likely not due to homology between targeted and nontargeted sequences as the alignment of the fragments used for the constructs and the corresponding CESA sequences did not contain any identical 21 nt sequence stretches (Figure S2). For example, our results show that the silencing of the clade 8 had an effect not only on *LusCESA8-B* and *LusCESA8-A* but also on other *CESA* genes expressed in the stem. Of interest is the observation that the down-regulation of nontargeted genes may be greater than that of the targeted gene. Such an effect was seen for the two *LusCESA8* genes in V_CS6 plants, as well as for the *LusCESA6-C* in V_CS8 plants. Although a cross-suppression effect cannot be totally excluded, the identity between the sequences in clade 6 and 8 does not exceed 55% and is unlikely to directly trigger the silencing pathway. A more plausible explanation could be that of *CESA* gene co-regulation. Previous studies on the VIGS down-regulation of *CESA* genes in barley showed that expression of the closely related *CSL* genes as well as the more distantly related *GT8* gene was reduced, suggesting that the expression of several cell wall biosynthetic genes is linked with that of *CESA* (Held *et al.*, 2008). In flax, it seems likely that clade 4 and clade 8 genes are involved in CWII synthesis, but it is difficult for the moment to determine the role of other genes with respect to the synthesis of the primary or secondary cell wall. In flax, the fact that significant silencing was observed for the 3 *LusCESA3* genes in V_CS1 plants, for the *LusCESA4* in V_CS6 and for both *LusCESA8* genes in V_CS6 already allows us to outline a number of possible interactions between these clades. The amount of flax transcriptomic data is currently growing and will lead to the identification of regulatory networks of genes involved in CWII synthesis.

VIGS-induced *CESA* down-regulation in flax also provoked a visible phenotype, and all plants infected with a TRV2-*CESA* construction were markedly shorter than plants infected with the empty control vector. In the stem, our VIGS approach showed that the down-regulation of CWII orthologous genes had no phenotypic effect on the cell walls, presumably because the residual gene activities were sufficient for cell construction. The most severe impact on tissue organization was observed in V_CS6 plants, and the strongest effect on bast fibre cells occurred in V_CS1 plants. In both cases, typical CWII organization was replaced by a structure of irregular thickness that resembled a primary wall. This strong phenotype can be related to the down-

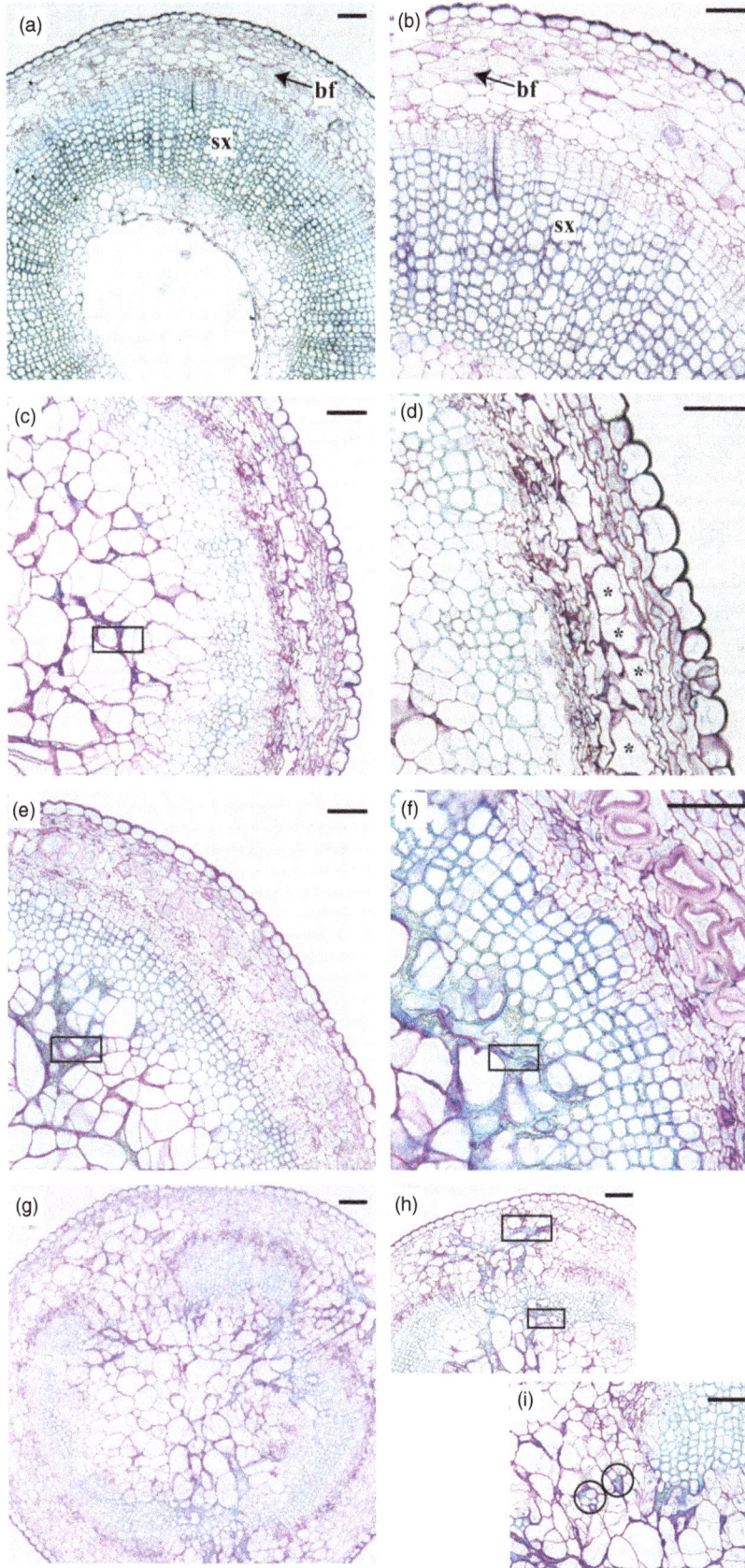

Figure 7 Cytological observation of CESA-silenced plant stems. (a–b): wild-type plants; (c–d): TRV2-*CESA1*-transformed plants; (e-f): TRV2-*CESA3*-transformed plants; (g-i): TRV2-*CESA6*-transformed plants. Square frames show collapsed/agglomerate cell walls; circles show residual primary xylem poles; asterisks shows large parenchyma cells without secondary walls in place of cellulosic fibres. sx: secondary xylem; bf: bast fibres. Bar: 50 μm.

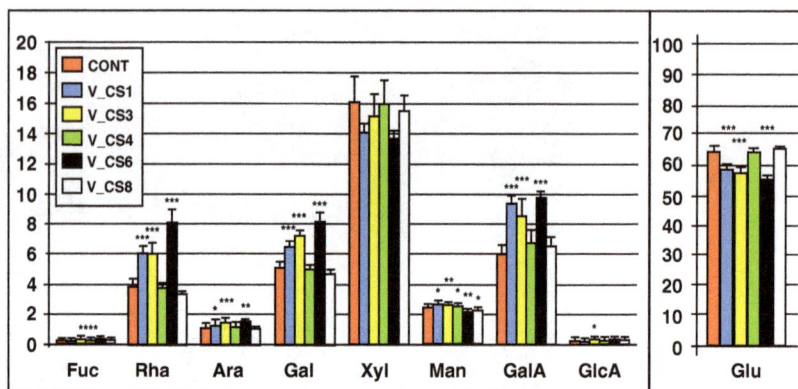

Figure 8 Sugar composition of cell wall polysaccharides in whole stems. The values are expressed as percentage ratios of each sugar to the total sugar content and are statistically compared to empty vector control plants. Significant differences between the transformed and the control plants were shown at $P < 0.001$ (***), $P < 0.010$ (**) and $P < 0.05$ (*). Mean expression values and standard deviation were presented.

regulation of all the identified *CESA* genes in these plants. Reduction in plant size was also observed when *CESA1* and *CESA3* genes were silenced by RNAi in *Arabidopsis* (Burn *et al.*, 2002) or when a *NtCesA-1* fragment was targeted by VIGS in tobacco (Burton *et al.*, 2000). In this later species, the size of the leaves was also reduced as observed in flax V_CS1, V_CS3 and V_CS6 plants. The cell shape modifications observed in the flax stem is also similar to that observed in the *Arabidopsis irx* (*cesA4-8*) mutants for Liepman *et al.* (2010). It still remains unclear whether the phenotypic modifications observed in the flax stem were only due to *CESA* down-regulation or to a global modification of co-regulated genes. Chemical analyses showed that VIGS down-regulation of *CESA* genes was associated with a slight but significant decrease in the proportion of glucose, suggesting that cellulose content was also reduced in these plants to the same extent as in VIGS silenced tobacco (Burton *et al.*, 2000) but lower than that observed in the *Arabidopsis prc-1* mutant (MacKinnon *et al.*, 2006) or the rice *Bc6* mutant (Kotake *et al.*, 2011). The observed increase in the proportion of other cell wall-related sugars also suggests that the structure/amount of noncellulosic polysaccharides (NCPs) was modified in these plants. The relatively low impact on cellulose content could be related to the propagation of heterogenous viral cDNA as previously suggested (Burton *et al.*, 2000). In the stem, virus particles are transported by the phloem and then move through the cells via the plasmodesma. It is possible that fibres located near the phloem are targeted more rapidly than the inner parts of the stem.

In conclusion, we have provided new transcriptomic and functional data on flax cellulose synthase genes. Their phylogenetic distribution is similar to that of poplar, which also belongs to the Malpighiale order. To provide functional information, we successfully developed a VIGS approach and demonstrated that the most important effect on bast fibres was obtained by targeting *CESA* genes previously described as actors of the primary cell wall formation. This may suggest that they can play an important role in the establishment of the fibre cell walls. These data will also contribute to our overall knowledge about the formation of bast fibres in other important fibre species such as ramie, jute and hemp. VIGS could also be used as a tool to explore the relationship between *CESA* genes and flax domestication by studying differential expression between cultivated flax (*L. usitatissimum*) and its wild progenitor (*L. bienne*). In addition, we are currently using this tool to target other cell wall genes identified in a mutant flax population (Chantreau *et al.*, 2013). The implementation of VIGS in this species will enhance the importance of flax as an emerging model system for studying cell wall metabolism.

Experimental procedures

Plant material

Flax plants (*Linum usitatissimum*) were grown in a greenhouse under 16 h/20 °C day and 8 h/18 °C night conditions. For microscopic, transcriptomic and sugar analyses, 43-day-old axillary stems from 3 silenced and control plants were immediately frozen in liquid nitrogen.

Light microscopy

The stem fragments were submitted to a series of alcohol baths and then embedded in methacrylate resin (Technovit 7100; Kulzer; Heraeus; Germany). Sections of 5 μm thickness were made using a microtome, stained with toluidine-blue-O and examined using a LEICA DM2000 (Leica Microsystems GmbH, Wetzlar, Germany).

Sugar analyses

Tissue samples from the pooled material were ground and repeatedly 80% ethanol extracted. Complete acid hydrolysis of cell wall polysaccharides into sugar monomers was performed using a two-step hydrolysis. Samples were swollen in 72% H_2SO_4 for 2 h at 20 °C followed by a second hydrolysis with 1 M H_2SO_4 for 2 h at 100 °C. The released monosaccharides were separated by high-performance anion-exchange chromatography (PA1 column, Dionex, Thermo Fisher Scientific, St Herblain, France using 2-deoxy-D-ribose as internal standard and standard solutions of neutral sugars and acidic sugars (Beaugrand *et al.*, 2004). One-way ANOVA was used to analyse the effect of VIGS transformation on the sugar composition at $P < 0.001$ (SigmaPlot, Systat Software, Erkrath, Germany). Multiple comparisons between transformed versus control samples (Holm–Sidak method) were then performed.

Molecular constructions

Cellulose synthase sequences from poplar and *Arabidopsis* were used to search for orthologs in the flax genome (Wang *et al.*, 2012) and in cDNA sequences obtained from the GenBank database (http://www.ncbi.nlm.nih.gov) and an EST database (Fenart *et al.*, 2010). The resulting sequences were then aligned and manually inspected. Fragments were PCR-amplified on cDNA using primers described in Table S3 and cloned into pCR2.1 plasmid using the TOPO TA cloning kit (Thermo Fisher Scientific, St Herblain, France). They were subsequently digested with *Eco*RI, purified on electrophoresis agarose gel using the QIAquick Gel Extraction kit (Qiagen, Courtaboeuf,

France) and cloned into TRV2 vector (Liu *et al.*, 2002) using T4 DNA ligase (Thermo Fisher Scientific, St Herblain, France).

Agrobacterium infiltration

TRV1 and TRV2 constructs were introduced into *Agrobacterium tumefaciens* strain GV2260 by electroporation, and the bacteria were grown in a selective LB medium supplemented with 10 mM MES and 20 mM acetosyringone and centrifuged and the pellet resuspended in an agroinfiltration medium containing 1 M MgCl$_2$, 10 mM MES and 150 μM acetosyringone. The agro-drench protocol (Ryu *et al.*, 2004) was conducted by depositing 1 mL of a 1 : 1 mixture of *Agrobacterium* solutions containing TRV1 and TRV2-PDS plasmids on the crown part of 15-day-old plants. Flax plants were also immersed in a mix of both bacterial suspensions supplemented with Silwet 0.05% and vacuum infiltrated for 5 min (Hileman *et al.*, 2005). The penetration of the agrobacterial suspensions was also maximized after sprinkling a small quantity of carborundum powder on 4 fully expanded leaves (Ruiz *et al.*, 1998). A viral multiplication was also done in *Nicotiana benthamiana* plants before infecting flax plants with a syringe. The leaf sap was collected by grinding the leaves and centrifugation of the extract followed by a precipitation step (Valentine *et al.*, 2004) or not (Brigneti *et al.*, 2004). The best results were obtained by syringe infiltration (Fu *et al.*, 2005). Using a 1-mL syringe, a 1 : 1 mixture of TRV1 and TRV2-*PDS Agrobacterium* suspension was infiltrated in the four first fully expanded leaves. The infection efficiency was further increased when the apical meristem was removed. The silencing then occurred in the two stems initiated from the lateral buds.

RNA analysis

Total RNA was extracted from isolated flax stems (three plants per experiments) as previously described (Huis *et al.*, 2010). RNA integrity was evaluated using the Experion electrophoresis system (Bio-Rad, Marnes-la-Coquette, France). For each sample, 500 ng of RNA was reverse-transcribed using the Iscript cDNA synthesis kit (Bio-Rad) according to the manufacturer's instructions. Primer pairs (Table S2) were specifically designed for the identified *CESA* gene models using Primer3 (Rozen and Skaletsky, 2000) and optimized for a Tm at 60 °C and for the amplification of 90–180 bp fragments. The qRT-PCRs were carried out in a reaction volume of 20 μL (5 μL diluted cDNAs, 10 μL of 2× SYBR Green mix and primer pairs at 0.4 μM). All PCRs were performed on three technical repetitions under the following conditions: 95 °C for 15 min, 40 cycles of 10 s at 95 °C and 30 s at 60 °C. The expression data were analysed as previously described (Huis *et al.*, 2010). We performed a generalized linear mixed model (GLMM) with the individuals as the random factor and the treatment as the fixed factors for each measured gene and for each organ separately. The procedure of model selection was followed, and the interpretation was performed as previously described (Zuur *et al.*, 2008).

Phylogenetic analysis

Phylogenetic tree of CESA proteins was made using neighbour-joining method conducted in MEGA5 (Tamura *et al.*, 2011). Bootstrap consensus tree inferred from 500 replicates. Branches corresponding to partitions reproduced <50% bootstrap replicates are collapsed. The evolutionary distances were computed using the *p*-distance method.

References

Aldaba, V.C. (1927) The structure and development of the cell wall in plants I. Bast fibers of *Boehmeria* and *Linum*. *Amer. J. Bot.* **14**, 16–22.

Babu, P.R., Rao, K.V. and Reddy, V.D. (2013) Structural organization and classification of cytochrome P450 genes in flax (*Linum usitatissimum* L.). *Gene*, **513**, 156–162.

Banttari, E.E. and Zeyen, R.J. (1972) Ultrastructure of flax with a simultaneous virus and mycoplasma-like infection. *Virology*, **49**, 305–308.

Barvkar, V.T., Pardeshi, V.C., Kale, S.M., Kadoo, N.Y. and Gupta, V.S. (2012) Phylogenomic analysis of UDP glycosyltransferase 1 multigene family in *Linum usitatissimum* identified genes with varied expression patterns. *BMC Genom.* **13**, 175.

Baulcombe, D.C. (2006) Short silencing RNA: the dark matter of genetics? *Cold Spring Harb. Symp. Quant. Biol.* **71**, 13–20.

Beaugrand, J., Cronier, D., Thiebeau, P., Schreiber, L., Debeire, P. and Chabbert, B. (2004) Structure, chemical composition, and xylanase degradation of external layers isolated from developing wheat grain. *J. Agric. Food Chem.* **52**, 7108–7117.

Brigneti, G., Martin-Hernandez, A.M., Jin, H., Chen, J., Baulcombe, D.C., Baker, B. and Jones, J.D. (2004) Virus-induced gene silencing in *Solanum* species. *Plant J.* **39**, 264–272.

Brunt, A.A., Crabtree, K., Dallwitz, M.J., Gibbs, A.J., Watson, L. and Zurcher, E.J. (1996). '*Plant Viruses Online: Descriptions and Lists from the VIDE Database.*' from http://pvo.bio-mirror.cn/refs.htm.

Burn, J.E., Hocart, C.H., Birch, R.J., Cork, A.C. and Williamson, R.E. (2002) Functional analysis of the cellulose synthase genes *CesA1*, *CesA2*, and *CesA3* in Arabidopsis. *Plant Physiol.* **129**, 797–807.

Burton, R.A., Gibeaut, D.M., Bacic, A., Findlay, K., Roberts, K., Hamilton, A., Baulcombe, D.C. and Fincher, G.B. (2000) Virus-induced silencing of a plant cellulose synthase gene. *Plant Cell*, **12**, 691–706.

Caillot, S., Rosiau, E., Laplace, C. and Thomasset, B. (2009) Influence of light intensity and selection scheme on regeneration time of transgenic flax plants. *Plant Cell Rep.* **28**, 359–371.

Carroll, A. and Specht, C.D. (2011) Understanding plant cellulose synthases through a comprehensive investigation of the cellulose synthase family sequences. *Front Plant Sci.* **2**, 5.

Chantreau, M., Grec, S., Gutierrez, L., Dalmais, M., Pineau, C., Demailly, H., Paysant-Leroux, C., Tavernier, R., Trouve, J.P., Chatterjee, M., Guillot, X., Brunaud, V., Chabbert, B., van Wuytswinkel, O., Bendahmane, A., Thomasset, B. and Hawkins, S. (2013) PT-Flax (phenotyping and TILLinG of flax): development of a flax (*Linum usitatissimum* L.) mutant population and TILLinG platform for forward and reverse genetics. *BMC Plant Biol.* **13**, 159.

Chantreau, M., Portelette, A., Dauwe, R., Kiyoto, S., Cronier, D., Morreel, K., Arribat, S., Neutelings, G., Chabi, M., Boerjan, W., Yoshinaga, A., Mesnard, F., Grec, S., Chabbert, B. and Hawkins, S. (2014) Ectopic lignification in the flax lignified bast fiber1 mutant stem is associated with tissue-specific modifications in gene expression and cell wall composition. *Plant Cell*, **26**, 4462–4482.

Charlet, K., Jernot, J.P., Eve, S., Gomina, M. and Breard, J. (2010) Multi-scale morphological characterisation of flax: from the stem to the fibrils. *Carbohydr. Polym.* **82**, 54–61.

Day, A., Addi, M., Kim, W., David, H., Bert, F., Mesnage, P., Rolando, C., Chabbert, B., Neutelings, G. and Hawkins, S. (2005a) ESTs from the fibre-bearing stem tissues of flax (*Linum usitatissimum* L.): expression analyses of sequences related to cell wall development. *Plant Biol (Stuttg).* **7**, 23–32.

Day, A., Ruel, K., Neutelings, G., Cronier, D., David, H., Hawkins, S. and Chabbert, B. (2005b) Lignification in the flax stem: evidence for an unusual lignin in bast fibers. *Planta*, **222**, 234–245.

Day, A., Fenart, S., Neutelings, G., Hawkins, S., Rolando, C. and Tokarski, C. (2013) Identification of cell wall proteins in the flax (Linum usitatissimum) stem. *Proteomics*, **13**, 812–825.

Delmer, D.P. (1999) CELLULOSE BIOSYNTHESIS: exciting times for a difficult field of study. *Annu. Rev. Plant Physiol. Plant Mol. Biol.* **50**, 245–276.

Fenart, S., Ndong, Y.P., Duarte, J., Riviere, N., Wilmer, J., van Wuytswinkel, O., Lucau, A., Cariou, E., Neutelings, G., Gutierrez, L., Chabbert, B., Guillot, X.,

Tavernier, R., Hawkins, S. and Thomasset, B. (2010) Development and validation of a flax (*Linum usitatissimum* L.) gene expression oligo microarray. *BMC Genom.* **11**, 592.

Fu, D.-Q., Zhu, B.-Z., Zhu, H.-L., Jiang, W.-B. and Luo, Y.-B. (2005) Virus-induced gene silencing in tomato fruit. *Plant J.* **43**, 299–308.

Galinousky, D.V., Anisimova, N.V., Raiski, A.P., Leontiev, V.N., Titok, V.V. and Khotyleva, L.V. (2014) Cellulose synthase genes that control the fiber formation of flax (*Linum usitatissimum* L.). *Russ. J. Genet.* **50**, 20–27.

Gorshkova, T. and Morvan, C. (2006) Secondary cell-wall assembly in flax phloem fibres: role of galactans. *Planta*, **223**, 149–158.

Gorshkova, T., Brutch, N., Chabbert, B., Deyholos, M., Hayashi, T., Lev-Yadun, S., Mellerowicz, E.J., Morvan, C., Neutelings, G. and Pilate, G. (2012) Plant fiber formation: state of the art, recent and expected progress, and open questions. *Crit. Rev. Plant Sci.* **31**, 201–228.

Harris, D.M., Corbin, K., Wang, T., Gutierrez, R., Bertolo, A.L., Petti, C., Smilgies, D.M., Estevez, J.M., Bonetta, D., Urbanowicz, B.R., Ehrhardt, D.W., Somerville, C.R., Rose, J.K., Hong, M. and Debolt, S. (2012) Cellulose microfibril crystallinity is reduced by mutating C-terminal transmembrane region residues CESA1A903V and CESA3T942I of cellulose synthase. *Proc. Natl. Acad. Sci. USA*, **109**, 4098–4103.

Held, M.A., Penning, B., Brandt, A.S., Kessans, S.A., Yong, W., Scofield, S.R. and Carpita, N.C. (2008) Small-interfering RNAs from natural antisense transcripts derived from a cellulose synthase gene modulate cell wall biosynthesis in barley. *Proc. Natl. Acad. Sci. USA*, **105**, 20534–20539.

Hileman, L.C., Drea, S., Martino, G., Litt, A. and Irish, V.F. (2005) Virus-induced gene silencing is an effective tool for assaying gene function in the basal eudicot species *Papaver somniferum* (opium poppy). *Plant J.* **44**, 334–341.

Hobson, N. and Deyholos, M.K. (2013) Genomic and expression analysis of the flax (*Linum usitatissimum*) family of glycosyl hydrolase 35 genes. *BMC Genom.* **14**, 344.

Huis, R., Hawkins, S. and Neutelings, G. (2010) Selection of reference genes for quantitative gene expression normalization in flax (*Linum usitatissimum* L.). *BMC Plant Biol.* **10**, 71.

Huis, R., Morreel, K., Fliniaux, O., Lucau-Danila, A., Fenart, S., Grec, S., Neutelings, G., Chabbert, B., Mesnard, F., Boerjan, W. and Hawkins, S. (2012) Natural hypolignification is associated with extensive oligolignol accumulation in flax stems. *Plant Physiol.* **158**, 1893–1915.

Jain, R., Lahiri, I. and Varma, A.J. (2000) Peanut stripe potyvirus: prevalence, detection and serological relationships. *Indian Phytopath.* **53**, 14–18.

Kotake, T., Aohara, T., Hirano, K., Sato, A., Kaneko, Y., Tsumuraya, Y., Takatsuji, H. and Kawasaki, S. (2011) Rice Brittle culm 6 encodes a dominant-negative form of CesA protein that perturbs cellulose synthesis in secondary cell walls. *J. Exp. Bot.* **62**, 2053–2062.

Kumagai, M.H., Donson, J., della-Cioppa, G., Harvey, D., Hanley, K. and Grill, L.K. (1995) Cytoplasmic inhibition of carotenoid biosynthesis with virus-derived RNA. *Proc. Natl. Acad. Sci. USA*, **92**, 1679–1683.

Kvavadze, E., Bar-Yosef, O., Belfer-Cohen, A., Boaretto, E., Jakeli, N., Matskevich, Z. and Meshveliani, T. (2009) 30 000-year-old wild flax fibers. *Science*, **325**, 1359.

Lei, L., Li, S. and Gu, Y. (2012) Cellulose synthase complexes: composition and regulation. *Front Plant Sci.* **3**, 75.

Li, A., Xia, T., Xu, W., Chen, T., Li, X., Fan, J., Wang, R., Feng, S., Wang, Y., Wang, B. and Peng, L. (2013) An integrative analysis of four CESA isoforms specific for fiber cellulose production between *Gossypium hirsutum* and *Gossypium barbadense*. *Planta*, **237**, 1585–1597.

Liepman, A.H., Wightman, R., Geshi, N., Turner, S.R. and Scheller, H.V. (2010) Arabidopsis - a powerful model system for plant cell wall research. *Plant J.* **61**, 1107–1121.

Liu, E. and Page, J.E. (2008) Optimized cDNA libraries for virus-induced gene silencing (VIGS) using tobacco rattle virus. *Plant Methods*, **4**, 5.

Liu, Y., Schiff, M. and Dinesh-Kumar, S.P. (2002) Virus-induced gene silencing in tomato. *Plant J.* **31**, 777–786.

Ma, M., Yan, Y., Huang, L., Chen, M. and Zhao, H. (2012) Virus-induced gene-silencing in wheat spikes and grains and its application in functional analysis of HMW-GS-encoding genes. *BMC Plant Biol.* **12**, 141.

MacKinnon, I.M., Sturcova, A., Sugimoto-Shirasu, K., His, I., McCann, M.C. and Jarvis, M.C. (2006) Cell-wall structure and anisotropy in procuste, a cellulose synthase mutant of *Arabidopsis thaliana*. *Planta*, **224**, 438–448.

Mokshina, N., Gorshkova, T. and Deyholos, M.K. (2014) Chitinase-like (CTL) and cellulose synthase (CESA) gene expression in gelatinous-type cellulosic walls of flax (*Linum usitatissimum* L.) bast fibers. *PLoS ONE*, **9**, e97949.

Nairn, C.J. and Haselkorn, T. (2005) Three loblolly pine CesA genes expressed in developing xylem are orthologous to secondary cell wall CesA genes of angiosperms. *New Phytol.* **166**, 907–915.

Neutelings, G. (2011) Lignin variability in plant cell walls: contribution of new models. *Plant Sci.* **181**, 379–386.

Oomah, B.D. (2001) Flaxseed as a functional food source. *J. Sci. Food Agric.* **81**, 889–894.

Persson, S., Wei, H., Milne, J., Page, G.P. and Somerville, C.R. (2005) Identification of genes required for cellulose synthesis by regression analysis of public microarray data sets. *Proc. Natl. Acad. Sci. USA*, **102**, 8633–8638.

Richmond, T.A. and Somerville, C.R. (2000) The cellulose synthase superfamily. *Plant Physiol.* **124**, 495–498.

Rozen, S. and Skaletsky, H. (2000) Primer3 on the WWW for general users and for biologist programmers. *Methods Mol. Biol.* **132**, 365–386.

Ruiz, M.T., Voinnet, O. and Baulcombe, D.C. (1998) Initiation and maintenance of virus-induced gene silencing. *Plant Cell*, **10**, 937–946.

Ryu, C.M., Anand, A., Kang, L. and Mysore, K.S. (2004) Agrodrench: a novel and effective agroinoculation method for virus-induced gene silencing in roots and diverse Solanaceous species. *Plant J.* **40**, 322–331.

Saxena, I.M., Brown, R.M. Jr and Dandekar, T. (2001) Structure-function characterization of cellulose synthase: relationship to other glycosyltransferases. *Phytochemistry*, **57**, 1135–1148.

Senthil-Kumar, M. and Mysore, K.S. (2011) New dimensions for VIGS in plant functional genomics. *Trends Plant Sci.* **16**, 656–665.

Somerville, C. (2006) Cellulose synthesis in higher plants. *Annu. Rev. Cell Dev. Biol.* **22**, 53–78.

Sveinsson, S., McDill, J., Wong, G.K., Li, J., Li, X., Deyholos, M.K. and Cronk, Q.C. (2014) Phylogenetic pinpointing of a paleopolyploidy event within the flax genus (*Linum*) using transcriptomics. *Ann. Bot.* **113**, 753–761.

Tamura, K., Peterson, D., Peterson, N., Stecher, G., Nei, M. and Kumar, S. (2011) MEGA5: molecular evolutionary genetics analysis using maximum likelihood, evolutionary distance, and maximum parsimony methods. *Mol. Biol. Evol.* **28**, 2731–2739.

Thompson, J.D., Higgins, D.G. and Gibson, T.J. (1994) CLUSTAL W: improving the sensitivity of progressive multiple sequence alignment through sequence weighting, position-specific gap penalties and weight matrix choice. *Nucleic Acids Res.* **22**, 4673–4680.

Tian, J., Pei, H., Zhang, S., Chen, J., Chen, W., Yang, R., Meng, Y., You, J., Gao, J. and Ma, N. (2014) TRV-GFP: a modified Tobacco rattle virus vector for efficient and visualizable analysis of gene function. *J. Exp. Bot.* **65**, 311–322.

Valentine, T., Shaw, J., Blok, V.C., Phillips, M.S., Oparka, K.J. and Lacomme, C. (2004) Efficient virus-induced gene silencing in roots using a modified tobacco rattle virus vector. *Plant Physiol.* **136**, 3999–4009.

Velasquez, A.C., Chakravarthy, S. and Martin, G.B. (2009) Virus-induced gene silencing (VIGS) in *Nicotiana benthamiana* and tomato. *J. Vis. Exp.* **28**, 1292.

Venglat, P., Xiang, D., Qiu, S., Stone, S.L., Tibiche, C., Cram, D., Alting-Mees, M., Nowak, J., Cloutier, S., Deyholos, M., Bekkaoui, F., Sharpe, A., Wang, E., Rowland, G., Selvaraj, G. and Datla, R. (2011) Gene expression analysis of flax seed development. *BMC Plant Biol.* **11**, 74.

Voinnet, O. (2009) Origin, biogenesis, and activity of plant microRNAs. *Cell*, **136**, 669–681.

Wang, L., Guo, K., Li, Y., Tu, Y., Hu, H., Wang, B., Cui, X. and Peng, L. (2010) Expression profiling and integrative analysis of the CESA/CSL superfamily in rice. *BMC Plant Biol.* **10**, 282.

Wang, Z., Hobson, N., Galindo, L., Zhu, S., Shi, D., McDill, J., Yang, L., Hawkins, S., Neutelings, G., Datla, R., Lambert, G., Galbraith, D.W., Grassa, C.J., Geraldes, A., Cronk, Q.C., Cullis, C., Dash, P.K., Kumar, P.A., Cloutier, S., Sharpe, A.G., Wong, G.K., Wang, J. and Deyholos, M.K. (2012) The genome of flax (*Linum usitatissimum*) assembled de novo from short shotgun sequence reads. *Plant J.* **72**, 461–473.

Wijdeveld, M.M., Goldbach, R.W., Meurs, C. and van Loon, L.C. (1992) Accumulation of viral 126 kDa protein and symptom expression in tobacco systemically infected with different strains of tobacco mosaic virus *Physio. Mol. Plant Pathol.* **41**, 437–451.

4

Heritable site-specific mutagenesis using TALENs in maize

Si Nian Char[1,†], Erica Unger-Wallace[1,†], Bronwyn Frame[2], Sarah A. Briggs[1], Marcy Main[2], Martin H. Spalding[1], Erik Vollbrecht[1], Kan Wang[2] and Bing Yang[1,*]

[1]Department of Genetics, Development and Cell Biology, Iowa State University, Ames, IA, USA

[2]Department of Agronomy, Iowa State University, Ames, IA, USA

*Correspondence

email byang@iastate.edu

[†]These two authors contribute equally to this work.

Keywords: TAL effector nuclease, gene editing, targeted mutagenesis, maize, Glossy2.

Summary

Transcription activator-like effector nuclease (TALEN) technology has been utilized widely for targeted gene mutagenesis, especially for gene inactivation, in many organisms, including agriculturally important plants such as rice, wheat, tomato and barley. This report describes application of this technology to generate heritable genome modifications in maize. TALENs were employed to generate stable, heritable mutations at the maize glossy2 (gl2) locus. Transgenic lines containing mono- or di-allelic mutations were obtained from the maize genotype Hi-II at a frequency of about 10% (nine mutated events in 91 transgenic events). In addition, three of the novel alleles were tested for function in progeny seedlings, where they were able to confer the glossy phenotype. In a majority of the events, the integrated TALEN T-DNA segregated independently from the new loss of function alleles, producing mutated null-segregant progeny in T1 generation. Our results demonstrate that TALENs are an effective tool for genome mutagenesis in maize, empowering the discovery of gene function and the development of trait improvement.

Introduction

Transcription activator-like effector nucleases (TALENs) are fusion proteins of native or artificial TAL effector DNA-binding domains and the DNA-cleavage domain of the restriction enzyme FokI (Christian et al., 2010; Li et al., 2011a,b). The modular TAL effector repeats can be custom-tailored into DNA recognizing domains for virtually any sequence in a genome (Boch et al., 2009). When expressed in cells, the paired TALENS recognize and bind to two adjacent, opposite subsites, enabling the FokI domains (homo- or heterodimeric) to dimerize and cleave the DNA double strands. The DNA double-strand breaks (DSBs) can be repaired in vivo through nonhomologous end-joining (NHEJ) repair or template-based homologous recombination, a process exploitable for site-specific genetic alteration (Joung and Sander, 2013; Miller et al., 2011). TALENs have emerged as a promising genetic tool for targeted gene mutagenesis as demonstrated in a plethora of organisms such as yeast, plant, nematode, insect, fish, mammal and human cells (Hockemeyer et al., 2011; Huang et al., 2011; Li et al., 2012; Sander et al., 2011; Tesson et al., 2011; Wang et al., 2014; Wood et al., 2011). Among plants, TALENs have been reported as effective genome mutagenesis tools in rice (Li et al., 2012), Arabidopsis (Christian et al., 2013), Brachypodium (Shan et al., 2013), barley (Wendt et al., 2013), wheat (Wang et al., 2014) and tomato (Lor et al., 2014).

Maize (Zea mays), as an agriculturally valued crop, has been a subject of genetic manipulation from varied approaches including selective breeding that harnesses natural genetic variation and methods that induce new variation such as chemical (EMS) mutagenesis or forward and reverse genetics that exploits endogenous transposon biology (Candela and Hake, 2008). Recently, methods to target specific sequences for mutagenesis in maize were reported (Gao et al., 2010; Liang et al., 2014; Shukla et al., 2009). Zinc finger nucleases (ZFNs) were used to modify the maize inositol-pentakisphosphate 2-kinase (IPK1) gene

generating altered phytate production in several maize lines (Shukla et al., 2009). An engineered homing nuclease based on I-CreI was demonstrated to successfully target the maize ligule-less-1 locus and facilitate NHEJ-based gene mutagenesis; site-specific insertions/deletions were detected in the primary (T0) transgenic plants at a frequency of 3%, and inheritance of several mutated alleles was demonstrated in the T1 generation (Gao et al., 2010). Most recently, both TALENs and CRISPR/Cas9 constructs were used to target four endogenous loci in maize protoplasts (Liang et al., 2014). While T0 maize plants containing TALEN-targeted somatic mutations were produced, the inheritance of TALEN-induced mutations in maize has yet to be demonstrated.

Here, we report the generation of heritable TALEN-induced mutations at the maize glossy2 (gl2) locus. We demonstrate TALEN-induced gl2 mutations that are heritable and segregate independently of the inserted TALEN T-DNA cassette, generating mutated null segregants in a single generation. In addition, several of the novel mutant alleles we generated condition the classical glossy phenotype associated with known mutations at this locus (Bianchi, 1978; Hayes and Brewbaker, 1928).

Results

Construction of TALENs targeting the maize endogenous gene glossy2

To apply TALEN technology for targeted gene mutagenesis in maize, we selected gl2 as the target and streamlined the procedures from construction of TALEN vectors to identification of intended mutations. We sought to exploit NHEJ-based repair of TALEN-induced DSBs for the generation of transgenic maize plants that contain site-specific mutations of gl2. An outline of the process including time frames of each step is presented in Supplemental Data (Figure S1). Using parameters defined in Li et al. (Li and Yang, 2013), a TALEN target site was identified in the reference genome sequence of the gl2 locus

(GRMZM2G098239). The target site is located in exon one, 110 bp downstream of the translation start site (Figure 1a). Deletion mutations at this site can result in a frame shift shortly after translation initiation, and thus have a high probability of generating a nonfunctional gene product and a mutant phenotype.

Using polymerase chain reaction (PCR) and DNA sequencing, we analysed the target site from the transformable maize genotypes Hi-II (Songstad et al., 1996), a hybrid and inbred line B104 (Frame et al., 2006) to ensure no sequence polymorphisms were present in the targeted genomic region. Modular repeats with repeat variable di-residues (RVDs) NI, NN, NG and HD were used to recognize the nucleotides A, G, T and C, respectively, in target sites (Figure S2a). A Golden Gate-based cloning strategy was employed to synthesize the TAL repeat domains corresponding to the two subsites of the gl2 target site using the modular assembly method as described (Li and Yang, 2013) (Figure S2b). We used a bacterial version of the TAL effector scaffold with the repeat region replaced by engineered repeats (Li et al., 2012) and used it to test two different TALEN configurations. Newly assembled TAL effector DNA-binding domains were fused to the bacterial homodimeric (Li et al., 2011a,b) and heterodimeric FokI endonuclease domain (Doyon et al., 2011), resulting in a pair of homodimeric and another pair of heterodimeric TALENs, respectively (Figure 1b).

Assessment of TALEN activity in yeast and maize embryo cells

We first validated the activities of TALEN constructs on the gl2 target site using a yeast single-strand annealing (SSA) assay (Li et al., 2011a,b). The assay depends on DNA repair to join two nonfunctional overlapping fragments of a LacZ gene and thus restoring the enzyme activity of its product, β-galactosidase, on the substrate X-gal (5-bromo-4-chloro-3-indolyl-beta-D-galacto-pyranoside) as a reporter of TALEN activity in yeast cells (See Methods). The gl2 target site sequence was inserted between the LacZ reporter gene fragments and that construct was co-expressed with the newly assembled TALEN genes. Homologous recombination-mediated restoration of the reporter LacZ gene

was observed, indicating the paired homodimeric TALENs were active when expressed in yeast (Figure 2a, b).

We next established a maize embryo cell-based transient assay to test the activity of TALENs in vivo. In this assay, TALEN activity was evaluated using a two-plasmid (reporter construct and TALEN construct) cotransformation system. Two reporter plasmids were constructed. First, a fluorescent reporter gene containing the gl2 target sequence (including the two TALEN-binding subsites and the spacer sequence) was generated. A fragment containing this sequence was cloned in frame with and immediately downstream of the endoplasmic reticulum (ER) localization signal of a 35S::ER-mCherry gene (Nelson et al., 2007). Thus, reporter plasmid p2813 contained a 120-bp insert that adds in-frame 40 amino acids to the mCherry coding region. This plasmid should encode a functional mCherry and serve as a positive control. The second reporter plasmid, p2825, contained a 121-bp fragment that adds the same 40 amino acids followed by one base pair that frame-shifts the mCherry coding region, rendering the reporter gene nonfunctional. The TALEN-expressing constructs, p467 (homodimeric Fok1 domain) and p555 (heterodimeric Fok1 domain), were generated by ligating the appropriate TALEN genes into the binary vector pTF101.1, which contains a CaMV 35S promoter-driven selectable marker gene [Figure 1b, (Paz et al., 2004)]). To help maximize the heterology of the overall TALEN cassette, the 35S promoter drives TALEN-L gene expression and the maize ubiquitin 1 promoter controls TALEN-R expression (Figure 1b).

The nonfunctional reporter plasmid p2825 was cobombarded with or without TALEN plasmid into 1.5–2 mm immature maize embryos. We speculated that TALEN-mediated DSBs at the spacer region of gl2 target sequence followed by NHEJ-based repair would in some cases correct the reading frame of the mCherry gene and confer production of a red fluorescent signal in maize embryo cells. Five days after bombardment, mCherry expression was examined by counting the number of red-fluorescing cell clusters (spots). While only a fraction of TALEN-induced modification would be expected to restore mCherry function, nevertheless, the number of fluorescing clusters may

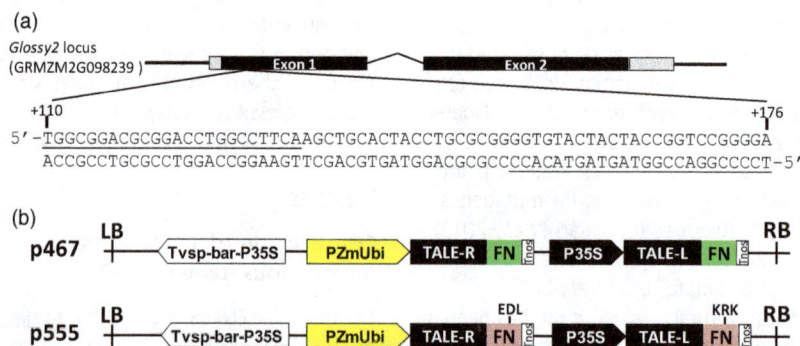

Figure 1 TALEN construct and the target maize Glossy2 sequences. (a) Gene structure of glossy2 containing the TALEN target sequences. The sequences in red are two subsites chosen for design and engineering of TALENs. The target site is located between 110 and 176 bp downstream of the translation initiation site of glossy2. The coding and untranslated regions are represented by solid black and grey bars, respectively. (b) Schematics of T-DNA construct for gene editing. The T-DNA region is defined by the two Agrobacterium-recognizable border sequences (left border-LB and right border-RB) and contains the selection marker bar gene under control of the CaMV 35S promoter, TALEN gene (TALEN-R) under control of the maize ubiquitin 1 promoter and the CaMV 35S promoter linked with another TALEN gene (TALEN-L). Construct p467 encodes the bacterial version of homodimeric (or identical) FokI nuclease domain (FN), while p555 contains the heterodimeric FokI nuclease domains with three amino acid mutations depicted as EDL and KRK, respectively (see Figure S2).

be used as reporter for TALEN activity. We also used this assay to evaluate the relative activity of homodimeric and heterodimeric FokI domains in the context of a single combination of TALE effector scaffold and RVD region. While these mutagenesis assays reported, as expected, low efficiency (Figure 2d, e) compared with mCherry expression derived from the control plasmid p2813 (Figure 2c), the results indicated that homodimeric FokI-containing TALENs (in construct p467) were nearly two times as active as the heterodimeric version (in construct p555) in this assay (Figure 2c–h, Table S1). This is consistent with several studies reporting compromised activity of several heterodimeric FokI designs leading to reduced efficiency of mutagenesis (Doyon et al., 2011; Miller et al., 2007; Szczepek et al., 2007). Thus, the homodimeric construct p467 was emphasized for the majority of the labour-intensive maize transformation experiments.

TALENs induce site-specific DNA changes at the maize glossy2 locus

TALEN construct p467 was mobilized into *Agrobacterium tumefaciens* EHA101 that was used to transform maize genotypes Hi-II and B104. Hi-II is a hybrid genotype that is amenable to *in vitro* tissue culture and genetic transformation; B104 is an inbred that is closely related to the sequenced inbred B73 but is more challenging than Hi-II in transformation (Frame *et al.*, 2006). We obtained 77 and 27 bialaphos-resistant callus lines from Hi-II and B104, respectively. T7 endonuclease I (T7E1) assay and DNA sequencing analysis on amplicons derived from genomic DNA from each callus line showed that eight Hi-II lines and one B104 line carried mutations at the *gl2* site (Figure 3). The nucleotide alterations at the *gl2* locus were all small deletions, ranging from 3 to 9 bp and resulting in either in-frame (3-, 6- or 9-bp deletion)

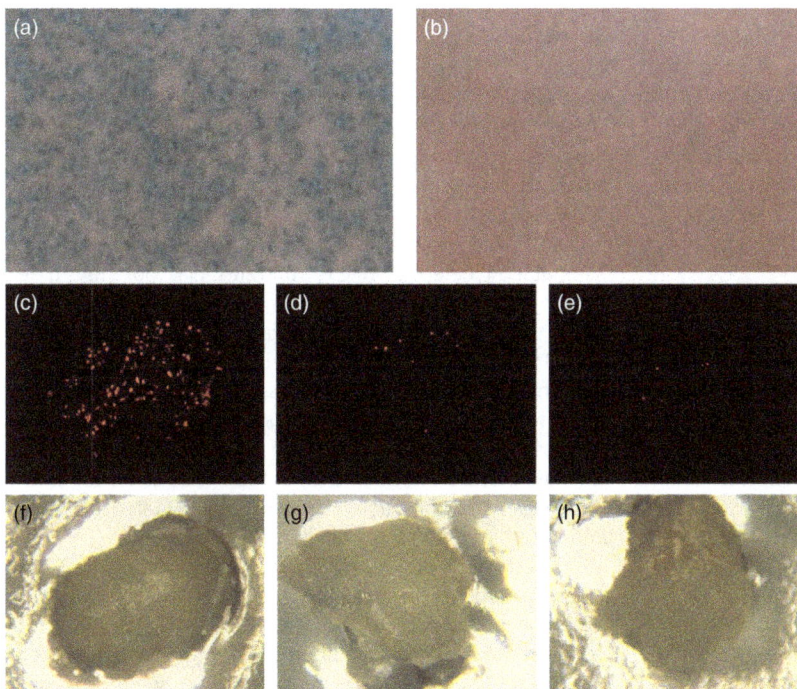

Figure 2 TALEN activity in yeast and maize embryo cells. Functionally restorable *LacZ* gene construct was expressed with (a) or without (b) homodimeric TALENs in yeast cells. Clones on the filter in (a) are blue due to restored LacZ enzymatic activity acting on the substrate X-gal, reflecting TALEN activity relative to the control in (b) that contains only white colonies. mCherry expression 5 days postbombardment. Images are of a representative embryo bombarded with (c, f) the positive control construct p2813, (d, g) with the reporter construct p2825+ p467 (d, g) and (e, h) with p2825+ p555. Paired images are of the same embryo, fluorescence (c–e) and bright field (f-h).

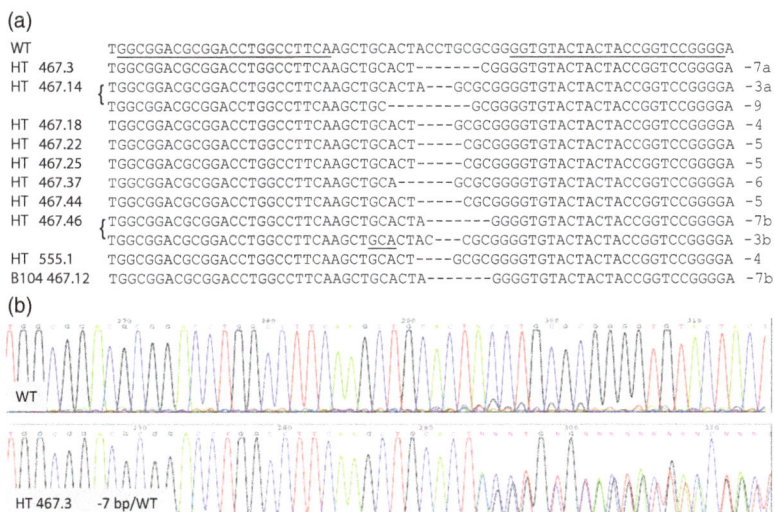

Figure 3 Site-specific nucleotide changes induced by TALENs in callus and plant lines of maize. (a) TALEN-induced deletion mutations at the *gl2* target site. Genomic DNA sequences at the relevant regions from each mutant are aligned with the corresponding wild-type sequence. The number of deleted nucleotides (dashes) is indicated to the right of each sequence. (b) An example of chromatographs showing the sequencing traces of two alleles, the wild type (WT) and 7-bp deletion (−7) at the *gl2* locus from a callus line.

or frame-shift (5- or 7-bp deletion) mutations of the *gl2* allele. Among those mutated callus lines, two lines (467.14 and 467.46) carried di-allelic (two, nonidentical chromosomal) mutations, while the rest contained mono-allelic mutations (Figure 3a). TALEN construct p555 (with heterodimeric FokI) was also introduced into a limited number of maize Hi-II embryos. Of 14 bialaphos-resistant callus lines, one mono-allelic 4-bp mutation (555.1) was obtained (Figure 3a). In summary, from ten callus lines with putative mutations at *gl2*, we identified twelve total mutations.

Moreover, examination of the DNA sequencing traces from the ten individual callus lines revealed two major overlapping sequence peaks starting at the deletion site in each line (a representative in Figure 3b), suggesting homogeneity of callus cells that contained the site-specific mutations and that TALEN-mediated gene mutagenesis occurred early during propagation of the callus line.

From each mutated callus line, multiple T0 plantlets (ranging from 5 to 21) were regenerated and brought to maturity. Genotype analysis using the T7E1 assay showed 70 of 88 T0 plants from the ten callus lines were positive for the mutated *gl2* alleles. DNA sequencing analysis on at least two T7E1-positive T0 plants of each line showed that plants derived from the same callus event carried identical nucleotide deletions and allele combinations, as predicted from assaying the progenitor callus lines. Therefore, the overall site-specific mutagenesis frequency in stable T0 plants from these experiments was 9.9% (9/91) for Hi-II and 3.7% (1/27) for B104.

To investigate whether the presence of TALEN constructs in the plant genome could lead to somatic mutagenesis during later growth stages of T0 plants, we also regenerated plants from callus lines that were transgenic but mutation negative at the *gl2* site. Reverse transcriptase (RT)-PCR analysis on T0 plants derived from six mutation-positive and three mutation-negative callus lines showed that the TALEN gene was expressed at a similar level in these plants (Figure S3). However, no somatic mutations at the *gl2* locus were detected in the mutation-negative callus lines, despite the expression of TALENs in these plants. Nor were new mutations detected from the mutation-positive lines as assayed by sequencing PCR amplicons of the relevant region. These results supported the approach to first screen the callus events for mutations prior to plant regeneration, as a majority of the T0 plants regenerated from mutated callus lines contained mutations, while no plant regenerated from TALEN-expressing, nonmutated callus lines contained an altered *gl2* allele. This observation suggests that mutagenesis

occurs at an early stage during the transformation process, similar to the situation observed using these same vectors in rice (Li *et al.*, 2012). It is possible that co-expression of the *gl2* targeting nuclease genes might be sufficient at early stages of embryogenic callus proliferation, but that reduced co-expression in differentiated plant tissues (leaves) results in a lack of detectable somatic mutations in these T0 plants. In contrast, mosaic patterns of mutations were reported in the T0 plants of barley (Wendt *et al.*, 2013) and maize (Liang *et al.*, 2014) when different TALEN scaffolds and promoters were used (Christian *et al.*, 2012).

Inheritance of TALEN-mediated mutations and the removal of T-DNA in T1 generation plants

Inheritance and segregation of the mutated *gl2* alleles and TALEN cassettes were examined in T1 progeny of six independent lines (Table 1). T0 plants from five mutated lines were pollinated by maize inbred line B104. In event 467.25, a T0 plant was crossed with a sibling, mutated T0 plant. Four of the six events (467.3, 467.18, 467.22 and 467.25) showed segregation of the TALEN cassette and of the mutated *gl2* locus that was supportive of a single-locus T-DNA integration genetically unlinked to *gl2*. In addition, the mutant *gl2* alleles were all inherited in a predicted Mendelian manner, further supporting the interpretation that the mutations detected in the T0 generation were not somatic in nature. In contrast, event 467.37 showed inheritance of the TALEN cassette, but not the mutation at the *gl2* locus suggesting the mutation detected in the T0 plants was not transmitted through the germ-line. Finally, segregation analysis of event 467.46 suggests multilocus insertions of the TALEN cassette, as 15 of 15 progeny plants examined by PCR amplification of the FokI region contained a TALEN cassette. In addition, this event was determined to be di-allelic at *gl2*, containing two different chromosomal mutations (3-bp and 7-bp deletions) in the T0 generation. While the inheritance of the TALEN genes suggests a multilocus T-DNA insertion, the two novel mutated *gl2* alleles both transmitted and segregated as expected.

To investigate more thoroughly the segregation of the T-DNA cassette and/or its components, the transformation selectable marker gene (35S-bar) and two TALEN genes, we designed four pairs of primers to detect the presence of six regions across the T-DNA and analysed by PCR the genomic DNA of some T1 progeny of the mutated line 467.18 (Figure 4a). Twenty-two T1 progeny plants of the line 467.18 were analysed by PCR amplification with T-DNA-specific primer pairs and genotyped at *gl2* using the T7E1 assay. Five plants (#5, 8, 18, 19 and 21) were identified as

Table 1 Inheritance of TALEN T-DNA and modified *gl2* alleles

Event	N	TALEN+ Edit+	TALEN+ Edit−	TALEN− Edit+	TALEN− Edit−	Segregation ratio	P
467.3	64	12	17	15	20	1 : 1 : 1 : 1	0.547
467.18	53	17	9	12	15	1 : 1 : 1 : 1	0.428
467.22	53	13	16	16	8	1 : 1 : 1 : 1	0.358
467.25	30	22	7	1	0	9 : 3 : 3 : 1	0.056
467.37	63	0	39	0	24	1 : 1 : 1 : 1	3.9×10^{-15}[†]
467.46*	15	15	NA	0	NA	1 : 1	1.0×10^{-4}[†]

*Di-allelic event, segregation of each mutated allele listed separately below.

[†]Deviates from predicted independent segregation for a single-locus T-DNA integration and *Gl2*.

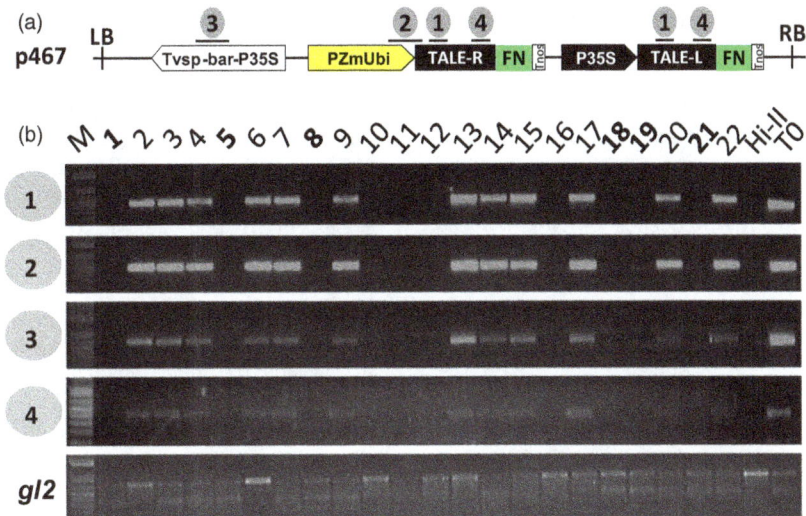

Figure 4 Genetic segregation produces null segregants of mutated *gl2* in the T1 generation. (a) Schematic of TALEN genes and transformation selection gene (*bar*) transgene cassette, as defined by the left (LB) and right border (RB) sequences. Pairs of site-specific primers detected four different regions of the TALEN T-DNA (lines with numbered, grey circles). Amplicon #1 is in the 5′ end of TAL effector domains, #2 spans the 3′ end of the maize ubiquitin 1 promoter to the 5′ end of TALEN-R, #3 is for *bar*coding region and #4 is for the 3′ end of TAL effector domains. (b) Images of agarose gels for electrophoresis of PCR amplicons derived from the primers depicted in A. 1 to 22 are T1 467.18 progeny plants; T0 is callus line 467.18; Hi-II genomic DNA from nontransformed Hi-II callus. The *gl2* panel reports the results from the T7E1 assay; double bands indicate *gl2* edit-positive. M, DNA ladder. Plants numbered in red and bold are null segregants with a *gl2* edit.

progeny that contain the mutated *gl2* alleles and lack T-DNA PCR amplicons attributable to the mutagenesis-inductive T-DNA (Figure 4b).

In summary, four of the six tested mutations at the *gl2* locus were inherited, and mutations could readily be segregated from the inductive TALEN cassette, resulting in stable *gl2* mutants lacking the TALEN transgene, or so-called null segregants.

Novel mutations at the *gl2* locus condition the classic phenotype

Glossy mutants in maize have been well characterized, and null mutations produce an easily observable seedling phenotype due to altered epicuticular wax composition in juvenile leaves (Bianchi, 1978; Hayes and Brewbaker, 1928). Alleles from three TALEN-mediated events (467.3, 467.18 and 467.22) were tested for their ability to condition a classic glossy phenotype and to complement existing *gl2* mutant alleles. Event 467.3 carries a mono-allelic 7-bp deletion of the *gl2* gene and was self-pollinated. The seedlings of this cross should segregate normal (Figure 5a) and glossy (Figure 5b) phenotypes in a 3 : 1 ratio, because *gl2* is a recessive mutation. Event 467.18 (mono-allelic 4-bp deletion) and event 467.22 (mono-allelic 5-bp deletion) were test-crossed with mutant alleles *gl2*-N239 and *gl2*-N718, respectively. If the alleles with mutations fail to complement the known *gl2* mutations, then the seedlings of these crosses should segregate 1 : 1 for the normal and glossy phenotypes. In all three tested events, progeny segregated the glossy phenotype at predicted ratios (Table 2), indicating that the mutations produced new, mutant *glossy2* alleles.

Discussion

In this report, we have developed an effective TALEN system suitable for *Agrobacterium* delivery of the nuclease genes and demonstrated efficacy of the system to generate heritable

Figure 5 Glossy phenotype conditioned by a TALEN-mutated *gl2* allele. Leaves of seedlings with a *Gl2* allele (a) and a *gl2* allele homozygous for 7-bp deletion recovered from line 467.3 (b) were misted with water. Water drops adhered to the surface of the mutant leaf (b) due to reduced epicuticular wax caused by loss of function of the *gl2* gene.

Table 2 Mutated alleles condition a glossy phenotype

Event	Edit	N	Normal	Glossy	Exp. segregation	χ^2	P
467.3	Δ7	30	20	10	3 : 1	1.111	0.292
467.18	Δ4	28	12	16	1 : 1	0.571	0.450
467.22	Δ5	31	13	18	1 : 1	0.806	0.369

gene mutagenesis in maize. The successful rate for TALEN-mediated mutagenesis reported here is about 10% for the hybrid Hi-II genotype and 3.7% for the inbred B104. A variety of factors influence the frequency of NHEJ-mediated mutagenesis, including for example the affinity of nuclease to recognize and bind to its genomic target site and the inherent activity of nuclease. In addition, both optimal nuclease activity and high levels of TALEN expression are important in achieving efficient mutagenesis. Variable expression of the TALEN genes could be attributable to the copy number or influence from the genomic site of T-DNA integration. The frequency of NHEJ-induced mutagenesis is also determined by the intrinsic error-prone nature of DSB repair in a eukaryote. For example, the NHEJ repair of DSBs in yeast is mostly accurate, resulting in low rates of gene disruption via its cryptic NHEJ repair pathway (Li *et al.*, 2011a,b; Moore and Haber, 1996; Wilson and Lieber, 1999). The frequency of mutagenesis obtained here (7–10%) using TALENs is slightly higher than the 3% mutagenesis frequency reported when the maize *Liguleless1* gene was targeted using an I-CreI-derived meganuclease (Gao *et al.*, 2010) and much lower than the 39% somatic mutagenesis rates reported in maize T0 leaves (Liang *et al.*, 2014). The TALEN-induced mutagenesis rates in general varied widely by species, transformation methods, TALEN architecture and construction. For example, TALENs produced about 48–63% mutant progeny in transgenic rice plants (Li *et al.*, 2012); mutation frequency in rice and *Brachypodium* callus lines ranged from 3.8 to 100% (Shan *et al.*, 2013); a pair of TALENs targeting a locus in wheat produced 6% NHEJ-based mutations (Wang *et al.*, 2014). In Arabidopsis, plants expressing the most active TALEN pairs transmitted mutations to the next generation at frequencies ranging from 1.5 to 12% (Christian *et al.*, 2013). While the TAL effector scaffolds, expression strategies, target genes and DNA delivery methods vary among these studies, it is difficult to draw direct comparisons of outcomes; however, it does clearly demonstrate that these designer nuclease-based tools can be effective in a diverse range of plant systems.

TALENs derived from the wild-type, homodimeric FokI nuclease domain could efficiently induce site-specific mutations in monocot plants including maize (Li *et al.*, 2012; Liang *et al.*, 2014; Wang *et al.*, 2014; Wendt *et al.*, 2013). However, the major drawback of the FokI nuclease domain is that the derived nucleases function as a homodimer, and in the case of a pair of TALENs, three combinations are possible: two left target site monomers, two right target site monomers or a pair of right/left site monomers. These homodimers can be as nearly functional as a nuclease composed of one left site and one right target site monomer. As a consequence of this, DSBs may be produced at sites recognized by either one or both nucleases potentially resulting in an overwhelming amount of unintended DSBs and collateral damage to the cells, leading to so-called off-target cytotoxicity (Wright *et al.*, 2014). To counter these potential off-target effects, several obligate heterodimeric FokI domains have been developed. The designs have reduced cytotoxic effects ascribed to off-target DSBs by limiting the production of functional nucleases to those generated by pairing the intended left and right target site monomers. The trade-off is that the enzymatic activity of many heterodimeric FokI designs is compromised, leading to reduced efficiency of mutagenesis (Doyon *et al.*, 2011; Miller *et al.*, 2007; Szczepek *et al.*, 2007). As a specific heterodimeric version (ELD/KKR) of FokI used in the context of a TALEN was reported to modify target loci in zebrafish as efficiently as a homodimer but with less toxicity (Cade *et al.*, 2012), we tested TALENs consisting of the same TAL effector DNA-binding domains fused with the homodimeric FokI domain or the obligate heterodimeric ELD/KKR FokI domains to compare

their efficiency of mutagenesis in maize cells. The results suggest that the heterodimeric TALENs were less active in inducing NHEJ-based mutagenesis in our transient assay than the homodimer, but generated the mutant *gl2* alleles at a frequency similar to the homodimeric counterparts (7.2% vs. 10%). As a possible read-out of cytotoxicity, we did not notice an obvious discrepancy of transformation efficiency associated with these two types of TALENs. Taken together, we conclude either homodimeric or heterodimeric TALENs are useful for targeted mutagenesis in maize.

To increase DNA heterology of the overall TALEN cassette, we used the cauliflower mosaic virus (CaMV) 35S gene promoter for TALEN-L and the maize ubiquitin 1 (UB1) gene promoter for TALEN-R. Both promoters are strong and constitutive promoters in monocotyledonous plants, although 35S is reported to be less active in monocots than in dicots (Battraw and Hall, 1990; Christensen *et al.*, 1992; Odell *et al.*, 1985). These promoters have also demonstrated expression in maize embryogenic callus enabling the selection of clones of mutagenized cells from which zygotic embryogenesis produces T0 plants (Zhao *et al.*, 2002). While it is possible that these two promoters might not be equally active to drive co-expression of two TALEN genes, this design has been successfully used to generate targeted genome modifications in rice (Li *et al.*, 2012). As targeted mutations were obtained with the construct design as described, other promoters or alternate strategies such as using a single promoter to express two TALEN genes separated by a T2A translational skipping sequence (Wang *et al.*, 2014; Zhang *et al.*, 2013) were not tested.

Similar to other nuclease-mediated gene mutagenesis technologies, TALENs are a powerful tool to generate target site-specific mutations within a relatively short time period. Compared to the fast emerging CRISPR/Cas9-based approach, TALEN-based gene mutagenesis has an advantage of conferring a low frequency generating nonspecific or off-target cleavage and mutagenesis (Reviewed by (Joung and Sander, 2013; Carroll, 2013)). Furthermore, certain genomic sites of interest for targeted mutagenesis (e.g. sequences corresponding to mature miRNA, small regulatory elements of genes) may not contain a PAM sequence required for CRISPR/Cas9 mutagenesis or the DNA sequence (e.g. a run of poly T causing immature transcriptional termination of guide RNA gene) not suitable for design of guide RNA.

In summary, our maize TALEN system expands the genetic toolbox for both basic biological research and agricultural applications intended to add value to maize by crop improvement.

Experimental procedures

Design and construction of TALEN genes

The paired TALEN (TALEN-L and TALEN-R) genes were designed and assembled using a Golden Gate cloning-based method as described (Li and Yang, 2013). Briefly, repeats from a library of single repeat units were digested with BsmBI and first ligated into three repeat fragments (each consisting eight repeats with RVDs corresponding the 8 nucleotides of target site). For each gene, the three fragments were released with restriction enzyme and assembled into TALEN scaffold to obtain plasmid pSK-TALEN-L or pSK-TALEN-R. The full length and AD (activation domain)-truncated backbone of TAL effector gene *avrXa7* were used for TALEN-L and TALEN-R, respectively. Three FokI endonuclease domain versions were used in this study: the bacterial homodimeric version was used for both TALEN-L and TALEN-R of construct p467, while a heterodimeric EDL version was used for TALEN-L and a KRK version for TALEN-R of construct p555. DNA sequences for the heterodimeric domains were rice-codon-optimized based on the amino acid sequences as reported by Doyon et al. (2011).

Yeast single-strand annealing assay

The paired homodimeric TALEN genes were released from pSK-TALEN-L and pSK-TALEN-R using the restriction enzymes BglII and SpeI and ligated into the BamHI and SpeI pretreated pCP3M and pCP4M, respectively, resulting in two yeast plasmids for TALEN gene expression in yeast cells. To construct the reporter plasmid, PCR amplicon from the Gl2 target site was digested with BamHI and SpeI and ligated into the BglII and SpeI pretreated pCP5 vector. The insertion of the TALEN-binding element containing Gl2 fragment separated the two halves of a nonfunctional LacZ gene, which encodes β-galactosidase, in the reporter construct pCP5-Gl2. The protocols for yeast transformation and β-galacto-sidase activity assay were followed as described (Li et al., 2011a, b). Briefly, the plasmids pCP3-TALEN-L and pCP4-TALEN-R were introduced into yeast mating strain YPH499. A combination of pCP3M and pCP4M empty vectors was used as negative control. The reporter plasmid pCP5-Gl2 was transferred into yeast YPH500. The YPH499 and YPH500 transformants were mixed, and cells were plated out on medium lacking appropriate amino acids to select and maintain the TALEN expression and reporter plasmids. Following a membrane lift, yeast colonies were stained with X-Gal (5-bromo-4-chloro-3-indolyl-beta-D-galacto-pyrano-side) for visualization of TALEN relative activity.

Construction of maize transformation plasmids

Plasmid pTF101.1, a binary vector (Frame et al., 2002), was modified to contain an expression cassette consisting of the cauliflower mosaic virus (CaMV) 35S promoter, the multiple cloning sites and the nopaline synthase (NOS) terminator, resulting in pTF-35S. The TALEN-L gene in pSK-TALEN-L was released by BglII and SpeI and cloned into the BamHI and SpeI pretreated pTF-35S, resulting in construct pTF-35S: TALEN-L. Meanwhile, TALEN-R gene in pSK-TALEN-R was similarly cloned into pEH3, a vector described previously (Li et al., 2012). The TALEN-R gene expression cassette (ZmUbi:TALEN-R) in pEH3 was further subcloned into pTF-35S:TALEN-L by HindIII, resulting in construct p467. Similarly, the heterodimeric TALEN genes were generated as construct p555.

Transient assay for TALEN activity in maize embryo cells

For the maize cell-based TALEN activity assay, a reporter construct was generated. The ER-mCherry gene was removed from CD3-959 (Arabidopsis Biological Resource Center; Nelson et al., 2007) by PCR and cloned into pTF101.1 at the EcoRV/AatII sites, replacing the phosphinothricin N-acetyltransferase coding gene (bar for Bialaphos resistance) with ER-mCherry. The 120-bp and 121-bp regions at Gl2 target site were amplified using primers EUO795 + EUO796 and EUO795 + EUO797, respectively. These fragments were cloned between the ER-localization signal and the mCherry encoding sequence (at the NcoI site) using infusion cloning reagents (Clontech, Mountain View, CA), generating reporter plasmids p2813 (in frame) and p2825 (out of frame). All PCR generated reading frames were confirmed by DNA sequence analysis.

Maize immature embryos (1.8–2 mm in length) were dissected, pre-incubated for 1–3 days at 28 °C and pretreated on osmotic media for 4 h prior to bombardment (Vain et al., 1993). Plasmid DNAs were coated on 0.6-μm gold particles, and embryos were bombarded with 650-psi rupture discs at 6 cm target distance as previously described (Frame et al., 2000). One day postbombardment, embryos were moved to the callus induction medium as described above, and mCherry expression was visualized 3–7 days later using a Leica dissecting fluorescence microscope (Chroma Technology 500–700; ET-mCherry/Texas Red filter).

Maize transformation

TALEN binary plasmids described above were introduced into Agrobacterium tumefaciens EHA101 through electroporation. Agrobacterium-mediated transformation of immature embryos of Hi-II and B104 maize lines was performed as previously described (Frame et al., 2002) (Frame et al., 2006) except that cocultivation media contained 300-mg/L L-cysteine and Agrobacterium liquid cultures used for embryo infection were shaken for 2 h prior to use. For Hi-II embryos, cocultivation medium was supplemented with 100-mg/L cefotaxime. Bialaphos-resistant callus events were screened for mutations at gl2 using the T7E1 assay.

T7E1 assay and sequencing to detect mutations at gl2

Callus cells and leaf tissues of the primary transgenic (T0) plants were used to extract genomic DNA using the CTAB method (Murray and Thompson, 1980). The genomic DNA was used to PCR-amplify the relevant regions centred on the TALEN target site with specific primers (see Table S2). The PCR products derived from the transgenic tissues were mixed with wild-type-derived amplicon, denatured (95 °C for 5 min) and re-annealed (ramp down to 25 °C at 5 °C/min), and then subjected to T7E1 digestion and agarose gel electrophoresis. The amplicons derived from the T7E1-positive samples were treated with ExoSAP-IT (Affymetrix, Santa Clara, CA) and subsequently sequenced using Sanger sequencing method. The sequencing chromatograms were carefully examined for patterns representative of mono-allelic or di-allelic mutations.

Progeny analysis

Genomic DNA was isolated from leaf tissue from individual progeny plants using the method described (Gao et al., 2010) and adding a second centrifugation step (3000 **g** for 10 min) to additionally clear the cell lysates prior to loading them onto the Whatman GF filter plates. The DNA was eluted from the plates in 200-μL water and 1–4 μL was used as template for PCR. The TALEN cassette was detected by standard PCR amplification using primers EUO772 + EUO773, and the gl2 locus was amplified using EUO750 + EUO751 and sequenced using EUO751. Four primer pairs (TALE.5′-F + TALE.5′-R, TALE.3′-F + TALE.3′-R, ZmU-bi.p-F + TALE.5′-R1 and BAR-F + BAR-R) were used to survey the complete T-DNA in 22 T1 progeny plants of 467.18. RNA was extracted from liquid nitrogen-frozen, powdered T0 and T1 leaves using 1.0 mL of TRIzol following the manufacturer's protocol (Invitrogen/Life technologies, Carlsbad, CA). An aliquot of 2.7 μg total RNA was treated with DNase I (Promega, Madison, WI), diluted to 20 μL and used for cDNA synthesis using the EcoDry oligo-dT reagents (Clontech). RT-PCR was performed using 1–2 μL input cDNA and standard conditions for 30 cycles. TALEN transcripts were amplified using EUO772 + EUO773; maize ubiquitin 1 was amplified using UQBF1 + UBQR1 as an internal control. Phenotype analysis of glossy mutants was performed by misting young seedlings at the 2-leaf stage with water. The gl2 mutant alleles, gl2-N239 and gl2-N718, were obtained from the Maize Genetics Cooperation Stock Center, Urbana, IL. Primer sequences are provided in Table S2.

Acknowledgements

The authors thank the US National Science Foundation (IOS-1238189 to B.Y.), the National Institute of Food and Agriculture of the US Department of Agriculture (2014-67013-21720 to B.Y.) and the Iowa State University Presidential Initiative for Interdisciplinary Research (M.S., K.W., E.V., B.Y.) for providing funding for this research. The authors are also grateful to the Maize Genetics Cooperation Stock Center, Urbana, IL for providing maize lines with *gl2* reference alleles, Xing Xu and Raven Saunders-Duckett for technical assistance, Katey Warnberg for greenhouse support and plant maintenance and Pete Lelonek for maintaining greenhouse and growth chamber spaces.

References

Battraw, M.J. and Hall, T.C. (1990) Histochemical analysis of CaMV 35S promoter-beta-glucuronidase gene expression in transgenic rice plants. *Plant Mol. Biol.* **15**, 527–538.

Bianchi, A. (1978) Glossy mutants: level of action and level of analysis. In *Maize Breeding and Genetics* (Walden, D.B., ed.), pp. 533–550. New York: Wiley & Sons.

Boch, J., Scholze, H., Schornack, S., Landgraf, A., Hahn, S., Kay, S., Lahaye, T., Nickstadt, A. and Bonas, U. (2009) Breaking the code of DNA binding specificity of TAL-type III effectors. *Science*, **326**, 1509–1512.

Cade, L., Reyon, D., Hwang, W.Y., Tsai, S.Q., Patel, S., Khayter, C., Joung, J.K., Sander, J.D., Peterson, R.T. and Yeh, J.R. (2012) Highly efficient generation of heritable zebrafish gene mutations using homo- and heterodimeric TALENs. *Nucleic Acids Res.* **40**, 8001–8010.

Candela, H. and Hake, S. (2008) The art and design of genetic screens: maize. *Nat. Rev. Genet.* **9**, 192–203.

Carroll, D. (2013) Staying on target with CRISPR-Cas. *Nat. Biotechnol.* **31**, 807–809.

Christensen, A.H., Sharrock, R.A. and Quail, P.H. (1992) Maize polyubiquitin genes: structure, thermal perturbation of expression and transcript splicing, and promoter activity following transfer to protoplasts by electroporation. *Plant Mol. Biol.* **18**, 675–689.

Christian, M., Cermak, T., Doyle, E.L., Schmidt, C., Zhang, F., Hummel, A., Bogdanove, A.J. and Voytas, D.F. (2010) Targeting DNA double-strand breaks with TAL effector nucleases. *Genetics*, **186**, 757–761.

Christian, M.L., Demorest, Z.L., Starker, C.G., Osborn, M.J., Nyquist, M.D., Zhang, Y., Carlson, D.F., Bradley, P., Bogdanove, A.J. and Voytas, D.F. (2012) Targeting G with TAL effectors: a comparison of activities of TALENs constructed with NN and NK repeat variable di-residues. *PLoS ONE*, **7**, e45383.

Christian, M., Qi, Y., Zhang, Y. and Voytas, D.F. (2013) Targeted mutagenesis of *Arabidopsis thaliana* using engineered TAL effector nucleases. *G3 (Bethesda)*, **3**, 1697–1705.

Doyon, Y., Vo, T.D., Mendel, M.C., Greenberg, S.G., Wang, J., Xia, D.F., Miller, J.C., Urnov, F.D., Gregory, P.D. and Holmes, M.C. (2011) Enhancing zinc-finger-nuclease activity with improved obligate heterodimeric architectures. *Nat. Methods* **8**, 74–79.

Frame, B., Zhang, H., Cocciolone, S., Sidorenko, L., Dietrich, C., Pegg, S., Zhen, S., Schnable, P. and Wang, K. (2000) Production of Transgenic maize from bombarded Type II callus: effect of gold particle size and callus morphology on transformation efficiency. *In Vitro Cell. Dev. Biol.* **36**, 21–29.

Frame, B.R., Shou, H., Chikwamba, R.K., Zhang, Z., Xiang, C., Fonger, T.M., Pegg, S.E., Li, B., Nettleton, D.S., Pei, D. and Wang, K. (2002) *Agrobacterium* tumefaciens-mediated transformation of maize embryos using a standard binary vector system. *Plant Physiol.* **129**, 13–22.

Frame, B.R., McMurray, J.M., Fonger, T.M., Main, M.L., Taylor, K.W., Torney, F.J., Paz, M.M. and Wang, K. (2006) Improved *Agrobacterium*-mediated

transformation of three maize inbred lines using MS salts. *Plant Cell Rep.* **25**, 1024–1034.

Gao, H., Smith, J., Yang, M., Jones, S., Djukanovic, V., Nicholson, M.G., West, A., Bidney, D., Falco, S.C., Jantz, D. and Lyznik, L.A. (2010) Heritable targeted mutagenesis in maize using a designed endonuclease. *Plant J.* **61**, 176–187.

Hayes, H.K. and Brewbaker, H.E. (1928) Glossy seedlings in maize. *Am. Nat.* **62**, 228–235.

Hockemeyer, D., Wang, H., Kiani, S., Lai, C.S., Gao, Q., Cassady, J.P., Cost, G.J., Zhang, L., Santiago, Y., Miller, J.C., Zeitler, B., Cherone, J.M., Meng, X., Hinkley, S.J., Rebar, E.J., Gregory, P.D., Urnov, F.D. and Jaenisch, R. (2011) Genetic engineering of human pluripotent cells using TALE nucleases. *Nat. Biotechnol.* **29**, 731–734.

Huang, P., Xiao, A., Zhou, M., Zhu, Z., Lin, S. and Zhang, B. (2011) Heritable gene targeting in zebrafish using customized TALENS. *Nat. Biotechnol.* **29**, 699–700.

Joung, J.K. and Sander, J.D. (2013) TALENs: a widely applicable technology for targeted genome editing. *Nat. Rev. Mol. Cell Biol.* **14**, 49–55.

Li, T. and Yang, B. (2013) TAL effector nuclease (TALEN) engineering. In *Methods in Molecular Biology*, vol. **978** (Samuelson, J.C., ed.), pp. 63–72. New York: Humana Press.

Li, T., Huang, S., Jiang, W.Z., Wright, D., Spalding, M.H., Weeks, D.P. and Yang, B. (2011a) TAL nucleases (TALNs): hybrid proteins composed of TAL effectors and FokI DNA-cleavage domain. *Nucleic Acids Res.* **39**, 359–372.

Li, T., Huang, S., Zhao, X., Wright, D.A., Carpenter, S., Spalding, M.H., Weeks, D.P. and Yang, B. (2011b) Modularly assembled designer TAL effector nucleases for targeted gene knockout and gene replacement in eukaryotes. *Nucleic Acids Res.* **39**, 6315–6325.

Li, T., Liu, B., Spalding, M.H., Weeks, D.P. and Yang, B. (2012) High-efficiency TALEN-based gene editing produces disease-resistant rice. *Nat. Biotechnol.* **30**, 390–392.

Liang, Z., Zhang, K., Chen, K. and Gao, C. (2014) Targeted mutagenesis in *Zea mays* using TALENs and the CRISPR/Cas system. *J. Genet. Genomics* **41**, 63–68.

Lor, V.S., Starker, C.G., Voytas, D.F., Weiss, D. and Olszewski, N.E. (2014) Targeted mutagenesis of the tomato PROCERA gene using TALENs. *Plant Physiol.* **166**, 1288–1291.

Miller, J.C., Holmes, M.C., Wang, J., Guschin, D.Y., Lee, Y.L., Rupniewski, I., Beausejour, C.M., Waite, A.J., Wang, N.S., Kim, K.A., Gregory, P.D., Pabo, C.O. and Rebar, E.J. (2007) An improved zinc-finger nuclease architecture for highly specific genome editing. *Nat. Biotechnol.* **25**, 778–785.

Miller, J.C., Tan, S., Qiao, G., Barlow, K.A., Wang, J., Xia, D.F., Meng, X., Paschon, D.E., Leung, E., Hinkley, S.J., Dulay, G.P., Hua, K.L., Ankoudinova, I., Cost, G.J., Urnov, F.D., Zhang, H.S., Holmes, M.C., Zhang, L., Gregory, P.D. and Rebar, E.J. (2011) A TALE nuclease architecture for efficient genome editing. *Nat. Biotechnol.* **29**, 143–148.

Moore, J.K. and Haber, J.E. (1996) Cell cycle and genetic requirements of two pathways of nonhomologous end-joining repair of double-strand breaks in *Saccharomyces cerevisiae*. *Mol. Cell. Biol.* **16**, 2164–2173.

Murray, M.G. and Thompson, W.F. (1980) Rapid isolation of high molecular weight plant DNA. *Nucleic Acids Res.* **8**, 4321–4325.

Nelson, B.K., Cai, X. and Nebenfuhr, A. (2007) A multicolored set of in vivo organelle markers for co-localization studies in Arabidopsis and other plants. *Plant J.* **51**, 1126–1136.

Odell, J.T., Nagy, F. and Chua, N.H. (1985) Identification of DNA sequences required for activity of the cauliflower mosaic virus 35S promoter. *Nature*, **313**, 810–812.

Paz, M.M., Shou, H., Guo, Z., Zhang, Z., Banerjee, A.K. and Wang, K. (2004) Assessment of conditions affecting *Agrobacterium*-mediated soybean transformation using the cotyledonary node explant. *Euphytica*, **136**, 167–179.

Sander, J.D., Cade, L., Khayter, C., Reyon, D., Peterson, R.T., Joung, J.K. and Yeh, J.R. (2011) Targeted gene disruption in somatic zebrafish cells using engineered TALENs. *Nat. Biotechnol.* **29**, 697–698.

Shan, Q., Wang, Y., Chen, K., Liang, Z., Li, J., Zhang, Y., Zhang, K., Liu, J., Voytas, D.F., Zheng, X., Zhang, Y. and Gao, C. (2013) Rapid and efficient gene modification in rice and *Brachypodium* using TALENs. *Mol. Plant* **6**, 1365–1368.

Shukla, V.K., Doyon, Y., Miller, J.C., DeKelver, R.C., Moehle, E.A., Worden, S.E., Mitchell, J.C., Arnold, N.L., Gopalan, S., Meng, X., Choi, V.M., Rock, J.M., Wu, Y.Y., Katibah, G.E., Zhifang, G., McCaskill, D., Simpson, M.A., Blakeslee, B., Greenwalt, S.A., Butler, H.J., Hinkley, S.J., Zhang, L., Rebar, E.J., Gregory, P.D. and Urnov, F.D. (2009) Precise genome modification in the crop species *Zea mays* using zinc-finger nucleases. *Nature*, **459**, 437–441.

Songstad, D.D., Armstrong, C.L., Peterson, W.L., Hairston, B. and Hinchee, M.A.W. (1996) Production of transgenic maize plants and progeny by bombardment of HI-II immature embryos. *In Vitro Cell. Dev. Biol. Plant*, **32**, 179–183.

Szczepek, M., Brondani, V., Buchel, J., Serrano, L., Segal, D.J. and Cathomen, T. (2007) Structure-based redesign of the dimerization interface reduces the toxicity of zinc-finger nucleases. *Nat. Biotechnol.* **25**, 786–793.

Tesson, L., Usal, C., Menoret, S., Leung, E., Niles, B.J., Remy, S., Santiago, Y., Vincent, A.I., Meng, X., Zhang, L., Gregory, P.D., Anegon, I. and Cost, G.J. (2011) Knockout rats generated by embryo microinjection of TALENS. *Nat. Biotechnol.* **29**, 695–696.

Vain, P., McMullen, M.D. and Finer, J.J. (1993) Osmotic treatment enhances particle bombardment-mediated transient and stable transformation of maize. *Plant Cell Rep.* **12**, 84–88.

Wang, Y., Cheng, X., Shan, Q., Zhang, Y., Liu, J., Gao, C. and Qiu, J.L. (2014) Simultaneous editing of three homoeoalleles in hexaploid bread wheat confers heritable resistance to powdery mildew. *Nat. Biotechnol.* **32**, 947–951.

Wendt, T., Holm, P.B., Starker, C.G., Christian, M., Voytas, D.F., Brinch-Pedersen, H. and Holme, I.B. (2013) TAL effector nucleases induce mutations at a pre-selected location in the genome of primary barley transformants. *Plant Mol. Biol.* **83**, 279–285.

Wilson, T.E. and Lieber, M.R. (1999) Efficient processing of DNA ends during yeast nonhomologous end joining. Evidence for a DNA polymerase beta (Pol4)-dependent pathway. *J. Biol. Chem.* **274**, 23599–23609.

Wood, A.J., Lo, T.W., Zeitler, B., Pickle, C.S., Ralston, E.J., Lee, A.H., Amora, R., Miller, J.C., Leung, E., Meng, X., Zhang, L., Rebar, E.J., Gregory, P.D., Urnov, F.D. and Meyer, B.J. (2011) Targeted genome editing across species using ZFNs and TALENs. *Science*, **333**, 307.

Wright, D.A., Li, T., Yang, B. and Spalding, M.H. (2014) TALEN-mediated genome editing: prospects and perspectives. *Biochem. J.* **462**, 15–24.

Zhang, Y., Zhang, F., Li, X., Baller, J.A., Qi, Y., Starker, C.G., Bogdanove, A.J. and Voytas, D.F. (2013) Transcription activator-like effector nucleases enable efficient plant genome engineering. *Plant Physiol.* **161**, 20–27.

Zhao, Z., Gu, W., Cai, T., Tagliani, L., Hondred, D., Bond, D., Schroeder, S., Rudert, M. and Pierce, D. (2002) High throughput genetic transformation mediated by *Agrobacterium tumefaciens* in maize. *Mol. Breeding* **8**, 323–333.

5

An *Agrobacterium*-delivered CRISPR/Cas9 system for high-frequency targeted mutagenesis in maize

Si Nian Char[1,†], Anjanasree K. Neelakandan[1,†], Hartinio Nahampun[2], Bronwyn Frame[2], Marcy Main[2], Martin H. Spalding[1], Philip W. Becraft[1], Blake C. Meyers[3], Virginia Walbot[4], Kan Wang[2,*] and Bing Yang[1,*]

[1]*Department of Genetics, Development and Cell Biology, Iowa State University, Ames, IA, USA*
[2]*Department of Agronomy, Iowa State University, Ames, IA, USA*
[3]*Donald Danforth Plant Science Center, St. Louis, MO, USA*
[4]*Department of Biology, Stanford University, Stanford, CA, USA*

*Correspondence
email
kanwang@iastate.edu and

email byang@iastate.edu
†These authors contribute equally to this work.

Keywords: *anthocyaninless*, *Argonaute*, CRISPR/Cas9, gene editing, maize, targeted mutagenesis.

Summary

CRISPR/Cas9 is a powerful genome editing tool in many organisms, including a number of monocots and dicots. Although the design and application of CRISPR/Cas9 is simpler compared to other nuclease-based genome editing tools, optimization requires the consideration of the DNA delivery and tissue regeneration methods for a particular species to achieve accuracy and efficiency. Here, we describe a public sector system, ISU Maize CRISPR, utilizing *Agrobacterium*-delivered CRISPR/Cas9 for high-frequency targeted mutagenesis in maize. This system consists of an *Escherichia coli* cloning vector and an *Agrobacterium* binary vector. It can be used to clone up to four guide RNAs for single or multiplex gene targeting. We evaluated this system for its mutagenesis frequency and heritability using four maize genes in two duplicated pairs: *Argonaute 18* (*ZmAgo18a* and *ZmAgo18b*) and *dihydroflavonol 4-reductase* or *anthocyaninless* genes (*a1* and *a4*). T_0 transgenic events carrying mono- or diallelic mutations of one locus and various combinations of allelic mutations of two loci occurred at rates over 70% mutants per transgenic events in both Hi-II and B104 genotypes. Through genetic segregation, null segregants carrying only the desired mutant alleles without the CRISPR transgene could be generated in T_1 progeny. Inheritance of an active CRISPR/Cas9 transgene leads to additional target-specific mutations in subsequent generations. Duplex infection of immature embryos by mixing two individual *Agrobacterium* strains harbouring different Cas9/gRNA modules can be performed for improved cost efficiency. Together, the findings demonstrate that the ISU Maize CRISPR platform is an effective and robust tool to targeted mutagenesis in maize.

Introduction

Clustered regularly interspaced short palindromic repeat/CRISPR-associated Cas9 (CRISPR/Cas9) constitutes an adaptive immune system in many proteobacteria and archaea (Bhaya *et al.*, 2011). These genes enable hosts to eliminate invading genetic parasites such as virus and plasmid DNA. Type I, II and III CRISPR/Cas systems with distinct characteristics of guide RNAs and Cas proteins have been documented. The Type II CRISPR/Cas system from *Streptococcus pyogenes* has been most widely adapted for site-specific genomic alteration or genome editing. Other CRISPR/Cas systems such as those derived from *Neisseria meningitidis* and *Streptococcus thermophiles* also have been adapted for genome editing in mammals (Hou *et al.*, 2013; Xu *et al.*, 2015). Modified CRISPR/Cas9 systems suitable for eukaryotes consist of a nuclear localized endonuclease and a guide RNA; this complex is referred to here as Cas9/gRNA. Unlike the predecessor zinc finger nuclease (ZFN) and TAL effector nuclease (TALEN), which involve dimerizing fusion proteins including the DNA binding domains of ZF and TAL and cleavage domains of *Fok*I endonuclease, Cas9/gRNA is a ribonucleoprotein active on target DNA. gRNA is a chimeric molecule of CRISPR RNA (crRNA) and transactivating crRNA (tracrRNA) preceded by a spacer sequence of 18–20 nucleotides complementary to the target DNA. Cas9 contains both RuvC and HNH DNA cleavage domains that cause DNA

double-strand breaks (DSB) predominantly located 3 bp upstream of the protospacer adjacent motif (PAM) sequence (5'-NGG or 5'-NAG for *S. pyogenes* Cas9) of the target DNA. Subsequently, host DSB DNA repair *in vivo* utilizes error-prone nonhomologous end-joining (NHEJ) or homology-directed repair (HDR). NHEJ often leads to random DNA insertions or deletions (indel mutations) at the cleavage site (so-called targeted mutagenesis), while HDR can be exploited for precise sequence or gene replacement or insertion by providing a donor DNA template with sequence homology to the predicted DSB region.

Different Cas9/gRNA systems have been tailored for targeted genomic alterations in both prokaryotes and eukaryotes. Tailored Cas9/gRNA systems have been successfully deployed into plants as DNA to generate site-specific mutagenesis in both mono-cotyledonous and dicotyledonous species including Arabidopsis (Feng *et al.*, 2014; Jiang *et al.*, 2014), tomato (Brooks *et al.*, 2014; Cermak *et al.*, 2015), potato (Wang *et al.*, 2015), soybean (Li *et al.*, 2015), rice (Feng *et al.*, 2013; Zhou *et al.*, 2014), sorghum (Jiang *et al.*, 2013), wheat (Wang *et al.*, 2014), barley (Lawrenson *et al.*, 2015) and maize (Liang *et al.*, 2014; Xing *et al.*, 2014; Svitashev *et al.*, 2015). The Cas9/gRNA can also be delivered into protoplasts of Arabidopsis, tobacco, lettuce and rice as a protein/RNA complex (a ribonucleoprotein, RNP) to induce mutations in cells, some of which can be regenerated into a gene-mutated plant (Woo *et al.*, 2015).

Although the CRISPR technology is simpler than ZFNs or TALENs, it must be optimized for each plant species to accommodate the type of tissue and the transformation delivery method. For example, different RNA polymerase II-based promoters (ubiquitin gene promoters, viral CaMV 35S promoter, etc.) suitable for the expression of Cas9, or RNA polymerase III-dependent promoters (U3, U6, etc.) for driving gRNA expression, need to be tested for efficacy in specific plant species. For DNA delivery, there are two major transformation methods: *Agrobacterium tumefaciens*-mediated or biolistic (gene gun)-mediated methods. Both methods are effective in transforming plant species that are not amenable to regeneration from single-cell protoplasts. The *Agrobacterium*-mediated method is more popular, because it has a propensity to insert single or a low copy number of transgenes and does not require an expensive particle gun apparatus and supplies.

Maize (*Zea mays*) supplies 25% of the world's calories. Genome editing protocols for this crop have been developed, including the application of ZFNs (Shukla *et al.*, 2009), TALENs (Char *et al.*, 2015) and Cas9/gRNA (Feng *et al.*, 2016; Liang *et al.*, 2014; Svitashev *et al.*, 2015; Xing *et al.*, 2014; Zhu *et al.*, 2016). The first report of Cas9/gRNA maize mutagenesis was by Liang *et al.* (2014) in protoplasts using polyethylene glycol (PEG) to mediate DNA uptake. By using the maize U3 promoter for gRNA and the CaMV 35S promoter for a rice codon-optimized *Cas9*, Liang *et al.* (2014) achieved 16.4% mutation frequency for one gRNA and 19.1% for the second gRNA in experiments targeting the *inositol phosphate kinase* gene (*ZmIPK*) in mesophyll protoplasts. Similarly, Xing *et al.* (2014) built a suite of Cas9/gRNA vectors; for maize, the Cas9/gRNA construct consisted of a maize codon-optimized *Cas9* under the maize *ubiquitin 1* gene promoter and gRNA under the rice U3 or wheat U3 promoters. A construct targeting *ZmHKT* was delivered by *Agrobacterium* into immature embryos of B73; twenty T_0 plants showed mutations of *ZmHKT*, although no explicit frequency was reported (Xing *et al.*, 2014). Svitashev *et al.* (2015) also reported CRISPR/Cas9-induced mutagenesis in maize, as well as gene replacement and gene insertion using biolistic-mediated transformation. The two key reagents of their protocol were the maize *ubiquitin 1* gene promoter joined to a maize codon-optimized *Cas9*, and a maize U6 promoter for gRNAs. A mixture of *Cas9*, gRNAs (in either the DNA gene or RNA form of gRNA), plus transformation selection and visual marker genes were co-bombarded into immature embryos of the maize Hi-II genotype. Additionally, *in vitro* transcribed guide RNAs were introduced into embryos expressing a stably integrated *Cas9* transgene (Svitashev *et al.*, 2015). Most recently, two groups demonstrated the feasibility of *Agrobacterium*-delivered Cas9/gRNA in targeted mutagenesis of the endogenous *phytoene synthase* at 13% frequency (Zhu *et al.*, 2016) or *Zmzb7* at 2% frequency (Feng *et al.*, 2016) assessed in T_0 Hi-II plants.

In this work, we present an easy-to-use binary vector system, ISU Maize CRISPR, for efficient site-specific mutagenesis in maize using *Agrobacterium*-mediated maize transformation. Our intention is to provide the public research community with an enabling platform for maize genome editing. We validated the Cas9/gRNA and *Agrobacterium*-mediated protocol using two maize gene families, *Argonaute 18* and *dihydroflavonol 4-reductase*. For each gene family, with members on two different chromosomes, we designed two gRNAs to target two sites within each allele. Here, we show that this vector can be used to insert up to four gRNAs for single or multiplex mutagenesis. We also show that the

Agrobacterium binary vector system achieves highly efficient and heritable site-specific mutagenesis for both maize hybrid genotype Hi-II and inbred B104. Because the preparation of staged maize embryos can be rate limiting and maize transformation process can be costly, we also demonstrated that efficient mutagenesis can be achieved by mixing two *Agrobacterium* strains for one infection experiment to generate transgenic plants independently mutated in each target by separate Cas9/gRNA construct. We confirm that the continuous presence of Cas9/gRNA in transgenic maize can cause mutagenesis of target genes of interest in subsequent generations. The Cas9/gRNA transgenic lines, therefore, can be used to convey the CRISPR-based mutagenesis by genetic cross to maize lines that are not amenable to genetic transformation.

Results

Targeted mutagenesis strategy

A schematic of the ISU Maize CRISPR plasmids used for *Agrobacterium*-mediated Cas9/gRNA introduction into maize is shown in Figure 1. The gRNA vectors are based on pENTR-gRNA1 and pENTR-gRNA2 described previously (Zhou *et al.*, 2014). In each intermediate vector, two different rice U6 small nuclear RNA gene promoters (PU6.1 and PU6.2) are used to express the gRNA genes. The first gRNA scaffold (85 nucleotides) is preceded by a cloning site containing two *BtgZI* sites in a tail-to-tail orientation downstream of PU6.1. The second gRNA scaffold follows a pair of tail-to-tail-oriented *BsaI* sequences downstream of PU6.2. Two sequential rounds of cloning permit the insertion of custom double-stranded gRNA spacer DNA sequences into these double *BtgZ*1 and double *Bsa*1 restriction enzyme sites in the vectors to generate intermediate constructs pgRNA-IM1 or pgRNA-IM2 (Figure 1).

As described in an earlier publication (Zhou *et al.*, 2014), these two vectors differ by one feature: pENTR-gRNA1 possesses two *Hind*III sites near the Gateway recombination sites attL1 and attL2, while pENTR-gRNA2 has only one *Hind*III site near the attL1 site (Figure 1). This feature allows pgRNA-IM2 to receive the gRNA cassettes from pgRNA-IM1 via *Hind*III digestion and subcloning. Therefore, this strategy can be used to construct up to four gRNAs, simultaneously targeting up to four DNA sequences in the maize genome.

The guide RNA spacer sequences were designed based on the maize B73 reference genome sequence (Schnable *et al.*, 2009) using the CRISPR Genome Analysis Tool (Brazelton *et al.*, 2015; http://cbc.gdcb.iastate.edu/cgat/). The relevant target regions in Hi-II and B104 genotypes were PCR-amplified and confirmed by sequencing. All pgRNA-IM constructs were confirmed for sequence accuracy at the insertion sites and flanking regions by Sanger sequencing. The confirmed gRNA cassette can be mobilized through Gateway recombination to the destination vector pGW-Cas9. The vector is built on the backbone of pMCG1005 (a gift from Dr. Vicki Chandler); this vector contains a rice codon-optimized *Cas9* with the maize *ubiquitin 1* gene promoter and the *bar* gene with a 4× CaMV 35S promoter used as transformation selectable marker (Figure 1). The binary plasmid is mobilized into *Agrobacterium* strain EHA101 for the transformation of maize immature embryos.

Agrobacterium-based maize transformation was previously described (Frame *et al.*, 2006). For a typical site-directed mutagenesis project, 20–30 bialaphos-resistant callus lines are identified for genotyping using the T7 endonuclease I (T7E1) assay (Char *et al.*, 2015). This assay uses PCR amplification of the target

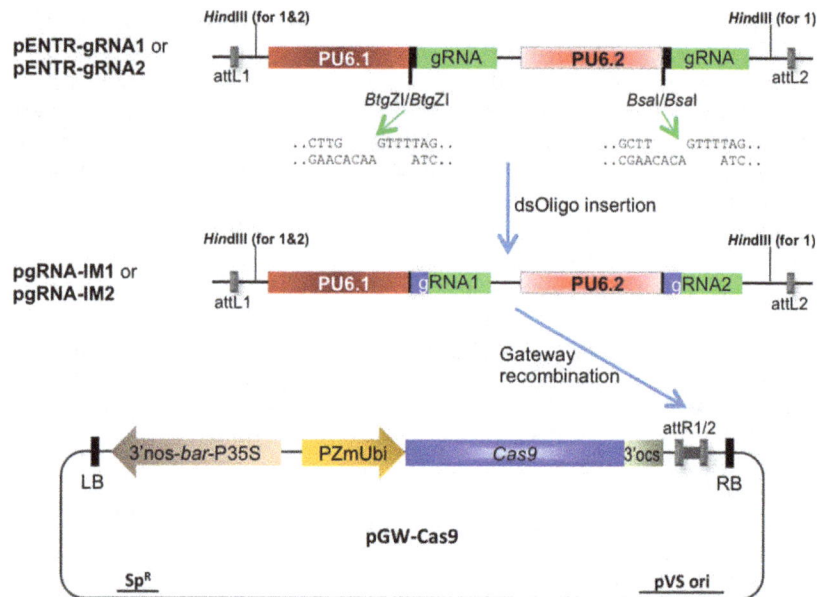

Figure 1 Schematic diagram of Cas9/gRNA construction. Cloning vectors pENTR-gRNA1 (with two HindIII sites) or pENTR-gRNA2 (with one HindIII site) were sequentially digested with BtgZI and BsaI restriction enzymes for the insertions of two double-stranded oligonucleotides. The subcloning resulted in two intermediate constructs, pgRNA-IM1 and pgRNA-IM2, each carrying two gRNA expression cassettes. The cassettes flanked by the Gateway recombination sequences attL1 and attL2 were mobilized to the binary vector pGW-Cas9 through Gateway recombination, resulting in a single plasmid Cas9/gRNA binary construct for Agrobacterium-mediated gene transfer.

gene followed by melting and reannealing the PCR products; homozygous individuals (two copies of the wild-type or two copies of the same mutant allele) yield a single duplex, while heterozygous individuals (diallelic mutants or wild-type allele plus mutated allele) yield multiple duplexes containing mismatches. The mismatched bases are targets for T7E1 cleavage, resulting in multiple fragments resolved by agarose gel electrophoresis. Comparison of restriction fragment patterns to wild-type size standards permits the classification of lines as diallelic mutants (DA), monoallelic mutants (MA) or nonmutant lines, with subsequent Sanger sequencing used to precisely describe the mutations (Char et al., 2015). Typically, ten independent callus lines with defined mutations are selected for plant regeneration, followed by self-crossing or crossing to a wild-type line to produce transgenic seeds. Multiple (usually two to five) plantlets are produced from each callus line. During T_0 growth, DNA from leaf samples are subjected to the T7E1 assay and sequencing of the site-specific PCR amplicons. The time frame from construct design to seed from a CRISPR maize line is approximately 7 months (Figure 2).

Cas9/gRNA constructs induce highly efficient mutagenesis in four genes

We first tested the platform for targeted mutagenesis on two closely related but polymorphic Argonaute (Ago) genes ZmAgo18a (GRMZM2G105250) and ZmAgo18b (GRMZM2G457370) that were implicated in 24-nt phasiRNA biogenesis in anthers (Zhai et al., 2015). To enhance mutagenesis success in the targeted exon, two closely located target sites in each Argonaute gene were selected for gRNA construction. ZmAgo18a was specifically targeted by gAgo18a-1/gAgo18a-2 in pgRNA-IM1, and ZmAgo18b was specifically targeted by gAgo18b-1/gAgo18b-2 in pgRNA-IM2 (Figure 3a). A third plasmid targeting both copies of Ago18 simultaneously was constructed by subcloning of the gAgo18a-1/gAgo18a-2 cassette from pgRNA-IM1 into pgRNA-IM2, which already contained gAgo18b-1/gAgo18b-2. The three gRNA constructs were subsequently moved to pGW-Cas9 through Gateway recombination. For simplicity, the three constructs are referred to as gAgo18a, gAgo18b and gAgo18a/b.

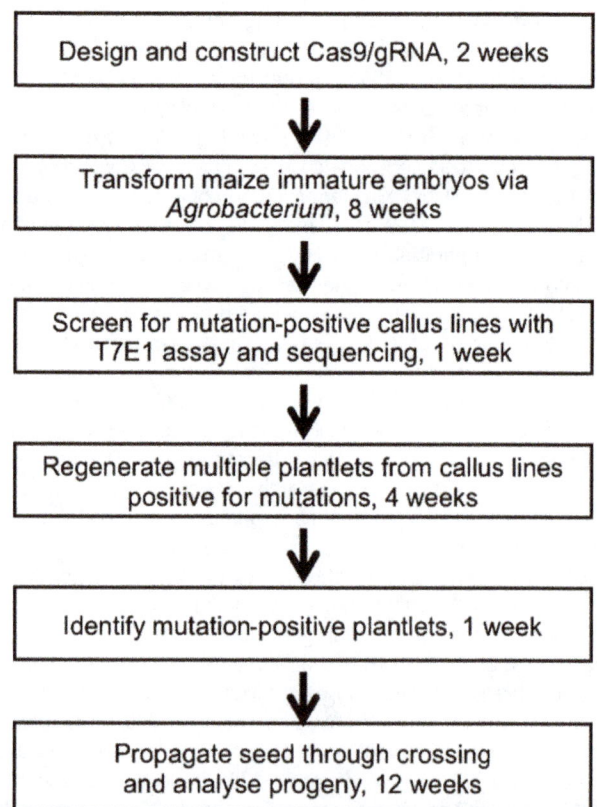

Figure 2 A flow chart of targeted mutagenesis in maize using Agrobacterium-mediated transformation illustrates the main steps in CRISPR-based mutagenesis. A minimum of 7 months is required from embryo transformation to production of mutant seeds.

We also constructed gRNAs that targeted the dihydroflavonol 4-reductase or anthocyanin biosynthesis gene a1 (anthocyaninless 1) and its homolog a4, two duplicated orthologs of the Arabidopsis Ben1 gene, encoding a dihydroflavonol 4-reductase that governs the levels of endogenous brassinosteroid hormones

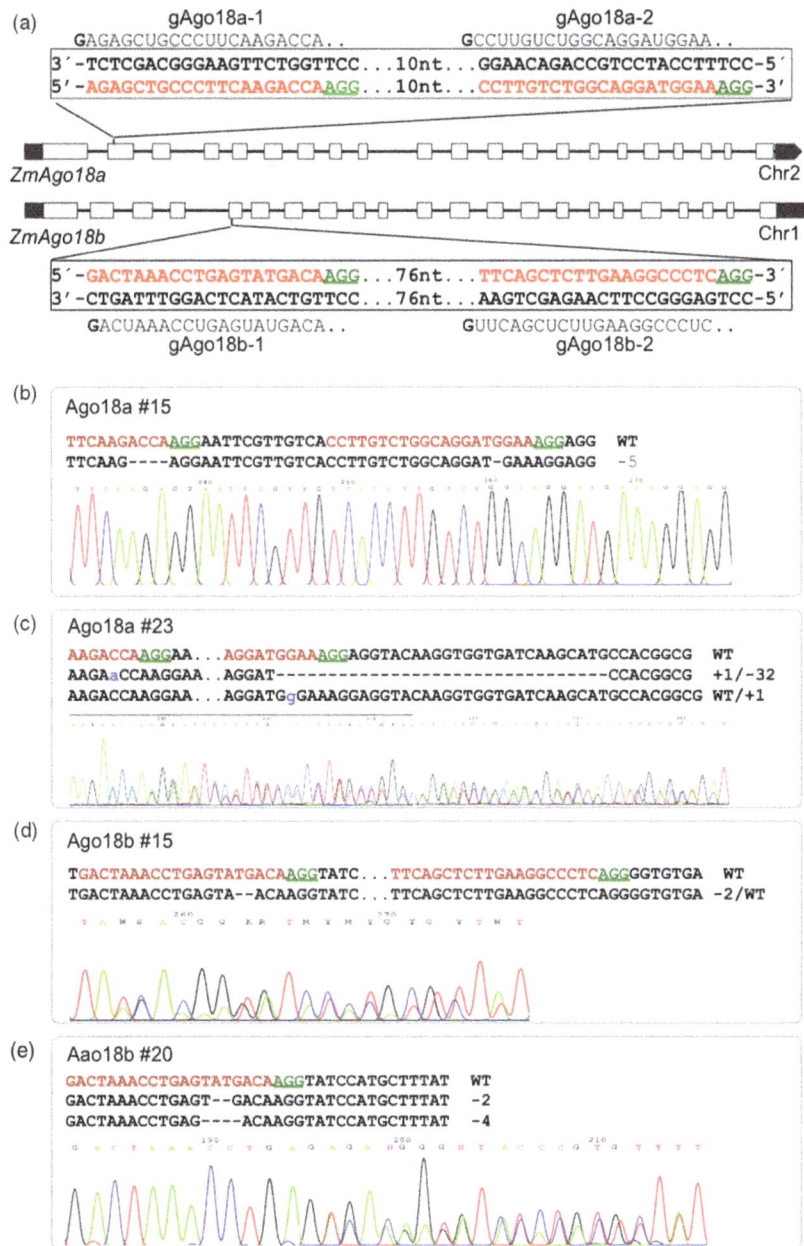

Figure 3 Cas9/gRNA-induced mutations in *ZmAgo18a* and *ZmAgo18b*. (a) Structure of the paralogous *Ago18* genes present on chromosomes 1 and 2 and the gRNAs designed to generate DSBs in exons (blank bars). gRNAs, gAgo18a-1 and gAgo18a-2 (above the double-strand box for *ZmAgo18a*), and gAgo18b-1 and gAgo18b-2 (below the double-strand box for *ZmAgo18b*). Nucleotides in red represent target sites, and green underlined nucleotides indicate PAM sequences for the gRNAs. 10 and 76 nt represent the numbers of nucleotides between the two target sites in each gene. (b–e) Sequences from selected T$_0$ plants with site-specific mutations accompanied by corresponding regions of the sequencing chromatograms. The nucleotide changes (dashes for deletion, lowercase letter for insertion and WT for unaltered) are also indicated to the right side of each sequence. Dots in Ago18a #23 and Ago18b #15 represent nucleotides not shown.

(Yuan *et al.*, 2007). The predicted protein sequences encoded by maize *a1* (GRMZM2G026930) and *a4* (GRMZM2G013726) share 88.3% similarity. The two gRNAs (gA1/A4-1 and gA1/A4-2) were designed to target conserved sites with a perfect match to *a4* but a mismatch to *a1* in position 3 at the 5′ end of each guide RNA (Figure 4a). Polymorphisms in the target regions between the two genes allowed specific amplification of the relevant regions for genotyping of individual genes.

The constructs were transferred to *Agrobacterium* strain EHA101 and used to infect immature Hi-II maize embryos. Bialaphos-resistant callus lines were identified, putative mutants were molecularly analysed, and mutation-positive callus lines were transferred to regeneration medium, yielding multiple plantlets per line. Plants were brought to maturity and self-pollinated or reciprocally crossed to a wild-type line for seeds. During plant growth in the greenhouse, successive leaf samples were taken from two randomly selected plants of each line, and

these were combined and used for genomic DNA (gDNA) extraction. PCR amplicons of *ZmAgo18a*, *ZmAgo18b*, *a1* and *a4* targeted regions were analysed by the T7E1 assay and sequenced.

Given the complexity of possible results, from zero to two target sites of one gene (or locus) and 0–4 target sites of two genes (or loci), we have adopted the following terminology to describe the allelic status of transformants. Monoallelic mutants, designated MA, have a mutation in one allele of the target gene (or locus) regardless of the site of the allele and an unmutated second allele. Diallelic mutants, designated DA, have mutations in both homologous copies of the target gene (or locus). For lines with two paralogous genes (or loci) targeted, we simply present the mutations in a combination of MA or DA for each locus. Interpretation of mono- and diallelic cases sometimes can be difficult in the T$_0$ generation; self-crossing or outcrossing to another line and the loss of Cas9/gRNA by segregation simplify

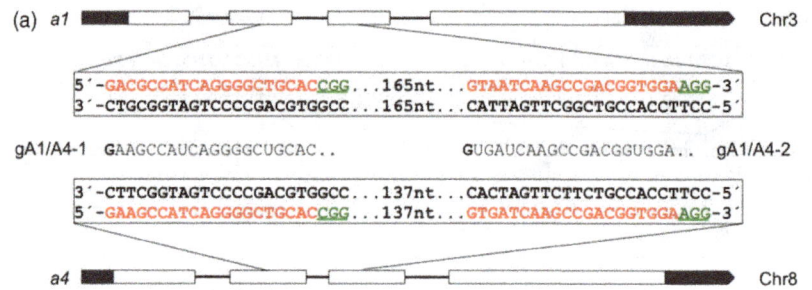

Figure 4 Cas9/gRNA-induced mutations at the *a1* (chromosome 3) and *a4* (chromosome 8) target sites. (a) Gene structures of *a1* and *a4* loci with gRNAs designed for DSBs in exons (blank bars). Nucleotides in red represent target sites, and green underlined nucleotides indicate PAM sequences for the gRNAs. gRNAs, gA1/A4-1 and gA1/A4-2 are between *a1* and *a4* gene boxes. 165 nt and 137 nt represent the numbers of nucleotides between the two target sites in each gene. (b) and (c) Sequences from selected T$_0$ plants containing the site-specific mutations. MT, mutant types; the nucleotide changes (dashes for deletion and lowercase letter in blue for insertion) are also indicated to the right side of each sequence, suffixed with a letter, if needed, to distinguish different alleles. Line, mutant line.

the interpretation of the T$_1$ DNA sequencing results and are used routinely for verification of the interpretation of T$_0$ sequences at the target sites.

The two single-gene targeting constructs gAgo18a and gAgo18b achieved similar transformation and mutagenesis frequencies (Table 1). In the T7E1 assay on selected bialaphos-resistant calli 74% (17 of 23) of gAgo18a lines and 70% (16 of 23) of gAgo18b lines were scored as mutated. Of the 17 mutated gAgo18a lines, 12 were MA and 5 were DA mutants. Similarly, nine MA mutants and seven DA mutants were identified among the 16 gAgo18b lines. The T7E1-positive PCR amplicons were subjected to Sanger sequencing and found to contain various combinations of mutations as illustrated for representative lines in Figure 3b, e. A majority of mutant plants tested (e.g. 7/10 of *ZmAgo18b*) contained mutations identical to those detected in the progenitor callus (Table 1). This result suggests that for most events, mutations occurred in a single cell from which each bialaphos-resistant callus was derived, rather than occurring sporadically during the subsequent callus growth.

As noted above in vector design, two gRNAs were used to mutate each *ZmAgo18* gene; the two gRNA targets were separated by 10 nucleotides (nt) in *ZmAgo18a* and by 76 nt for *ZmAgo18b* (Figure 3a). Interestingly, mutations were detected at both target sites for *ZmAgo18a* (Figure 3b, c). However, only one of two gRNAs for *ZmAgo18b* was effective, because all mutations detected in *ZmAgo18b* were located in the target site 1

(Figure 3d, e). It is unknown whether the distance between the two gRNAs or an aspect of gAgo18b-2 structure contributed to the lack of mutations from this gRNA.

Table 1 summarizes two duplex targeting experiments for *ZmAgo18a* and *ZmAgo18b*, as well as parallel experiments for *a1* and *a4*. In the gAgo18a/b experiment, a total of 22 mutated lines were identified from 26 bialaphos-resistant callus lines, an 85% frequency of mutagenesis. Among the 22 lines, 12% contained only mutations in *ZmAgo18a*, representing one MA and two DA mutants. Fifteen per cent involved only *ZmAgo18b*, with three MA mutants and one DA mutant. Most mutated lines (58%) had mutations in both *ZmAgo18* genes: 11 MA and 4 DA for *ZmAgo18a* and 10 MA and 5 DA for *ZmAgo18b*.

The design of the *a1/a4* duplex targeting experiment was different, with each gRNA targeting sites conserved between the two genes. We achieved a 79% mutagenesis frequency with 37 callus mutants confirmed from 47 bialaphos-resistant callus lines. Thirty-five of the 37 callus lines were regenerated and produced plants (Table 1). Similar to gAgo18a/b, gA1/A4 produced three groups of callus-mutant lines: *a1* single, *a4* single and *a1* and *a4* double mutants. Fifteen per cent of mutations involved only *a1*; 49% were in *a4* only, three times higher than that in *a1*. The lower mutation efficiency in *a1* is likely attributable to the 1-bp mismatch between the target sequence of *a1* and each of two gRNAs (Figure 4a). Nine lines (15%) involved both the *a1* and *a4*

Table 1 Summary of CRISPR mutagenesis frequencies on four genes in maize Hi-II genotype

gRNA	Target gene	# bar+ callus line analysed	# Mutation+ callus line	% Mutation frequency	# Monoallelic mutant	# Diallelic mutant	# Mutation+ line regenerated
gAGO18a	ZmAgo18a	23	17	74	12	5	17
gAGO18b	ZmAgo18b	23	16	70	9	7	16
gAGO18a/b	ZmAgo18a	26	3	12	1	2	22
	ZmAgo18b		4	15	3	1	
	ZmAgo18a&18b		15	58	11 (*18a*), 10 (*18b*)	4 (*18a*),5 (*18b*)	
gA1/A4	a1	47	7	15	1	6	35
	a4		23	49	1	20	
	a1 & a4		7	15	0 (*a1*), 0 (*a4*)	7 (*a1*), 7 (*a4*)	

genes. Various combinations of MA and DA mutations were observed. Figure 4b, c shows the mutations identified for selected lines.

Inheritance of Cas9/gRNA-mediated mutations

Individual T_0 plants from selected mutant lines were self-pollinated or cross-pollinated with the maize inbred line B73. Once mutated, target sites should no longer be recognized by gRNA and therefore not subject to further rounds of mutagenesis. To avoid the complications of continuing action by the Cas9/gRNA transgene in B73 alleles introduced by cross-pollination, we chose populations derived from selfing of T_0 lines carrying homogeneous or heterogeneous DA mutations for transmission analysis. For genotyping T_1 plants, 20 seeds from independent transgenic events were germinated and grown in the greenhouse. Genomic DNA was extracted and PCR amplicons from the targeted region were examined using the T7E1 assay and, for a subset of samples, Sanger sequencing of amplicons was performed.

Inheritance of the mutated ZmAgo18a and ZmAgo18b alleles was analysed in independent lines. Table 2 shows inheritance results for four selected lines, two ZmAgo18a mutants (Ago18a #2 and #15) and two ZmAgo18b mutants (Ago18b #19 and #20). For all four lines, mutations in the T_0 plants were transmitted to the T_1 generation. Ago18a #2 was a DA mutant: one mutated allele had a 51-bp deletion and the second allele had a 35-bp deletion. Ago18a#15 was a homogenous DA mutant with a 5-bp deletion (Figure 3b); thus, there was no segregation for mutations among the progeny (Table 2). Ago18b #19 was DA of 1-bp deletion and 1-bp insertion, while Ago18b #20 was DA for heterogeneous 2- and 4-bp deletions (Figure 3e). For the inheritance of Cas9/gRNA, Cas9 was analysed using a PCR approach with gene-specific primers. If T_0 plants had a single transgene

locus, Cas9 would be expected to segregate 3 : 1 in the T_1 progeny from the self-pollination; this Mendelian expectation was confirmed in Ago18a #2, Ago18b #19 and #20. However, all 20 seedlings from Ago18a #15 carried the Cas9/gRNA transgene (Table 2). This type of non-Mendelian segregation could result from transgenic lines with multiple transgene copies integrated on different chromosomes and could also occur if a sterility mutation eliminates noncarrier pollen, that is via an unselected mutation from transformation or tissue culture in repulsion to the Cas9/gRNA transgene.

Inherited Cas9 and gRNA expression induces new mutations in progeny plants

To investigate whether the Cas9/gRNA transgene cassette remains active and can induce mutations after the carrier plant is crossed to a wild-type inbred line, we chose one a1/a4 mutation-positive line (#20) for analysis. This line was chosen because it has DA mutations in both the a1 and a4 genes (Figure 5a, c, blue T_0). Female flowers of T_0 plants were crossed by using pollen from B73 carrying wild-type a1 and a4 alleles. As shown in Figure 5a, the T_1 progeny population can be divided into Cas9-positive (transgenic lines) and Cas9-negative (transgene null) segregants. Two of eight T_1 plants screened (25%) carried the Cas9/gRNA transgene. DNA sequencing analysis of the two Cas9-positive individuals indicated that both had novel mutations in a4, but not in a1 (Table 3). The novel mutations were verified from three independent leaves of the same plant, indicating that the mutations did not reflect mosaicism (Figure 5a, c). This observation suggests that the novel mutations in the wild-type B73 allele occurred early in development, perhaps in the zygote.

To further confirm the heritable activity of the Cas9/gRNA transgene, Cas9-positive plants from the two T_1 lines, designated

Table 2 Transmission of Cas9/gRNA-induced mutations in T1 progeny

Lines*	# analysed	Cas9 positive			Cas9 negative			Cas9 pos vs Cas9 neg	P-value[†]
		aa	ab	bb	aa	ab	bb		
AGO18a #2	20	6	7	5	1	1	0	18 : 2	0.121
AGO18a #15	20	20	0	0	0	0	0	20 : 0	0.010
AGO18b #19	18	1	12	3	0	0	2	16 : 2	0.174
AGO18b #20	16	3	5	4	2	0	2	12 : 4	1.000

aa & bb, homozygous a and b; ab, heterozygous mutant ab.

*All mutant lines were self-pollinated; expected segregation ratio is 3 : 1 (Cas9 positive: Cas9 negative).

[†]Chi-square probability with one degree of freedom.

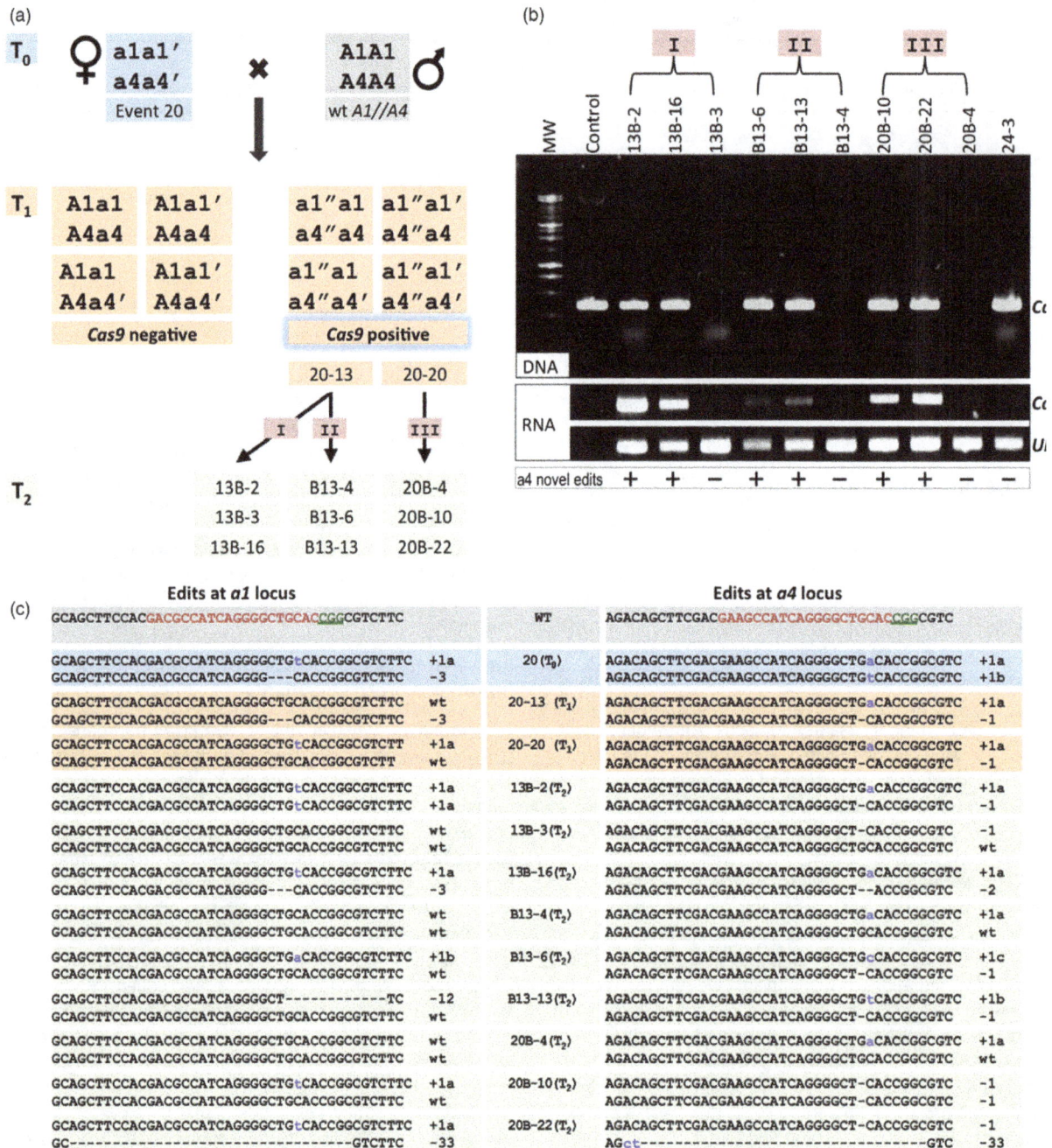

Figure 5 Characterization of sexually heritable new alleles after targeted mutagenesis by Cas9/gRNA in maize cells. (a) Schematic diagram showing the inheritance and segregation of original edited alleles as well as the generation of new mutations from *a1//a4* CRISPR event 20. The T_0 has DA mutations for both *a1* and *a4*. The T_1 and T_2 progeny were derived by crossing mutants to recipient lines with wild-type *a1* and *a4* loci. The wild-type allele is represented as 'A', the T_0 mutated alleles as 'a' and 'a''. 'a'''' indicates alleles that potentially contain novel mutations. For the development of the T_2 generation, T_1 plant 20-13 was used as a female (cross I), or male (cross II), and T_1 plant 20-20 was used as a female (cross III). (b) The top panel shows the presence of *Cas9* in genomic DNA in the T_2 progeny plants of event 20 as assayed by PCR. The control lane represents a plasmid-positive control with cloned *Cas9*. The bottom panel shows *Cas9* transcript levels by RT-PCR. The control lane represents -RT (negative control), and *ubiquitin 1* gene expression (Ubi) serves as the positive control. '+' stands for the presence of novel mutation in T_2, and '−' stands for its absence. (c) Sequence information at the *a1* and *a4* targeted loci for the T_0, T_1 and T_2 plants from event 20. Nucleotides in red represent target sites and green underlined nucleotides indicate PAM sequences for the gRNAs. The nucleotide variations (dashes for deletion and lowercase letter in blue for insertion) are marked on the right side of each sequence with a number, suffixed with a letter, if needed, to distinguish different alleles. Line names are listed in the middle column.

as 20-13 and 20-20, were further outcrossed to B73 wild-type plants to generate T_2 progeny. Plant 20-13 was crossed with B73 either as female (Figure 5a, cross I) or as male (Figure 5a, cross II) to assess the efficiency of novel mutation generation and of Cas9

activity in reciprocal crosses. Plant 20-20 was the ear parent in crosses with B73 (Figure 5a, cross III).

A total of 60 T_2 generation plants from lines 20-13 and 20-20 were screened for the presence of *Cas9*, and novel mutations

Table 3 Novel mutations in Cas9/gRNA-positive progenies

Outcross with wild-type a1//a4		Total # screened	Cas9 pos	Novel mutations (%)	
				a1	a4
T₁ generation					
I	Event 20 as female	8	2	0 (0%)	2 (100%)
T₂ generation					
I	Line 20–13 as female	17	9	7 (78%)	9 (100%)
II	Line 20–13 as male	19	12	10 (83%)	12 (100%)
III	Line 20–20 as female	24	12	7 (58%)	12 (100%)
T₂ total		60	33	24 (73%)	33 (100%)

were detected at the *a1* and *a4* loci, which revealed continued mutagenesis at high frequencies (Table 3). As expected, segregation for the *Cas9* transgene was approximately 1 : 1, with 33 of the 60 plants carrying *Cas9*. All *Cas9*-positive plants carried novel mutations in the *a4* gene, but only a subset (58%–83%) also carried mutations in *a1*. Preferential mutation of *a4* was also observed in the T₁ generation (Table 3), an observation consistent with the T₀ generation analysis (Table 1).

RT-PCR-based expression analysis of *Cas9* in the T₂ progeny demonstrated a positive correlation between the presence of *Cas9* mRNA and the occurrence of novel mutations. As can be seen in Figure 5b, c, all T₂ lines that showed continued mutagenesis contained an actively transcribed *Cas9*, whereas no further mutations in *a1* or *a4* genes were detected in lines lacking the *Cas9* transgene (13B-3, B13-4, 20B-4). Notably, continued mutating was not detected in one line (24-3) that contained a *Cas9* transgene. However, no *Cas9* expression was detectable in this plant (Figure 5b, Figure S1), indicating that inheritance of the *Cas9* transgene is not sufficient and that *Cas9* expression is also required for mutagenesis.

Co-infection of two *Agrobacterium* strains harbouring different Cas9/gRNA constructs produces respective mutations

Given the high efficiency of our Cas9/gRNA system in the initial transformations, we explored the feasibility of mutating two genes (or two groups of genes) in one infection procedure by mixing *Agrobacterium* strains harbouring different gRNA constructs for co-transformation. The motivation for this was to reduce the number of embryos required and the cost of plant transformation while still producing an adequate number of mutants for each gene or group of genes. This represents an alternative strategy for multiplex targeting which can also be achieved with multiple gRNAs in one construct. As a proof-of-concept experiment, two *Agrobacterium* EHA101 strains, one containing Cas9/gAgo18a and another Cas9/gAgo18b, were mixed to obtain equal bacterial cell density. The mixed bacterial culture was then used to infect a similar number of maize B104 immature embryos as is usually used for a single *Agrobacterium*-mediated transformation.

A total of twenty-two independent bialaphos-resistant calli were identified. These lines were first screened for the presence of Cas9/gRNA transgenes using PCR. Twenty-one of twenty-two

lines (95%) were positive for the Cas9/gRNA transgenes. Of these twenty-one calli, nine lines (43%) were positive only for the gAgo18a transgene, nine lines (43%) were positive only for gAgo18b and three lines (14%) were positive for both gAgo18a and gAgo18b transgenes (Table 4). The results indicate that one *Agrobacterium* infection experiment can produce two major groups of transgenic callus lines with a small portion of double transformation.

Transformed calli were further analysed for mutations in the targeted genes using both the T7E1 assay and Sanger sequencing. Among those lines carrying Cas9/gRNA transgenes, five were positive for mutations of *ZmAgo18a*, two for *ZmAgo18b* and one for both genes. On the other hand, neither callus line carrying both transgenes (lines 4 and 6) produced mutations in both genes. This discrepancy could be attributed to issues related to false negatives in PCR analysis or incomplete sampling of representative callus cells.

Multiple plantlets (3–5) per callus line were regenerated from the 21 transgenic lines. Two to three plantlets derived from the same transformed callus line were randomly selected and individually analysed for mutations in the target genes. As can be seen in Table 4, seven (47%) of 15 lines that were mutation positive in the callus tissue for *ZmAgo18a* and *18b* were also mutation positive in plantlets. These include five lines positive to *ZmAgo18a* (lines 1, 6, 15, 18 and 19) and one line positive for *Ago18b* (line 12); no line was positive for both genes. One line (line 13) was positive for *Ago18b* in callus assay, but the mutation was not detected in plants. On the other hand, eight lines (lines 4, 5, 8, 9, 14, 16, 17 and 20) tested negative for mutations in the

Table 4 Summary of transformation and mutagenesis frequencies in co-infection experiment

Event ID	Transgene (callus)		Mutant (callus)		Mutant (plant)	
	gAgo18a	gAgo18b	Ago18a	Ago18b	Ago18a	Ago18b
1	+	−	+	−	+	−
2	−	+	−	−	−	−
3	+	−	−	−	−	−
4	+	+	−	−	−	+
5	−	+	−	−	−	+
6	+	+	+	−	+	−
7	−	+	−	−	−	−
8	+	−	−	−	+	−
9	+	−	−	−	+	−
10	+	+	+	+	+	−
11	−	+	−	−	−	−
12	−	+	−	+	−	+
13	−	+	−	+	−	−
14	−	+	−	−	−	+
15	+	−	+	−	+	−
16	+	−	−	−	+	−
17	+	−	−	−	+	−
18	+	−	+	−	+	−
19	+	−	+	−	+	−
20	−	+	−	−	−	+
21	−	+	−	−	−	−
Total	12	12	6	3	10	5
Efficiency	Transformation		Mutagenesis		Mutagenesis	
	57%	57%	29%	14%	48%	24%

callus, but the plants were found to contain mutations in their respective genes. These results indicate that the initial callus assays are reasonable predictors of plant genotype; however, assay improvements may be warranted to increase the accuracy or analysis of multiple regions of calli to determine whether there is unexpected chimerism.

These results from this co-infection experiment indicate that mixing two *Agrobacterium* strains generates mutation frequencies in individual target genes of respective Cas9/gRNA similar to single strain infections. This is consistent with earlier findings showing that when two *Agrobacterium* strains were used for co-infection, the majority of transgenic plants obtained carry only one type of T-DNA in their genome (De Buck *et al.*, 2009; De Neve *et al.*, 1997). Hence, the co-infection approach can be exploited to effectively induce mutations in individual target genes (or groups of conserved genes) and produce mutant plants, thus reducing transformation costs and increasing throughput.

Discussion

We report a high-efficiency CRISPR platform, ISU Maize CRISPR, consisting of Cas9/gRNA utilized with *Agrobacterium*-mediated transformation for targeted mutagenesis of maize in a 7-month process. Efficacy was tested with four maize genes: combining all results, 60% of transgenic callus lines contained site-specific mutations that persisted in regenerated plants and were heritable in the T_1 generations. The Cas9/gRNA transgene, when mobilized to the B73 inbred via genetic crosses, could induce new heritable mutations in wild-type alleles. Finally, co-infection of two *Agrobacterium* strains harbouring distinct Cas/gRNA constructs generated mutations in the respective target genes with frequencies similar to those observed in single transformations. We expect that this highly efficient and cost-effective CRISPR platform will become an enabling genomic tool for the public research community.

gRNA-directed Cas9, like other types of engineered nucleases (e.g. meganucleases, ZFNs and TALENs), has been engineered to induce site-specific DSBs in a host genome, wherein NHEJ is the predominant repair mechanism. The propensity of NHEJ for introducing small indels leads to NHEJ-based genomic mutagenesis as the major application in genome editing (Sander and Joung, 2014). In contrast to the endonucleases first deployed, Cas9/gRNA transgenes are simpler to construct and are more efficient mutagens. Our experience with the same maize transformation platform, but a different gene target than either *Ago18* or *a1/a4*, indicates that mutagenesis frequency by TALENs is about 10% in Hi-II and 3.7% in B104 (Char *et al.*, 2015), which is at least sixfold lower than reported here for Cas9/gRNA-mediated events. On the other hand, the reported mutagenesis efficiencies of different species or even different maize CRISPR platforms differ significantly. For example, relatively low mutagenesis frequency (5.6%) was reported in wheat, a cereal crop with a large and complex genome (Wang *et al.*, 2014). In contrast, much higher mutagenesis rates (up to 100%) for rice have been reported by a number of laboratories (Zhou *et al.*, 2014; Zhang *et al.*, 2014; Xie *et al.*, 2015; Ma *et al.*, 2015). In dicotyledonous tomato and soybean, CRISPR systems produced more than 50% mutated T_0 plants (Brooks *et al.*, 2014; Li *et al.*, 2015). Among the reported maize CRISPR systems, frequencies of 2%–100% were reported in the T_0 Cas9-positive plants (Feng *et al.*, 2016; Liang *et al.*, 2014; Svitashev *et al.*, 2015; Xing *et al.*, 2014; Zhu *et al.*, 2016). These differences reflect a variety of

factors that influence the frequency of NHEJ-mediated mutagenesis. The main factors impacting this efficiency are different promoters that direct the spatiotemporal expression of *Cas9* and gRNAs, methods of Cas9/gRNA delivery that result in different transgene copy numbers and thus variation in the host cell abundance of the Cas9/gRNA ribonucleoprotein complex and, finally, the specific target gene sequence and chromosomal location, which affects the accessibility the target gene to Cas9/gRNA ribonucleoprotein to cause DSBs (Wu *et al.*, 2014; Horlbeck *et al.*, 2016). Nevertheless, our Cas9/gRNA system showed similarly high efficiency with all four genes tested and with an additional 27 constructs targeting 30 maize genes (unpublished data).

The current necessity for plant transformation makes maize genome editing an expensive, laborious and time-consuming process. We adapted three strategies to improve efficiency in terms of the cost of consumables and labour. First, we designed and constructed two gRNAs targeting each gene to increase (presumably double) the success rate or improve the possibility that at least one gRNA will be active for mutagenesis. Our experience in CRISPR-mediated mutagenesis in plants indicates that not every gRNA constructed is active or highly active *in planta* to induce DSBs and subsequently targeted mutagenesis. For example, in this work, only one (gAgo18b-1) of the two gRNAs constructed for targeting *ZmAgo18b* was mutagenic. Until a simple and rapid assay to identify the most effective gRNA candidate is available prior to maize transformation, the approach described here of using two gRNAs for one gene would improve the odds of achieving targeted mutagenesis. Another important utility of the 2-gRNAs-for-1-gene approach is to enable large deletion mutation in the targeted gene. Different vector systems and cloning approaches can be used to enhance the efficiency of the multiple gRNA construction process. For example, the Golden Gate assembly technology can be adapted to make multiple gRNA expression units in one reaction and in either the gRNA intermediate vector or directly in the destination Cas9 vector (Engler *et al.*, 2008).

Second, we incorporated a genotyping procedure to screen bialaphos-resistant callus lines and retain only mutation-positive lines for regeneration. The callus screening is an important step for resource conservation especially when performing the transformation of inbred B104. Compared to the Hi-II transformation, regeneration of transgenic B104 callus is technically demanding, time-consuming and requires extended use of growth chambers. Therefore, the early identification of mutation-positive callus lines will maximize the number of CRISPR plants for each mutagenesis project. In our work, the majority of sequencing chromatograms showed two predominant peaks starting at the expected mutation sites, indicating uniform genotypes within calli, including some lines containing homogenous or heterogeneous DA mutations. Over 70% of edit-positive callus lines remained positive for mutations in the regenerated plants. Third, we tested the feasibility of combining two independent mutageneses through co-incubation of two *Agrobacterium* strains with the standard number of starting embryos and showed that each independent mutagenesis occurred at frequencies similar to those for individual infections. The possibility of combining more than two different Cas9/gRNA-containing *Agrobacterium* strains needs to be further tested for efficacy.

In this study, we analysed the progeny between integrated Cas9/gRNA in T_0 and T_1 plants crossed with B73 for heritable activity of the mutagenesis reagents. In both populations, novel mutations at specific genomic sites were detected at frequencies ranging from 0% to 100%. This continuing action of Cas9/gRNA is in agreement with previous reports of new mutations (Brooks *et al.*, 2014; Svitashev *et al.*, 2015; Wang *et al.*, 2014; Zhang *et al.*, 2014). Heritable Cas9/gRNA action points to the prospect of performing intergenotype targeted mutagenesis for some applications. For example, transgenic B104 carrying a specific Cas9/gRNAs might be crossed with transformation-recalcitrant maize genotypes to generate desired mutagenesis in specific alleles in the Cas9/gRNA recipient maize. This approach could also facilitate other genetic procedures, such as introgression of a recessive trait generated by Cas9/gRNA into specific germplasm, or performance of double-mutant analyses. Furthermore, once a *Cas9*-expressing plant exists, transient introduction of gRNAs by any means could permit gene modification for any chosen target without the necessity of generating a new maize transformant.

With our efficient and robust CRISPR platform, ISU Maize CRISPR, transgenic maize plantlets with gene-specific mutations can be generated as early as 16 weeks from the day of construct delivery after which seed can be produced in an additional 3–4 months. Transgenic plants with mutations are typically pollinated with wild-type donor pollen to produce segregants that have gene-specific mutations, but are free of the Cas9/gRNA transgene. These null-segregant mutant seeds containing no foreign DNA sequences can then be used for further fundamental and applied research, with minimal or no regulatory and containment requirements. We expect that this *Agrobacterium*-delivered ISU Maize CRISPR system will empower the public research community and accelerate the exploration of both gene function and trait improvement.

Experimental procedures

Plasmids and bacterial strains

Molecular cloning and the construction of maize transformation plasmids were performed as previously described (Ausubel *et al.*, 1993). The transformation vector was based on pMCG1005 (kindly provided by Dr. Vicki Chandler), a binary vector for *Agrobacterium*-mediated maize transformation containing four copies of an enhanced CaMV 35S promoter driving *bar* gene expression for bialaphos resistance. pMCG1005 also contains a cassette of the maize *ubiquitin 1* gene promoter (Christensen and Quail, 1996) coupled with the first intron of maize alcohol dehydrogenase (*Adh1*) gene (Callis *et al.*, 1987) and the terminator of octopine synthase gene of *Agrobacterium tumefaciens* (Koncz *et al.*, 1983). This vector was modified by replacing the first intron of the rice waxy gene with a linker sequence to facilitate the cloning of *Cas9*. The rice codon-optimized *Cas9* was as described (Jiang *et al.*, 2013). Additionally, the Gateway recombination cassette of attR1-ccdB-attR2 was cloned into pMCG1005, resulting in the destination vector, pGW-Cas9. Sequence information is available upon request.

For the construction of guide RNA genes, the intermediate vectors pENTR-gRNA1 and pENTR-gRNA2 that each can express two gRNAs were used (Zhou *et al.*, 2014). Briefly, in each gRNA vector, a cloning site with 2×*BtgZI* downstream of one rice U6 promoter and another cloning site with 2×*BsaI* downstream of the second rice U6 promoter were used for sequential insertions of two gRNA spacer sequences (Zhou *et al.*, 2014). To construct the gRNA gene targeting a specific genomic locus, two complementary oligonucleotides (21–25 nt) were annealed to produce a double-stranded DNA oligonucleotide (dsOligo). For

the *BtgZI* cloning site, the sense strand contains a 5′ 4-nt overhang of TGTT and the antisense strand contains a AAAC 5′ overhang. For the *BsaI* cloning site, the double-stranded oligonucleotides were designed with a GTGT 5′ overhang on the sense strand and AAAC 5′ overhang on the antisense strand. All oligonucleotides were synthesized at Integrated DNA Technology (Coralville, IA). The first dsOligo was inserted into *BtgZI* restriction site and the second dsOligo was sequentially inserted at the *BsaI* restriction site followed by sequencing to confirm the accuracy of construction. To construct a gRNA cassette expressing four guide RNA genes, the gRNA cassette from pgRNA-IM1 was cut out using *HindIII* and subcloned into pgRNA-IM2 that was already constructed to contain two gRNA sequences. The gRNA cassettes were finally mobilized to pGW-Cas9 by using the Gateway LR Clonase (Thermo Fisher Scientific, Waltham, MA).

Escherichia coli strain XL1-Blue was used for molecular cloning of Cas9/gRNA constructs and *Agrobacterium tumefaciens* strain EHA101 for maize transformation. *Escherichia coli* cells were grown in Luria–Bertani (LB) medium at 37 °C with a standard culture technique (Ausubel *et al.*, 1993), and *Agrobacterium* was grown at 28 °C in YEP medium (yeast extract 5 g/L, peptone 10 g/L, NaCl$_2$ 5 g/L) with appropriate antibiotics.

Maize tissue culture and transformation

Maize (*Zea mays*) hybrid genotype Hi-II and inbred line B104 were used. *Agrobacterium*-mediated transformation of the immature embryos of Hi-II and B104 maize genotypes was performed at the Iowa State University Plant Transformation Facility as described (Frame *et al.*, 2002, 2006). The plants were grown in greenhouses with controlled temperatures of 26 °C/22 °C and a photoperiod of 16 h/8 h (day/night).

Genotyping maize callus lines and plants

Genomic DNA from maize calli and leaves of transgenic plants was extracted using the cetyltrimethyl ammonium bromide (CTAB) method (Murray and Thompson, 1980). Genomic DNA was used for PCR amplification of relevant regions with specific primers flanking the target sites (Table S1). PCR reaction conditions were optimized for each primer pair and are available upon request. PCR amplicons were assessed for mutations using the T7 endonuclease I (T7E1) assay and Sanger sequencing. PCR amplicons obtained from the transgenic tissues were mixed with the respective amplicon derived from wild type, denatured (95 °C for 5 min) and reannealed (ramp down to 25 °C at 5 °C/min), then subjected to T7E1 digestion and agarose gel electrophoresis as previously described (Char *et al.*, 2015). The amplicons derived from the T7E1-positive samples were treated with ExoSAP-IT (Affymetrix, Santa Clara, CA) and subsequently evaluated by the Sanger sequencing method by the Iowa State University DNA facility (http://www.dna.iastate.edu/). The sequencing chromatograms were carefully examined for exact patterns that might indicate monoallelic or diallelic mutations.

Expression analysis in progeny plants by RT-PCR

The expression of Cas9 mRNA in progeny plants was evaluated by RT-PCR on an Eppendorf Mastercycler (Eppendorf, Hamburg, Germany). Total RNA was isolated from one-week-old seedlings using RNeasy Plant Mini Kit (Qiagen, Valencia, CA) as per the manufacturer's instructions. The concentration and purity of the isolated RNA was confirmed by NanoDrop ND-1000

Spectrophotometer. One microgram of RNA was subjected to DNase treatment with Promega RQ1 RNase-free DNase I (Promega Corporation, Madison, WI) followed by reverse transcription mediated by SuperScript™ III Reverse Transcriptase (Thermo Fisher Scientific, Waltham, MA) following the manufacturer's protocol. The cDNA was amplified using OsCas9 primers (Table S1) for 26 cycles, and the products were separated on agarose gels, visualized by SYBR Safe DNA gel stain (Thermo Fisher Scientific) and photographed. The expression of maize *ubiquitin* cDNA was also determined in the same sample set, as an endogenous positive control.

Acknowledgements

Research support was from the National Institute of Food and Agriculture (NIFA) of the US Department of Agriculture (USDA) (2014-67013-21720 to B.Y.), the Iowa State University Presidential Initiative for Interdisciplinary Research (M.H.S., P.W.B., K.W., B.Y.) and the US National Science Foundation (PGRP 1339229 to B.C.M. and V.W.). Additional support was provided by NIFA-USDA Hatch project number # IOW05162, by State of Iowa funds (K.W., B.F., M.M.) and by Charoen Pokphand Indonesia (H.N.). The authors are grateful to Michael Muszynski for providing greenhouse and growth chamber space and thank Daniel Little and Pete Lelonek for greenhouse support and plant maintenance.

References

Ausubel, F., Brent, R., Kingston, R., Moore, D., Seidman, J. and Struhl, K. (1993) *Current Protocols in Molecular Biology*, 1st ed. New York: Canada John Wiley and Sons.

Bhaya, D., Davison, M. and Barrangou, R. (2011) CRISPR-Cas systems in bacteria and archaea: versatile small RNAs for adaptive defense and regulation. *Annu. Rev. Genet.* **45**, 273–297.

Brazelton, V.A. Jr., Zarecor, S., Wright, D.A., Wang, Y., Liu, J., Chen, K., Yang, B., et al. (2015) A quick guide to CRISPR sgRNA design tools. *GM Crops Food*, **6**, 266–276.

Brooks, C., Nekrasov, V., Lippman, Z.B. and Van Eck, J. (2014) Efficient gene editing in tomato in the first generation using the clustered regularly interspaced short palindromic repeats/CRISPR-associated 9 system. *Plant Physiol.* **166**, 1292–1297.

Callis, J., Fromm, M. and Walbot, V. (1987) Introns increase gene expression in cultured maize cells. *Genes Dev.* **1**, 1183–1200.

Cermak, T., Baltes, N.J., Cegan, R., Zhang, Y. and Voytas, D.F. (2015) High-frequency, precise modification of the tomato genome. *Genome Biol.* **16**, 232.

Char, S.N., Unger-Wallace, E., Frame, B., Briggs, S.A., Main, M., Spalding, M.H., Vollbrecht, E., et al. (2015) Heritable site-specific mutagenesis using TALENs in maize. *Plant Biotechnol. J.* **13**, 1002–1010.

Christensen, A.H. and Quail, P.H. (1996) Ubiquitin promoter-based vectors for high-level expression of selectable and/or screenable marker genes in monocotyledonous plants. *Transgenic Res.* **5**, 213–218.

De Buck, S., Podevin, N., Nolf, J., Jacobs, A. and Depicker, A. (2009) The T-DNA integration pattern in Arabidopsis transformants is highly determined by the transformed target cell. *Plant J.* **60**, 134–145.

De Neve, M., De Buck, S., Jacobs, A., Van Montagu, M. and Depicker, A. (1997) T-DNA integration patterns in co-transformed plant cells suggest that T-DNA repeats originate from co-integration of separate T-DNAs. *Plant J.* **11**, 15–29.

Engler, C., Kandzia, R. and Marillonnet, S. (2008) A one pot, one step, precision cloning method with high throughput capability. *PLoS ONE*, **3**, e3647.

Feng, Z., Zhang, B., Ding, W., Liu, X., Yang, D.L., Wei, P., Cao, F., et al. (2013) Efficient genome editing in plants using a CRISPR/Cas system. *Cell Res.* **23**, 1229–1232.

Feng, Z., Mao, Y., Xu, N., Zhang, B., Wei, P., Yang, D.L., Wang, Z., et al. (2014) Multigeneration analysis reveals the inheritance, specificity, and patterns of CRISPR/Cas-induced gene modifications in Arabidopsis. *Proc. Natl Acad. Sci. USA*, **111**, 4632–4637.

Feng, C., Yuan, J., Wang, R., Liu, Y., Birchler, J.A. and Han, F. (2016) Efficient targeted genome modification in maize using CRISPR/Cas9 system. *J. Genet. Genom.* **43**, 37–43.

Frame, B.R., Shou, H., Chikwamba, R.K., Zhang, Z., Xiang, C., Fonger, T.M., Pegg, S.E., et al. (2002) *Agrobacterium tumefaciens*-mediated transformation of maize embryos using a standard binary vector system. *Plant Physiol.* **129**, 13–22.

Frame, B.R., McMurray, J.M., Fonger, T.M., Main, M.L., Taylor, K.W., Torney, F.J., Paz, M.M., et al. (2006) Improved *Agrobacterium*-mediated transformation of three maize inbred lines using MS salts. *Plant Cell Rep.* **25**, 1024–1034.

Horlbeck, M.A., Witkowsky, L.B., Guglielmi, B., Replogle, J.M., Gilbert, L.A., Villalta, J.E., Torigoe, S.E., et al. (2016) Nucleosomes impede Cas9 access to DNA in vivo and in vitro. *Elife*, **5**, e12677.

Hou, Z., Zhang, Y., Propson, N.E., Howden, S.E., Chu, L.F., Sontheimer, E.J. and Thomson, J.A. (2013) Efficient genome engineering in human pluripotent stem cells using Cas9 from *Neisseria meningitidis*. *Proc. Natl Acad. Sci. USA*, **110**, 15644–15649.

Jiang, W., Zhou, H., Bi, H., Fromm, M., Yang, B. and Weeks, D.P. (2013) Demonstration of CRISPR/Cas9/sgRNA-mediated targeted gene modification in Arabidopsis, tobacco, sorghum and rice. *Nucleic Acids Res.* **41**, e188.

Jiang, W., Yang, B. and Weeks, D.P. (2014) Efficient CRISPR/Cas9-mediated gene editing in *Arabidopsis thaliana* and inheritance of modified genes in T2 and T3 generations. *PLoS ONE*, **9**, e99225.

Koncz, C., De Greve, H., Andre, D., Deboeck, F., Van Montagu, M. and Schell, J. (1983) The opine synthase genes carried by Ti plasmids contain all signals necessary for expression in plants. *EMBO J.* **2**, 1597–1603.

Lawrenson, T., Shorinola, O., Stacey, N., Li, C., Ostergaard, L., Patron, N., Uauy, C., et al. (2015) Induction of targeted, heritable mutations in barley and *Brassica oleracea* using RNA-guided Cas9 nuclease. *Genome Biol.* **16**, 258.

Li, Z., Liu, Z.B., Xing, A., Moon, B.P., Koellhoffer, J.P., Huang, L., Ward, R.T., et al. (2015) Cas9-guide RNA directed genome editing in soybean. *Plant Physiol.* **169**, 960–970.

Liang, Z., Zhang, K., Chen, K. and Gao, C. (2014) Targeted mutagenesis in *Zea mays* using TALENs and the CRISPR/Cas system. *J. Genet. Genom.* **41**, 63–68.

Ma, X., Zhang, Q., Zhu, Q., Liu, W., Chen, Y., Qiu, R., Wang, B., et al. (2015) A robust CRISPR/Cas9 system for convenient, high-efficiency multiplex genome editing in monocot and dicot Plants. *Mol. Plant.* **8**, 1274–1284.

Murray, M.G. and Thompson, W.F. (1980) Rapid isolation of high molecular weight plant DNA. *Nucleic Acids Res.* **8**, 4321–4325.

Sander, J.D. and Joung, J.K. (2014) CRISPR-Cas systems for editing, regulating and targeting genomes. *Nat. Biotechnol.* **32**, 347–355.

Schnable, P.S., Ware, D., Fulton, R.S., Stein, J.C., Wei, F., Pasternak, S., Liang, C., et al. (2009) The B73 maize genome: complexity, diversity, and dynamics. *Science*, **326**, 1112–1115.

Shukla, V.K., Doyon, Y., Miller, J.C., DeKelver, R.C., Moehle, E.A., Worden, S.E., Mitchell, J.C., et al. (2009) Precise genome modification in the crop species *Zea mays* using zinc-finger nucleases. *Nature*, **459**, 437–441.

Svitashev, S., Young, J.K., Schwartz, C., Gao, H., Falco, S.C. and Cigan, A.M. (2015) Targeted mutagenesis, precise gene editing, and site-specific gene insertion in maize using Cas9 and guide RNA. *Plant Physiol.* **169**, 931–945.

Wang, Y., Cheng, X., Shan, Q., Zhang, Y., Liu, J., Gao, C. and Qiu, J.L. (2014) Simultaneous editing of three homoeoalleles in hexaploid bread wheat confers heritable resistance to powdery mildew. *Nat. Biotechnol.* **32**, 947–951.

Wang, S., Zhang, S., Wang, W., Xiong, X., Meng, F. and Cui, X. (2015) Efficient targeted mutagenesis in potato by the CRISPR/Cas9 system. *Plant Cell Rep.* **34**, 1473–1476.

Woo, J.W., Kim, J., Kwon, S.I., Corvalan, C., Cho, S.W., Kim, H., Kim, S.G., et al. (2015) DNA-free genome editing in plants with preassembled CRISPR-Cas9 ribonucleoproteins. *Nat. Biotechnol.* **33**, 1162–1164.

Wu, X., Scott, D.A., Kriz, A.J., Chiu, A.C., Hsu, P.D., Dadon, D.B., Cheng, A.W., *et al.* (2014) Genome-wide binding of the CRISPR endonuclease Cas9 in mammalian cells. *Nat. Biotechnol.* **32**, 670–676.

Xie, K., Minkenberg, B. and Yang, Y. (2015) Boosting CRISPR/Cas9 multiplex editing capability with the endogenous tRNA-processing system. *Proc. Natl Acad. Sci. USA*, **112**, 3570–3575.

Xing, H.L., Dong, L., Wang, Z.P., Zhang, H.Y., Han, C.Y., Liu, B., Wang, X.C., *et al.* (2014) A CRISPR/Cas9 toolkit for multiplex genome editing in plants. *BMC Plant Biol.* **14**, 327.

Xu, K., Ren, C., Liu, Z., Zhang, T., Zhang, T., Li, D., Wang, L., *et al.* (2015) Efficient genome engineering in eukaryotes using Cas9 from *Streptococcus thermophilus. Cell. Mol. Life Sci.* **72**, 383–399.

Yuan, T., Fujioka, S., Takatsuto, S., Matsumoto, S., Gou, X., He, K., Russell, S.D., *et al.* (2007) Ben1, a gene encoding a dihydroflavonol 4-reductase (DFR)-like protein, regulates the levels of brassinosteroids in *Arabidopsis thaliana. Plant J.* **51**, 220–233.

Zhai, J., Zhang, H., Arikit, S., Huang, K., Nan, G.L., Walbot, V. and Meyers, B.C. (2015) Spatiotemporally dynamic, cell-type-dependent premeiotic and meiotic phasiRNAs in maize anthers. *Proc. Natl Acad. Sci. USA* **112**, 3146–3151.

Zhang, H., Zhang, J., Wei, P., Zhang, B., Gou, F., Feng, Z., Mao, Y., *et al.* (2014) The CRISPR/Cas9 system produces specific and homozygous targeted gene editing in rice in one generation. *Plant Biotechnol. J.* **12**, 797–807.

Zhou, H., Liu, B., Weeks, D.P., Spalding, M.H. and Yang, B. (2014) Large chromosomal deletions and heritable small genetic changes induced by CRISPR/Cas9 in rice. *Nucleic Acids Res.* **42**, 10903–10914.

Zhu, J., Song, N., Sun, S., Yang, W., Zhao, H., Song, W. and Lai, J. (2016) Efficiency and inheritance of targeted mutagenesis in maize using CRISPR-Cas9. *J. Genet. Genom.* **43**, 25–36.

6

Tuber-specific silencing of *asparagine synthetase-1* reduces the acrylamide-forming potential of potatoes grown in the field without affecting tuber shape and yield

Rekha Chawla, Roshani Shakya and Caius M. Rommens*

Simplot Plant Sciences, J. R. Simplot Company, Boise, ID, USA

*Correspondence

email crommens@simplot.com

Keywords: potato, acrylamide, tubers, metabolic engineering, asparagine synthetase.

Summary

Simultaneous silencing of *asparagine synthetase* (*Ast*)*-1* and *-2* limits asparagine (ASN) formation and, consequently, reduces the acrylamide-forming potential of tubers. The phenotype of silenced lines appears normal in the greenhouse, but field-grown tubers are small and cracked. Assessing the effects of silencing *StAst1* and *StAst2* individually, we found that yield drag was mainly linked to down-regulation of *StAst2*. Interestingly, tubers from untransformed scions grafted onto intragenic *StAst1/2*-silenced rootstock contained almost the same low ASN levels as those in the original silenced lines, indicating that ASN is mainly formed in tubers rather than being transported from leaves. This conclusion was further supported by the finding that overexpression of *StAst2* caused ASN to accumulate in leaves but not tubers. Thus, ASN does not appear to be the main form of organic nitrogen transported from leaves to tubers. Because reduced ASN levels coincided with increased levels of glutamine, it appears likely that this alternative amide amino acid is mobilized to tubers, where it is converted into ASN by StAst1. Indeed, tuber-specific silencing of *StAst1*, but not of *StAst2*, was sufficient to substantially lower ASN formation in tubers. Extensive field studies demonstrated that the reduced acrylamide-forming potential achieved by tuber-specific *StAst1* silencing did not affect the yield or quality of field-harvested tubers.

Introduction

Amino acids are the major building blocks for proteins (Andrews, 1986) and represent only ~2% of the total amino-nitrogen (N) content of potato leaves (Koch *et al.*, 2003). They are released upon senescence-associated hydrolysis to be transported to tubers (Hörteinsteiner and Feller, 2002; Taylor *et al.*, 2010), where they ultimately represent half of the total amino-N content (Koch *et al.*, 2003). If the long-distance transport of amino acids is hindered, for instance through silencing of the Aap1 transporter, the amino acid content in tubers is reduced by 32%–50% (Koch *et al.*, 2003). Asparagine (ASN) and glutamine (GLN) are thought to play a particularly important role in the transport and storage of nitrogen (Lehmann and Ratajczak, 2008) because of their relatively stable nature and high N/C ratio (Ireland and Lea, 1999; Masclaux-Daubresse *et al.*, 2006). ASN is the predominant free amino acid in *Solanum tuberosum* (potato) tubers and constitutes up to 25% of the total free amino acid pool in tubers (Golan-Goldhirsh, 1986; Koch *et al.*, 2003).

Asparagine synthetase (Ast) catalyses the last step in ASN formation. Plants contain multiple Ast isoenzymes, each of which functions during a specific process in development, such as the mobilization of N during seed germination, the recycling of N in vegetative organs in response to stress, the mobilization of N from source to sink and the storage of N in seeds. However, the regulation of a specific Ast-encoding gene is not necessarily conserved across plant species (Duff *et al.*, 2011; Herrera-Rodríguez *et al.*, 2002; Lam *et al.*, 1998; Tsai and Coruzzi, 1991). Molecular and genetic analyses indicate that distinct *Ast* genes are regulated differentially by environmental stimuli (light/dark, stress), metabolic status (sugar and N levels), developmental cues and tissue/cell-type specificity (Gaufichon *et al.*, 2010). Further studies are required to determine whether any correlation exists between phylogenetic classification and the diverse physiological roles of *Ast* genes during the life cycle of plants. The predominantly expressed *Ast* genes in potato, *StAst1* and *StAst2*, are poorly characterized. However, as *StAst1* has high sequence similarity with Arabidopsis *Asn1*, it may play a similar role in assimilating, transporting and storing N (Lam *et al.*, 2003). On the other hand, *StAst2* is homologous to *Arabidopsis Asn2* and may thus play a role in ammonium metabolism (Wong *et al.*, 2004).

As ASN is oxidized into a toxic compound, acrylamide, during cooking of starchy foods in a low-water environment, high tuber concentrations of ASN are undesirable. Both acrylamide and its reactive metabolite, glycidamide, are neurotoxins and probable carcinogens (Parzefall, 2008; Tareke *et al.*, 2002), and the consumption of certain heat-processed foods, including French fries, potato chips, bread and coffee, results in an average dietary intake of 0.3–0.7 µg acrylamide/kg per day (Dybing *et al.*, 2005). This intake may be linked to the development of certain degenerative diseases, including cancer (Tardiff *et al.*, 2010; Tareke *et al.*, 2002).

In an effort to reduce the acrylamide-forming potential of potatoes by minimizing tuber concentrations of ASN, we previously generated transgenic potato plants, in which *StAst1* and *StAst2* were simultaneously silenced (Rommens *et al.*, 2008). These lines exhibited a ~20-fold reduction in ASN and, when processed into French fries, contained just 5% of the acrylamide present in the control. In the light- and humidity-controlled conditions of a greenhouse, the transgenic potato plants appeared normal and had normal yields. However, we subsequently found that the *StAst1/2*-silenced plants did not grow well in the field, producing small, deformed tubers and fewer tubers per plant. We thus aimed to generate transgenic potato lines that had reduced levels of ASN, but grew normally in the field. By silencing *StAst1* and *StAst2* individually, we established that StAst1 is mainly involved in ASN formation in tubers, whereas *StAst2* has a larger impact on agronomic traits in field-grown plants. Silencing of only *StAst1* in tubers reduced ASN content by 80% without compromising the quality or yield of field-grown plants.

Results

Simultaneous silencing of *StAst1* and *StAst2* in tubers leads to cracking

To reduce the levels of free ASN in tubers and thereby limit the precursors of the Maillard reaction, we previously generated transgenic potato lines, in which *StAst1* and *StAst2* were simultaneously silenced, particularly in the tubers (Rommens *et al.*, 2008). Greenhouse trials showed that tubers of these plants contained less ASN than the control and produced less acrylamide when prepared as French fries (Rommens *et al.*, 2008). These plants exhibited a normal phenotype in the greenhouse, producing similar yields of tubers as the control. To our surprise, upon subsequent field trials, tubers of all 23 *StAst1/2*-silenced lines were small and cracked, while also developing secondary growth (Figure 1a). These abnormal tubers maintained reduced levels of ASN when compared with wild type (WT) and empty vector tissue culture (TC) controls (Figure 1b). Previous studies had shown that tuber cracking in potato is generally associated with environmental stress (Hochmuth *et al.*, 2001; Jefferies and Mackerron, 1987). However, because our control tubers did not show any signs of stress, the altered phenotype of silenced tubers was because of simultaneous silencing of *StAst1* and *StAst2*. These unexpected results prompted us to more thoroughly analyse the role of the two ASN biosynthetic genes in potato.

StASt1 silencing reduces yield in the field

To constitutively silence the *StAst1* gene alone, an inverted repeat carrying two copies of a small fragment of this gene was operably linked to the promoter of the potato ubiquitin-7 (*Ubi7*) gene (Garbarino *et al.*, 1995) and cloned into pSIM1714. pSIM401, which only contains the selectable marker gene *nptII*, was used as empty vector control. The plants were transformed using a standard protocol (Richael *et al.*, 2008) and rooted on kanamycin-containing medium. Plant lines positive for the presence of the transgene, as determined by PCR, were transferred to the greenhouse 1 month after rooting. RNA gel blot analysis demonstrated that *StAst1* was silenced in most lines, but lines 3, 9, 11, 13, 17 and 19–21 did not contain reduced transcript levels, possibly because of position integration effects (Figure 2a).

In the greenhouse, these plants were phenotypically similar to the controls (Figure 2b,c) while also displaying similar yields. Both the leaves and tubers of *StAst1*-silenced plants had reduced levels of ASN. The GLN levels were higher, especially in the tubers of the silenced lines, indicating that the precursor of ASN accumulated in the absence of ASN (Figure 2c). Lines 1714-9 and 1714-13 were not silenced and served as internal controls (Figure 2c).

The tubers of greenhouse-grown *StAst1*-silenced lines were stored at 38 °C for 5 months to break dormancy and then planted in the field. In a five-hill trial, transgenic lines were agronomically similar to the controls (Figure 3a), but their leaves had slightly reduced ASN levels (Figure 3b). No significant change was observed in total NH_4^+ and N content in the leaves (Figures 3c and 5d). However, the yield of tubers (Figure 6a) and the concentration of ASN in the tubers were reduced (Figure 6b), as were the total N and total protein content in the tubers on a per plant basis (Figure 6c,d).

Therefore, although the aerial and below-ground parts of *StAst1*-silenced lines resembled the controls, constitutive silencing of *StAst1* resulted in reduced ASN levels in both the leaves and tubers, which in turn affected tuber yield and thus reduced the total N and total protein content in the tubers.

Loss of StAst2 function leads to aberrant phenotype and decreased yield in the field

StAst2 alone was constitutively silenced using the pSIM1715 construct, in which a *StAst2* silencing cassette was placed under

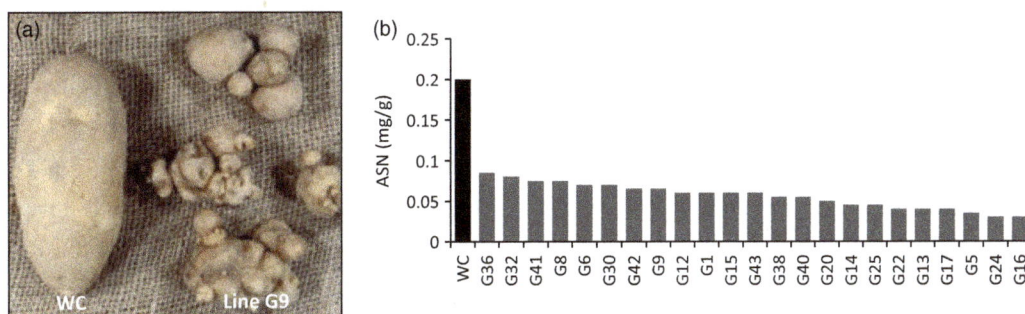

Figure 1 Analysis of field-grown transgenic potato lines containing construct pSIM1256 that simultaneously silenced *StAst1* and *StAst2* in the tubers. (a) Phenotypes of field-grown Russet Ranger tubers silenced for both *StAst1* and *StAst2* (G9) compared with the controls (TC and WT). (b) asparagine levels (mg/g FW) in three randomly pooled tubers/line from a single 5-plant plot.

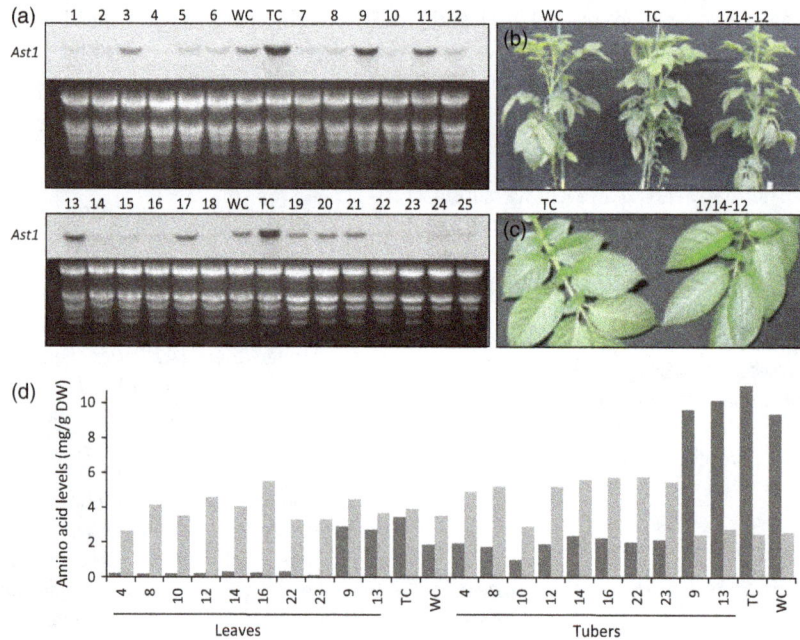

Figure 2 Analysis of greenhouse-grown lines containing construct pSIM1714 that constitutively silenced *StAst1*. (a) RNA blot analysis was performed on leaf tissue to identify *StAst1*-silenced lines. (b) Morphology of a *StAst1*-silenced line (1714-12) as compared to the wild type and empty vector control (TC). (c) Close-up of a leaf of the silenced line 1714-12 and TC lines. (d) asparagine (dark grey bars) and glutamine (light grey bars) levels in three pooled leaves/line and three pooled tubers/line. It should be noted that *StAst1* was not silenced in transgenic lines pSIM1714-9 and 13, which means that these lines represent internal controls.

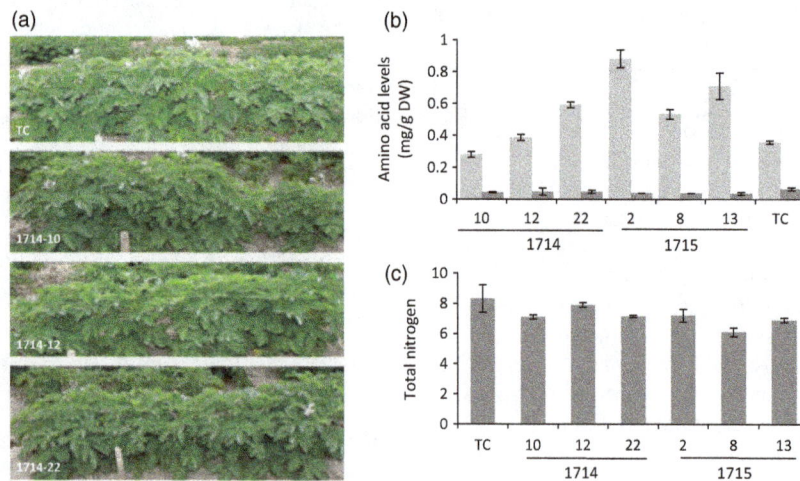

Figure 3 Characterization of *StAst1*- and *StAst2*-silenced potato lines grown in the field. (a) Phenotypes of *StAst1*-silenced, field-grown Russet Ranger plants in a five-hill field trial. TC is the empty vector tissue culture control, and 1714-10, 1714-12 and 1714-22 are three transgenic lines silenced for *StAst1*. (b) asparagine (dark grey) and glutamine (light grey) levels in leaves of *StAst1* (1714-10, 12 and 22)- and *StAst2* (1715-2, 8 and 13)-silenced lines. (c) Total N content in the leaves of *StAst1* (1714-10, 12 and 22)- and *StAst2* (1715-2, 8 and 13)-silenced lines. All data represent the average of three measurements ± SD.

the control of the potato *Ubi7* promoter. Upon transformation, lines silenced for *StAst2* were identified by RNA blot analysis. The blots were also probed with a *StAst1* probe to ensure that *StAst1* alone was not cross-silenced (Figure 4a).

In the greenhouse, the phenotypes and yield of these plants were similar to those of the controls (Figure 4b). Neither the leaves nor the tubers of the *StAst2*-silenced plants exhibited any changes in ASN/GLN levels in the greenhouse (Figure 4c). Lines

1715-15 and 1715-23 were not silenced and served as internal controls (Figure 4c).

These *StAst2*-silenced lines and their controls were stored for 5 months and then planted in the field. Surprisingly, all *StAst2*-silenced lines that looked normal in the greenhouse exhibited a conditional phenotype in the field. The plants were stunted and exhibited reduced vigour, narrow leaf blades with wavy margins and a chlorophyll deficiency (Figure 5a–c). Biochemical analysis

Figure 4 Analysis of greenhouse-grown lines, in which *StAst2* was constitutively silenced (pSIM1715). (a) RNA blot analysis was performed on leaf tissue to identify *StAst2*-silenced lines. The blots were stripped and re-probed with *StAst1* probes to confirm that *StAst1* was not silenced. (b) Phenotype of the entire plant, WT is the wild type; TC is the empty vector control; 1715-4 is a representative transgenic line. (c) Leaf morphology of *StAst2*-silenced lines compared with that of the WT and empty vector control, TC. (d) asparagine (dark grey) and glutamine levels (light grey) in the leaves and tubers of *StAst2*-silenced lines. *StAst2* was not silenced in lines 15 and 23, which served as internal controls.

of the leaves showed a significant reduction in total N (*P* value <0.05) (Figure 3c). ASN levels were lower than in the controls, whereas GLN levels were higher (Figure 3b). In agreement with findings in Arabidopsis lines silenced in *AtAst2* expression (Wong *et al.*, 2004), NH_4^+ levels were significantly (*P* < 0.05) elevated in the leaves of *StAst2*-silenced lines (Figure 5d). Thus, *StAst2*, like *Arabidopsis AtAsn2*, may be involved in NH_4^+ detoxification in the leaves.

Although the tubers had normal morphologies, the yield per plant appeared drastically decreased in all transgenic lines containing pSIM1715 (Figure 6a). ASN levels were comparable to those in the controls (Figure 6b). Therefore, *StAst2* does not contribute significantly towards tuber ASN levels. However, total N and total protein content was significantly reduced in the tubers and was even lower than in the constitutive *StAst1*-silenced lines (Figure 6c,d). No significant changes in NH_4^+ level were observed in the tubers (data not shown). The decrease in the yield of *StAst2*-silenced lines could be due to an overall reduction in growth and a change in aerial morphology, as seen in these field-grown plants (Figure 5a). Therefore, *StAst2* function is important for maintaining a normal plant phenotype, overall growth, and yield in potato.

To determine the effect of *StAst2* on tuber ASN levels, we designed a construct (pSIM1716) that drives the tuber-specific silencing of *StAst2*. The AGP promoter was used to drive the tuber-specific silencing of *StAst2*. RNA blot analysis identified

lines 1, 2, 3, 4, 5, 8, 12, 13, 14, 15, 16, 18, 19, 20, 21, 22, 23, 24 and 25 as being silenced for *StAst2* (Figure 7a). Biochemical analysis of the tubers of these plants grown in the greenhouse revealed no change in tuber ASN levels compared with the controls (Figure 7b). Therefore, neither constitutive nor tuber-specific silencing of *StAst2* influenced tuber ASN levels. Thus, *StAst2* does not affect tuber ASN levels, and its constitutive silencing is deleterious in the field.

Increased ASN formation in leaves does not increase ASN content in the tubers

To further assess the effect of individual asparagine synthetase proteins on the ASN pool in tubers, we constructed pSIM1704, in which the *StAst2* coding region is driven by the 35S promoter.

StAst2 expression was tested in the leaves of the transgenic plants to identify overexpressing lines (Figure 8a). Based on their high levels of expression, lines 1704-3, 13, 15, 18, 19, 20 and 23 were chosen for biochemical analysis. Surprisingly, we found that ASN levels were elevated up to threefold in the leaves of *35S:StAst2* plants, while the level of ASN in tubers was unaffected (Figure 8b,c). Thus, *StAst2* is predominantly active in leaves. Similarly, *Arabidopsis AtAsn2* was predominantly expressed in leaves and was found to be regulated by light (Wong *et al.*, 2004). We thus propose that *StAst2*, like *AtAsn2*, is involved in NH_4^+ detoxification in the

Figure 5 Characterization of field-grown *StAst2*-silenced lines containing pSIM1715. (a) Phenotypes of *StAst2*-silenced field-grown Russet Ranger plants in a five-hill field trial. TC is the empty vector control, and 1715-2, 1714-8 and 1714-13 are transgenic lines silenced for *StAst2*. (b) Phenotype of an individual plant (1715-8) silenced for *StAst2* as compared to TC and a *StAst1*-silenced (1714-22) plant. (c) Leaf morphology of *StAst2*-silenced lines (1715-2, 8 and 13) grown in the field. (d) Levels of $NH_4^+/\mu g$ of protein in the leaves of *StAst2*- (1715-2, 8 and 13) as compared with *StAst1* (1714-10, 12 and 22)-silenced lines. All data represent the average of three measurements ± SD.

leaves. The morphology and yield of *35S:StAst2* lines grown in the greenhouse were comparable with those of WT plants (data not shown).

A key finding of this analysis is that increased (up to three-fold) ASN levels in leaves do not correlate with increased levels of ASN in tubers. Foliar ASN appears to have a limited impact on ASN levels in the potato tuber.

Contribution of long-distance transport to ASN accumulation in tubers

To reduce ASN levels to a minimum in the tuber, we needed to establish the contribution of foliar ASN to the tuber pool. ASN is the major nitrogenous compound in the phloem of several legumes (up to 30 mM) (Atkins *et al.*, 1975), whereas in maize, ASN has been reported to represent as much as 14% of the free amino acids in phloem (Lohaus *et al.*, 1998; Weiner *et al.*, 1991).

We designed a construct (pSIM1256) to simultaneously silence *StAst1* and *StAst2* expression under the control of the *AGP* and *GBSS* promoters, which are much more active in tubers than in the aerial parts (Rommens *et al.*, 2008). Both of these genes were silenced in line B26, which was used to study the extent of ASN transport from the leaves to the tubers.

We performed a reciprocal grafting experiment between line B26 and the WT. Whereas line B26 contained *StAst1* and *StAst2* transcripts in the leaves, but not in the tubers, the WT contained transcripts of both genes in the leaves and tubers. Transcripts of both genes were also detected in grafted tubers

whenever the WT parent was used as stock (Table 1). Three-week-old plants intended to be used as the scion were cut two to three nodes beneath the apical meristem and were secured and sealed to the desired stock as described in Methods. New leaves that emerged from the stock were removed to ensure that all N was transported from the scion. A control grafting of WT/WT and B26/B26 was performed to study the effect of grafting itself on transcript and amino acid levels. The tubers of control B26/B26 grafts had low levels of ASN (2.8%), as did the parental B26 lines, and the WT/WT graft exhibited levels similar to those in WT tubers (100%). Whereas ASN levels in the tubers of B26/WT grafts were similar to those in WT tubers, the tubers of WT/B26 grafts had low levels of ASN (4%; Table 1). Hence, transport of ASN from leaves to tubers is limited, and ASN does not appear to be the primary amino acid that transports N to sink tissue in potato. In the grafted lines that exhibited reduced levels of ASN (i.e. B26/B26 and WT/B26), GLN levels were increased two- to threefold (Table 1). Therefore, a lack of *StAst1* in tubers leads to GLN accumulation, which is the precursor of ASN (Table 1). In potato, GLN seems to be the major amino acid transported from the leaves to tubers, where it is converted to ASN by Ast.

Potato quality can be improved by tuber-specific silencing of *StAst1*

We next silenced *StAst1* alone, using two convergently oriented tuber-specific promoters from the ADP-glucose pyrophosphorylase (*Agp*) gene (Muller-Rober *et al.*, 1994) and

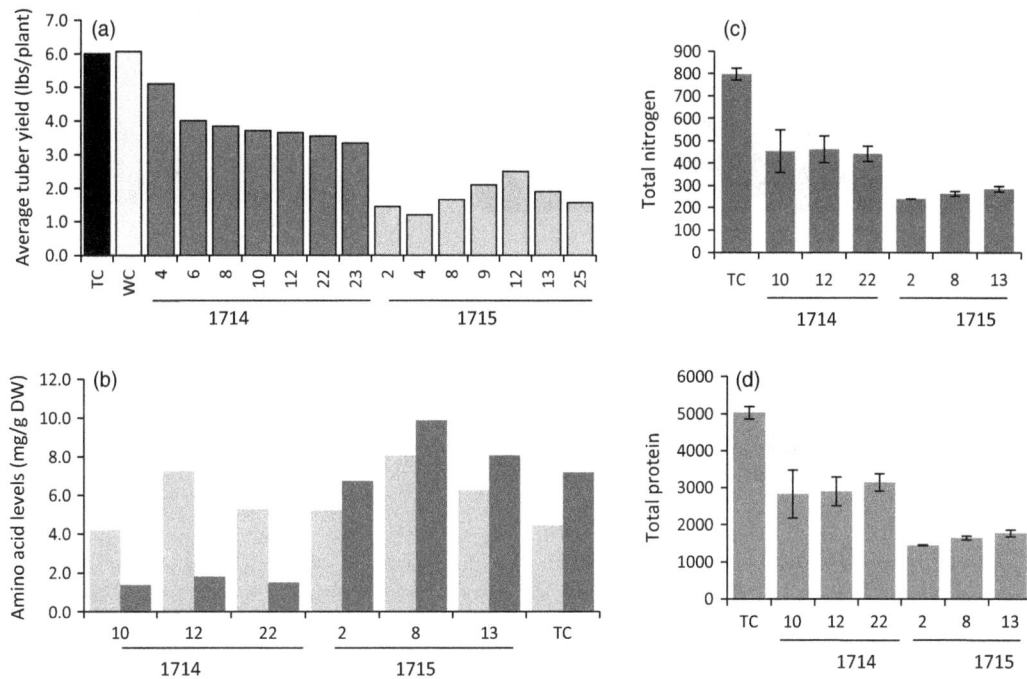

Figure 6 Tuber yield, amino acid levels, nitrogen and protein content of *StAst1*- and *StAst2*-silenced plants grown in the field. (a) Average tuber yield per plant (lb/plant) from field trials. The wild type is an untransformed control, TC is an empty vector control; 1714-4, 6, 8, 10, 12, 22 and 23 are *StAst1*-silenced lines, and 1715-2, 4, 8, 9, 12, 13 and 25 are *StAst2*-silenced lines. (b) asparagine (dark grey) and glutamine (light grey) levels in the tubers of *StAst1* (1714-10, 12, and 22)- and *StAst2* (1715-2, 8 and 13)-silenced lines. (c) Total nitrogen (mg/g fresh weight) and (d) total stored protein (mg/g fresh weight) in the tubers of *StAst1*-silenced lines (1714-10, 12 and 22) and three *StAst2*-silenced lines (1715-2, 8 and 13). TC is an empty vector control. Values shown for yield and amino acid levels are from a single measurement, the total nitrogen and protein levels represent the average of three measurements ± SD.

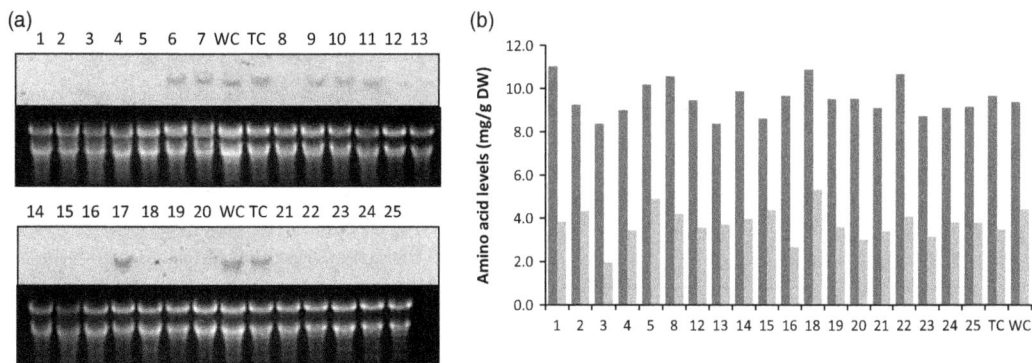

Figure 7 Tuber-specific silencing of *StAst2* (pSIM1716). (a) RNA blot analysis of tuber RNA to identify lines silenced for *StAst2*. It should be noted that lines 1716-6, 7, 9, 10, 11 and 17 are not silenced. (b) asparagine (dark grey) and glutamine (light grey) levels in the tubers of the transgenic lines. Wild type is untransformed control, and TC is empty vector control.

granule-bound starch synthase (*Gbss*) gene (Visser *et al.*, 1991), respectively. Five silenced lines were identified by RNA blot analysis. The leaf and tuber morphology of these plants were similar to controls grown in the greenhouse. Sixty tubers from each of the selected lines were planted in three locations (three random plots per field with one row of 20 tubers per plot). No differences were observed between the controls and the *StAst1*-silenced lines in agronomic traits monitored such as emergence and stand counts (early-season), date of flowering, date of row closure, plant vigour, foliage colour, leaflet size, leaflet curl, incidental disease, and

incidental insects (mid-season), and date of vine kill, vine maturity, and incidental disease (late-season). RNA gel blot analyses were performed to confirm that, despite the lack of observable effects (Figure 9a,b), the intragenic tubers were still silenced for StAst1 (Figure 9c). The yield of plants silenced for *StAst1* was similar to that of WT plants under field conditions ($P > 0.05$) (Figure 9d). The tubers were also indistinguishable from the WT in terms of size profile, specific gravity by size, colour analysis and internal analysis during grading. However, *StAst1*-silenced tubers showed a considerable reduction in ASN levels (ranging from 60% to 80%),

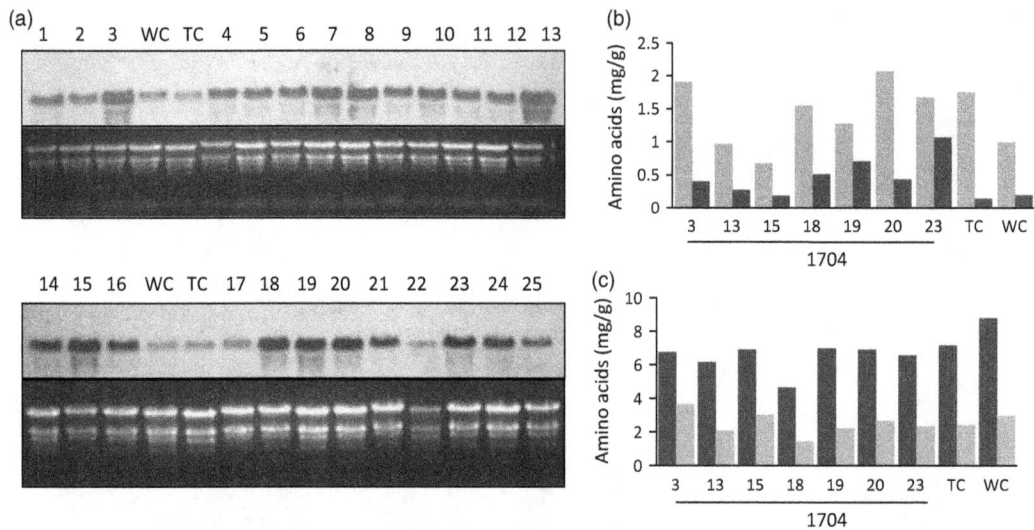

Figure 8 Transgenic lines constitutively expressing *StAst2* (pSIM1704). (a) Expression analysis in the leaves of transgenic lines harbouring *35S:StAst2*. Based on *StAst2* transcript levels, 1704-13, 15, 18, 19, 20 and 23 were chosen for biochemical analysis. Wild type is untransformed control, and 401 is empty vector control. (b) Free asparagine (ASN) (dark grey) and glutamine (GLN) (light grey) levels in leaves of *35S:StAst2* plants. (c) Free ASN (dark grey) and GLN (light grey) levels in tubers of *35S:StAst2* transgenic plants.

leading us to conclude that *StAst1* is the major enzyme responsible for ASN synthesis in tubers (Figure 9e). Data from two potato varieties Russet Ranger (R) and Russet Burbank (B) showed a similar trend. We propose that StAst1 enzyme can be used as a tool to lower acrylamide levels in heat-processed potato products (Figure 9f).

Discussion

In this study, we sought to limit acrylamide production in French fries by reducing the amount of ASN in potato tubers. The choice to target asparagine synthetase was clear from the start of our investigation, based on successful greenhouse

Table 1 Relative *StAs1* and *StAs2* transcript levels and amino acid levels for ASN and GLN in leaves and tubers of grafted plants.* The column headings show the scion/stock genotype

	Wild type / Wild type	Ast1⁻/Ast2⁻ (B26) / Wild type	Wild type / Ast1⁻/Ast2⁻ (B26)	Ast1⁻/Ast2⁻ (B26) / Ast1⁻/Ast2⁻ (B26)
Leaves				
StAst1 RNA	100	90	100	90
StAst2 RNA	100	95	100	90
ASN†	100	100	100	100
GLN‡	100	100	100	100
Tubers				
StAst1 RNA	100	75	0	0
StAst2 RNA	100	75	0	0
ASN§	100	97.9	4.36	2.79
GLN¶	100	107.9	252.8	317.6

ASN, asparagine; GLN, glutamine.

*Data are shown as percentage of wild type.

†100% = 0.169 mg/g FW.

‡100% = 1.374 mg/g FW.

§100% = 8.25 mg/g FW.

¶100% = 2.59 mg/g FW.

trials conducted previously (Rommens *et al.*, 2008). However, as this approach affected the quality of potatoes grown in the field, we set out to determine whether (i) one of the ASN synthetases had a greater impact on tuber ASN levels and could be fine-tuned to minimize the negative impact on the agronomic characteristics of the plant, (ii) the asparagine synthetase should be silenced constitutively or specifically in tubers, and (iii) the strategy to silence asparagine synthetase only in tubers is sufficient, or if significant amounts of ASN are transported from foliar tissues to the potato via other mechanisms. The following paragraphs discuss our findings.

StAst1/2 double mutants exhibit a synergistic phenotype

Tuber folding was only observed when both *StAst1* and *StAst2* were silenced in the tubers (Figures 1a and 10). As the tuber phenotypes appeared normal under field conditions when only one of the genes was silenced (Figure 10), the tuber cracking phenotype has a synergistic nature. The deleterious effect of silencing both enzymes in tubers can be attributed to the fact that ASN is critical for potato tubers. Levels of ASN that are below a certain threshold are perceived as stress and thus lead to the cracking of the tubers in the field (Hochmuth *et al.*, 2001; Jefferies and Mackerron, 1987). Tuber cracking could be because of global alterations in ASN levels and metabolism, which is an important intermediate in C/N portioning within the plant and thus affects other metabolites. It is also possible that significant phenotypic effects on tuber could be as a result of modified tuber water relations as we could mimic tuber cracking of StAst1/StAst2 plants in greenhouse by overwatering stress.

Increased foliar ASN levels do not result in increased ASN in potato tubers

In this study, we explored the extent of ASN transport by reciprocally grafting low ASN stock with WT scions. Interestingly, there was no increase in the level of free ASN in tubers of these grafted lines (Table 1). Also, lines overexpressing

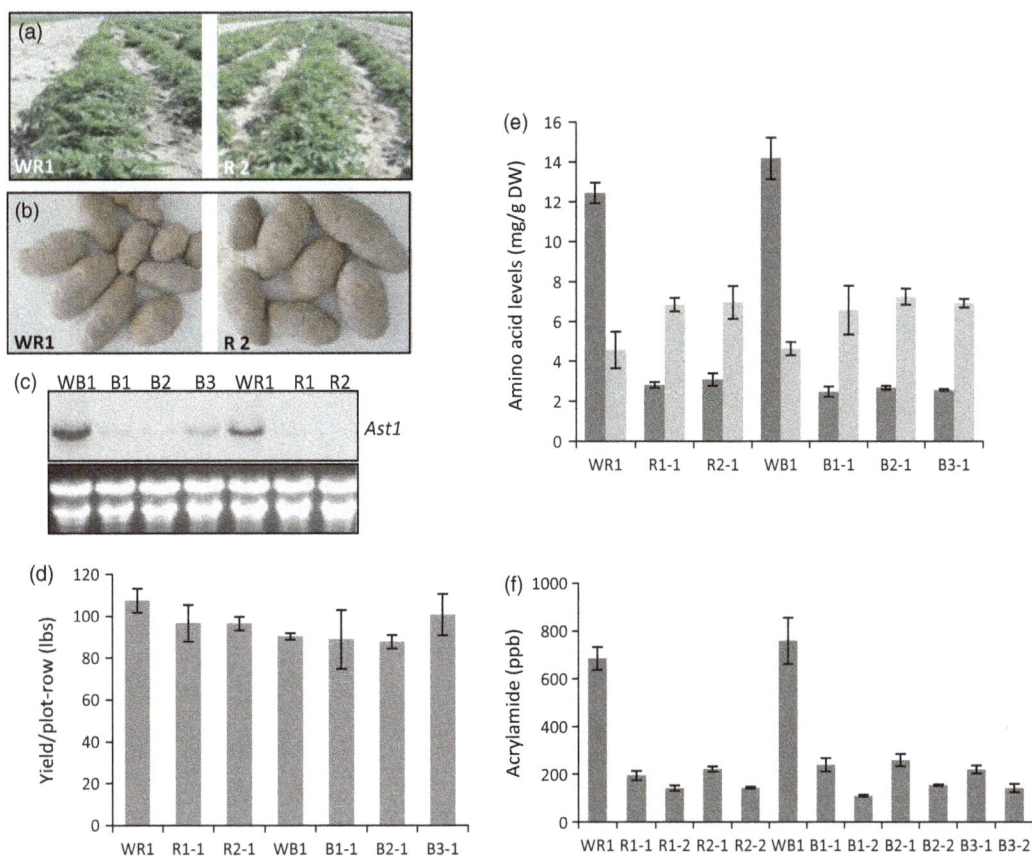

Figure 9 Phenotypic analysis of tuber-specific *StAst1*-silenced lines grown in the field. (a) Foliar and (b) tuber phenotypes of a Russet Ranger tuber-specific *StAst1*-silenced line (R2) compared with those of the WR (control). (c) Expression analysis of *StAst1* in the tubers of transgenic lines. (d) Yield/-plot-row consisting of 20 plants. Each line was represented three times in the field. Bars represent SD. (e) Free asparagine (dark grey bars) and glutamine (light grey bars) levels in the tubers of the *StAst1*-silenced lines. WR1 is the control for Russet Ranger variety, R1 and R2 are two independent Russet Ranger lines, WB1 is the Russet Burbank control, and B1, B2 and B3 are three independent Burbank lines silenced for *StAst1*. The data show standard deviation of three replicates. (f) Acrylamide levels (ppb) in the *StAst1*-silenced lines. WR1 and WB1 are the untransformed controls. Acrylamide data for two replicates of each transgenic line are shown. All data shown represent the average of three measurements ± SD.

StAst2 had increased *StAst2* transcript and ASN levels in the leaves, but did not have increased levels of ASN in the tubers (Figure 8b,c). This result indicates that ASN is not the major amino acid that transports N from source to sink in potato. This finding differs from observations in 35S-*AtAsn1* Arabidopsis lines, where sink tissues such as flowers and developing siliques exhibited higher levels of free ASN than source tissues, such as leaves and stems, despite the presence of significantly higher levels of *AtAsn1* mRNA in the source tissues. This was partially attributed to enhanced ASN transport from source to sink tissues via the phloem, based on increased levels of ASN in the phloem exudates of 35S:*AtAsn1* plants (Lam et al., 2003). This difference may be due to physiological differences between potato tubers and Arabidopsis siliques or may be a side effect of the constitutive promoters used.

However, our results are in agreement with the results of an aphid-feeding study that found that phloem exudates of 5-week-old potato plants contained significantly more GLN (30.1 mol%) than ASN (6.1 mol%; Karley et al., 2002). A similar observation was made in *Nicotiana tabacum* (tobacco), where GLN was implicated as the major N-transporting amino acid in the phloem sap (Masclaux-Daubresse

et al., 2006). Furthermore, GLN is the major amino acid entering the *Zea mays* (maize) cob just after silking, where it is further metabolized to ASN, a process that is required for optimal kernel growth and accumulation of storage proteins (Canas et al., 2010; Seebauer et al., 2004). Phloem sap analysis confirmed that GLN is the major form of N transported in maize (Oaks, 1992). GLN concentration was considerably reduced in the *gln1-3*, *gln1-4* and *gln1-3/gln1-4* maize mutants (Martin et al., 2006). Whereas ASN accumulated in the leaves of these plants during kernel filling, it was not sufficiently transported to the developing ears and was present in the phloem sap at concentrations below those found in WT plants. Furthermore, the concentration of soluble ASN was lower in the kernels of these mutants than in those of the WT (Martin et al., 2006). These findings support the hypothesis that GLN is the major amino acid involved in N transport in potato and maize.

Loss of StAst2 function is deleterious because of its important role in ammonium detoxification

StAst2 is required for overall vigour and yield in the field. Like *AtASN2*, *StAst2* is involved in ammonium detoxification, and it

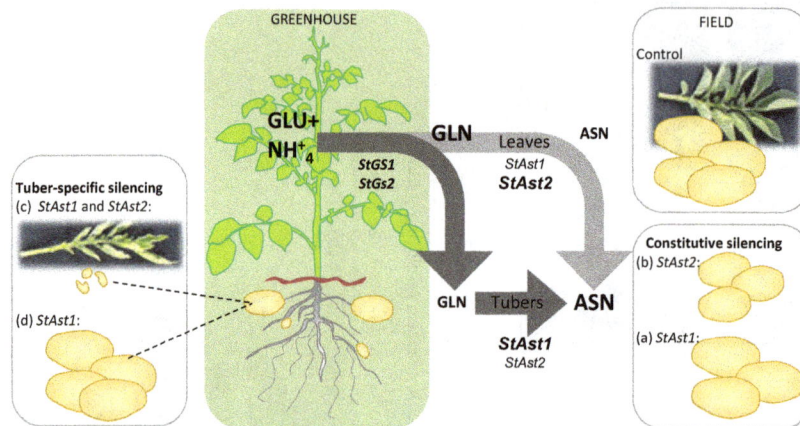

Figure 10 Functional characterization of StAst1 and StAst2 in potato. Majority of nitrogen is transported from senescing leaves as glutamine to the tubers where it is converted to asparagine (ASN) dominantly by *StAst1*, while *StAst2* is more important in ammonium detoxification in leaves of potato. (a) Constitutive *StAst1* silencing leads to a decreased yield in the field. (b) Constitutive *StAst2* silencing imparts a conditional phenotype on the foliage in the field and results in overall stunted plants with yield reduction. (c) Tuber silencing of both *StAst1* and *StAst2* results in tuber cracking in the field, while phenotype and yield are normal in the greenhouse. (d) Tuber-specific *StAst1* silencing is successful in decreasing ASN levels by up to 70%–80% and results in a normal phenotype and yields in the field.

is the accumulation of NH_4^+ in the leaves of *StAst2*-silenced plants that cause the severe phenotype (Figure 5). Although NH_4^+ toxicity is a universal phenomenon in both plant and animal systems (Britto and Kronzucker, 2002), the threshold at which symptoms develop differs widely among species. Chlorosis and the overall suppression of growth have been observed in species that are sensitive to NH_4^+. Solanaceae has been ranked as the family that is most sensitive to NH_4^+ toxicity (Britto and Kronzucker, 2002), with symptoms of toxicity (i.e. decreased dry weights) occurring earlier than in other plants grown in growth chambers in the presence of 8–12 mM NH_4^+ (Cao and Tibbits, 1998).

Therefore, *StAst2* appears to have a major role in NH_4^+ detoxification in potato (Figure 5d). NH_4^+ is generated during nitrate reduction, photorespiration and senescence. Stress factors, including drought, overwatering and infection, enhance the formation of ASN (Kanamori and Matsumoto, 1974; Lea et al., 2007; Ta and Joy, 1986; Wong et al., 2004), and NH_4^+ accumulation is generally viewed as an index of stress in plants (Barker, 1999a,b; Britto and Kronzucker, 2002). Although the physiological significance of up-regulating the biosynthesis of ASN is not fully understood, it is likely that an inability to detoxify NH_4^+ would, directly or indirectly, affect stress tolerance in the field. Based on phylogenetic analysis and on the observation that the expression of *AtAsn1* and *AtAsn2* is reciprocally regulated by light and metabolites, it has been proposed that these genes have different physiological roles in N metabolism in plants (Lam et al., 2003 and Wong et al., 2004). In Arabidopsis, light induction of *AtAsn2* is dependent on the presence of NH_4^+ in the medium (Wong et al., 2004). Accumulation of ASN in response to abiotic stresses could be a mechanism to detoxify NH_4^+ and a means of storing N when protein synthesis is inhibited by stress conditions. ASN also accumulates when high amounts of ammonium are generated by deamination of soluble amino acids released by proteolysis during leaf senescence and thereby prevents ammonium toxicity (Herrera-Rodríguez et al., 2007).

Furthermore, this study illustrates that field conditions are vastly different from those in the greenhouse and emphasizes the importance of field trials. Whereas *StAst1*- and *StAst2*-silenced plants appeared to be normal in the greenhouse, a variety of aberrations emerged in field trials (Figure 10). Such phenotypic variations would be hard to identify in model species, for example Arabidopsis, grown under the carefully controlled conditions in a growth chamber. Thus, field trials are critical when evaluating the phenotypes of genetically modified crops.

StAst1 is the major enzyme catalysing ASN production both in the leaves and in the tubers of potato

Most ASN in the potato plant is produced by *StAst1*, which contributes to both foliar and tuber ASN levels (Figures 9e and 10). The majority of N/C transport in potato occurs as GLN, which is converted to ASN by *StAst1* in the tuber. In agreement with results obtained for Arabidopsis (Lam et al., 2003), changes in total free ASN in tubers parallel the expression of *StAst1*, and not *StAst2*. Although StAST1 is the major enzyme in the production of ASN, it was not possible to silence this enzyme altogether in the plant. Constitutive silencing of *StAST1* affects the overall yield and the total N and protein content, indicating that some basal level of ASN transport from the leaves to the tubers is required to maintain the overall tuber yield. However, tuber-specific silencing of *StAst1* was sufficient to reduce levels of ASN in tubers without having a negative effect on yield in field trials (Figure 9a,b,d,e). Furthermore, this strategy was sufficient to reduce acrylamide levels by up to 70% of those in the control (Figure 9f). This study thus highlights the importance of using a tissue-specific promoter for fine-tuning the expression in a particular organ and is a key feature of the novel genetic modification strategy presented here (Figure 10).

Despite the central role of ASN in N storage in potato, most N is transported to the potato tuber as GLN. In tubers, the majority of GLN is converted to ASN by StAST1 (Figure 10). It is likely that StAST2 is involved in stress-induced ASN synthesis, although a more detailed study would enhance our understanding of this process. The present investigation shows that silencing of *StAst1* in tubers decreases the levels of free ASN, while increasing GLN amounts, without affecting plant growth or tuber phenotype and can be safely used as strategy to lower acrylamide levels in French fries. The increased GLN levels did not alter the quality of potato as food because they were still within the normal range for potato (Davies, 1977; Lisinska and Leszczynski, 1989). Furthermore, GLN is a nonessential amino acid that can be synthesized by the body and does not represent a significant precursor for acrylamide (Stadler, 2005). Consumption of processed potato products contributes to approximately one-third of the average dietary exposure to acrylamide (Boettcher *et al.*, 2005). Thus, an eventual replacement of existing potatoes by low-StAst1 varieties would lower the acrylamide intake by ~30%.

In the light of our new understanding of the role of *StAst1*, it might be tempting to try and knock this gene out through methods other than genetic engineering. However, the creation of such a recessive mutant would be difficult to achieve because backcrossing triggers inbreeding depression in potato (Golmirzaie *et al.*, 1998). Furthermore, constitutive downregulation of *StAst1* expression was shown here to negatively affect the plant's agronomic characteristics.

Experimental procedures

Plasmid construction

A multigene silencing construct pSIM1256 containing 348-bp fragments of *StAst1* and *StAst2* was cloned into a basic P-DNA vector, as previously described (Rommens *et al.*, 2004, 2008). For tuber-specific expression, the gene fragments were driven by the Gbss promoter (coordinates 1138–1823 of GEN-BANK accession X83220), and the Agp promoter is identical to coordinates 2183–4407 of GENBANK accession X96771 as described in Rommens *et al.*, 2008;. RNA isolation was carried out using Plant RNA easy kit (Qiagen), and RT-PCRs were performed with the one-step RT-PCR kit (Qiagen). Amplified products were gel purified and cloned into pGEM-Teasy (Promega) for subsequent sequence analysis. For constitutive expression of *StAst2*, pSIM1704 the complete ORF of *StAst2* was amplified using primers for RT-PCR (1704F GGA GCA TCG TGG AAA AAG AAG ACG and 1704R GAA CAA AAG AGA ACA TCC CAT CC) placed downstream of the *35S* CaMV promoter. For *StAst1* silencing, inverted repeat corresponding to 403 bp (coordinates 355–757 of accession DM 178284) was cloned into binary vector under the control of potato ubiquitin-7 (ubi7) gene promoter (Garbarino *et al.*, 1995) and the ubiquitin-3 gene terminator (Garbarino and Belknap, 1994) using a standard restriction enzyme method. The primers used for RT-PCR of *StAst1* are forward TCA TGT GTG GAA TTT TGG CTT TG and reverse AAA CAT CAA GTA TGG ACA GCA AGA. pSIM 1715 silencing targeted 250 bp of *StAst2* (coordinates of accession DM 178281). Primers used to amplify this region were forward TGG TCT TGT CCT CAT TAT ACA GC and reverse TGG TCT TGT CCT CAT TAT ACA GC. The same region was silenced in pSIM1716 using the ubi7 promoter and ubi3 terminator. Control binary vector pSIM401 (Rommens *et al.*, 2005) was used to produce empty vector tissue culture (TC) control plants. The binary vectors were transferred into *Agrobacterium tumefaciens* strain LBA4404 for plant transformation.

Plant transformation and genotyping

Stock plants were maintained in magenta boxes containing 40 mL half-strength M516 medium (PhytoTechnology, Shawnee Mission, KS) with 3% sucrose and 2 g/L gelrite. Plants were transformed using *Agrobacterium* as described previously (Richael *et al.*, 2008). Transformed plants were genotyped for the presence of P-DNA, T-DNA and backbone DNA using a robust and reliable PCR method, as described in Xin *et al.* (2003).

Field plot design and maintenance

Field trials using select *StAst1*- and *StAst2*-silenced lines were conducted at the University of Idaho Research and Extension Center. Field trials using *StAst1*-silenced lines were conducted at two additional sites in Idaho and Michigan, respectively. Seeds were sown 3–4 inches deep, at 10-inch spacing within rows and 36-inch spacing between rows. Pre-plant and in-season fertilizer applications were based on soil tests and petiole analyses, respectively, following standard protocols for russet potatoes. Each plot was evaluated qualitatively, in some cases using standardized monitoring scales for differential responses to insect, disease and environmental stresses. Herbicides, insecticides and fungicides were applied as needed. Vines were removed by a flail-type mower at Parma or desiccated with 6,7-dihydrodipyrido(1,2-a:2',1'-c)pyrazinediium dibromide (Reglone) (Syngenta, Greensboro, NC) 2 weeks prior to harvest. Tubers were harvested using a single-row mechanical harvester, and ~40 kg of tubers/line/site (equal amounts from each plot) was scored for defects, about 2 weeks later, according to standard industry practices.

Plant growth and grafting

Plants were grown in a controlled greenhouse environment with average day temperatures of 21 ± 3 °C and night temperatures of 16 ± 3 °C and 16-h days, which were facilitated by supplemental high-pressure sodium lights. The plants were grown in SunGro Mix 1 (SunGro Horticulture, Bellevue, WA) (8 L/pot) for 3 months. Fertilizations were carried out at 2, 4 and 6 weeks after planting with All-Purpose MiracleGro 24-8-16 (Scotts Company, Marysville, OH) according to the manufacturer's recommendations.

Plants were grafted using a technique similar to that described by others (Clayber, 1975; Marcotrigiano and Gouin, 1984). Seedlings were grown in a greenhouse in individual 1-gallon pots for about 3 weeks or until the stems were about 5 mm in diameter. The stem that was to be used as the scion was cut about two to three nodes below the apical meristem. Using a sharp scalpel, all leaves but the youngest were removed from the scion stem, and the bottom was trimmed on either side to form a 'V', with a greatest width of 1 mm. All leaves were removed from the stock. The stem that was to be used as the rootstock was cut widthwise to about the third node. The stem was sliced lengthwise down the centre to form a cut that was as long as the 'V' of the scion. The scion was then placed into the cut of the rootstock and secured with a strip of parafilm, and the graft was enclosed in a plastic bag fastened with a twist tie. After 7 days, a hole was punctured in the plastic bag to allow the scion to acclimate to a lower humidity, and the bag was removed 2 days later.

RNA isolation and RNA blot analysis

RNA was extracted from 1 g of leaf and tuber tissue using TRI-ZOL and Plant RNA Reagent (Invitrogen), respectively, according to the manufacturer's protocol. For RNA blot analysis, 20 µg RNA was blotted to the nylon membrane using the standard protocol. DIG-labelled DNA probes were synthesized using StAst1 and StAst2 primers, according to the protocol from Roche. Prehybridization and hybridization were performed in DIG-Easy Hyb buffer, and washing and developing were carried out as per the protocol (Roche). Nonradioactive digoxigenin RNA gel blot hybridization was performed according to the manufacturer's recommendations (Roche Applied Science). A 1.1-kb labelled probe for StAst2 was derived from a gene fragment amplified with the primer pair 5'-CTT GCT C AT CAA CGA TTG GCA ATA G and 5'-AGG TCG GAT CAT TTT CCA TTC TG. The primers used to produce a 1.1-kb probe for StAst1 were 5'-GGT TGA TGACTG ATG TCC CCT TTG and 5'-AGT TA G CTC CTT ATT GTG AGC TC. After exposure, the film was developed using a Konica SRX-101A developing machine.

Total ammonium, protein and nitrogen and acrylamide quantification assays

Protein and total nitrogen levels were determined according to the Dumas method as described in Official Methods of analysis of AOAC International (2005, 18th edn, Official Method 992.15, Gaithersburg, MD). The amount of ammonium/µg of protein was quantified at Covance Inc. (Madison, WI) according to protocols described in Official Methods of analysis of AOAC International (2000, 17th edn, Official Method 920.03, Gaithersburg, MD). Acrylamide levels in French Fries were measured as described previously (Rommens et al., 2008).

Determination of ASN and other amino acid levels

Asparagine and other amino acids were extracted by homogenizing 250 mg of ground freeze-dried tissue in 5 µmol sarcosine (internal standard) in 3.0 mL of a 0.03 M triethylamine HCl buffer. Next, 150 µL of 3.2% potassium hexacyanoferrate trihydrate, 150 µL of 7.2% zinc sulphate and 250 µL 0.1 N NaOH with 3.0 mL 0.03 M TEA buffer were added, with vortexing after each addition. The extract was centrifuged for 15 min at 4 °C, 2147 g, and the supernatant was transferred to a new tube. The pellet was re-suspended in 5 mL nanopure water and centrifuged. Supernatant was pooled with the first tube, and the volume was adjusted to 12.5 mL with water. The extracted free amino acids were derivatized using the EZ:faast method (Phenomenex) according to the user's manual. Derivatized samples were analysed by liquid chromatography–mass spectrometry (LC-MS) using an Agilent 1200 series HPLC system coupled with a 6300 series ion trap. For HPLC, an EZ:faast AAA-MS column (250 × 3.0 mm) was used, and the mobile phase was 10 mM ammonium formate in water (buffer A) and 10 mM ammonium formate in methanol (buffer B), with a gradient elution of 68%–83% buffer B over a period of 0–11 min at a flow of 0.25 mL/min and a column temperature of 35 °C. The mass spectrometer was equipped with an electrospray ionization source and was operated in the positive mode with auto MSn. The source was operated using drying gas (N_2; 350 °C; 10 L/min) and nebulizer gas (N_2; 30 psi), and the source voltage had a scan range of 50–550 m/z. Automated MS/MS analysis was conducted using Agilent's SmartFrag software, with a Smart ICC Target of 50 000 and a maximum accumulation

time of 200 ms. Bruker's quant analysis software was used for quantification.

Statistical analyses

Data are presented as the means of the results of at least three experiments, and the error bars shown represent the standard deviation (SD) of the mean. Significance was determined using the Student's two-tailed t-test.

Acknowledgements

We thank Robert Chretien and Michele Krucker for plant transformations and grafting/greenhouse maintenance of plants. Craig Richael and Mike Thornton are acknowledged for managing field trials.

References

Andrews, M.C. (1986) The partitioning of nitrate assimilation between the root and shoot of higher plants. *Plant, Cell Environ.* **9**, 511–519.

Atkins, C.A., Pate, J.S. and Sharkey, P.J. (1975) Asparagine metabolism-key to the nitrogen nutrition of developing legume seeds. *Plant Physiol.* **56**, 807–12.

Barker, A.V. (1999a) Ammonium accumulation and ethylene evolution by tomato infected with root-knot nematode and grown under different regimes of plant nutrition. *Commun. Soil Sci. Plant Anal.* **30**, 175–182.

Barker, A.V. (1999b) Foliar ammonium accumulation as an index of stress in plants. *Commun. Soil Sci. Plant Anal.* **30**, 167–174.

Boettcher, M.I., Schettgen, T., Kütting, B., Pischetsrieder, M. and Angerer, J. (2005) Mercapturic acids of acrylamide and glycidamide as biomarkers of the internal exposure to acrylamide in the general population. *Mutat. Res.* **580**, 167–176.

Britto, D.T. and Kronzucker, H.J. (2002) NH_4^+ toxicity in higher plants: a critical review. *J. Plant Physiol.* **159**, 567–584.

Canas, R.A., Quillere, I., Lea, P.J. and Hirel, B. (2010) Analysis of amino acid metabolism in the ear of maize mutants deficient in two cytosolic glutamine synthetase isozymes highlights the importance of asparagine for nitrogen translocation within sink organs. *Plant Biotech. J.* **8**, 966–978.

Cao, W. and Tibbits, T.W. (1998) Response of potatoes to nitrogen concentrations differ with nitrogen forms. *J. Plant Nutr.* **21**, 615–623.

Clayber, C.D. (1975) Insect resistance in a graft-induced periclinal chimera of tomato. *HortScience*, **10**, 13–15.

Davies, A.M.C. (1977) The free amino acids of tubers of potato varieties grown in England and Ireland. *Potato Res.* **20**, 9–21.

Duff, S.M.G., Qi, Q., Reich, T., Wu, X., Brown, T., Crowley, J.H. and Fabbri, B. (2011) A kinetic comparison of asparagine synthetase isozymes from higher plants. *Plant Phys. Biochem.* **49**, 251–256.

Dybing, E., Farmer, P.B., Andersen, M., Fennell, T.R., Lalljie, S.P., Müller, D.J., Olin, S., Petersen, B.J., Schlatter, J., Scholz, G., Scimeca, J.A., Slimani, N., Törnqvist, M., Tuijtelaars, S. and Verger, P. (2005) Human exposure and internal dose assessments of acrylamide in food. *Food Chem. Toxicol.* **43**, 365–410.

Garbarino, J.E. and Belknap, W.R. (1994) Isolation of a ubiquitin ribosomal protein gene (ubi3) from potato and expression of its promoter in transgenic plants. *Plant Mol. Biol.* **24**, 119–127.

Garbarino, J.E., Oosumi, T. and Belknap, W.R. (1995) Isolation of a polyubiquitin promoter and its expression in transgenic potato plants. *Plant Physiol.* **109**, 1371–1378.

Gaufichon, L., Reisdorf-Cren, M., Rothstein, S.J., Chardon, F. and Suzuki, A. (2010) Biological functions of asparagine synthetase in plants. *Plant Sci.* **179**, 141–153.

Golan-Goldhirsh, A. (1986) Effect of the add-back process on the free amino acid pool of potatoes. *Z. Lebensm. Unters. Forsch.* **182**, 29–32.

Golmirzaie, A.M., Bretschneider, K. and Ortiz, R. (1998) Inbreeding and true seed in tetrasomic potato. II. Selfing and sib-mating in heterogeneous hybrid populations of *Solanum tuberosum*. *Theor. Appl. Genet.* **97**, 1129–1132.

Herrera-Rodríguez, M.B., Carrasco-Ballesteros, S., Maldonado, J.M., Pineda, M., Aguilar, M. and Pérez -Vicente, R. (2002) Three genes showing distinct regulatory patterns encode the asparagine synthetase of sunflower (*Helianthus annuus*). *New Phytol.* **155**, 33–45.

Herrera-Rodríguez, M.B., Pérez-Vicente, R. and Maldonado, J.M. (2007) Expression of asparagine synthetase genes in sunflower (*Helianthus annuus*) under various environmental stresses. *Plant Physiol. Biochem.* **45**, 33–38.

Hochmuth, G.J., Hutchinson, C.M., Maynard, D.N., Stall, W.M., Kucharek, T.A., Webb, S.E., Taylor, T.G., Smith, S.A. and Simonne, E.H. (2001) Potato production in Florida. In *Vegetable Production Guide for Florida* (Maynard, D.N. and Olson, S.M., eds), pp. 223–230. Vance Publishing, Lincolnshire.

Hörteinsteiner, S. and Feller, U. (2002) Nitrogen metabolism and remobilization during senescence. *J. Exp. Bot.* **53**, 927–937.

Ireland, R.J. and Lea, P.J. (1999) The enzymes of glutamine, glutamate, asparagine, and aspartate metabolism. In *Plant Amino Acids: Biochemistry and Biotechnology* (Singh, B.K., ed.), pp. 49–109. New York: Marcel Dekker.

Jefferies, R.A. and Mackerron, D.K.L. (1987) Observations on the incidence of tuber growth cracking in relation to weather patterns. *Potato Res.* **30**, 613–623.

Kanamori, T. and Matsumoto, H. (1974) Asparagine biosynthesis by *Oryza sativa* seedlings. *Phytochemistry*, **13**, 1407–1412.

Karley, A.J., Douglas, A.E. and Parker, W.E. (2002) Amino acid composition and nutritional quality of potato leaf phloem sap for aphids. *J. Exp. Biol.* **205**, 3009–3018.

Koch, W., Kwart, M., Laubner, M., Heineke, D., Stransky, H., Frommer, W.B. and Tegeder, M. (2003) Reduced amino acid content in transgenic potato tubers due to antisense inhibition of the leaf H+/amino acid symporter StAAP1. *Plant J.* **33**, 211–220.

Lam, H.-M., Hsieh, M.-H. and Coruzzi, G. (1998) Reciprocal regulation of distinct asparagine synthetase genes by light and metabolites in *Arabidopsis thaliana*. *Plant J.* **16**, 345–353.

Lam, H.-M., Wong, P., Chan, H.K., Yam, K.M., Chen, L., Chow, C.M. and Coruzzi, G.M. (2003) Overexpression of the ASN1 gene enhances nitrogen status in seeds of Arabidopsis. *Plant Physiol.* **132**, 926–935.

Lea, P.J., Sodek, L., Parry, M.A.J., Shewry, P.R. and Halford, N.G. (2007) Asparagine in plants. *Ann. Appl. Biol.* **150**, 1–26.

Lehmann, T. and Ratajczak, L. (2008) The pivotal role of glutamine dehydrogenase (GDH) in the mobilization of N and C from storage material to asparagine in germinating seeds of yellow lupine. *J. Plant Physiol.* **165**, 149–158.

Lisinska, G. and Leszczynski, W. (1989) *Potato Science and Technology. Chapter 2: Potato Tubers as a Raw Material for Processing and Nutrition.* London: Elsevier Applied Science, pp. 11–128.

Lohaus, G., Büker, M., Hubmann, M., Soave, C. and Heldt, H.W. (1998) Transport of amino acids with special emphasis on the synthesis and transport of asparagine in the Illinois Low Protein and Illinois High Protein strains of maize. *Planta*, **205**, 181–188.

Marcotrigiano, M. and Gouin, F.R. (1984) Experimentally synthesized plant chimera 2. A comparison of *in vitro* and in vivo techniques for the production of interspecific *Nicotiana* chimeras. *Ann. Bot.* **54**, 513–521.

Martin, A., Lee, J., Kichey, T., Gerentes, D., Zivy, M., Tatout, C., Dubois, F., Balliau, T., Valot, B., Davanture, M., Tercé-Laforgue, T., Quilleré, I., Coque, M., Gallais, A., Gonzalez-Moro, M., Bethencourt, L., Habash, D.Z., Lea, P.J., Charcosset, A., Perez, P., Murigneux, A., Sakakibara, H., Edwards, K.J. and Hirel, B. (2006) Two cytosolic glutamine synthetase isoforms of maize are specifically involved in the control of grain production. *Plant Cell*, **18**, 3252–3274.

Masclaux-Daubresse, C., Reisdorf-Cren, M., Pageau, K., Lelandais, M., Grandjean, O., Kronenberger, J., Valadier, M.H., Feraud, M., Jouglet, T. and Suzuki, A. (2006) Glutamine synthetase-glutamate synthase pathway and glutamate dehydrogenase play distinct roles in the sink-source nitrogen cycle in tobacco. *Plant Physiol.* **140**, 444–456.

Muller-Rober, B., La Cognata, U., Sonnewald, U. and Willmitzer, L. (1994) A truncated version of an ADP-glucose pyrophosphorylase promoter from potato specifies guard cell-selective expression in transgenic plants. *Plant Cell*, **6**, 601–612.

Oaks, A. (1992) A re-evaluation of nitrogen assimilation in roots. *Bioscience*, **42**, 103–110.

Parzefall, W. (2008) Minireview on the toxicity of dietary acrylamide. *Food Chem. Toxicol.* **46**, 1360–1364.

Richael, C.M., Kalyaeva, M., Chretien, R.C., Yan, H., Adimulam, S., Stivison, A., Weeks, J.T. and Rommens, C.M. (2008) Cytokinin vectors mediate marker-free and backbone-free plant transformation. *Transgenic Res.* **5**, 905–917.

Rommens, C.M., Humara, J.M., Ye, J., Yan, H., Richael, C., Zhang, L., Perry, R. and Swords, K. (2004) Crop improvement through modification of the plant's own genome. *Plant Physiol.* **135**, 421–431.

Rommens, C.M., Bougri, O., Yan, H., Humara, J.M., Owen, J., Swords, K. and Ye, J. (2005) Plant-derived transfer DNAs. *Plant Physiol.* **139**, 1338–1349.

Rommens, C.M., Yan, H., Swords, K., Richael, C. and Ye, J. (2008) Low-acrylamide French fries and potato chips. *Plant Biotech. J.* **6**, 843–853.

Seebauer, J.R., Moose, S.P., Fabbri, B.J., Crossland, L.D. and Below, F.E. (2004) Amino acid metabolism in maize earshoots. Implications for assimilate preconditioning and nitrogen signaling. *Plant Physiol.* **136**, 4326–4334.

Stadler, R.H. (2005) Acrylamide formation in different foods and potential strategies for reduction. *Adv. Exp. Med. Biol.* **561**, 157–169.

Ta, T.C. and Joy, K.W. (1986) Separation of amino acid and amide nitrogen from plant extracts for 15^N analysis. *Anal. Biochem.* **154**, 564–569.

Tardiff, R.G., Gargas, M.L., Kirman, C.R., Carson, M.L. and Sweeney, L.M. (2010) Estimation of safe dietary intake levels of acrylamide for humans. *Food Chem. Toxicol.* **48**, 658–667.

Tareke, E., Rydberg, P., Karlsson, P., Eriksson, S. and Törnqvist, M. (2002) Analysis of acrylamide, a carcinogen formed in heated foodstuffs. *J. Agric. Food Chem.* **50**, 4998–5006.

Taylor, L., Nunes-Nesi, A., Parsley, K., Leiss, A., Leach, G., Coates, S., Wingler, A., Fernie, A. and Hibberd, J.M. (2010) Cytosolic pyruvate, orthophosphate dikinase functions in nitrogen remobilization during leaf senescence and limits individual seed growth and nitrogen content. *Plant J.* **62**, 641–652.

Tsai, F.Y. and Coruzzi, G.M. (1991) Light represses transcription of asparagine synthetase genes in photosynthetic and non-photosynthetic organs of plants. *Mol. Cell. Biol.* **11**, 4966–4972.

Visser, R.G., Stolte, A. and Jacobsen, E. (1991) Expression of a chimaeric granule-bound starch synthase-GUS gene in transgenic potato plants. *Plant Mol. Biol.* **17**, 691–699.

Weiner, H., Blechschmidt-Schneider, S., Mohme, H., Eschrich, W. and Heldt, H.W. (1991) Phloem transport of amino acids. Comparison of amino acid contents of maize leaves and of the sieve tube exudates. *Plant Physiol. Biochem.* **29**, 19–23.

Wong, H.K., Chan, H.K., Coruzzi, G.M. and Lam, H.M. (2004) Correlation of ASN2 gene expression with ammonium metabolism in *Arabidopsis*. *Plant Physiol.* **134**, 332–338.

Xin, Z., Velten, J.P., Oliver, M.J. and Burke, J.J. (2003) High-throughput DNA extraction method suitable for PCR. *Biotechniques*, **34**, 820–824.

7

Transcriptome-wide sequencing provides insights into geocarpy in peanut (*Arachis hypogaea* L.)

Xiaoping Chen[1†], Qingli Yang[2,3†], Haifen Li[1†], Heying Li[4], Yanbin Hong[1], Lijuan Pan[2], Na Chen[2], Fanghe Zhu[1], Xiaoyuan Chi[2], Wei Zhu[1], Mingna Chen[2], Haiyan Liu[1], Zhen Yang[2], Erhua Zhang[1], Tong Wang[2], Ni Zhong[1], Mian Wang[2], Hong Liu[4], Shijie Wen[1], Xingyu Li[1], Guiyuan Zhou[1], Shaoxiong Li[1], Hong Wu[4], Rajeev Varshney[1,5], Xuanqiang Liang[1]* and Shanlin Yu[2]*

[1]*Crops Research Institute, Guangdong Academy of Agricultural Sciences (GAAS), South China Peanut Sub-center of National Center of Oilseed Crops Improvement, Guangdong Key Laboratory for Crops Genetic Improvement, Guangzhou, China*
[2]*Shandong Peanut Research Institute, Shandong Academy of Agricultural Sciences, Qingdao, China*
[3]*College of Food Science and Engineering of Qingdao Agricultural University, Qingdao, China*
[4]*South China Agricultural University, Guangzhou, China*
[5]*International Crops Research Institute for the Semi-Arid Tropics (ICRISAT), Patancheru, India*

*Correspondence

email liang-804@163.com

email yshanlin1956@163.com
[†]These authors contributed equally to this work.

Keywords: peanut (*Archis hypogaea* L.), RNA-Seq, Geocarpy, fruit development.

Summary

A characteristic feature of peanut is the subterranean fructification, geocarpy, in which the gynophore ('peg'), a specialized organ that transitions from upward growth habit to downward outgrowth upon fertilization, drives the developing pod into the soil for subsequent development underground. As a step towards understanding this phenomenon, we explore the developmental dynamics of the peanut pod transcriptome at 11 successive stages. We identified 110 217 transcripts across developmental stages and quantified their abundance along a pod developmental gradient in pod wall. We found that the majority of transcripts were differentially expressed along the developmental gradient as well as identified temporal programs of gene expression, including hundreds of transcription factors. Thought to be an adaptation to particularly harsh subterranean environments, both up- and down-regulated gene sets in pod wall were enriched for response to a broad array of stimuli, like gravity, light and subterranean environmental factors. We also identified hundreds of transcripts associated with gravitropism and photomorphogenesis, which may be involved in the geocarpy. Collectively, this study forms a transcriptional baseline for geocarpy in peanut as well as provides a considerable body of evidence that transcriptional regulation in peanut aerial and subterranean fruits is complex.

Introduction

Peanut or groundnut (*Arachis hypogaea* L.) is an important food and cash crop for edible oil and protein production in tropical, subtropical and warm regions of the world. Despite a member of the Fabaceae family, its fruit is actually the subterranean pod. Peanut shows a distinctive pattern of fruit development, 'aerial flower and subterranean fruit' (Smith, 1950), quite different from those of other legume species. Following fertilization, the gynophore elongates to form a specialized geotropic organ (peg) (Smith, 1950). The peg-harbouring embryo continues to grow and pushes the developing pod into the soil for the subsequent pod development underground. After penetration into soil, pod formation and embryo differentiation occur, and a seed is produced (Feng *et al.*, 1995). The subterranean fructification is the most prominent characteristic of seed production in peanut and thus has the biologically important value for studying organogenesis and evolution. More importantly, the study of peanut pod is of significance for understanding mechanisms controlling plant reproductive development and crop improvement under dark conditions. Thought to be an adaptation to particularly harsh environments, geocarpy is severely influenced by subterranean environmental factors. Pod wall tissues in peanut provide functions not only in serving to protect the seeds, but also

in delivering nutrients to seed as part of source–sink pathway, as well as producing metabolized storage products, like other legumes (Setia *et al.*, 1987; Thorne, 1979; Wang and Grusak, 2005). Thus, a comprehensively transcriptomic foundation should help in this effort.

Transcriptomics resources for peanut research have emerged in the past decade. The peanut transcriptomes have been surveyed by cDNA sequencing, such as expressed sequence tags (EST) (Bi *et al.*, 2010; Guo *et al.*, 2008, 2009; Luo *et al.*, 2005). These EST sequencing efforts have allowed the identification of functional genes resistant to abiotic and biotic stresses in peanut. Microarrays have also been used to investigate the global transcription profiles of different peanut varieties (Chen *et al.*, 2012) as well as expression patterns in a variety of peanut tissues under various conditions (Guo *et al.*, 2011; Li *et al.*, 2013; Payton *et al.*, 2009; Wang *et al.*, 2012). Using such approaches, however, it is difficult to define the transcriptome at single-base resolution. Recently, RNA sequencing (RNA-Seq) using next-generation sequencing (NGS) is becoming a revolutionary tool for transcriptomics (Wang *et al.*, 2009) and has been widely applied in plant biology, both in model species such as *Arabidopsis* (Lister *et al.*, 2008) and in crop plants including rice (Lu *et al.*, 2010) and maize (Li *et al.*, 2010), as well as recently in peanut (Chen *et al.*, 2013; Xia *et al.*, 2013; Yin *et al.*, 2013; Zhang *et al.*,

2012). Transcriptomics has matured to the point where complex gene regulatory networks comprising mRNA expression and transcription factors aid in elucidating complex developmental processes (Scanlon and Timmermans, 2013). Understanding the global expression profiling of peanut pods and defining their transcriptomes will provide crucial information for understanding geocarpy in peanut.

Here, we yielded the first transcriptional map of the pod developmental transcriptome at single-base resolution using RNA-seq. We sequenced two stages of whole pods (aerial pod and not swelling subterranean pod) and nine stages of isolated pod walls representing 11 distinct stages of pod development with the following objectives: (i) to catalogue gene expression patterns during peanut pod development across the aerial and underground developmental stages; (ii) to characterize the major biological processes like gravitropism and subterranean fruit development. Taken together, our data could serve as a valuable resource for transcriptomics studies related to peanut pod for developmental biologists, who are interested in studying fruit development under dark conditions, as well as for the general transcriptomics community.

Results and Discussions

Transcriptome sequencing of aerial pod and subterranean pod wall

For a comprehensive understanding of subterranean fructification in peanut, we selected two stages of whole pods (aerial pod, P0 and not swelling subterranean pod, P1) and nine stages of isolated pod walls (P2SH-P10SH) (Table S1) for RNA-Seq analysis. Using Illumina HiSeq2000 platform, between 38 and 72 million 101-nt reads were generated for each sample (Table S2 and S3). The sequence data are deposited in NCBI Sequence Read Archive (SRA, http://www.ncbi.nlm.nih.gov/sra) under the accession number SRP033292. A reference transcriptome sequence was assembled using a genome-guided strategy using Trinity (Grabherr et al., 2011). We aligned each individual library's data against the reference transcriptome sequence using Bowtie2 (Langmead et al., 2009). Tolerances were set to allow up to five mismatches in each alignment, and reads that aligned to multiple reference transcripts were ignored. By these criteria, 90.14%–93.15% of paired-end reads (PEs) were uniquely mapped to the reference transcriptome and 4.02%–5.83% of PEs was filtered as reads with bad matches (Table S4 and Figure 1a). The remaining reads (2.83%–4.03%) were defined as unmapped reads. The mean number of mapped reads in different libraries ranged from 98 (P1) to 214 (P4SH) (Figure 1b).

We estimated the expression levels of transcripts using fragments per kilobase of transcript per million mapped fragments (FPKM) (Trapnell et al., 2010). A gene was considered expressed in a sample if the FPKM was greater than one, and the lower value of the FPKM 95% confidence interval was greater than zero (Hansey et al., 2012). By this criterion, we found that 110 217 transcripts were expressed in at least one of the 11 samples, ranging from 51 293 to 79 518 (Figure 1c) in individual samples. The largest portion of transcripts showed medium expression (2 ≤ FPKM < 10), followed by low expression (FPKM < 2). The proportions of genes at three expression levels were similar in all stages. The number of transcripts expressed in single or multiple samples tended to shape a reverse parabolic distribution with stage-specific and shared expressed transcripts being the largest group (Figure 1d). Among the 11 stages, aerial

pod (P0) contained the smallest number of transcripts (271) specifically detected in this stage, while 16 724 transcripts were detected only in subterranean pod swelling stages (P2–P6). About 43 788 transcripts were shared in the four periods (Figure 1e). Despite the similar number of expressed transcripts in each sample, Pearson correlation coefficients indicated that underlying expression dynamics were greatly diverse during pod development (Figure S1). We identified 109 063 transcripts that were differentially expressed between samples pairwise (Figure S2 and Table S5). Strikingly, up-regulated transcripts in aerial pod (P0) accounted for only ~19% of differentially expressed (DE) transcripts compared with those in subterranean samples. We employed qRT-PCR to validate the expression levels based on RNA-Seq. A high correlation between two approaches was observed (Figure S3).

Identification of temporally up- and down-regulated genes in pod wall throughout pod development

To characterize the temporal expression patterns of genes expressed throughout pod development, we used StepMiner (Sahoo et al., 2007) to identify four temporal expression patterns including one-step-up (expression level transitions from low to high in two consecutive developmental stages), one-step-down (from high to low), two-step-up/down (transitions from low to high and then back down in a series of developmental stages) and two-step-down/up transitions (from high to low and then back up) (Figure 2a). We identified 7,702 genes with one or two transition points in expression during P0–P10 stages (Figure 2 and Table S6). One-step patterns accounted for the majority of the identified genes, including 39.3% of genes with the one-step-up pattern and 32.2% of genes with the one-step-down pattern, while 28.5% of identified genes showed two transition points (two-step-up/down and two-step-down/up genes) (Table S6). With genes exhibiting a one-step-down transition point, the transition in expression level occurred mostly at P10SH stage in which desiccation occurs.

According to material collection information, pod expansion started at P2 stage and the pod size reached its maximum (~16.5 mm) at P6 stage (Table S1). Pod enlargement occurred during P2–P6 stages, suggesting that up-at-P2 genes, showing lower expression at P0–P1 stages and higher expression during P2SH–P6SH, may be associated with pod swelling. We further classified these one-step and two-step genes based on phenotypic difference during pod swelling underground. For one-step-up genes, the major expression transition occurs at P2SH (Figure 2), indicating a dramatic reprogramming of pod wall transcriptome started from P2SH, consistent with phenotypic differentiation that pod swelling starts at this time point. With genes showing two-step-up-down transition point, transition occurred more frequently during P2SH-P6SH. Approximately 49% (920/1879) of two-step-up-down genes may be involved in the pod enlargement. To confidently identify a set of transcripts accompanying with pod enlargement, we further filtered the 920 genes using a more stringent criteria, which required that all genes showed lower expression at P0–P1, higher expression consistently during P2SH to P6SH and then lower expression during P7SH-P10SH. Accordingly, we obtained 43 transcripts that sufficed the criteria (Table S7). This gene set contained a number of known genes involved in cell wall expansion; for example, endoglucanase can induce extension of cell walls and play roles in the assembly of the cellulose–hemicellulose network in the expanding cell wall (Klose et al., 2015; Nicol et al., 1998; Yuan

(a)

PEs with unique and good matches (91.8%)

PEs with unique but bad matches (4.7%)

PEs with one end unmapped (3.06%)

PEs with both ends unmapped (0.44%)

(b)

log2 (Number of reads mapped to a transcript)

(c)

(d)

Number of samples

(e)

Aerial peg (P0)

Not swelling underground (P1)

Swelling underground (P2-P6)

Mature and desiccation (P7-P10)

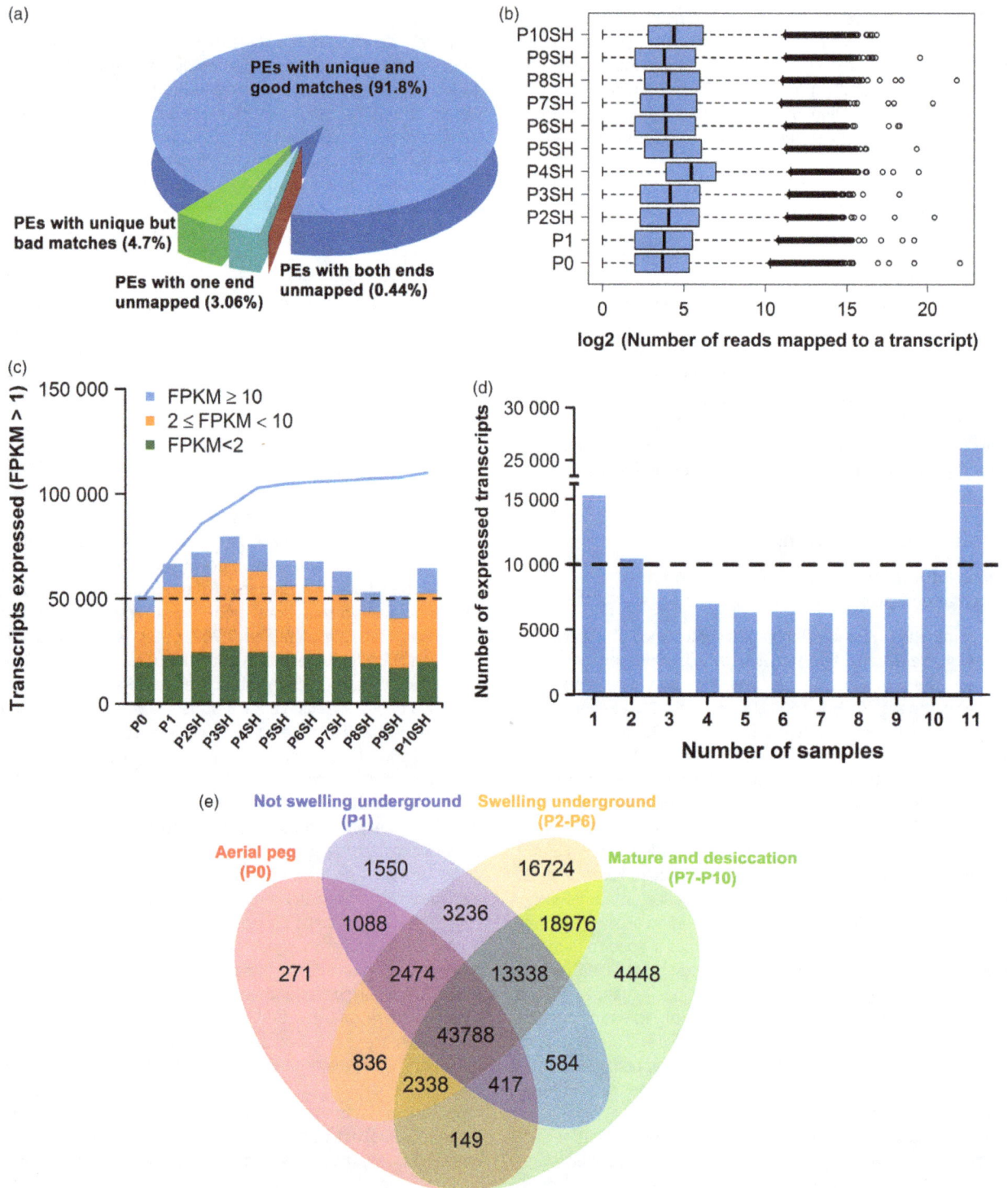

Figure 1 Number of genes expressed in aerial peg and subterranean pod walls. (a) Overall mapping results of paired-end reads (PEs) for all libraries referring to the reference transcriptome. (b) Distribution of the number of uniquely mapped paired-end reads (PEs) for each library. (c) Proportion of genes expressed at different levels (based on FPKM) during the 11 developmental stages. The bars indicate the number of transcripts expressed in each sample, and the line indicates the cumulative number of expressed transcripts. (d) Number of specific and shared transcripts expressed in single sample (1), multiple samples (2–10) or all samples (11) during pod development. (e) Venn diagram of the numbers of expressed transcripts in aerial pod (P0), whole pod not swelling underground (P1) and pod walls during pod swelling underground (P2–P6) as well as pod walls during mature and desiccation stages (P7–P10).

et al., 2001), consistent with cross section of pod wall (Figure S4). In addition, approximately 63% (27/43) of genes in this group are unknown (uncharacterized genes or no homologs found in the UniProt database), suggesting that pod enlargement underground is a complex biological process regulated by a series of genes not yet identified in addition to previously described genes.

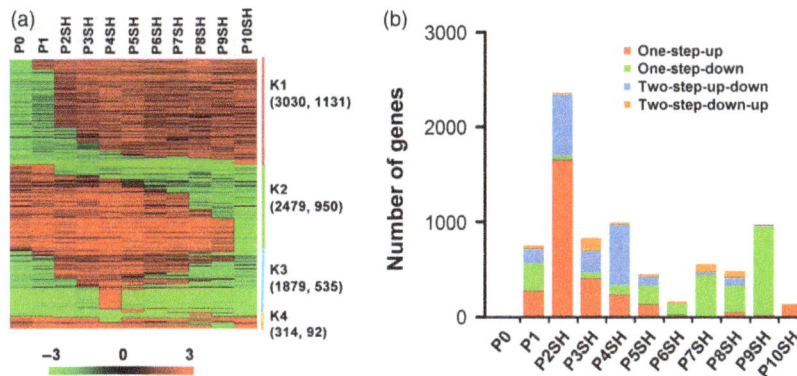

Figure 2 Identification of temporal up- and down-regulated gene sets. (a) Identification of the one-step-up (K1), one-step-down (K2), two-step-up/down (K3) and two-step-down/up (K4) transitions for all transcripts across pod development using StepMiner. The number of all transcripts (left) and transcription factors (right) is indicated in parentheses for each cluster. The scale colour bar is shown on the bottom. (b) Number of transcripts in one-step and two-step gene sets.

Biological processes enriched in temporally up- and down-regulated gene sets of pod wall

The identified temporal up- and down-regulated gene sets are likely associated with specific biological processes or pathways involved in pod development. We used Cytoscape (Su et al., 2014) with plugin BiNGO (Maere et al., 2005) to identify the major biological processes that were significantly enriched (false discovery rate < 0.01) in temporally up- and down-regulated gene sets.

The one-step-up gene sets were shown to be enriched for 'response to stimulus', 'developmental process', 'transport' and 'metabolic process' (Figure S5). The response to stimulus category included a broad range of stimuli like abiotic and biotic, external and endogenous stimuli (Figure 3a), suggesting that subterranean developing pods suffered from various adverse environmental factors. Pod wall has evolved the capacity to adapt to particularly harsh subterranean environments as a protective organ. We found that functional overrepresentations included 'response to abscisic acid stimulus'. Significant quantities of abscisic acid were once found in legume pod walls as maturity approached (Eeuwens and Schwabe, 1975). In addition to response to endogenous stimulus, genes in peanut pod walls were shown to be enriched for response to fungus, bacterium and nematode which are soilborne major pathogens causing damages to peanut (Holbrook et al., 2000; Singh et al., 1997; Starr and Simpson, 2006). As expected, developmental process was enriched in pod walls during pod development (Figure 3b). In addition, transport was overrepresented in pod wall, consistent with its function in not only serving to protect the seeds but also delivering nutrients to the seed as part of the source–sink pathway (Wang and Grusak, 2005).

The one-step-down genes were enriched for less diverse GO categories in comparison with one-step-up genes. Key functional overrepresentations included 'response to stimulus' and 'metabolic process' (Figure S6). Under the category of 'response to stimulus', child categories showed dramatic difference between the one-step-up and one-step-down gene sets, indicating that the two gene sets may play different roles in response to various stimuli (Figure 3c). The one-step-down genes were enriched for 'response to auxin' and 'response to light stimulus' as well as 'response to gravity' which are important to the peanut promi-

nent feature 'aerial flower and subterranean fruit'. Although 'developmental process' category was also found to be enriched in one-step-down genes, only few child categories were identified to be enriched, indicating developmental processes of pod walls underground were regulated mainly by up-regulated genes but not down-regulated genes.

The two-step gene sets covered a narrow range of GO categories with enrichment in comparison with one-step gene sets (Figure S7 and S8). Two-step-up-down genes were enriched for 'cell wall organization or biogenesis' and its child categories like 'plant-type-cell wall organization or biogenesis' and 'plant-type cell wall loosening' (Figure 3d), indicating that pod wall was gradually swelling during P2-P6 stages and then ceased swelling during P7-P10 stages.

Collectively, due to the geocarpy feature, the fructification of peanut is affected not only by terrestrial environmental factors, but also by subterranean environmental factors (Ono, 1979; Varaprasad et al., 1999, 2000), resulting in that the four temporal gene sets are enriched for 'response to stimulus', indicating that peanut subterranean pod develops under various endogenous and exogenous stimuli.

Identification of stage-specific gene sets across pod development

To discover the gene expression programs that characterize pod wall during pod development, mRNA-specific accumulation at each stage was measured using a specificity index (τ) (Yanai et al., 2005) to the genes with FPKM $\geq$ 2 in at least one sample. Here, τ values, varying between 0 for completely housekeeping genes and 1 for strictly one-stage-specific genes, shaped a parabolic distribution with intermediate specificities ($0.5 \leq \tau \leq 0.55$) being the largest group (Figure 4a). In this study, genes with $\tau \geq 0.9$ were defined as being expressed in a stage-specific pattern. By this criterion, 5474 stage-specific genes were identified (Figure 4 and Table S8). The number of identified stage-specific genes showed dramatic difference among the 11 stages, ranging from a low of 12 to a high of 2611 (Figure 4b), quite different from the similarity of the overall mRNA profiles detected among the stages (Figure 1c). The highest number of stage-specific genes was identified in P4SH (2611 genes), followed by P10SH (1120 genes), two times more than the numbers of those genes detected in other stages. In terms of TFs, the highest number of stage-specific

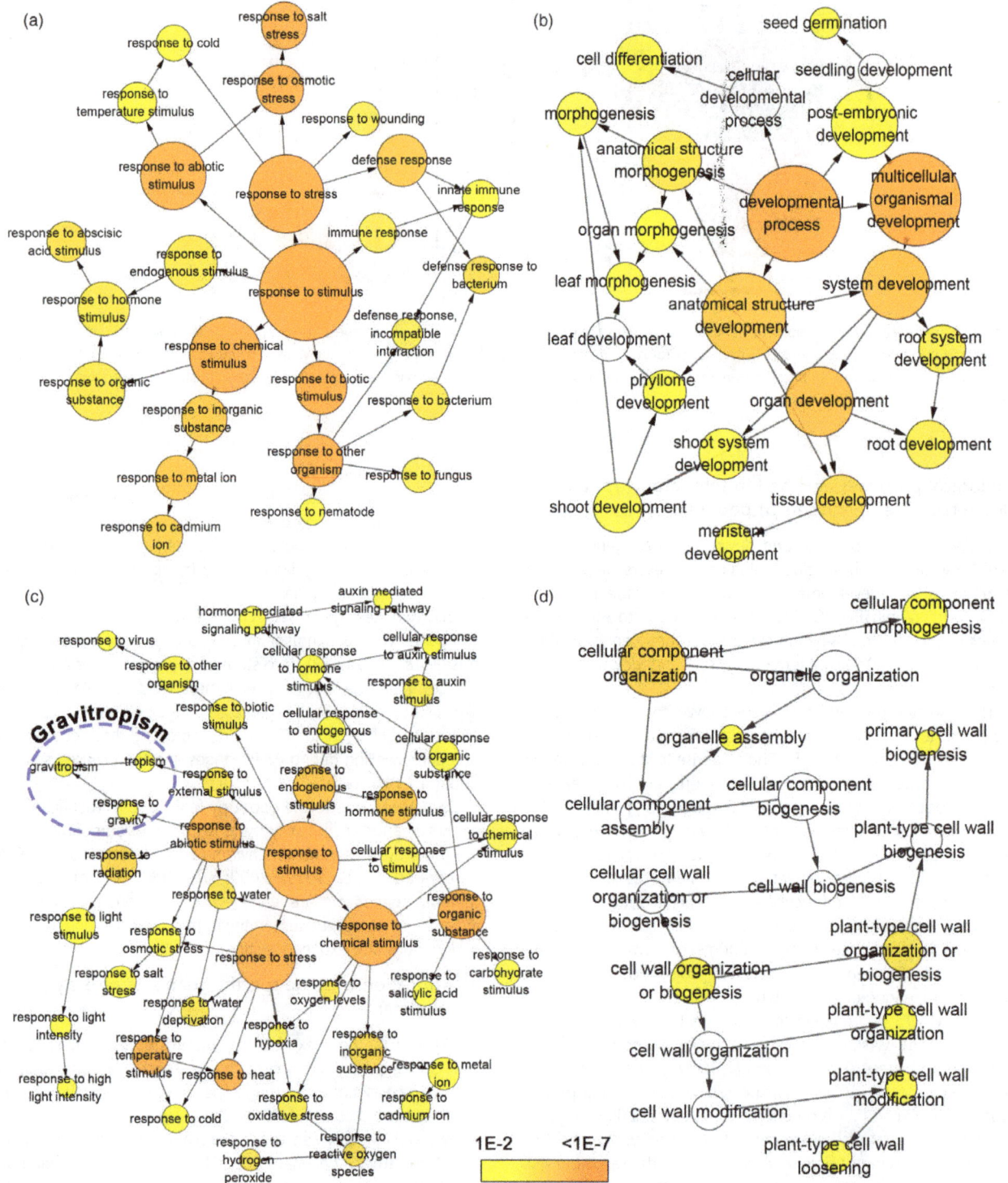

Figure 3 Enrichment analysis of gene ontology and functional annotation of one- and two-step gene sets. Overrepresented gene ontology categories in one-step-up (a,b), one-step-down (c) and two-step-up-down (d) gene sets. The circles are shaded based on significance level (yellow, false discovery rate < 0.01), and the radius of each circle denotes the number of genes in each category.

TFs was detected in P10SH (475), followed by P4SH (324). Pod enlargement started at P2 stage, and then, pod size reached its maximum (~16.5 mm) at P6 stage in which the seed size remained ~ 2 mm (**Table S1**). The P4SH sample was collected from pod walls at the middle time point of pod enlargement, at which pod wall expanded rapidly relative to at other stages. Given

desiccation occurring at P10SH stage, large numbers of genes and TFs were required to function in pod desiccation. Taken together, the variable number and proportion of stage-specific genes and TFs suggested the more complex gene expression profiles in pod enlargement and desiccation, implying the complexity of the peanut geocarpy.

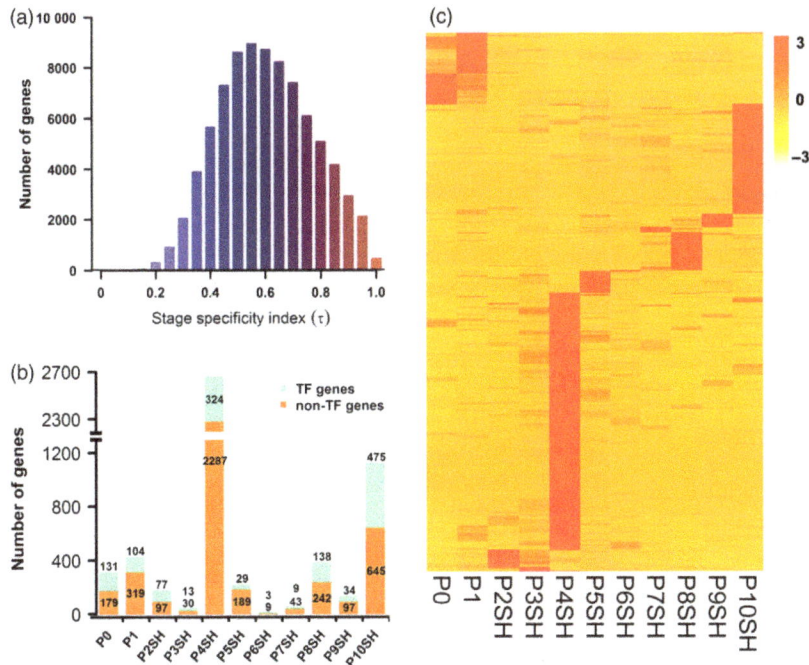

Figure 4 Stage-specific gene sets in aerial peg and subterranean pod wall across pod development. (a) Distribution of τ values of expressed genes from all samples. τ values varied between 0 for completely housekeeping genes and 1 for strictly stage-specific genes. In this study, τ values were divided into twenty bins for histogram. (b) Numbers of TF and non-TF genes in each stage-specific gene set. (c) Heatmap of scaled FPKM values of the stage-specific genes. Red, high expression; white, low expression.

Identification of gravitropism- and photomorphogenesis-related genes during pod development

The most salient feature of peanut fruit production is that the pod does not swell until it penetrates into the soil. Penetrating into the soil and then swelling of gynophore ('peg') tips are essential processes described for peanut pod development, which is the crucial determinant of peanut yield. In the previous study, we elucidated why aerially developing pods could not swell normally compared with those that penetrated into soil (Chen et al., 2013). Here, we attempt to focus on the downward outgrowth and pod formation in subterranean darkness.

We searched expressed transcripts in pod wall against the gene set falling under the GO category 'gravitropism' (GO:0009630) or experimentally identified in Arabidopsis using an E-value cut-off of 1E-10, and identified 151 gravitropism-related genes expressed during pod development. GO enrichment analysis showed that these genes were enriched for 'gravitropism', 'response to gravity', 'positive gravitropism' and 'negative gravitropism' (Figure 5a) with the GO category including experimentally identified gravitropic genes, like Auxin influx carrier (AUX1), Altered Response to Gravity (ARG1) and Gravitropism Defective 2 (GRV2/SGR8) genes, which have been well characterized to function in gravitropism (Bennett et al., 1996; Sedbrook et al., 1999; Silady et al., 2004). Further analysis identified 25 genes likely to be gravitropic including 4 involved in gravity perception, eight in signal transduction and 13 in organ response (Table S9) based on previous functional studies (Bennett et al., 1996; Caspar and Pickard, 1989; Friml et al., 2002; Harrison and Masson, 2008; Kato et al., 2002; Morita et al., 2006; Noh et al., 2003; Sedbrook et al., 1999; Silady et al., 2004; Swarup et al., 2004; Withers et al., 2013; Yano et al., 2003; Young et al., 2006). Of

them, 24 were identified in the peanut progenitor A genome (Arachis duranensis). This suggests that bona fide gravitropic genes not only were identified in the Arachis genome, but also were truly expressed during pod development.

In general, fruit development is genetically controlled usually under light conditions, while peanut fruit development occurs in darkness (subterranean fructification) with peg elongation responding conversely to light/dark conditions (Shlamovitz et al., 1995; Smith, 1950). In this study, we identified 245 expressed genes associated with photomorphogenesis using similarity search against 274 photomorphogenesis-related genes previously identified in Arabidopsis or falling under the GO category 'photomorphogenesis'(GO:0009640). These genes were identified to be enriched for a broad array of GO categories associated with photomorphogenesis (Figure 5a). We also found two genes involved in skotomorphogenesis and six in de-etiolation. Expression of identified photomorphogenesis-related genes was grouped around four patterns (Figure 5b). Majority of genes were up-regulated in P2SH and P10SH, suggesting that these genes may play roles in the regulation of peanut pod morphogenesis in subterranean darkness.

Conclusion

Collectively, through generating a highly resolved and extensive transcriptome map of developing peanut pod, we have set up a solid framework for a systemic approach to understand fruit development underground. We have shown that the peanut fruit, with its prominent characteristics of subterranean development, would provide great potential to elucidate this distinctive fruit development – 'aerial flower, subterranean fruit', as well as explore the fruit development under dark conditions.

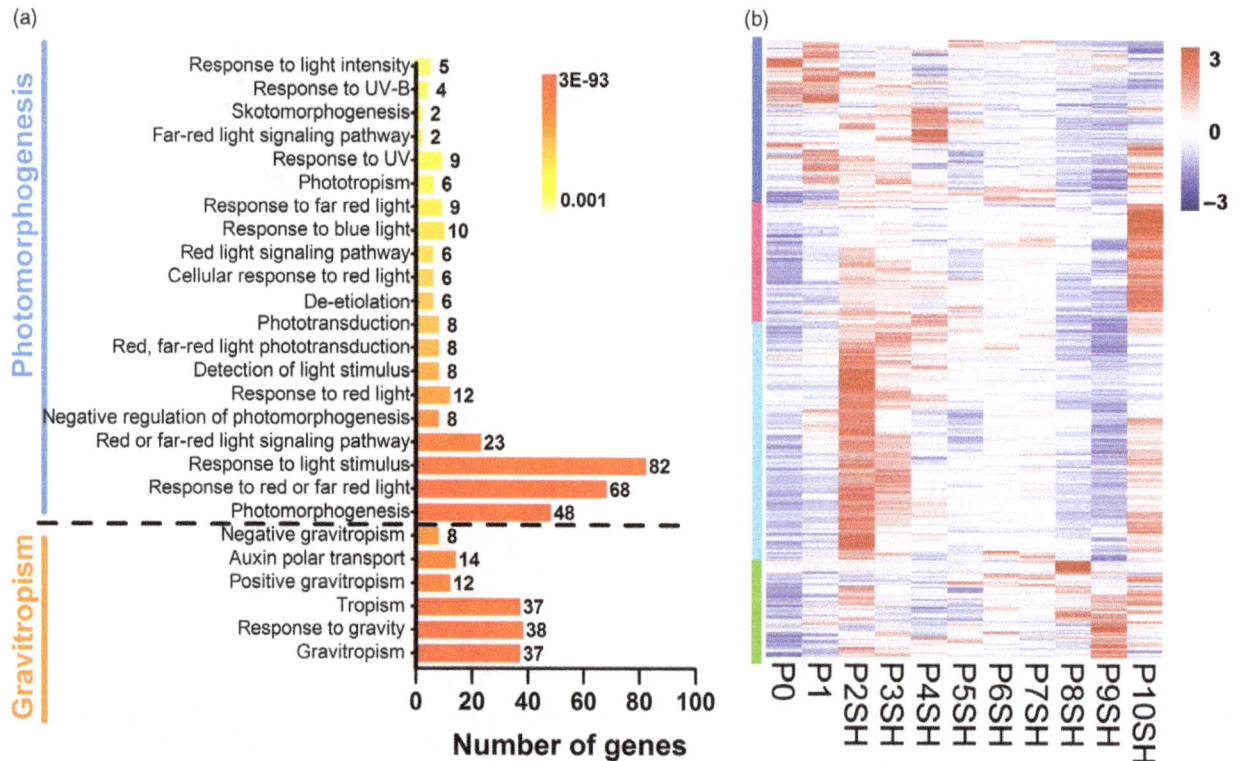

Figure 5 Expressed genes related to gravitropism and photomorphogenesis. (a) GO enrichment analysis of identified genes. Bars with numbers showed the number of genes involved in a particular GO category. Colours indicated P-value of enrichment analysis (yellow, $P < 0.001$; orange, $P < 3E-93$). (b) Heatmap showing distinct expression profiles of identified genes related to gravitropism and photomorphogenesis.

Materials and Methods

Plant materials and RNA isolation

Plants of 'Yueyou 7', a widespread cultivar in the southern China, were grown in fields in the summer (March–July, 2011) at the experimental station of Guangdong Academy of Agricultural Sciences. Aerial and subterranean pods were collected from plants grown in the field. Selfed flowers were identified with coloured plastic thread, and elongating aerial pegs were tied with coloured tags on the eighth day after flowering (DAF). The samples were from two stages of whole pods (aerial pod and subterranean pod, not swelling) and nine stages of isolated pod walls. The stages were mainly defined by pod diameter and seed diameter with days after flowering as reference. The pod diameter was determined at the base of the pod adjacent to the gynophore, and the seed located at the base of the pod was used to measure the diameter. The fruits were classified based on size rather than time after flowering because it is more reproducible in various growing conditions. The detailed information for sample collection was provided in Table S1. Total RNA was extracted using the RNeasy Plant Mini Kit (Qiagen). The quality and quantity of each RNA sample were assayed using NanoDrop (Thermo Scientific) and the Agilent 2100 Bioanalyzer (Agilent).

Illumina sequencing and reference transcriptome

RNA sequencing was performed at MacroGen Inc. (www.macrogen.com) using Illumina HiSequation 2000 platform. Library construction and sequencing followed the standard sequencing protocols recommended by Illumina. DNA fragments in the size for each library were listed in Table S2. We employed a combined

assembly strategy (Martin and Wang, 2011) to yield a peanut reference transcriptome sequence. First, reads were aligned to the peanut progenitor (*Arachis duranensis*) genome (unpublished data) using TopHat (Trapnell *et al.*, 2009) with default arguments. Second, aligned reads were assembled using Trinity with a genome-guided strategy (–genome_guided_max_intron 10000 – max_memory 64G –CPU 16). Third, *de novo* assembly was performed on unaligned reads using Trinity (–max_memory 64G – CPU 16 –bflyCPU 16). Fourth, both assemblies by Trinity were used for a super assembly using TGICL (version 2.0; http://sourceforge.net/projects/tgicl). Finally, redundancy and isoforms were removed using CD-HIT-EST (Li and Godzik, 2006) and the longest transcripts were retained, generating a reference transcriptome sequence.

RNA-seq data alignment and transcriptome profile

We used Bowtie 2 (Langmead and Salzberg, 2012) to map paired-end reads (PEs) to the reference transcriptome, allowing at most five mismatches. Reads with multiple matches are removed from the primary search results. For each pair of forward and reverse reads, we required that both ends should uniquely map to the same transcript. After these filtering, we collected a set of uniquely mapped pairs for the subsequent abundance estimation. Using the uniquely mapped read pairs, we estimated the expression levels of transcripts using Fragments Per Kilobase of transcript per Million mapped fragments (FPKM) (Trapnell *et al.*, 2010) in a way similar to reads per kilobase of exons per million mapped reads (RPKM) (Mortazavi *et al.*, 2008). The upper and lower boundary values of a FPKM 95% confidence interval were calculated using the R stats package. A gene was defined as expressed in a given sample if its low confidence boundary value was greater than zero. A gene was considered lowly expressed if the FPKM value was ≤2, moderately expressed if the FPKM value was >2 and ≤10, and highly expressed if the FPKM value was >10.

Identification of stage-specific gene sets

The genes specifically expressed at each stage were identified using a single statistical analysis, τ value, which was developed by Yanai et al. (2005) for tissue specificity index:

$$\tau_i = \sum_{i=1}^{N} (1 - R_{(i,j)}/R_{(i,max)})/(N-1)$$

where N is the number of samples, R(i, j) is the expression value of i gene in j sample, and R(i, max) is the maximal value of gene i in all samples examined. The τ values range from 0 to 1, and the higher the τ value of a gene for a stage, the more likely the genes is specifically expressed at that stage.

Assessment of gene expression

In this study, we employed two programs for the identification of DE transcripts. Transcripts that were defined as DE genes by both programs were considered to be DE transcripts. First, DE transcripts were calculated from different samples using an R package (DEGseq) proposed by Wang et al. (2010). The DEGseq analysis was performed on read counts of expressed genes as defined above. For each gene, the P-value and Q-value were calculated. The significant threshold to control the FDR at a given value was computed. Then, we identified DE transcripts using the GFOLD package (Feng et al., 2012). GFOLD assigned reliable statistics for expression changes based on the posterior distribution of log fold change. We used an in-house Perl script to extract DE transcripts from the output files generated by both programs. DE transcripts that were detected by only one program were ignored and not used for further analysis.

Functional annotation and analysis

Expressed genes were subjected to BLASTX analysis against the following databases: (i) Uniprot Viridiplantae database for deducing putative function; (ii) the Arabidopsis protein database (TAIR10_pep_20101214_update ftp://ftp.arabidopsis.org) for GO enrichment analysis. An E-value threshold of 1E-5 was used to determine the significant hits. The putative functions of query transcripts were defined by the first subject hits. An in-house Perl script was used to perform gene ontology (GO) annotation based on UniProtKB GOA file (ftp://ftp.ebi.ac.uk). Identification of significantly enriched gene ontology (GO) categories (P-value ≤ 0.05) was performed using the Cytoscape (version 3.0) with a plugin BiNGO (Maere et al., 2005; Su et al., 2014).

qRT-PCR analysis

To validate RNA-seq results, quantitative real-time RT-PCRT was conducted as previously described (Chen et al., 2013). All assays for a particular gene were performed in triplicate synchronously under identical conditions. All qRT-PCR experiments were run in a 25-μl volume with the Roche LightCyler 480 system (Roche). The actin gene was used as a reference. Relative quantification analyses of all target genes were performed using the E (Efficiency)-method from Roche Applied Science (Tellmann and Geulen, 2006). The expression level of each target gene was normalized to the level of the reference gene. The relative expression values were then validated for the RNA-seq data.

Acknowledgements

This research was funded by grants from Science and Technology Planning Project of Guangdong Province (No. 2012B050700007, 2011B010500019, 2013B050800021), Pearl River Science and Technology Nova of Guangzhou (No. 2011J2200035, 2013J2200088), National Natural Science Foundation of China (No. 31200155, 31501246, 31271767) and Natural Science Foundation of Guangdong Province (S2013020012647) and supported by the earmarked fund for Modern Agro-industry Technology Research System (CARS-14). The founders had no role in study design, data collection and analysis, decision to publish or preparation of the manuscript. We declare no conflict of interests.

References

Bennett, M.J., Marchant, A., Green, H.G., May, S.T., Ward, S.P., Millner, P.A., Walker, A.R. et al. (1996) Arabidopsis AUX1 gene: a permease-like regulator of root gravitropism. Science, **273**, 948–950.

Bi, Y.P., Liu, W., Xia, H., Su, L., Zhao, C.Z., Wan, S.B. and Wang, X.J. (2010) EST sequencing and gene expression profiling of cultivated peanut (Arachis hypogaea L.). Genome, **53**, 832–839.

Caspar, T. and Pickard, B.G. (1989) Gravitropism in a starchless mutant of Arabidopsis : Implications for the starch-statolith theory of gravity sensing. Planta, **177**, 185–197.

Chen, X., Hong, Y., Zhang, E., Liu, H., Zhou, G., Li, S., Zhu, F. et al. (2012) Comparison of gene expression profiles in cultivated peanut (Arachis hypogaea) under strong artificial selection. Plant Breeding, **131**, 620–630.

Chen, X., Zhu, W., Azam, S., Li, H., Zhu, F., Li, H., Hong, Y. et al. (2013) Deep sequencing analysis of the transcriptomes of peanut aerial and subterranean young pods identifies candidate genes related to early embryo abortion. Plant Biotechnol. J. **11**, 115–127.

Eeuwens, C.J. and Schwabe, W.W. (1975) Seed and pod wall development in Pisum sativum, L. in relation to extracted and applied hormones. J. Exp. Bot. **26**, 1–14.

Feng, Q.L., Stalker, H.T., Pattee, H.E. and Isleib, T.G. (1995) Arachis hypogaea plant recovery through in vitro culture of peg tips. Peanut Sci. **22**, 129–135.

Feng, J., Meyer, C.A., Wang, Q., Liu, J.S., Shirley Liu, X. and Zhang, Y. (2012) GFOLD: a generalized fold change for ranking differentially expressed genes from RNA-seq data. Bioinformatics, **28**, 2782–2788.

Friml, J., Wisniewska, J., Benkova, E., Mendgen, K. and Palme, K. (2002) Lateral relocation of auxin efflux regulator PIN3 mediates tropism in Arabidopsis. Nature, **415**, 806–809.

Grabherr, M.G., Haas, B.J., Yassour, M., Levin, J.Z., Thompson, D.A., Amit, I., Adiconis, X. et al. (2011) Full-length transcriptome assembly from RNA-Seq data without a reference genome. Nat. Biotechnol. **29**, 644–652.

Guo, B., Chen, X., Dang, P., Scully, B.T., Liang, X., Holbrook, C.C., Yu, J. et al. (2008) Peanut gene expression profiling in developing seeds at different reproduction stages during Aspergillus parasiticus infection. BMC Dev. Biol. **8**, 12.

Guo, B., Chen, X., Hong, Y., Liang, X., Dang, P., Brenneman, T., Holbrook, C. et al. (2009) Analysis of gene expression profiles in leaf tissues of cultivated peanuts and development of EST-SSR markers and gene discovery. Int. J. Plant Genomics, **2009**, 715605.

Guo, B., Fedorova, N.D., Chen, X., Wan, C., Wang, W., Nierman, W.C., Bhatnagar, D. et al. (2011) Gene expression profiling and identification of resistance genes to Aspergillus flavus Infection in peanut through EST and microarray strategies. Toxins, **3**, 737–753.

Hansey, C.N., Vaillancourt, B., Sekhon, R.S., de Leon, N., Kaeppler, S.M. and Buell, C.R. (2012) Maize (Zea mays L.) genome diversity as revealed by RNA-sequencing. PLoS ONE, **7**, e33071.

Harrison, B.R. and Masson, P.H. (2008) ARL2, ARG1 and PIN3 define a gravity signal transduction pathway in root statocytes. Plant J., **53**, 380–392.

Holbrook, C.C., Kvien, C.K., Rucker, K.S., Wilson, D.M. and Hook, J.E. (2000) Preharvest aflatoxin contamination in drought tolerant and intolerant peanut genotypes. Peanut Sci., **27**, 45–48.

Kato, T., Morita, M.T., Fukaki, H., Yamauchi, Y., Uehara, M., Niihama, M. and Tasaka, M. (2002) SGR2, a phospholipase-like protein, and ZIG/SGR4, a SNARE, are involved in the shoot gravitropism of Arabidopsis. *Plant Cell*, **14**, 33–46.

Klose, H., Gunl, M., Usadel, B., Fischer, R. and Commandeur, U. (2015) Cell wall modification in tobacco by differential targeting of recombinant endoglucanase from Trichoderma reesei. *BMC Plant Biol.* **15**, 54.

Langmead, B. and Salzberg, S.L. (2012) Fast gapped-read alignment with Bowtie 2. *Nat. Methods*, **9**, 357–359.

Langmead, B., Trapnell, C., Pop, M. and Salzberg, S.L. (2009) Ultrafast and memory-efficient alignment of short DNA sequences to the human genome. *Genome Biol.* **10**, R25.

Li, W. and Godzik, A. (2006) Cd-hit: a fast program for clustering and comparing large sets of protein or nucleotide sequences. *Bioinformatics*, **22**, 1658–1659.

Li, P., Ponnala, L., Gandotra, N., Wang, L., Si, Y., Tausta, S.L., Kebrom, T.H. *et al.* (2010) The developmental dynamics of the maize leaf transcriptome. *Nat. Genet.* **42**, 1060–1067.

Li, H., Chen, X., Zhu, F., Liu, H., Hong, Y. and Liang, X. (2013) Transcriptome profiling of peanut (Arachis hypogaea) gynophores in gravitropic response. *Funct. Plant Biol.* **40**, 1249–1260. doi:10.1071/FP13075.

Lister, R., O'Malley, R.C., Tonti-Filippini, J., Gregory, B.D., Berry, C.C., Millar, A.H. and Ecker, J.R. (2008) Highly integrated single-base resolution maps of the epigenome in Arabidopsis. *Cell*, **133**, 523–536.

Lu, T., Lu, G., Fan, D., Zhu, C., Li, W., Zhao, Q., Feng, Q. *et al.* (2010) Function annotation of the rice transcriptome at single-nucleotide resolution by RNA-seq. *Genome Res.* **20**, 1238–1249.

Luo, M., Dang, P., Guo, B.Z., He, G., Holbrook, C.C., Bausher, M.G. and Lee, R.D. (2005) Generation of expressed sequence tags (ESTs) for gene discovery and marker development in cultivated peanut. *Crop Sci.*, **45**, 346–353.

Maere, S., Heymans, K. and Kuiper, M. (2005) BiNGO: a Cytoscape plugin to assess overrepresentation of gene ontology categories in biological networks. *Bioinformatics*, **21**, 3448–3449.

Martin, J.A. and Wang, Z. (2011) Next-generation transcriptome assembly. *Nat. Rev. Genet.* **12**, 671–682.

Morita, M.T., Sakaguchi, K., Kiyose, S., Taira, K., Kato, T., Nakamura, M. and Tasaka, M. (2006) A C2H2-type zinc finger protein, SGR5, is involved in early events of gravitropism in Arabidopsis inflorescence stems. *Plant J.* **47**, 619–628.

Mortazavi, A., Williams, B.A., McCue, K., Schaeffer, L. and Wold, B. (2008) Mapping and quantifying mammalian transcriptomes by RNA-Seq. *Nat. Methods*, **5**, 621–628.

Nicol, F., His, I., Jauneau, A., Vernhettes, S., Canut, H. and Hofte, H. (1998) A plasma membrane-bound putative endo-1,4-beta-D-glucanase is required for normal wall assembly and cell elongation in Arabidopsis. *EMBO J.* **17**, 5563–5576.

Noh, B., Bandyopadhyay, A., Peer, W.A., Spalding, E.P. and Murphy, A.S. (2003) Enhanced gravi- and phototropism in plant mdr mutants mislocalizing the auxin efflux protein PIN1. *Nature*, **423**, 999–1002.

Ono, Y. (1979) Flowering and fruiting of peanut plants. *Jap. Agric. Res. Quart.* **13**, 226–229.

Payton, P., Kottapalli, K.R., Rowland, D., Faircloth, W., Guo, B., Burow, M., Puppala, N. *et al.* (2009) Gene expression profiling in peanut using high density oligonucleotide microarrays. *BMC Genom.*, **10**, 265.

Sahoo, D., Dill, D.L., Tibshirani, R. and Plevritis, S.K. (2007) Extracting binary signals from microarray time-course data. *Nucleic Acids Res.* **35**, 3705–3712.

Scanlon, M. and Timmermans, M. (2013) Growth and development: from genes to networks and a mechanistic understanding of plant development. *Curr. Opin. Plant Biol.* **16**, 1–4.

Sedbrook, J.C., Chen, R. and Masson, P.H. (1999) ARG1 (altered response to gravity) encodes a DnaJ-like protein that potentially interacts with the cytoskeleton. *Proc. Natl. Acad. Sci. USA*, **96**, 1140–1145.

Setia, R.C., Setia, N. and Malik, C.P. (1987) The pod wall structure and function in relation to seed development in some legumes. *Phyton*, **27**, 205–220.

Shlamovitz, N., Ziv, M. and Zamski, E. (1995) Light, dark and growth regulator involvement in groundnut (Arachis hypogaea L.) pod development. *Plant Growth Regul.* **16**, 37–42.

Silady, R.A., Kato, T., Lukowitz, W., Sieber, P., Tasaka, M. and Somerville, C.R. (2004) The gravitropism defective 2 mutants of Arabidopsis are deficient in a

protein implicated in endocytosis in Caenorhabditis elegans. *Plant Physiol.* **136**, 3095–3103; discussion 3002.

Singh, A.K., Mehan, V.K. and Nigam, S.N. (1997) *Sources of resistance to groundnut fungi and bacterial diseases: an update and appraisal.* Information Bulletin No 50 International Crops Research Institute for the Semi-Arid Tropics, Patancheru, Andhra Pradesh, India.

Smith, B.W. (1950) *Arachis hypogaea.* Aerial flower and subterranean fruit. *Am. J. Bot.* **37**, 802–815.

Starr, J.L. and Simpson, C.E. (2006) Improving the utility of nematode resistance in groundnut. *Commun. Agric. Appl. Biol. Sci.*, **71**, 647–651.

Su, G., Morris, J.H., Demchak, B. and Bader, G.D. (2014) Biological network exploration with cytoscape 3. *Curr. Protoc. Bioinformatics*, **47**, 8.13.1–8.13.24.

Swarup, R., Kargul, J., Marchant, A., Zadik, D., Rahman, A., Mills, R., Yemm, A. *et al.* (2004) Structure-function analysis of the presumptive Arabidopsis auxin permease AUX1. *Plant Cell*, **16**, 3069–3083.

Tellmann, G. and Geulen, O. (2006) LightCycler 480 Real-Time PCR system: Innovative solutions for relative quantification. *Biochemica*, **4**, 16–17.

Thorne, J.H. (1979) Assimilate redistribution from soybean pod walls during seed development. *Agron. J.* **71**, 812–816.

Trapnell, C., Pachter, L. and Salzberg, S.L. (2009) TopHat: discovering splice junctions with RNA-Seq. *Bioinformatics*, **25**, 1105–1111.

Trapnell, C., Williams, B.A., Pertea, G., Mortazavi, A., Kwan, G., van Baren, M.J., Salzberg, S.L. *et al.* (2010) Transcript assembly and quantification by RNA-Seq reveals unannotated transcripts and isoform switching during cell differentiation. *Nat. Biotechnol.* **28**, 511–515.

Varaprasad, P.V., Craufurd, D.Q. and Summerfield, R.J. (1999) Sensitivity of peanut to timing of heat stress during reproductive development. *Crop Sci.* **539**, 1352–1357.

Varaprasad, P.V., Craufurd, D.Q. and Summerfield, R.J. (2000) Effect of high air and soil temperature on dry matter production, pod yield and yield components of groundnut. *Plant Soil*, **222**, 231–239.

Wang, H.L. and Grusak, M.A. (2005) Structure and development of Medicago truncatula pod wall and seed coat. *Ann. Bot.* **95**, 737–747.

Wang, Z., Gerstein, M. and Snyder, M. (2009) RNA-Seq: a revolutionary tool for transcriptomics. *Nat. Rev. Genet.* **10**, 57–63.

Wang, L., Feng, Z., Wang, X. and Zhang, X. (2010) DEGseq: an R package for identifying differentially expressed genes from RNA-seq data. *Bioinformatics*, **26**, 136–138.

Wang, T., Chen, X., Li, H., Liu, H., Hong, Y., Yang, Q., Chi, X. *et al.* (2012) Transcriptome identification of the resistance-associated genes (RAGs) to *Aspergillus flavus* infection in pre-harvested peanut (Arachis hypogaea). *Funct. Plant Biol.* **40**, 292–303.

Withers, J.C., Shipp, M.J., Rupasinghe, S.G., Sukumar, P., Schuler, M.A., Muday, G.K. and Wyatt, S.E. (2013) Gravity Persistent Signal 1 (GPS1) reveals novel cytochrome P450s involved in gravitropism. *Am. J. Bot.* **100**, 183–193.

Xia, H., Zhao, C., Hou, L., Li, A., Zhao, S., Bi, Y., An, J. *et al.* (2013) Transcriptome profiling of peanut gynophores revealed global reprogramming of gene expression during early pod development in darkness. *BMC Genom.* **14**, 517.

Yanai, I., Benjamin, H., Shmoish, M., Chalifa-Caspi, V., Shklar, M., Ophir, R., Bar-Even, A. *et al.* (2005) Genome-wide midrange transcription profiles reveal expression level relationships in human tissue specification. *Bioinformatics*, **21**, 650–659.

Yano, D., Sato, M., Saito, C., Sato, M.H., Morita, M.T. and Tasaka, M. (2003) A SNARE complex containing SGR3/AtVAM3 and ZIG/VTI11 in gravity-sensing cells is important for Arabidopsis shoot gravitropism. *Proc. Natl. Acad. Sci. USA*, **100**, 8589–8594.

Yin, D., Wang, Y., Zhang, X., Li, H., Lu, X., Zhang, J., Zhang, W. *et al.* (2013) *De novo* assembly of the peanut (Arachis hypogaea L.) seed transcriptome revealed candidate unigenes for oil accumulation pathways. *PLoS ONE*, **8**, e73767.

Young, L.S., Harrison, B.R., Narayana Murthy, U.M., Moffatt, B.A., Gilroy, S. and Masson, P.H. (2006) Adenosine kinase modulates root gravitropism and cap morphogenesis in Arabidopsis. *Plant Physiol.* **142**, 564–573.

Yuan, S., Wu, Y. and Cosgrove, D.J. (2001) A fungal endoglucanase with plant cell wall extension activity. *Plant Physiol.* **127**, 324–333.

Zhang, J., Liang, S., Duan, J., Wang, J., Chen, S., Cheng, Z., Zhang, Q. *et al.* (2012) *De novo* assembly and characterisation of the transcriptome during seed development, and generation of genic-SSR markers in Peanut (Arachis hypogaea L.). *BMC Genom.*, **13**, 90.

8

Discovery of rare mutations in extensively pooled DNA samples using multiple target enrichment

Xu Chi[1,2], Yingchun Zhang[1], Zheyong Xue[1], Laibao Feng[1], Huaqing Liu[3], Feng Wang[3] and Xiaoquan Qi[1,*]

[1]Key Laboratory of Plant Molecular Physiology, Institute of Botany, Chinese Academy of Sciences, Beijing, China
[2]Graduate University of Chinese Academy of Sciences, Beijing, China
[3]Fujian Provincial Key Laboratory of Genetic Engineering for Agriculture, Fujian Academy of Agricultural Sciences, Fuzhou, China

*Correspondence

email xqi@ibcas.ac.cn

Keywords: NGS, induced mutation detection, large samples pooling, multiplexed target enrichment.

Summary

Chemical mutagenesis is routinely used to create large numbers of rare mutations in plant and animal populations, which can be subsequently subjected to selection for beneficial traits and phenotypes that enable the characterization of gene functions. Several next-generation sequencing (NGS)-based target enrichment methods have been developed for the detection of mutations in target DNA regions. However, most of these methods aim to sequence a large number of target regions from a small number of individuals. Here, we demonstrate an effective and affordable strategy for the discovery of rare mutations in a large sodium azide-induced mutant rice population (F_2). The integration of multiplex, semi-nested PCR combined with NGS library construction allowed for the amplification of multiple target DNA fragments for sequencing. The $8 \times 8 \times 8$ tridimensional DNA sample pooling strategy enabled us to obtain DNA sequences of 512 individuals while only sequencing 24 samples. A stepwise filtering procedure was then elaborated to eliminate most of the false positives expected to arise through sequencing error, and the application of a simple Student's t-test against position-prone error allowed for the discovery of 16 mutations from 36 enriched targeted DNA fragments of 1024 mutagenized rice plants, all without any false calls.

Introduction

Genetic variations that are caused by radiation, chemical mutagens and errors in the process of DNA replication are the basis of genetic diversity. Chemical mutagens, such as ethyl methanesulfonate (EMS), ethylnitrosourea (ENU) and sodium azide, have been widely used in animal and plant mutagenesis (Natarajan, 2005). The alkylating agents, for example EMS, can alkylate guanine bases, leading to the misplacing of a thymine residue over a cytosine residue opposite to the O-6-ethyl guanine by DNA polymerase during replication. Also, sodium azide was believed to be metabolized by the cells to form the mutagenic agent, presumably azidoalanine (Owais and Klein-hofs, 1988), which preferentially create A to G and T to C mutations (Cooper et al., 2008; Olsen et al., 1993; Suzuki et al., 2008; Till et al., 2007). The high frequency of point mutations in the chemical-induced mutagenized population often leads to a large phenotypic variation. A reverse genetic approach, namely target-induced local lesions in genome (TILLING), was developed to identify mutations from this type of mutated population (McCallum et al., 2000). Basically, point mutations of the heteroduplex DNA are recognized and nicked by Cel-I endonuclease (Yang et al., 2000), resulting in a mixture of two shorter DNA fragments in addition to the original PCR products (Colbert et al., 2001). High-performance liquid chromatography (Caldwell et al., 2004), polyacrylamide gel electrophoresis (PAGE) (Raghavan et al., 2007), capillary electrophoresis (Suzuki et al., 2008) and matrix-assisted laser desorption ionization time-of-flight (MALDI-TOF) (Van Den Boom and Ehrich, 2007) have been applied to effectively separate and detect those nicked DNA fragments from highly

mutated mutant populations (Wang et al., 2010). TILLING has been applied for target gene functional analysis in all major crops such as rice (Leung et al., 2001), barley (Caldwell et al., 2004), wheat (Slade et al., 2005), maize (Till et al., 2004), pea (Triques et al., 2007) and soybean (Cooper et al., 2008). But, detection of point mutations from a mutant population consisting of several thousand M_2 individuals by TILLING technology is labour-intensive and time-consuming.

The next-generation sequencing (NGS) technology enables scientists to obtain huge amount of DNA sequence data from a single experiment, providing a potential way to directly detect point mutations in a highly mixed DNA samples in a large-scale population. As NGS remains too costly to allow for routine whole-genome sequencing of large numbers of individuals within a given species (Mamanova et al., 2010), the focus has been to develop target enrichment methods to allow sequencing to be directed at a small number of specific genomic regions. Such strategies are relevant for assessing levels of genetic variation and for developing diagnostic marker assays, where nontarget genomic regions are irrelevant (O'Roak et al., 2012). The three leading NGS target enrichment strategies involved (i) a PCR-based approach, (ii) the use of molecular inversion probes (MIP) (O'Roak et al., 2012) and (iii) microarray-based capture (Nijman et al., 2010). Both the MIP- and microarray-based methods are particularly well suited to the parallel sequencing of large numbers of target DNA fragments, but given the complexity of the enrichment process and the need to construct an NGS library, these methods are not cost- and/or labour-efficient enough to deal with the sort of sample numbers needed for the detection of rare mutations. Instead, a TILLING by sequencing strategy has been proposed, in which NGS was applied to amplicons obtained from a template of pooled DNA samples (Tsai *et al.*, 2011). Such a PCR-based approach is difficult to multiplex, and the many cycles of PCR required to initially achieve the amplification of the target DNA sequences, then subsequently constructing the NGS library, risk the introduction of many artefactual nucleotide changes, leading to an unacceptably high background error rate. Moreover, the uneven pooling strategy of the tridimensional pool ($12 \times 16 \times 16$) lowers the comparability of bulked samples from different dimensions, which resulted in the unresolved false-negative problem.

We present here a strategy in which multiplexed semi-nested PCR is combined with NGS library construction for the sequencing of amplicons derived from tridimensional, evenly pooled DNA samples (Figure 1). The application of semi-nested PCR, which is similar to a nested PCR (Bej *et al.*, 1991; Haqqi

et al., 1988) except that one of the primers in the first PCR is reused in the second PCR, ensures high specificity and efficiency in target enrichment. Integration of semi-nested PCR with NGS library construction allows for a reduction in the number of PCR cycles, thereby greatly diminishing the background mutation rate. A stepwise filtering procedure was then elaborated to eliminate most of the false positives expected to arise through sequencing error while including most of the genuine mutations, and a simple Student's *t*-test was applied to further eliminate position-prone errors. The even number of samples pooled in each bulk ensured the comparability of bulks from different dimensions, which enhanced the accuracy of mutation calling. As a result of this novel strategy, the detection of rare mutations via direct sequencing became both cost-effective and easily manageable in terms of the complexity of the data handling procedures.

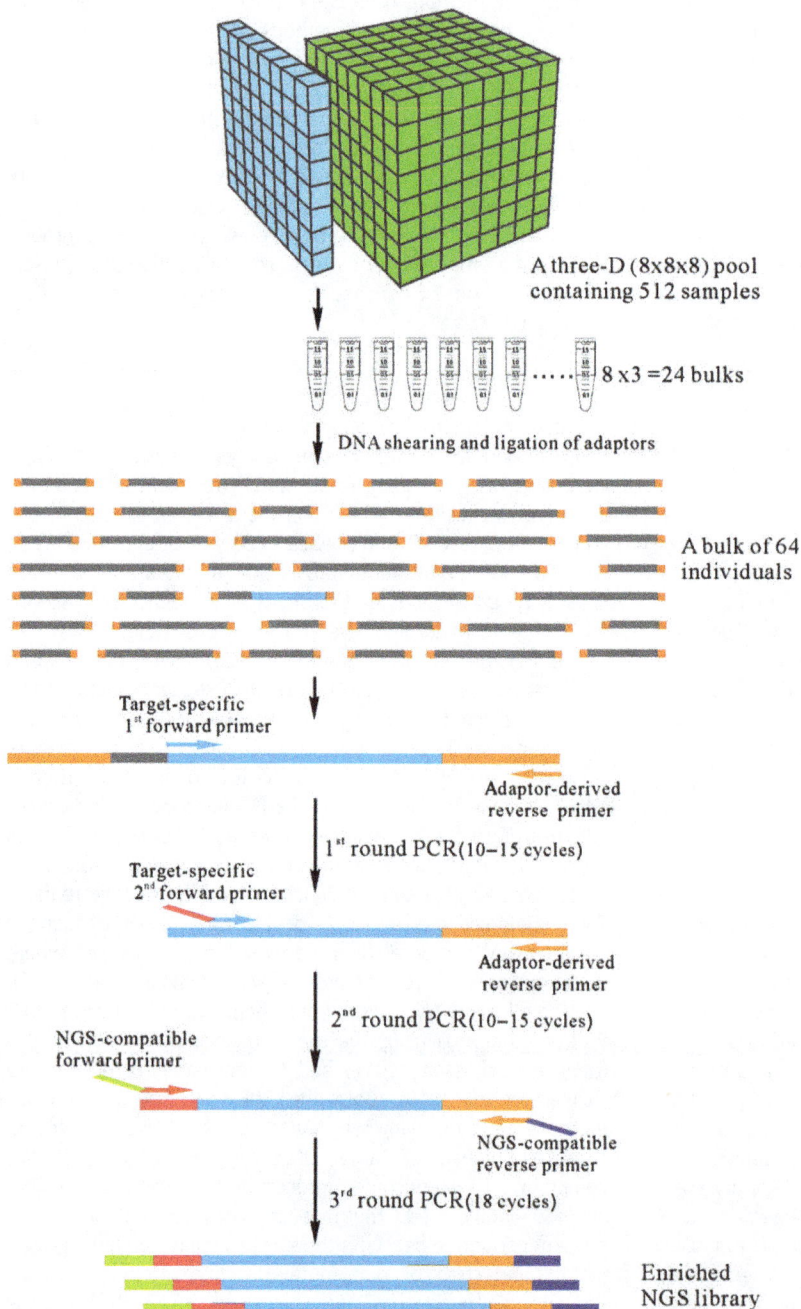

Figure 1 Schematic illustration of the semi-nested PCR-based multiple target enrichment method, based on tridimensional pooling of DNA. Two independent sets of 512 DNA samples were arranged in the form of $8 \times 8 \times 8$ arrays, from each of which 24 DNA bulks, each comprising 64 individuals, were created by pooling equimolar amounts of DNA taken from the samples arrayed in each plane in all three orthogonal directions. The DNA was then sheared, ligated to adaptors and amplified by three rounds of PCR. The first PCR was based on a target-specific forward primer and an adapter-derived reverse primer. The second PCR was based on a target-specific nested primer with a 5′ overhang and the nested adapter-derived reverse primer. The overhang and the adapter that anchored the NGS-compatible forward and reverse primers were used for the third round PCR. Bar codes were embedded within the adaptors. The resulting NGS library consisted of a population of fragments of variable length, each having one fixed and one variable end.

Results

Multiplex enrichment of target DNA from pooled samples

In a pilot experiment, a model template was generated by mixing DNA extracted from five known mutants (C4151T, C7973T, C9835T, C9880T and G9818A) for the rice gene *Os08g12740* with DNA from wild type cv. Zhonghua 11 in a ratio of 1:63, which simulated a mutant allele frequency of 1/64 where the mutation was in the homozygous state or1/128 if heterozygous. The subpools were enriched for the targets by a semi-nested PCR based on a variable number (3, 5, 7, 10 or 13) of target-specific primers (Table S1). When NGS was applied to these five libraries, a bar coding method (Table S1) showed that of the 2.04 Gbp of sequence generated, 19.21% related to C4151T, 18.71% to C7973T, 22.75% to C9835T, 22.82% to C9880T and 16.50% to G9818A. Overall, 54–72% of the high-quality reads are mapped to the targeted sequences (Figure 2a). Encouragingly, the semi-nested PCRs involving ten or fewer primers achieved an even enrichment of all of the target DNA fragments (Figure 2a). The highest on-target ratio (72%) was achieved by the ten primer reaction. With 13 primers, only 11 of the target fragments were enriched, and an uneven enrichment of the target fragments was observed (varying from 3% to 20%). An uneven depth coverage pattern ('cliff' shape), which resulted from overlapping reads derived from both the fixed and random amplified ends, featured in nearly all of the enriched regions (Figure 2b). The length of enriched target fragments ranged from 208 to 431 bp, when a minimum sequencing depth of 512 was applied. In a 10-plex semi-nested PCR, 3563 bp of the target DNA fragments were amplified (Figure 2b). This model experiment demonstrated that combining semi-nested PCR with NGS library construction provided an effective means of enriching for multiple target DNA fragments.

Reducing the number of PCR cycles, for the most part, resulted in a reduction in sequencing error (0.19–0.62×10^{-3}, see Figure 2c), as compared to the rate (0.08–1.3×10^{-3}) observed in a rice TILLING population in a previous study (Tsai *et al.*, 2011). To our surprise, the GGT/ACC motif appeared to be particularly prone to sequencing error (6.21×10^{-3} and 13.53×10^{-3}, respectively) (Figure 2d). As Illumina HiSeq2000-based sequencing uses the same laser wavelength to excite the fluorophore attached to both G and T (Minoche *et al.*, 2011), this may simply be a contamination artefact. The estimated mean frequencies of the known heterozygous and homozygous mutations (5.37–17.32×10^{-3}) were close to their expected values of 7.81 and 15.63×10^{-3} (Figure 2c). Where the sequencing depth was particularly high (i.e. >10 000), all five mutations could be unambiguously distinguished (Figure 2e). It was not possible to discriminate between genuine mutations and sequencing errors at sequencing depths <10 000, even though the true mutation rate was about one order of magnitude higher than the average sequencing error rate (Figure 2f).

Estimation of the mutation rate variation

To discriminate between genuine mutations and sequencing errors, the variation of the mutation rate was analysed based on the data from the pilot experiment. The minimum mutation rate (MiMR, see Experimental procedures section) was proposed to distinguish the genuine mutations from sequencing errors. This criterion was estimated utilizing the genuine mutation data from the pilot experiment. The two heterozygous mutations, C7973T and C9880T, were each sequenced five times, which gave five m/n values (Figure 2e). The respective mean m/n values of C7973T and C9880T were unbiased estimators of X; therefore, they were used to calculate the σ_X (0.0010). The five m/n values of C7973T and C9880T, respectively, were used to calculate their own $\sigma_{m/n}$ (Figure 3). This quantity was assumed to be independent of X, so the two separate $\sigma_{m/n}$ estimates were then averaged, giving $\sigma_{m/n} = 2.8 \times 10^{-4}$. The probability density function of the normal distributions of both X and m/n is presented in Figure 3. The sample size of X was eight (two in the same subpool were obviously higher than expected; therefore, they were excluded from the calculation); therefore, we used $t_{= 0.025, \ df = 7}$ to calculate the one-tail leftborder of X (3.7×10^{-3}). The one-tail left border representing 97.5% of m/n lies at 3.02×10^{-3}. This value was considered as the MiMR threshold, indicating that any position with an m/n below this value was due to sequencing error. Theoretically, 95% (97.5% × 97.5%) of the genuine mutations will remain following the application of this threshold filter.

Sequencing the enriched target DNA from extensively pooled tridimensional samples

In total, 70 468 200 high-quality reads (6.95 Gbp) were obtained from the 48 bulked templates from two pools (Pool A and Pool B), derived from 1024 M_2 individuals of a rice sodium azide-induced mutated population. The mean proportion of on-target reads across all 48 bulks was 57.2%, rising to 73.9% in one case (Table S3); thus, the enrichment achieved was similar to that attained in the pilot experiment (Figure 2a). The average length of the 36 target fragments was 146 bp, ranging from 27 to 289 bp in Pool A and from 11 to 292 bp in Pool B (Figure S1a,b, Tables S4 and S5). The average length of the screened region was 2603 bp for each individual in Pool A and 3162 bp in Pool B (Tables S4 and S5). The number of enriched target fragments in the 24 bulks of Pool A was 732 (84.7%) and 785 (90.9%) for Pool B (Figure S1a, b). The total enriched fragment length of the 512 M_2 progeny was 1.33 Mbp in Pool A and 1.62 Mbp in Pool B. The average sequencing depth was 7002 in Pool A and 8402 in Pool B (see Figure S1c,d).

Application of multiple criteria for rare mutation discovery

As the mutations induced by sodium azide are typically C to T or G to A (Qi *et al.*, 2006), our focus was directed onto the 117 602 Cs and 120 148 Gs which were mapped to the reference sequences (Figure 4a; Tables S6 and S7). The set of criteria, a minimum sequencing depth (MiSD) of 512 and a minimum mutation rate (MiMR) of 3.02×10^{-3}, were applied sequentially, which succeeded in excluding most of the likely sequencing errors (Figure 4a). What remained after the application of the 1X1Y1Z criterion, that is, a candidate mutation was defined by having exactly one base position's mutation rates above 3.02×10^{-3} in each of tridimensional bulks, was a set of 36 Cs and 45 Gs, corresponding to 12 C to T and 15 G to A candidate mutations. Application of a Student's *t*-test enabled the elimination of position-prone errors at the same base position and resulted in a set of 16 highly likely candidate mutations (Figure 4a). Sanger sequencing confirmed that the identified 16 candidates were all the genuine mutations and that the selected 12 false mutations which were eliminated by the *t*-test were indeed false positives (Figure 4b; Table S8).

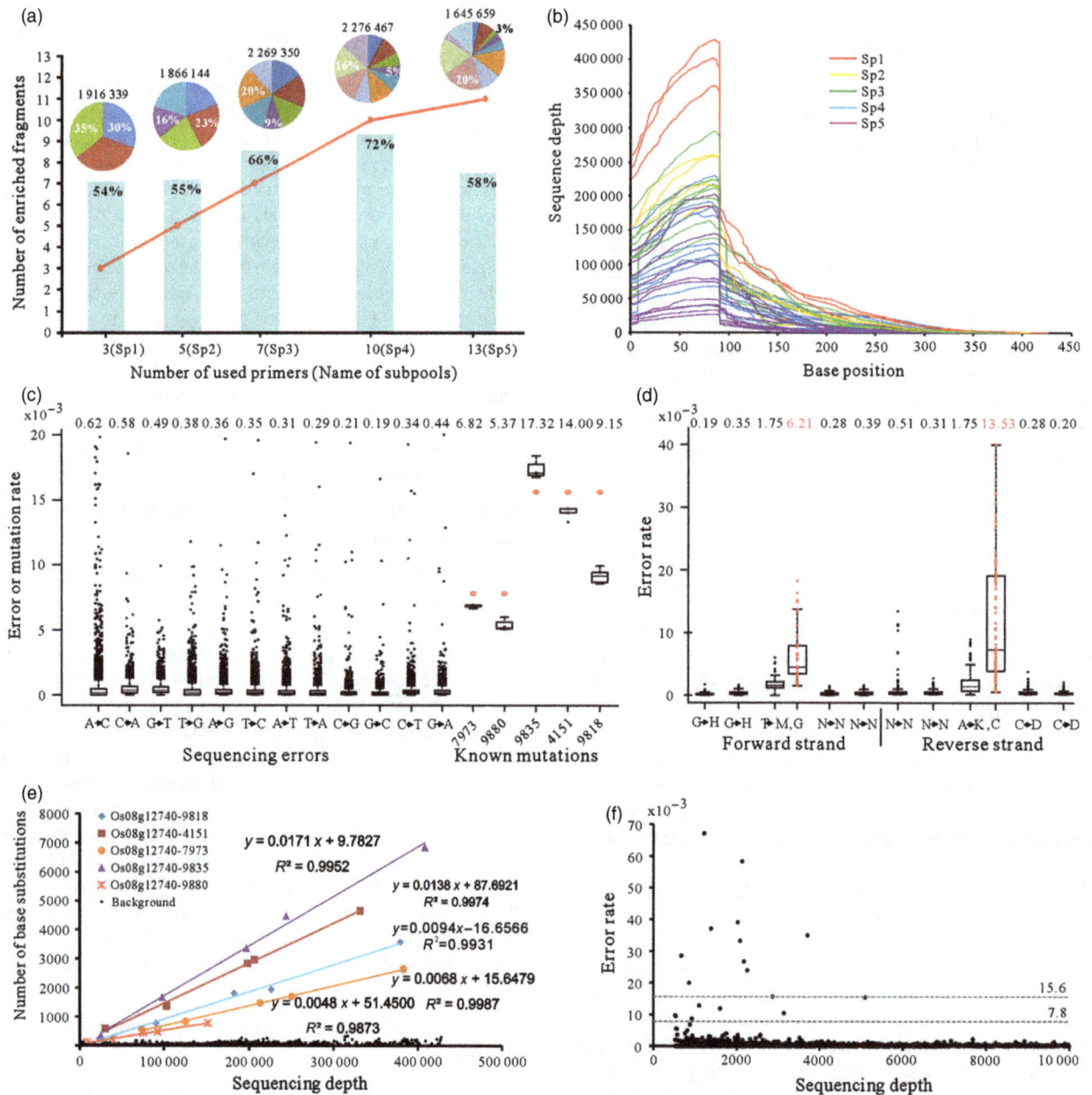

Figure 2 Summary of the pilot experiment. (a) The number of enriched target fragments and the enrichment efficiency. The pie chart at the top shows the total number of reads and the maximum and minimum proportions of reads in the five enriched libraries produced by varying the number of target-specific semi-nested PCR primers. The red line indicates the number of enriched target fragments when 3–13 target-specific semi-nested PCR primers were used to amplify templates of the corresponding subpools. The histogram in cyan indicates the ratio of on-target sequences. (b) Sequencing depth of the enriched target DNA fragments. (c) A box plot showing the rate of both sequencing errors and genuine mutations in the five subpools. The mean rate of each nucleotide substitution type is shown above the box plot. (d) The sequencing error rate in the vicinity of GGT/ACC trinucleotides. (e) The detected base substitution rates of C to T or G to A plotted against sequencing depth. (f) The background sequencing error rate plotted against sequencing depth. The dotted lines indicate expected allele frequencies of 15.6×10^{-3} and 7.8×10^{-3} in a pooled sample comprising, respectively, a single homozygous and a single heterozygous mutant along with 63 wild-type individuals.

We used a very stringent criterion of the 1X1Y1Z to define the candidate mutations. It is possible that more than one genuine mutation is included at the same base position in a bulk which contains 64 highly mutated genomic DNA samples from F_2 individuals, or a few extra-high error rates may have been undistinguishable from genuine mutations. When a maximum cumulative bulk data (MaCBD) of 4, 5 and 6 were applied, three more genuine mutations were identified, but 12 false-positive mutations were included (Table 1). Usage of higher MaCBD

values more than three results in the substantial increase of false-positive ratios to 62% and 46%. The exclusion of these three genuine mutations by the 1X1Y1Z criterion was due to the extra-high error rates of the same base positions in other bulks.

Discussion

The strategy that we used for rare mutation discovery relies on three key attributes. Firstly, the combination of multiplex semi-nested PCR with NGS library construction allowed a 12-plex amplification of target DNA fragments and greatly reduced the total number of PCR cycles. This strategy reduces not only the effort required in the experimental process, but also resulted in the reduction in the background mutation rate. Secondly, the application of a $8 \times 8 \times 8$ tridimensional pooling strategy allowed us to simultaneously screen 512 individuals within each 3-D pool while only actually needing to sequence 24 bulked samples. Thirdly, the development of multiple criteria and the stepwise filtering strategy that gradually increases filtering stringency enabled us to eliminate most of the false positives expected to arise through sequencing errors. We were thus able to unambiguously identify genuine rare mutations.

Multiplexing PCR often results in uneven and unspecific amplification of the targets, which may be due to the different efficiencies in primer binding and extension. Our method reduces the differences of primer binding and increases the amplification specificity by introducing semi-nested PCR. Furthermore, the method minimizes the extension difficulty by fragmenting the template. Without additional instrumentation or procedures, we achieved multiplexing of at least 11 primers with minimum 5% to maximum 16% proportion of enriched targets and enriched 3563-bp target DNA fragments by a single 10-plex nested PCR following with NGS library construction reaction (Figure 2b). The multiplex semi-nested PCR strategy used in this study generated relative short enriched target DNA fragments (208–431 bp). This method is particularly well suited to targeting exon sequences of genes with many introns, which can greatly increase chances of screening coding sequences. However, the length of a screened region is determined by the shortest amplified fragments of the tridimensional (X, Y and Z) pools. Indeed, in our mutation identification experiments, much shorter target fragments (2603 bp for each individual in Pool A and 3162 bp in Pool B by three 12-plex PCR enrichments) were screened. To further increase the screening efficiency, techniques for obtaining longer enriched fragments in all tridimensional pools are required.

Recently, two studies (Chen *et al.*, 2013; Tsai *et al.*, 2011) have reported practical efforts in sequencing extensively pooled samples. In these two papers, single targeted DNA fragment was amplified by PCR, and the amplicons were then used for NGS library construction. The singleplex PCR strategy mainly involves eight steps, that is, PCR amplification of targeted DNA fragments, amplicons purification, quantification and fragmentation, adaptor ligation following column purification, one-round PCR and a gel purification. The strategy used in this study integrates target selection into NGS library construction and includes nine major steps: genomic DNA fragmentation, adaptor ligation, three rounds of PCR, column purification of the ligation and the first two PCR products, and gel purification of the third PCR product. Reducing the number of purification and PCR will simplify the experiment procedure and reduce the cost.

Previous works (Chen *et al.*, 2013; and Tsai *et al.*, 2011) used 3-D pooling strategy in rather complex ways. Chen *et al.* (2013) claimed that they could identify each singleton's pooling pattern

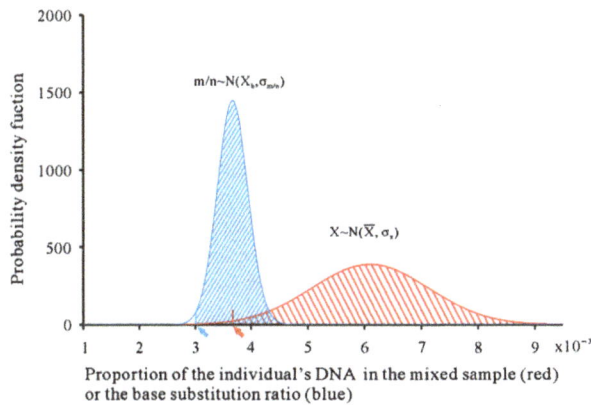

Figure 3 The probability density function of the normal distribution of DNA proportions of genuine mutations in the bulked samples (red area) and the ratio of base substitution numbers in total sequencing depth (the *m/n* ratio, blue area). The red and blue arrows indicate the one-tail left-hand borders equivalent to 97.5% of the DNA proportion and the *m/n* ratio, respectively.

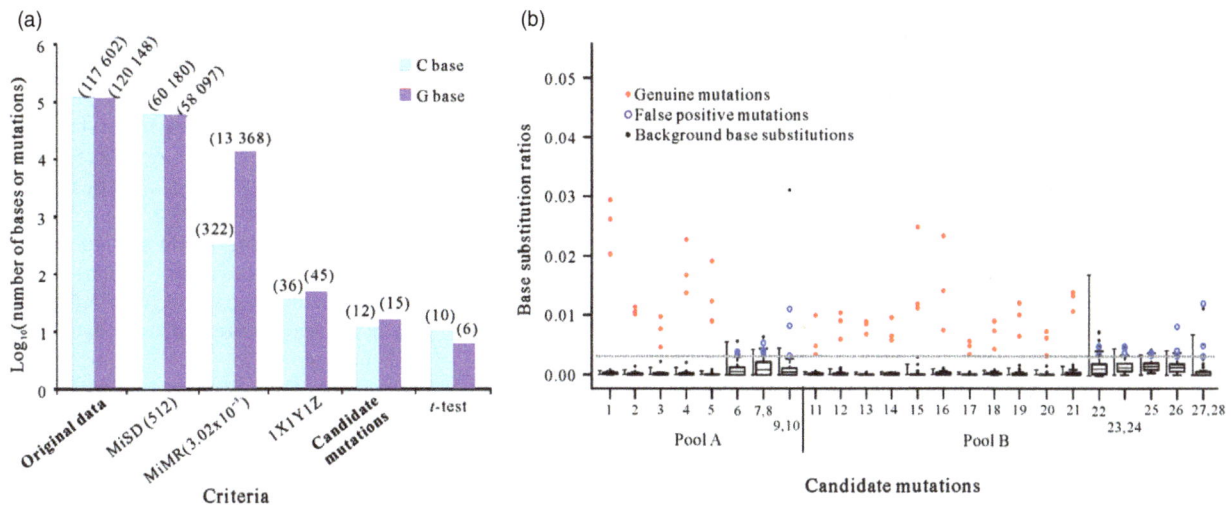

Figure 4 The putative and confirmed candidates detected in the tridimensional pooled DNA samples. (a) The $\log_{10}$ of the number of presumptive mutations remaining after each filtering step (original numbers shown above the relevant column). (b) Validation of the candidate mutations. The grey dotted lines indicate the one-tail left-hand border, equivalent to 95% of the 'true mutations' base substitution rate (3.02×10^{-3}). The confirmed genuine mutations are marked by red dots. Sequencing errors are marked by blue circles.

Table 1 The number of genuine mutations identified using different MaCBD values

| MaCBD | Pool A | | | Pool B | | |
	Number of genuine mutations	Number of false-positive mutations	False-positive rate (%)	Number of genuine mutations	Number of false-positive mutations	False-positive rate (%)
4	6	4	40	12	3	20
5	6	6	50	13	6	32
6	6	10	62	13	11	46

MaCBD, Maximum cumulative bulk data.

without any further experiments. However, there are actually 192 pairs of singletons in their pooling design that have the same pooling pattern (due to the symmetrical positions of each pair in their Pooling 3), which cannot be distinguished from the sequencing result alone. Moreover, the number of mixed samples in each pool did not reach the full pooling capacity; that is, for a $12 \times 12 \times 12$ tridimensional pool, there were only 96 samples in each bulk (if orthogonal pooling was used, there would be $12 \times 12 = 144$ samples in each pool). Therefore, the number of possible patterns of singletons is less than $10 \times 10 \times 10 = 1000 < 384 \times 3 = 1152 < 12 \times 12 \times 12 = 1728$ (pooling of 96 samples in each bulk resulted in less pooling patterns than $10 \times 10 = 100$ samples in each bulk). This is less than the 24^3 claimed by Chen et al., which might be the reason why they cannot actually distinguish every singleton's pooling pattern with their method. It is worth noting that full pooling capacity was not achieved in the Tsai et al. method either. They screened 768 samples by sequencing 44 bulked samples ($12 + 16 + 16 = 44$), using $12 \times 16 \times 16$ tridimensional pooling. In contrast, with our simple $8 \times 8 \times 8$ orthogonal pooling strategy, we sequenced 48 bulked samples to screen 1024 samples of the population, with 9% more bulked samples and 33% more samples screened overall.

The major challenge in detection of rare mutations from large populations is to distinguish genuine mutations from sequencing errors. Therefore, the model for sequencing errors is extraordinarily important for reducing false positives. As pointed out by many previous researches, the sequence context of a given base position can greatly influence the sequencing error rate of that site. However, the algorithm in Chen et al. (2013) did not take sequence context into account (the noise data for the null model was drawn randomly from a pool of all positions' noise data), which resulted in low repeatability of the variant detection (two experimental repeats, one detected 138 putative mutations, the other 161, with a total of 118 overlapped detected putative mutations). Tsai et al. (2011) and Missirian et al. (2011) took the sequence context into account and developed a complex algorithm, but due to their uneven pooling strategies, bulked samples from different dimensions were not comparable, which resulted in an unresolved false-negative problem. In contrast to these particular failings, our stepwise filtering of the data took the advantage of our even number of multidimensional pooling, by comparing the predicted mutation's base substitution rate to all the other 21 predicted noises' sequencing error rates at the same base position (hence fixed the influence of the sequence context) using a simple Student's

t-test. This eliminated all of the false-positive predictions from the former filtering step.

In this study, five genuine mutations were identified through direct sequencing of 1.33 Mbp enriched target DNA fragments among the 512 M_2 progeny in Pool A, and 11 genuine mutations were identified in Pool B (1.62 Mbp enriched target sequences) of a sodium azide-induced rice mutant population, equivalent to a mutation rate of one per 267 and 147 Kbp, respectively. This represents a higher mutation frequency than the one mutation per 500 Kbp reported elsewhere (Till et al., 2007) and showed that our NGS method is likely more effective/sensitive than traditional TILLING.

Compared with most of the current target re-sequencing strategies (such as Rain-dance PCR (Tewhey et al., 2009), array-based hybridization (Nijman et al., 2010) and molecular inverse probe (O'Roak et al., 2012)) that aim at parallel sequencing of tens of thousands of target regions in a small number of individuals, our method is particularly suitable for sequencing of less than a hundred target DNA regions in thousands of individuals. As such, it is particularly well suited to applications such as the identification of target gene mutants from ten thousand individuals in chemical-induced mutant populations of plants or animals. It is important to note that the extensive tridimensional pooling strategy is better used in populations containing random rare mutations, such as the chemical-induced plant F_2 populations. For applications focusing on populations containing inherited rare mutations, such as the detection of disease rare mutations in large human populations, it is recommended to use a more limited pooling strategy that bulks only a few samples, and re-confirmation of rare mutations in the individual samples is required (Comai et al., 2004; Till et al., 2006).

Reduction in sequencing error rate is particularly important for rare mutation detection in the extensively pooled samples. Lowering the sequencing error rate will enable the detection of mutations more accurately and may henceforth allow us to use an even more extensively pooled tridimensional strategy (e.g. $12 \times 12 \times 12$) for sequencing more individuals. Lowering the sequencing error rate would also reduce the required sequencing depth. Coupling our method with the recently proposed duplex sequencing strategy (Schmitt et al., 2013), which eliminates most of the point mutations induced in PCRs, would achieve more cost-effective/labour-effective methods for rare mutation detection.

In summary, through thorough consideration of the critical factors that affect the accuracy of rare mutation detection and carefully designed pooling and multiplexed enrichment strategies, our method achieved a higher efficiency in rare mutation detection in large populations as compared to existing methods. The reductions in cost and improvements in dimensionality (number of individuals screened) will be particularly attractive to research groups who are seeking ways to economically employ and benefit from NGS methods.

Experimental procedures

Generation of rice mutant population

Batches of 25 g of mature caryopses of rice (Oryza sativa L. ssp. japonica) cultivar Zhonghua 11 (obtained from the Chinese Academy of Agricultural Science, Beijing, China) were imbibed in water and then immersed in 2 mM sodium azide for 6 h. The germinating caryopses were rinsed three times for 5 min in water and left to soak for a further 13 h at 30–35 °C, before being planted in a field at Shangzhuang in Beijing. DNA was isolated from each self-fertile individual (M_2 plant), using a standard CTAB extraction method.

Establishment of pooled DNA samples

In the pilot experiment, a 3 µL aliquot of 200 ng/µL DNA extracted from each of the five known *Os08g12740* mutants (C4151T, C7973T, C9835T, C9880T, G9818A) was mixed with 177 µL of 200 ng/µL cv. Zhonghua 11 DNA. A 30 µL aliquot of this subpool DNA was ligated to adaptors and amplified sequentially using appropriate primers (Table S1).

A tridimensional pooling strategy (as shown in Figure 1), where 512 samples were arrayed as a cube (8 × 8 × 8 = 512), was applied in this study. The three dimensions, or directions, were designated as *X*, *Y* and *Z*, respectively. Each sample in this pool was represented as a unique coordinate, (X_i, Y_j, Z_k) [$i, j, k \in (1, 2, 3, ..., 8)$]. The samples located on the same plane were mixed together; for example, samples (X_1, Y_j, Z_k) [$j, k \in (1,2,3,..8)$] containing 64 individuals' DNA samples were bulked together and named as *X*1. As such, samples (X_i, Y_1, Z_k) [$i, k \in (1,2,3,..8)$] were pooled and named as *Y*1, and so on. For each tridimensional pool, there were 24 bulked samples in total. Each 8 bulked samples of the 24 comes from one of three directions (*X*1, *X*2, ..., *X*8 for direction *X*, *Y*1, *Y*2, ..., *Y*8 for direction *Y* and *Z*1, *Z*2,..., *Z*8 for direction *Z*). Therefore, each individual's DNA samples must be present once in each of the three directional bulked samples. For example, an individual (X_3, Y_5, Z_6) would only be present in the bulked samples *X*3, *Y*5 and *Z*6. In this study, two unique tridimensional pools, each containing the DNA of 512 individuals, were established.

Multiple target enrichment and NGS library construction

A 6 µg aliquot of each of the five subpool DNA samples of the pilot experiment (or the 48 bulk DNA of the TILLING population) was fragmented by a 90 s exposure to sonication (40 kHz,80W), and then, the DNA was purified using a QIAquick PCR purification kit using protocols recommended by the manufacturer (Qiagen, Hilden, Germany). The fragments were blunted in a reaction mixture containing 20 µL of 50 ng/µL sonicated DNA, 55 µL ddH$_2$O, 10 µL 10 × T4 DNA ligase buffer containing 10 mM ATP, 4 µL 10 mM dNTPs, 5 µL 3 U/µL T4 DNA polymerase, 1 µL 5 U/µL Klenow fragment and 5 µL 10 U/µL T4 PNK. The reaction mixtures were held at 20 °C for 30 min. The resulting DNA was purified using a Qiagen QIAquick PCR Purification kit using protocols recommended by the manufacturer. An adenine was added to the 3′ end of each fragment by mixing 32 µL of the blunting reaction mixture with 25 µL 10 × NEBuffer4, 10 µL 1 mM dATP and 3 µL 5 U/µL Klenow fragment. This mixture was held at 37 °C for 30 min. To prepare the adaptors for ligation, each oligomer ('long' and 'short', Tables S1 and S2) was diluted to 100 µM, and 25 µL of each dilution was added to 10 µL 10 × NEBuffer4 and 40 µL ddH$_2$O. The DNA was denatured by holding at 95 °C for 2 min, after which the temperature was reduced by 1 °C and held for 36 s for each additional cycle until the temperature reached 24 °C, for the formation of double-stranded DNA. For the ligation reaction, 12 µL of the DNA fragment solution was mixed with 25 µL 2 × quick ligase buffer, 6 µL 25 µM treated adaptors, 5 µL T4 DNA quick ligase and 2 µL ddH$_2$O and held at 20 °C for 15 min. The reaction was finally purified using a MinElute PCR Purification kit, using the protocols recommended by the manufacturer (Qiagen).

The primers for the first two PCRs were presented in an equimolar mixture made up to 2 µM with 1 × TE buffer. In the first round of PCR of the pilot experiment, the forward primers of each subpool consisted of different number of TSPs (target-specific primer, Table S2). The reverse primer was AP1 (adaptor-derived primer), which was used for each of the first round PCRs. In the second round PCRs of the pilot experiment, the forward primers for each subpool were similar to the first round PCR primers, with TSP2-1 instead of TSP 1-1, TSP2-2 instead of TSP1-2,..., TSP2-13 instead of TSP1-13. The reverse primer used in the second round was AP2. In the study of the mutagenized rice population, we used 12 primers in each multiplexed PCR. Therefore, the 36 TSPs were grouped in three sets, with each set containing 12 TSPs. The third round PCR primers were AP3F and AP3R, which were universal for all of the third round PCRs. For the first round PCRs, a 6 µL aliquot of adaptor-ligated DNA was added to 2 µL primer mixture, 0.4 µL of first step universal primer (AP1), 15 µL 2 × Phusion High Fidelity Polymerase PCR Master Mix and 6.6 µL ddH$_2$O; the PCR program comprised 95 °C/60 s denaturation, followed by 12 cycles of 95 °C/20 s, 60 °C/60 s, 68 °C/120 s and an extension step of 68 °C/5 min. The amplicons were purified using a QIAquick PCR purification kit, using protocols recommended by the manufacturer. For the second round PCRs, the template was 6 µL of the amplicon produced by the first PCR, 3 µL primer mixture, 0.6 µL of the second step universal primer (AP2), 15 µL 2 × Phusion High Fidelity Polymerase PCR Master Mix and 5.4 µL ddH$_2$O, and the PCR program comprised 95 °C/60 s denaturation, followed by 13 cycles of 95 °C/20 s, 60 °C/60 s, 68 °C/120 s and followed by an extension of 68 °C/5 min. The resulting amplicons were purified as above. For the third round PCRs, the template was 2 µL of the purified amplicon produced by the second round PCR, 0.6 µL 25 µM of AP3F and AP3R each, 15 µL 2 × Phusion High Fidelity Polymerase PCR Master Mix and 11.8 µL ddH$_2$O. The PCR program comprised 98 °C/30 s denaturation, followed by 18 cycles of 98 °C/10 s, 65 °C/30 s, 72 °C/30 s and then by an extension of 72 °C/5 min. The amplicons were purified using a QIAquick Gel Extraction kit, using protocols recommended by the manufacturer, applying a size selection of 300–500 bp.

Next-generation sequencing and data processing

The resulting library was sequenced using an Illumina HiSeq 2000 platform (paired-end, 2 × 100 bp read length). Low-quality reads (phred quality score lower than 20, that is, error rate >1%) and reads including undefined nucleotides were rejected. The remaining reads were sorted according to the incorporated bar codes (Tables S1 and S2). Then, the bar code sequences were removed by a customized perl script, and the resulting sequences were aligned to the reference sequences of the relevant rice genes using Bowtie 2.0 (a software tool which aligns NGS data with reference sequences, http://bowtie-bio.sourceforge.net/bowtie2/index.shtml). The data files were converted into 'bam' format to allow sorting and pile-up analyses using the software package SAMtools (http://samtools.sourceforge.net/). A customized *perl* script was written to calculate sequencing depth and the number of base substitutions for all possible base substitutions at a given position.

Establishment of multiple criteria for rare mutation discovery

A set of criteria was established to distinguish between genuine mutations and those arising from sequencing errors. This set consisted of the following: (i) a minimum sequencing depth

(MiSD), (ii) a minimum mutation rate (MiMR), (iii) the feature of the 3-D pool that a genuine mutation must be presented in each of the tridimensional bulks, X, Y and Z (defined as 1X 1Y 1Z) and (iv) a Student's t-test that was used to discriminate genuine mutations from position-prone sequencing errors. Our strategy for the selection of these criteria was to maximize the retention of genuine mutations through a stepwise exclusion of sequencing errors.

The MiSD threshold of 512

In a bulk of 64 DNA samples, a heterozygous allele is represented by a variant base in one of 128 of the DNA samples. To balance error-free base calling with the depth of sequencing, the aim was to identify the mutant sequence four times. This required a minimum sequencing depth of $128 \times 4 = 512$.

The MiMR threshold of 3.02×10^{-3}

The data for each base position of one of the 48 bulks consisted of two major elements: the number of sequencing depth (the number of times for this base to be sequenced, here designated as n) and the number of base substitutions (here designated as m). Theoretically, for a heterozygous mutation site, the m/n ratio should be 1/128, equalling the proportion of the mutated individual's DNA. But in practice, there were biases in DNA mixing, which could cause a deviation to the mutation rate. For a heterozygous mutation site, if we designated the proportion of the mutated individual's DNA in the mixture to be X, then it was reasonable to assume that X follows a normal distribution (caused by DNA handling), that is,

$$X \sim N\left(\frac{1}{128}, \sigma_X\right)$$

and σ_X was unknown. After the DNA samples were mixed, X of each mutated individual was fixed, respectively, but the library construction step caused the mutation rate (m/n) to vary from X. Assuming m/n follows a normal distribution, then

$$\frac{m}{n} \sim N(X, \sigma_{m/n})$$

As the library construction and sequencing process were conducted in parallel for each individual's DNA, it was logical to assume that $\sigma_{m/n}$ stays constant for each individual. Estimation of σ_X, X and $\sigma_{m/n}$ allowed us to calculate a minimum mutation rate under a certain confident level (in this case, $97.5\% \times 97.5\% = 95\%$).

The 1X1Y1Z

A consequence of the three-dimensional pooling strategy was that a genuine mutation (as opposed to a sequencing error) should be represented once in each of the three-dimensional bulks, that is, one in bulk X, one in bulk Y and one in bulk Z, of the total 24 (8 Xs + 8 Ys + 8 Xs) bulks of the tridimensional pool. The 1X 1Y 1Z criterion is a high-stringency filter that allows only exactly matched positions to pass through, prior to further statistical analysis.

Student's t-test of significance

The DNA sequence context generally has certain influences on sequencing error rates. A few error-prone motifs, such as the GGT/ACC motif, that were identified in this study can be excluded in the analysis when it was predicted or identified. However, it is still hard to deal with a base position where the error rate is close to that of a genuine mutation. For each identified candidate mutation, a nucleic acid base position-based t-test was applied. The 24 m/n ratios of the tridimensional pool at this particular base position were grouped as that of three from candidate genuine mutations and 21 from randomized sequencing error. The t-test was introduced with the null hypothesis that one of the m/n ratios from the identified candidate mutation was due to a sequencing error. The one-tail right-hand border of 99% sequencing errors' m/n ratio was calculated for each of the candidate positions. Any m/n ratio that was smaller than this value was not considered to be significantly different from a sequencing error and was excluded, while whose base positions having mutation rates significantly different from the cut-off values were used to identify the candidate genuine mutations.

Acknowledgement

We thank Xuan Li, John Hugh Snyder and Saleha Bakht for discussion of the manuscript, Yan Yan, Yongzhen Sun and Haiyin Wang for the help on the statistics and the *perl* script.

Funding

This work was supported by the fund of National Transgenic Megaproject of China [2013ZX08009001] and the Key Project of Chinese National Programs for Fundamental Research and Development [2013CB127000].

References

Bej, A.K., Mahbubani, M.H. and Atlas, R.M. (1991) Amplification of nucleic acids by polymerase chain reaction (PCR) and other methods and their applications. *Crit. Rev. Biochem. Mol. Biol.* **26**, 301–334.

Caldwell, D.G., McCallum, N., Shaw, P., Muehlbauer, G.J., Marshall, D.F. and Waugh, R.A. (2004) Structured mutant population for forward and reverse genetics in Barley (*Hordeum vulgare* L.). *Plant J.* **40**, 143–150.

Chen, C.T., McDavid, A.N., Kahsai, O.J., Zebari, A.S. and Carlson, C.S. (2013) Efficient identification of rare variants in large populations: deep re-sequencing the CRP locus in the CARDIA study. *Nucleic Acids Res.* **41**, e85.

Colbert, T., Till, B.J., Tompa, R., Reynolds, S., Steine, M.N., Yeung, A.T., McCallum, C.M., Comai, L. and Henikoff, S. (2001) High-ghroughput screening for induced point mutations. *Plant Physiol.* **126**, 480–484.

Comai, L., Young, K., Till, B.J., Reynolds, S.H., Greene, E.A., Codomo, C.A., Enns, L.C., Johnson, J.E., Burtner, C., Odden, A.R. and Henikoff, S. (2004) Efficient discovery of DNA polymorphisms in natural populations by Ecotilling. *Plant J.* **37**, 778–786.

Cooper, J.L., Till, B.J., Laport, R.G., Darlow, M.C., Kleffner, J.M., Jamai, A., El-Mellouki, T., Liu, S., Ritchie, R., Nielsen, N., Bilyeu, K.D., Meksem, K., Comai, L. and Henikoff, S. (2008) TILLING to detect induced mutations in soybean. *BMC Plant Biol.* **8**, 9.

Haqqi, T.M., Sarkar, G., David, C.S. and Sommer, S. S. (1988) Specific amplification with PCR of a refractory segment of genomic DNA. *Nucleic Acids Res.* **16**, 11844.

Leung, H., Wu, C., Baraoidan, M., Bordeos, A., Ramos, M., Madamba, S., Cabauatan, P., Vera Cruz, C., Portugal, A., Reyes, G., Bruskiewich, R., Mclaren, G., Lafitte, R., Gregorio, G., Bennett, J., Brar, D., Khush, G., Schnable, P, Wang, G. and Leach, J. (2001) Deletion mutants for functional genomics: progress in phenotyping, sequence assignment, and database development. *Rice Genet.* **4**, 239–251.

Mamanova, L., Coffey, A.J., Scott, C.E., Kozarewa, I., Turner, E.H., Kumar, A., Howard, E., Shendure, J. and Turner, D.J. (2010) Target-enrichment strategies for next-generation sequencing. *Nat. Methods,* **7**, 111–118.

McCallum, C.M., Comai, L., Greene, E.A. and Henikoff, S. (2000) Targeted screening for induced mutations. *Nat. Biotechnol.* **18**, 455–457.

Minoche, A.E., Dohm, J.C. and Himmelbauer, H. (2011) Evaluation of genomic high-throughput sequencing data generated on Illumina HiSeq and Genome Analyzer systems. *Genome Biol.* **12**, R112.

Missirian, V., Comai, L. and Filkov, V. (2011) Statistical mutation calling from sequenced overlapping DNA pools in TILLING experiments. *BMC Bioinformatics*, **12**, 287.

Natarajan, A.T. (2005) Chemical mutagenesis: from plants to human. *Curr. Sci. India*, **89**, 2.

Nijman, I.J., Mokry, M., Boxtel, R., Toonen, P., Bruijn, E. and Cuppen, E. (2010) Mutation discovery by targeted genomic enrichment of multiplexed barcoded samples. *Nat. Methods*, **7**, 913–915.

Olsen, O., Wang, X. and Von Wettstein, D. (1993) Sodium azide mutagenesis: preferential generation of A·T→G·C transitions in the barley *Ant18* gene. *Proc. Natl Acad. Sci. USA*, **90**, 8043–8047.

O'Roak, B.J., Vives, L., Fu, W., Egertson, J.D., Stanaway, I.B., Phelps, I.G., Carvill, G., Kumar, A., Lee, C., Ankenman, K., Munson, J., Hiatt, J.B., Turner, E.H., Levy, R., O'Day, D.R., Krumm, N., Coe, B.P., Martin, B.K., Borenstein, E., Nickerson, D.A., Mefford, H.C., Doherty, D., Akey, J.M., Bernier, R., Eichler, E.E. and Shendure, J. (2012) Multiplex targeted sequencing identifies recurrently mutated genes in autism spectrum disorders. *Science*, **338**, 1619–1622.

Owais, W.M. and Kleinhofs, A. (1988) Metabolic activation of the mutagen azide in biological systems. *Mutat. Res.* **197**, 313–323.

Qi, X., Bakht, S., Qin, B., Leggett, M., Hemmings, A., Mellon, F., Eagles, J., Werck-Reichhart, D., Schaller, H., Lesot, A., Melton, R. and Osbourn, A. (2006) A different function for a number of an ancient and highly conserved cytochrome P450 family: from essential sterols to plant defense. *Proc. Natl Acad. Sci. USA*, **103**, 18848–18853.

Raghavan, C., Naredo, M.E.B., Wang, H., Atienza, G., Liu, B., Qiu, F., McNally, K.L. and Leung, H. (2007) Rapid method for detecting SNPs on agarose gels and its application in candidate gene mapping. *Mol. Breed.* **19**, 87–101.

Schmitt, M.W., Kennedy, S.R., Salk, J.J., Fox, E.J., Hiatt, J.B. and Loeb, L.A. (2013) Detection of ultra-rare mutations by next-generation sequencing. *Proc. Natl Acad. Sci. USA*, **109**, 14508–14513.

Slade, A.J., Fuerstenberg, S.I., Loeffler, D., Steine, M.N. and Facciotti, D. (2005) A reverse genetic nontransgenic approach to wheat crop improvement by TILLING. *Nat. Biotechnol.* **23**, 75–81.

Suzuki, T., Eiguchi, M., Kumamaru, T., Satoh, H., Matsusaka, H., Moriguchi, K., Nagato, Y. and Kurata, N. (2008) MNU-induced mutant pools and high performance TILLING enable finding of any gene mutation in rice. *Mol. Genet. Genom.* **279**, 213–223.

Tewhey, R., Warner, J.B., Nakano, M., Libby, B., Medkova, M., David, P.H., Kotsopoulos, S.K., Samuels, M.L., Hutchison, J.B., Larson, J.W., Topol, E.J., Weiner, M.P., Harismendy, O., Olson, J., Link, D.R. and Frazer, K.A. (2009) Microdroplet-based PCR enrichment for large-scale targeted sequencing. *Nat. Biotechnol.* **27**, 1025–1031.

Till, B.J., Reynolds, S.H., Weil, C., Springer, N., Burtner, C., Young, K., Bowers, E., Codomo, C.A., Enns, L.C., Odden, A.R., Greene, E.A., Comai, L. and Henikoff, S. (2004) Discovery of induced point mutations in maize genes by TILLING. *BMC Plant Biol.* **4**, 12.

Till, B.J., Zerr, T., Bowers, E., Greene, E.A., Comai, L. and Henikoff, S. (2006) High-throughput discovery of rare human nucleotide polymorphisms by Ecotilling. *Nucleic Acids Res.* **34**, e99.

Till, B.J., Cooper, J., Tai, T.H., Colowit, P., Greene, E.A., Henikoff, S. and Comai, L. (2007) Discovery of chemically induced mutations in rice by TILLING. *BMC Plant Biol.* **7**, 19.

Triques, K., Sturbois, B., Gallais, S., Dalmais, M., Chauvin, S., Clepet, C., Aubourg, S., Rameau, C., Caboche, M. and Bendahmane, A. (2007) Characterization of *Arabidopsis thaliana* mismatch specific endonucleases: application to mutation discovery by TILLING in pea. *Plant J.* **51**, 1116–1125.

Tsai, H., Howell, T., Nitcher, R., Missirian, V., Watson, B., Ngo, K.J., Lieberman, M., Fass, J., Uauy, C., Tran, R.K., Khan, A.A., Filkov, V., Tai, T.H., Dubcovsky, J. and Comai, L. (2011) Discovery of rare mutations in populations: TILLING by sequencing. *Plant Physiol.* **156**, 1257–1268.

Van Den Boom, D. and Ehrich, M. (2007) Discovery and identification of sequence polymorphism and mutations with MALDI-TOF MS. *Methods Mol. Biol.* **366**, 287–306.

Wang, T., Uauy, C., Till, B. and Liu, C.M. (2010) TILLING and associated technologies. *J. Integr. Plant Biol.* **52**, 1027–1030.

Yang, B., Wen, X., Kodali, N.S., Oleykowski, C.A., Miller, C.G., Kulinsi, J., Besack, D., Yeung, J.A., Kowalski, D. and Yeung, A.T. (2000) Purification, cloning, and characterization of the CEL I nuclease. *Biochemistry*, **39**, 3533–3541.

9

Targeted mutagenesis of a conserved anther-expressed P450 gene confers male sterility in monocots

A. Mark Cigan[1,*], Manjit Singh[1], Geoffrey Benn[1,†], Lanie Feigenbutz[1], Manish Kumar[1], Myeong-Je Cho[2], Sergei Svitashev[1] and Joshua Young[1]

[1]Trait Technologies, DuPont Pioneer, Johnston, IA, USA
[2]Trait Technologies, DuPont Pioneer, Hayward, CA, USA

*Correspondence
email mark.cigan@genusplc.com
present address: Genus plc, 1525 River Road, DeForest, WI, USA.
†Present address: Department of Plant Biology, University of California, One Shields Avenue, Davis, CA, USA.

Keywords: endonuclease, male sterile, rice, sorghum, wheat.

Summary

Targeted mutagenesis using programmable DNA endonucleases has broad applications for studying gene function *in planta* and developing approaches to improve crop yields. Recently, a genetic method that eliminates the need to emasculate the female inbred during hybrid seed production, referred to as Seed Production Technology, has been described. The foundation of this genetic system relied on classical methods to identify genes critical to anther and pollen development. One of these genes is a P450 gene which is expressed in the tapetum of anthers. Homozygous recessive mutants in this gene render maize and rice plants male sterile. While this P450 in maize corresponds to the male fertility gene *Ms26*, male fertility mutants have not been isolated in other monocots such as sorghum and wheat. In this report, a custom designed homing endonuclease, Ems26+, was used to generate *in planta* mutations in the rice, sorghum and wheat orthologs of maize *Ms26*. Similar to maize, homozygous mutations in this P450 gene in rice and sorghum prevent pollen formation resulting in male sterile plants and fertility was restored in sorghum using a transformed copy of maize *Ms26*. In contrast, allohexaploid wheat plants that carry similar homozygous nuclear mutations in only one, but not all three, of their single genomes were male fertile. Targeted mutagenesis and subsequent characterization of male fertility genes in sorghum and wheat is an important step for capturing heterosis and improving crop yields through hybrid seed.

Introduction

Crop improvement on a global scale is required to meet growing food demands (Parry and Hawkesford, 2012; Ray et al., 2013). The expansion of yield gains made by developing hybrid crops, rather than varieties, would be an important step to achieve such progress where harnessing the advantages of heterosis is the main objective of hybrid breeding (Duvick, 1999; Melchinger, 1999; Tester and Landrige, 2010). Genetic divergence between parent lines has been demonstrated to correlate with the extent to which heterosis of a hybrid increases. Hybrid seed production requires that a male parent inbred pollinates a genetically distinct female inbred parent that is either male sterile or has been emasculated. This is straightforward in monoecious plants such as maize that have separate male (tassel) and female (ear) flowers on the same plant, where detasseling can emasculate the female parent during hybrid seed production. While detasseling has been used for hybrid maize seed production, physical emasculation is not a scalable solution for crop species with perfect flowers such as rice, sorghum, barley, millet and wheat. Historically, cytoplasmic male sterility (CMS) has been used in these crops for hybrid seed production; however this method is not effective in all germplasm due to the dependence on nuclear restorer genes specific for a given cytoplasm (Dwivedi et al., 2008; Li and Yuan, 1999; Reddy et al., 2005; Whitford et al., 2013). Recently, a novel hybrid seed production system has been described for maize that is not limited by the genetic background (Wu et al., 2015). This Seed Production Technology (SPT) system involves restoring pollen formation while maintaining a homozygous nuclear recessive mutation in the maize male fertility gene *Ms45*. Although homozygous *ms45* maize plants are male sterile, pollen formation can be restored in plants that contain a transformed version of the *Ms45* gene. In the SPT system, only the *ms45* allele and not the Ms45 restoration cassette is transmitted through pollen. The resultant *ms45* progeny are used as females during hybrid production; thus, emasculation by detasseling is not necessary as these females are genetically male sterile. Although this system originally incorporated the maize *Ms45* gene, it is applicable to other crops and to other nuclear male sterile mutations, whether dominant or recessive.

Extending this hybrid seed production system to other monoecious crops could afford the ability to maximize yield potentials through heterosis despite important differences such as asynchronous flower development and architecture. This is particularly relevant for rice, sorghum and wheat. However, deployment of a genetically based hybrid seed production system in these crops requires the identification and isolation of male fertility genes. While male fertility genes and mutants have been described and isolated in maize (Cigan et al., 2001; Neuffer et al., 1997) and rice (Guo and Liu, 2012; Yamagata et al., 2007) by classical genetic approaches, there is a paucity of available mutants in crops such as sorghum and wheat. As opposed to the use of chemicals or irradiation to generate random mutations and the associated time-consuming screening of mutagenized plants, recent progress in DNA double-strand break (DSB) technologies holds great promise for the systematic generation of male sterility mutants in these agriculturally important crops. A variety of nucleases can be employed for the generation of targeted DSBs; zinc-finger nucleases (ZFN), I-CreI-based homing endonucleases (meganucleases), transcription activator-like effector nucleases (TALEN) and the Cas9-guide RNA system (often referred to as 'CRISPR-Cas') have been used to produce mutations in various plant species (Kumar and Jain, 2015; Osakabe and Osakabe, 2015; Svitashev et al., 2015; Voytas and Gao, 2014). DSBs are repaired through endogenous cellular processes of nonhomologous end-joining (NHEJ) or homology-directed repair; the NHEJ repair pathway can often be inexact resulting in mutations at the site of the DSB and in this way targeted mutagenesis is achieved. Such a directed mutagenesis strategy utilizing a redesigned I-CreI homing endonuclease to target the Ms26 male fertility gene in maize has recently been reported (Djukanovic et al., 2013). As demonstrated in rice, the rice Ms26 ortholog encodes for a cytochrome P450 mono-oxygenase enzyme (OsCYP704B2) and is expressed in the tapetum and microspores in the developing anther (Hobo et al., 2008; Li et al., 2010). The homozygous recessive ms26 mutants in maize and rice are defective in both tapetum and microspore development resulting in male sterility (Li et al., 2010; Loukides et al., 1995). The redesigned endonuclease described by Djukanovic et al. (2013), Ems26+, targets a 22-nt sequence found in the 5th exon of Ms26. DSB generated by Ems26+ within the predetermined chromosomal recognition site and NHEJ repair resulted in mutations in the Ms26 gene and loss of function. Maize plants containing homozygous recessive ms26 mutant alleles were phenotypically normal, producing tassels that contained developed spikelets, with the exception that anthers did not shed pollen. These mutant ms26 alleles were transmitted to progeny at the expected segregation ratio (Djukanovic et al., 2013). The male sterile phenotype generated by targeted gene disruption using the Ems26+ endonuclease was indistinguishable from ms26 maize mutants generated by transposon-tagging (Loukides et al., 1995; Fox, T., DuPont Pioneer, unpublished observation).

This report demonstrates that the Ems26+ endonuclease can generate targeted mutations in the orthologs of Ms26 genes of rice (Oryza sativa, Os), sorghum (Sorghum bicolor, Sb) and common wheat (Triticum aestivum, Ta). Orthologous Ms26 mutations in rice and sorghum plants, such as maize (Zea mays L., Zm), confer a recessive male sterile phenotype, and restoration of fertility in these mutant sorghum plants was achieved by a transformed copy of maize Ms26. In addition, using a conditionally regulated Ems26+, mutations in the wheat Ms26 ortholog were recovered in all three genomes of hexaploid spring wheat. These results demonstrate the utility of nuclease-facilitated gene editing to produce male fertility gene mutants for the purpose of unlocking yield potential through exploiting heterosis in these major food crops.

Results

The maize Ms26 gene is a member of a P450 protein family whose expression is limited to developing anthers (Cigan et al., 2005). Biochemical characterization indicated putative roles of the rice Ms26 ortholog (OsCYP704B2) catalysing ω-hydroxylated C16 and C18 fatty acids as well as synthesis and transport of sporopollenin precursor components important to pollen wall formation (Li et al., 2010). Maize and rice plants carrying homozygous recessive mutations disrupting the coding-frame of the Ms26 and OsCYP704B2 genes, respectively, resulted in an inability of these plants to generate fertile pollen grains

(Djukanovic et al., 2013; Li et al., 2010). Consistent with the importance of this P450 for pollen development, protein sequences with high homology to the maize Ms26 were identified in sorghum and wheat by searching plant sequence databases. The sorghum and wheat Ms26-like sequences along with the rice OsCYP704B2 gene were, therefore, referred to herein as SbMs26, TaMs26 and OsMs26, respectively. Although there is significant sequence identity across these crop species (Figure S1 and Text S1), nothing is known regarding the functional conservation of this P450 in sorghum and wheat. As shown in Figure 1, the 22-nt Ems26+ recognition site is conserved across these monocots and resides in the last exon of each gene, and this exon contains the important haem-binding loop common to haem-thiolate cytochrome P450s (Werck-Reichhart and Feyereisen, 2000). To determine whether the Ms26 gene is required for male fertility, several directed mutagenesis approaches using the Ems26+ endonuclease were developed and tested in rice and then used to determine whether this gene is required for male fertility in sorghum and wheat.

Ems26+-generated mutations in rice

Young rice callus [(Oryza sativa L. ssp. Japonica (cv. Kitaake)] initiated from germinating seed was used as transformation targets for biolistic- or Agrobacterium-mediated delivery of vectors that contained the custom endonuclease, Ems26+. For biolistic transformation, two vectors were constructed: (i) a vector containing a codon-optimized Ems26+ under the transcriptional regulation of the maize ubiquitin promoter (UBIpro:Ems26+) enabling constitutive expression, and (ii) a vector containing a herbicide resistance selectable marker phosphinothricin acetyltransferase (PAT) fused to the red fluorescence protein (RFP) gene and placed under the regulation of the maize END2 promoter. This vector allowed transformed rice cells to grow in the presence of the herbicide glufosinate while monitoring red fluorescence, when codelivered into 2-week-old, seed-derived rice callus. Ems26+ contains a nuclear localization signal at the N-terminus and an intron to eliminate expression in bacterial cells used for vector propagation (Djukanovic et al., 2013). Approximately 6 weeks after bombardment, callus sectors that grew on media containing glufosinate were screened for mutations at the 22-nt Ms26 target site. Due to the presence of a BsiWI restriction site (5'-CGTACG-3') across the centre of the Ems26+ recognition sequence and at the site of Ems26+-mediated DNA cleavage (Figure 1), PCR amplification and digestion by BsiWI was used as an initial screening tool to detect Ms26 mutations. PCR amplicons resistant to BsiWI digestion would be indicative of sequence changes at this recognition site due to cutting by Ems26+ and imprecise repair by NHEJ (Djukanovic et al., 2013). Genomic DNA isolated from these glufosinate-resistant callus sectors was amplified using the primer pair UNIMS26 5'-2 and UNIMS26 3'-1 to generate a 653-bp amplicon that was then subjected to digestion with BsiWI. Products of these reactions were electrophoresed on 1% agarose gels and screened; amplicons containing wild-type sequence across the Ems26+ recognition site would produce 376- and 277-bp DNA digestion products (Figure S2, lanes 2, 4). Screening of 292 glufosinate-resistant events generated by cobombardment with the Ems26+ vector identified 22 callus sectors that contained PCR products resistant to BsiWI digestion. Subcloning and DNA sequence analysis of these PCR products revealed a variety of mutations across the Ms26 target site, which included single- and multiple-nucleotide insertions and deletions (Figure 2).

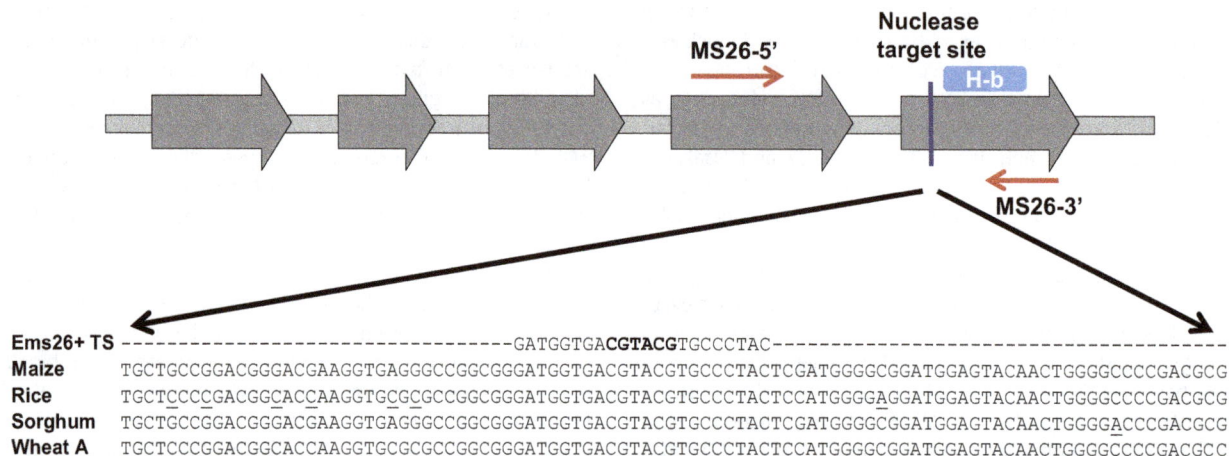

Figure 1 Ems26+ endonuclease target site is conserved across maize, rice, sorghum and wheat. Gene structure illustration of maize *Ms26*; exons (grey arrows) and the haem-binding (H-b) domain are shown. The haem-binding domain resides in the 5th exon of the maize *Ms26* (Gene ID: 100191749) gene and sorghum (Gene ID: 8082128) ortholog. This domain is also found in the last exon of the rice (Gene ID: 4331756) and the wheat (Fielder A-genome) *Ms26* orthologs, which each consists of four exons. The Ems26+ target site is depicted as a vertical line upstream of the H-b domain. Sequence differences in rice, sorghum and wheat A-genome as compared to the maize *Ms26* gene are indicated as underlined nucleotides. The UNIMS26 primer pair (arrows labelled MS26-5′ and MS26-3′) was used to amplify the region of interest from isolated genomic DNA to screen for *Ms26* mutations. A *Bsi*WI restriction enzyme site (red font) present in the 22-nt Ems26+ target site allowed for *Ms26* mutation screening by determining resistance to *Bsi*WI digestion.

To examine the male fertility phenotype of rice plants containing *OsMs26* mutations generated by Ems26+, plants were regenerated from glufosinate-resistant callus sectors, analysed for *OsMs26* mutations and allowed to set self-pollinated seed (T1 seed). T1 seed from three plants containing nonidentical *OsMs26* mutations, but lacking the randomly integrated transformation vectors, were advanced for progeny testing. Using the *Bsi*WI digestion assay, genomic DNA collected from progeny plantlets was amplified with the UNIMS26 primer pair and screened for mutations at the Ms26 target site. Heterozygous *OsMs26/Osms26* (Figure S2, lane 4) and homozygous *Osms26/Osms26* (Figure S2, lane 6) mutant plants were advanced and scored for their ability to form pollen by analysing seed set upon

self-fertilization. Throughout vegetative growth, these plants appeared phenotypically normal and flowered at the same time as wild-type plants with the exception that dehisced anthers were not observed in the homozygous recessive *Osms26* plants (Figure 3a). Microscopic examination of anthers from *OsMs26/Osms26* plants staged at late uninucleate microspore development revealed normal appearing microspores, which continued to develop into mature, fertilization-competent pollen. However, examination of anthers from *Osms26/Osms26* plants revealed that microspores arrested development immediately after tetrad release, whereupon these plants contained anthers devoid of mature pollen (Figure 3b). These observations were similar to those reported by Loukides *et al.* (1995) and Li *et al.* (2010).

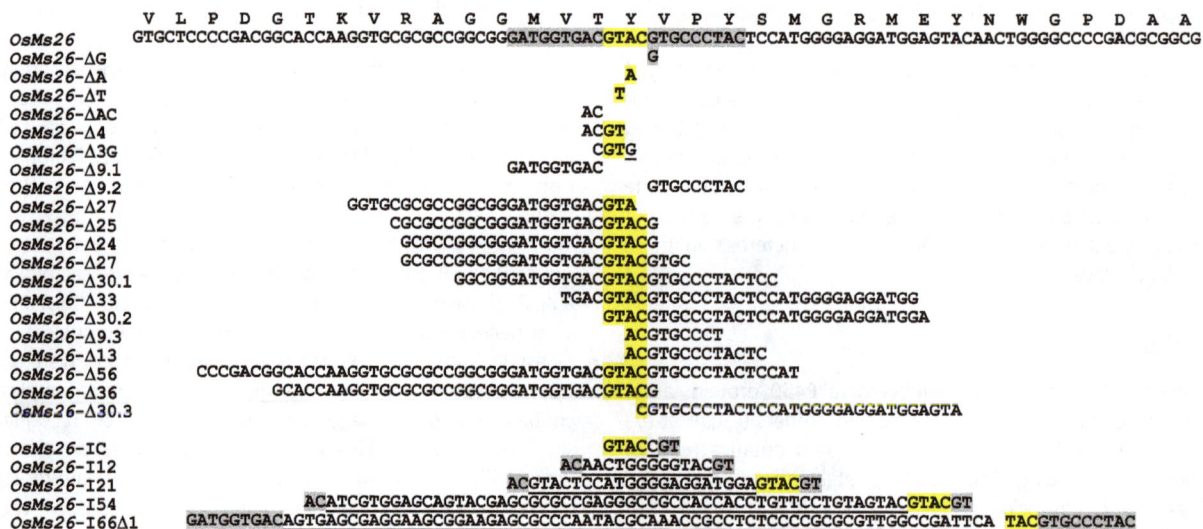

Figure 2 Ems26+-generated *OsMs26* mutations in rice. Shown in the WT reference sequence, the 22-nt Ems26+ recognition sequence is highlighted in grey and yellow. The latter indicates the 4 bp overhang, 5′-GTAC-3′, produced by nuclease cutting and is highlighted throughout for orientation purposes. For each mutant, deleted nucleotides are shown and inserted nucleotides are shown as underlined. Amino acid sequence corresponding to the *OsMs26* gene is provided as single letters above the WT nucleotide reference sequence.

(a)

(b)

OsMs26/Osms26-Δ9.2   **Osms26-Δ9.2/Osms26-Δ9.2**

Figure 3 Male sterile rice generated from targeted mutagenesis of the *OsMs26* gene via bombardment delivery of Ems26+. (a) Rice panicles from male fertile (left; *OsMs26/Osms26-Δ36*) and male sterile (right; *Osms26-Δ36/Osms26-Δ36*) plants. Dehisced anthers are shown with white arrows. (b) Longitudinal anther squash from heterozygous *OsMs26/Osms26-Δ9.2* male fertile plants (left) containing microspores (white arrows) and homozygous *Osms26-Δ9.2/Osms26-Δ9.2* male sterile plants (right). Magnification bar = 100 μM.

Surprisingly, as shown in Figure 3b, the in-frame deletion of nine nucleotides in *Osms26-Δ9.2* conferred a male sterile phenotype. In summary, all *OsMs26/Osms26* plants were male fertile; and all *Osms26/Osms26* plants were male sterile (Table 1). In addition, homozygous *Osms26/Osms26* plants in this study were female fertile as demonstrated by their ability to set seed when fertilized with wild-type rice pollen.

Inducible Ems26+ for producing double-strand breaks

Given the ability to successfully direct mutations to the rice *Ms26* ortholog that confer a male sterile phenotype, the requirement of *Ms26* for pollen development in sorghum and wheat was investigated. Relative to maize and rice, transformation of sorghum and wheat is challenging (Casas *et al.*, 1997; Zhao *et al.*, 2000; Zhu *et al.*, 1998). Moreover, gene editing in wheat is complicated due to the size and polyploid nature of the genome. Thus, as an alternative to biolistic delivery of Ems26+, an *in planta* method to stimulate targeted mutagenesis in sorghum and wheat was devised. This strategy depended on the generation of plants containing a stably integrated, temperature-regulated Ems26+ expression cassette delivered on a T-DNA vector by *Agrobacterium*-mediated transformation. Ems26+ was placed under the transcriptional control of a maize temperature-regulated promoter, MDH (ZmMDHpro:Ems26+) (Svitashev *et al.*, 2015), which allowed for temperature-inducible expression of the nuclease in plant tissues. In this strategy, immature embryos from plants containing conditionally expressed Ems26+ would be heated at 37 °C for 24 h to generate mutations. This T-DNA also contained a blue fluorescence gene (Cyan) regulated by an embryo-preferred END2 promoter (ZmEnd2pro:Cyan). This reporter served as a visible marker for selection of transformed callus, fertilized immature embryos and mature seed. A disrupted copy

of the DsRED gene transcriptionally regulated by a maize Histone 2B promoter was also present in this T-DNA to monitor endonuclease activity. This nonfunctional colour marker consisted of duplicated DsRED gene fragments (RF-FP) with a 369-bp overlap separated by a 136-bp spacer containing the Ems26+ recognition site (Text S2). DSBs in RF-FP would either be repaired by NHEJ or intramolecular recombination within the repeated RF-FP fragments (Figure 4a). Duplicated gene coding regions interrupted by a DSB target site(s) have been demonstrated to be useful visual reporters to detect nuclease activity (Puchta and Hohn, 2012). As such, one outcome of a DSB at the target site would be to stimulate intramolecular recombination within the repeated RF-FP restoring DsRED gene function as revealed by the appearance of red-fluorescing callus cells against a background of blue fluorescence. Thus in contrast to control callus where no red fluorescence would be expected, the appearance of red-fluorescing cells in treated samples would demonstrate activity of Ems26+ endonuclease and suggest potential cutting and, therefore, mutagenesis within the endogenous *Ms26* gene.

To test this scheme, rice plants were generated to contain a stable, randomly integrated ZmMDHpro:Ems26+ expression cassette as described above in the rice genome. Four independently transformed plants were allowed to self-pollinate, and blue-fluorescing seed (ZmEnd2pro:Cyan) were selected to initiate rice calli (Hervé and Kayano, 2006; Toki, 1997). Calli from these four events were allowed to grow at 26 °C (control) for 30 days, whereupon a portion of callus from each event was then incubated at 37 °C (treated) for 48 h. The temperature-treated calli were returned to 26 °C, upon which control and treated samples were examined for the appearance of red-fluorescing sectors. Over the course of 2 weeks, red sectors were readily visible in three of the four independent callus isolates treated at 37 °C (Figure 4b) and none in the control calli. Detection of red fluorescence indicated DSBs as a result of active Ems26+; therefore, genomic DNA was isolated from several red-sectoring callus events and screened for mutations at the endogenous *OsMs26* gene. A DNA fragment spanning a portion of the *OsMs26* gene was amplified from control and temperature-treated calli and screened using the *BsiWI* digestion assay. Analysis of amplified products from the three callus events demonstrated RF-FP repair and revealed *BsiWI*-resistant fragments present only in genomic DNA derived from these callus sectors incubated at 37 °C. Uncut amplified PCR products from control and temperature-treated callus DNA were sequenced.

Table 1 Male fertility phenotypes of T1 progeny plants with heterozygous or homozygous *OsMs26* mutations

Mutation	Plants	OsMs26/Osms26		Osms26/Osms26	
		Male fertile	Male sterile	Male fertile	Male sterile
Osms26-Δ9.2	18	10	0	0	8
Osms26-Δ36	9	5	0	0	4
Osms26-I66Δ1	15	11	0	0	4

(a)

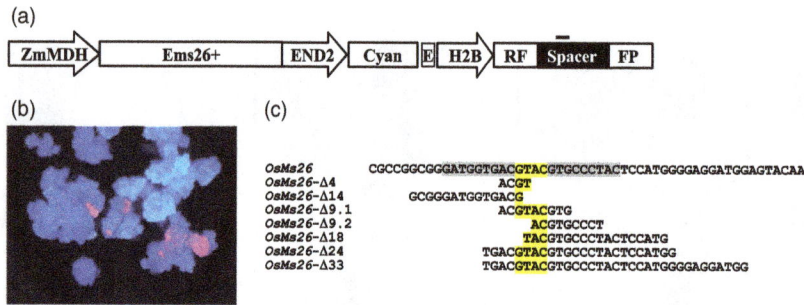

(b)

(c)

OsMs26	CGCCGGCGGGGATGGTGACGTACGTGCCCTACTCCATGGGGAGGATGGAGTACAA
OsMs26-Δ4	ACGT
OsMs26-Δ14	GCGGGATGGTGACG
OsMs26-Δ9.1	ACGTACGTG
OsMs26-Δ9.2	ACGTGCCCT
OsMs26-Δ18	TACGTGCCCTACTCCATG
OsMs26-Δ24	TGACGTACGTGCCCTACTCCATG
OsMs26-Δ33	TGACGTACGTGCCCTACTCCATGGGGAGGATGG

Figure 4 *OsMs26* mutations in rice generated using a stably integrated cassette containing Ems26+ under transcriptional control of a maize temperature-regulated promoter. (a) *Agrobacterium* vector for stable integration of the ZmMDHpro:Ems26+. The vector also contained a blue fluorescence gene (Cyan) as a visible marker regulated by the maize END2 promoter (END2). The disrupted DsRED gene (H2Bpro:RF-FP) contained duplicated DsRED gene fragments (RF, FP) separated by a 136-bp spacer containing the Ems26+ recognition site (position identified by the black bar above spacer sequence). (b) Blue and red fluorescence overlay image of rice calli containing stably integrated ZmMDHpro:Ems26+ after treatment at 37 °C for 48 h. Images captured 7 days after treatment. (c) Deletions in *OsMs26* achieved via targeted mutagenesis. Shown in the reference sequence, the 22-nt Ems26+ recognition sequence is highlighted in grey and yellow. The latter indicates the 4 bp overhang, 5'-GTAC-3', produced by nuclease cutting and is highlighted throughout for orientation purposes. Nucleotides shown are deleted sequence in seven of eight identified mutations, one larger deletion not shown. ZmMDH, Maize Mannitol dehydrogenase promoter; Ems26+, homing endonuclease targeting Ms26 target site; E, CaMV 35S promoter; H2B, maize Histone 2B promoter.

Analysis of PCR products amplified from callus sectors maintained at 26 °C revealed only wild-type *OsMs26* sequences. In contrast, only four of the 16 PCR products derived from callus exposed to 37 °C contained wild-type *OsMs26* sequences. The remaining 12 products revealed mutations across the Ems26+ recognition site, which consisted of several independent deletions (Figure 4c). These observations demonstrated that a stably integrated and conditionally regulated endonuclease could be used in rice to generate site-specific mutations and, thus, it has the potential to perform similarly in other monocots such as sorghum and wheat.

Mutations in the sorghum *Ms26* ortholog confer male sterility

To determine whether the ortholog of maize *Ms26* in *S. bicolor* was required for pollen formation, Ems26+ was used to direct mutations to the conserved recognition site in the 5th exon of this single-copy gene (*SbMs26*) present on Chromosome 1 (Sb01g045960). Immature embryos from *S. bicolor* [Tx430 (Miller, 1984)] were transformed with the ZmMDHpro:Ems26+ T-DNA using *Agrobacterium*-mediated delivery. Six independent blue-fluorescing callus sectors, indicative of the presence of a stably integrated copy of ZmMDHpro:Ems26+ T-DNA, were used to generate sorghum plants (approximately 2–12 plants per callus sector). Intactness and copy number of the T-DNA insertion were determined in these plants by DNA hybridization analysis (Southern, 1975) while amplification and sequencing of genomic DNA confirmed the presence of only wild-type *SbMs26* alleles in these plants. In addition, no red fluorescence was detected when progeny seed were screened visually. The presence of only wild-type *SbMs26* alleles and the absence of red-fluorescing seed suggested that the Ems26+ was not activated in these T0 plants. Several plants containing single-copy ZmMDHpro:Ems26+ T-DNA insertions were allowed to self-pollinate, and blue-fluorescing immature embryos (10 plants, 16 embryos per plant, representing four independent T-DNA insertions) were harvested 10–15 days postfertilization to be used for *SbMs26* mutational analysis. Visual inspection of these harvested embryos revealed that there were no red-fluorescing sectors or red-fluorescing embryos present. These blue-only-fluorescing embryos were incubated at 26 °C or 37 °C for 48 h and examined for RF-FP repair. In contrast to

control embryos maintained at 26 °C, red-fluorescing sectors were readily detected to varying degrees in all embryos incubated at 37 °C indicating the presence of active Ems26+. Calli from embryos treated at 37 °C were grown for 4 weeks and used to regenerate plants that were subsequently screened for mutations at the *SbMs26* target site on Chromosome 1. In total, approximately 20% of the plants (n = 440) regenerated from these four independent T-DNA insertion events revealed heterozygous mutations within the *SbMs26* gene when screened by PCR amplification and digestion with *BsiWI*. Although expected at a low frequency, biallelic mutations were not detected in these sorghum experiments. Similar to the experiments in rice, DNA sequence analysis of these PCR products (n = 50 independent *BsiWI*-resistant PCR fragments) revealed a variety of mutations across the *SbMs26* region that consisted of deletions and insertions. In total, 16 nonidentical mutations were identified in these regenerated sorghum plants (Figure S3 and Text S3).

Sorghum plants containing heterozygous mutations at *SbMs26* were allowed to flower, and seed from self-pollinated plants was advanced to examine male fertility phenotype in progeny plants. T1 progeny containing heterozygous *SbMs26* mutations [36-bp deletion (*Sbms26-Δ36*), 59-bp deletion with 11-bp insertion (*Sbms26-Δ59 + I11*) and 78-bp deletion (*Sbms26-Δ78*)] were phenotypically identical to wild-type sorghum Tx430 plants and were male fertile (Table 2). As shown in Figure 5a, panicles from *SbMs26/Sbms26-Δ78* plants extruded anthers that contained pollen and had begun to set seed as demonstrated by the appearance of collapsed stigmas after anthesis. Homozygous *Sbms26* plants were also normal in stature and development with the exception of being completely male sterile: *Sbms26-Δ78* plants extruded small shrivelled anthers, did not shed pollen and did not set seed upon self-pollination (Figure 5b and c). Upon closer examination, pollen was easily detected in anthers from *SbMs26/Sbms26-Δ78* plants (Figure 5d and g), while pollen was not observed in anthers from *Sbms26-Δ78/Sbms26-Δ78* plants (Figure 5e and h). In summary, all *SbMs26/Sbms26* plants were male fertile, while all *Sbms26/Sbms26* plants were male sterile (Table 2). These male sterile plants were female fertile as demonstrated by their ability to set seed when pollinated with wild-type sorghum pollen.

Table 2 Male fertility phenotypes associated with sorghum mutants and complementation of male sterility using maize *Ms26* gene

Mutation	Plants	SbMs26/Sbms26		Sbms26/Sbms26	
		Male fertile	Male sterile	Male fertile	Male sterile
Sbms26-Δ36	6	3	0	0	3
Sbms26-Δ78	6	3	0	0	3
Sbms26-Δ59 + I11	6	3	0	0	3

Complementation data

Mutation	Plants	Sbms26/Sbms26		Sbms26/Sbms26 + ZmMs26	
		Male fertile	Male sterile	Male fertile	Male sterile
Sbms26-Δ36	10[†]	0	4	6	0
Sbms26-Δ78	15[‡]	0	10	5	0

SbMs26, sorghum *Ms26* ortholog; *ZmMs26*, maize *Ms26* gene.

[†]Includes events #3 and #4.

[‡]Includes events #1, #2, and #3.

To further investigate that the male sterile phenotype observed in these sorghum mutant plants was directly due to deletions at *SbMs26*, restoration of fertility was tested using a copy of the maize *Ms26* gene (also referred to as *ZmMs26*) introduced by transformation. Wild-type sorghum plants were generated to contain a single copy of the *ZmMs26* gene introduced on a T-DNA expression cassette and crossed with plants carrying either the *Sbms26-Δ36* or the *Sbms26-Δ78* deletion mutation. From these crosses, F1 plants were screened and those that contained both the *Sbms26* mutation and *ZmMs26* were allowed to self-pollinate to generate F2 seeds. F2 plants were grown under greenhouse conditions and genotyped for the *Sbms26* mutation and presence of *ZmMs26*. A set of F2 plants homozygous for the *Sbms26* mutation but segregated for the *ZmMs26* expression cassette were selected. Plants homozygous for *Sbms26-Δ36* or *Sbms26-Δ78* but lacking *ZmMs26* were male sterile; anthers did not contain pollen, and plants did not set seed upon self-pollination (Table 2 and Table S1). However, when *ZmMs26* was present, these homozygous *Sbms26* mutant plants were male fertile (Table 2) with pollen development comparable to wild type (Figure 5f and i). The requirement of Sb01g045960 for pollen development in sorghum suggests that this gene is the ortholog of the maize male fertility gene *Ms26*.

Ems26+-targeted mutations in the wheat *Ms26* ortholog

As described above, the maize Ms26 polypeptide sequence was used to search the translated hexaploid wheat genome database (Chinese spring) (Consortium, 2014) for homologous sequences. A 4.2-kb contig derived from Chromosome 4AS was found capable of encoding a 550 amino acid polypeptide having 91% sequence similarity to maize Ms26 protein (Figure S1 and Text S4). Additional BLAST (NCBI, Bethesda, MD, USA) search yielded wheat 4BL and 4DL contigs that only partially overlapped the region corresponding to the 2nd (4DL) and last exons (4BL and 4DL) of the 4AS sequence. To ascertain whether the Ems26+

recognition site was conserved across all three wheat genomes, UNIMS26 primers were used to amplify genomic DNA from the hexaploid spring wheat cultivar Fielder. DNA sequence analysis of these amplified products, when compared to the contigs described above, suggested that the 22-nt Ems26+ recognition site was conserved. The A-, B- and D-genomic copies of the wheat *Ms26* ortholog (referred to as *TaMs26*) were assigned to the Fielder genome (Figure S4). Distinguishing features within this amplified region include a six-nucleotide, two-amino acid gap; functional relevance of this sequence difference is not yet known, but interestingly is found in a region that is diverse across maize, sorghum, rice and wheat Ms26 protein alignments (Figure S5).

To examine whether the Ems26+ endonuclease was competent to generate targeted mutations at the *TaMs26* gene orthologs in hexaploid wheat, Fielder plants containing the ZmMDHpro:Ems26+ T-DNA were generated by *Agrobacterium*-mediated transformation. Immature Fielder embryos were used as transformation targets. Blue-fluorescing callus sectors indicative of T-DNA integration were selected to initiate plant regeneration. Young wheat plantlets grown under greenhouse conditions were analysed for ZmMDHpro:Ems26+ T-DNA copy number by qPCR. Four independently transformed wheat plants containing single- or low-copy ZmMDHpro:Ems26+ T-DNA were allowed to self-pollinate and embryos were harvested 15 days postfertilization. Blue-fluorescing embryos ($n = 12$) isolated from each plant were either held in the dark at 26 °C (control, $n = 4$) or incubated at 37 °C for 24 h (treated, $n = 8$), whereupon the treated embryos were returned to 26 °C and allowed to grow in the dark. As observed for the rice and sorghum experiments, approximately 72 h after treatment, wheat embryos incubated at 37 °C began to develop red-fluorescing sectors, consistent with the presence of active Ems26+ endonuclease (Figure S6). No red-fluorescing sectors were detected in control, nontreated wheat embryos. To ascertain whether Ems26+ generated mutations at endogenous *TaMs26* sequences, total genomic DNA was isolated from temperature-treated and control embryos and amplified using UNIMS26 primers. Deep sequencing analysis of these amplicons (approximately 5 million reads per sample) revealed that 1.6% of the amplicon reads from the temperature-treated embryos contained mutations as compared to <0.007% from control embryos (Table S2 and Figure S7). Callus from embryos treated at 37 °C and containing ZmMDHpro:Ems26+ was used to regenerate 121 plants that were then screened for mutations using the *Bsi*WI digestion assay; ten wheat plants contained *Bsi*WI digestion-resistant amplicons. DNA sequencing of these resistant PCR products identified seven different mutations within the Ems26+ recognition site consisting of a single-nucleotide addition and deletions of various sizes; a nonoverlapping subset of these mutations was found within each genome copy of *TaMs26* (Figure 6). Plants containing *TaMs26* mutations were allowed to self-pollinate for progeny analysis. Inheritance analysis determined that, with the exception of the single-C-nucleotide-insertion mutant (no seed set), mutations in all three genomes were sexually transmitted to progeny plants. These progeny plants containing single-genome, homozygous mutations in the A-, B- or D-genomic copies of *TaMs26* were phenotypically normal and male fertile (Table S3). Together these experiments demonstrated that a conditionally expressed Ems26+ was capable of generating targeted mutations at the *TaMs26* gene *in planta* and that single-genome mutants in this P450 are not sufficient to confer a male sterile phenotype in the spring wheat

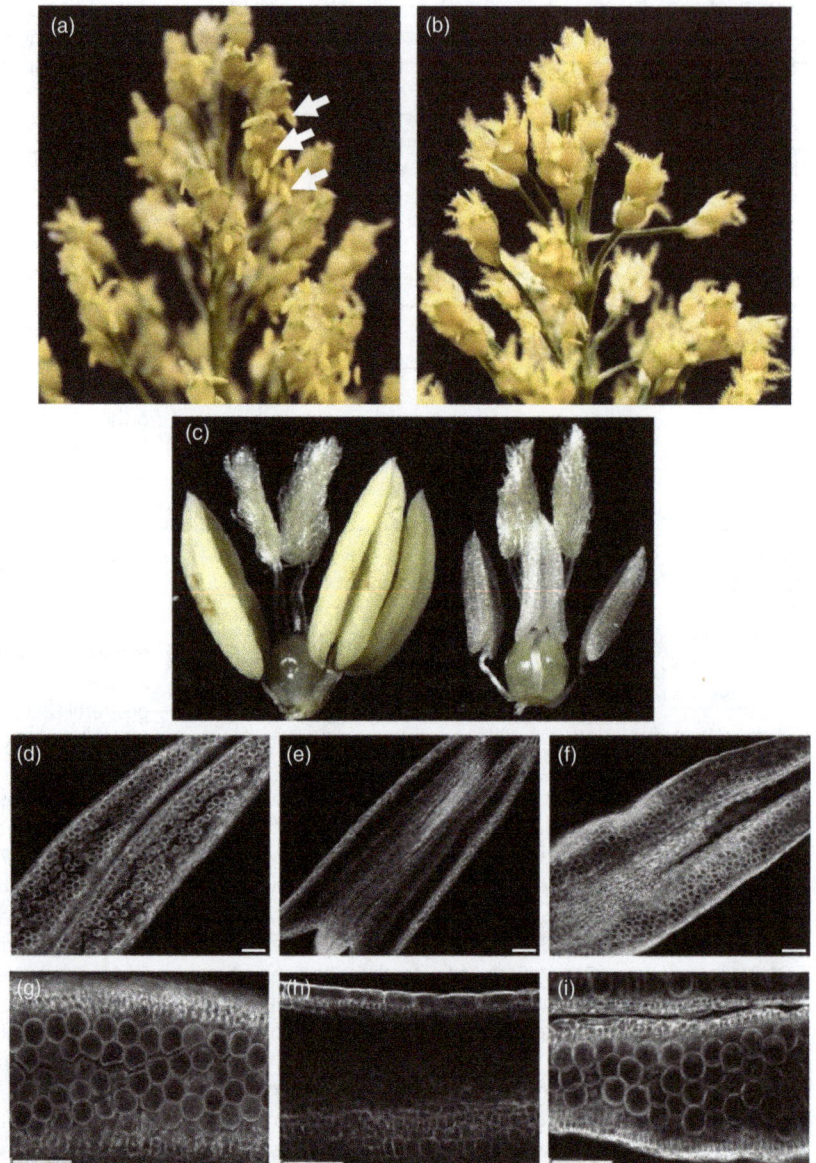

Figure 5 Male sterile sorghum generated from targeted mutagenesis of *SbMs26* gene. (a) Panicles from male fertile sorghum plants (*SbMs26/Sbms26-Δ78*) with extruded anthers (white arrows). (b) Panicles from male sterile sorghum plants (*Sbms26-Δ78/Sbms2-Δ78*). (c) Floret from male fertile (left) and from male sterile (*Sbms26/Sbms26*) sorghum plants (right). (d-i) Images of sorghum anthers derived from wild-type plants (d and g), homozygous *Sbms26/Sbms26* plants (e and h), and *Sbms26/Sbms26* plants containing ZmMs26 complementation vector (f and i). Magnification bar = 100 μM.

cultivar Fielder. Studies that genetically combine these different A-, B- and D-genome mutations are in progress to determine the requirement of *Ms26* for pollen development in wheat.

Discussion

The ability to perform targeted mutagenesis in plants is a valuable tool for gene functional analysis and generation of genetic diversity. In this report, targeted mutagenesis of the monocot male fertility gene, *Ms26*, was accomplished in three different plant species using either a constitutively or conditionally expressed custom designed I-*Cre*I endonuclease, Ems26+. In rice, approximately 7% of the primary plants contained mutations across the highly conserved Ems26+ target site; these mutations ranged from single-nucleotide deletions to large insertions. Similar to mutations identified by classical mutagenesis of maize and rice, targeted mutations to the rice *Ms26* ortholog also resulted in complete male sterility by preventing the formation of pollen in homozygous recessive mutant plants. During these studies, it was determined that plants carrying a constitutively

expressed Ems26+ endonuclease did not promote mutagenesis in subsequent generations of plants suggesting that mutations occurred early in the plant transformation process and functional endonucleases were not stably transmittable. In contrast, when Ems26+ was introduced into transformed plants and placed under the transcriptional control of a conditional promoter, the endonuclease was functional in progeny sorghum and wheat plants.

In previous reports, the low frequency of stable event recovery has been suggested to be an indicator of cytotoxicity due to off-target recognition for ZFNs in human cell lines and plants (Cornu *et al.*, 2007; Szczepek *et al.*, 2007; Townsend *et al.*, 2009). This suggestion is supported by the observation that in maize and rice experiments, in contrast to control experiments that did not introduce UBIpro:Ems26+, event recovery was lower when Ems26+ was delivered by biolistic transformation. Although it cannot be ruled out that reduced transformation frequency in these experiments was simply a consequence of the high concentration of vectors containing Ems26+ delivered by particle bombardment when compared to *Agrobacterium*

```
FIELDER_A_GENOME_2        TGATCGCCGGGCGGGACACCACGGCCACGACGCTCTCCTGGTTCACCTACATGGCCATGACGCACCCGGCCGTGGCCGAGAAGCTCCGCCGCGAGCTGGCCGCCTTCGAGGCGGACCGCGCCCCGCGAGGATGGCGTCGCCGCTGGTCCCCT
PLANT14_4_BPDEL           TGATCGCCGGGCGGGACACCACGGCCACGACGCTCTCCTGGTTCACCTACATGGCCATGACGCACCCGGCCGTGGCCGAGAAGCTCCGCCGCGAGCTGGCCGCCTTCGAGGCGGACCGCGCCCCGCGAGGATGGCGTCGCCGCTGGTCCCCT
PLANT45_C_INS            TGATCGCCGGGCGGGACACCACGGCCACGACGCTCTCCTGGTTCACCTACATGGCCATGACGCACCCGGCCGTGGCCGAGAAGCTCCGCCGCGAGCTGGCCGCCTTCGAGGCGGACCGCGCCCCGCGAGGATGGCGTCGCCGCTGGTCCCCT
FIELDER_B_GENOME_3        TGATCGCCGGGCGGGACACGACGGCCACGACGCTCTCCTGGTTCACCTACATGGCCATGACGCACCCGGCCGTGGCCGAGAAGCTCCGCCGCGAGCTGGCCGCCTTCGAGGCGGACCGCGCCCCGCGAGGATGGCGTCGCTCTGGTCCCCT
PLANT09_81_BPDEL          TCATCGCCGGGCGGGACACGACGGCCACGACGCTCTCCTGGTTCACCTACATGGCCATGACGCACCCGGCCGTGGCCGAGAAGCTCCGCCGCGAGCTGGCCGCCTTCGAGTCCGAGCGGCGCCGCGAGGATGGCGTCGCTCTGGTCCCCT
PLANT30_9_BPDEL           TCATCGCCGGGCGGGACACGACGGCCACGACGCTCTCCTGGTTCACCTACATGGCCATGACGCACCCGGCCGTGGCCGAGAAGCTCCGCCGCGAGCTGGCCGCCTTCGAGTCCGAGCGGCGCCGCGAGGATGGCGTCGCTCTGGTCCCCT
PLANT121_23_BPDEL         TCATCGCCGGGCGGGACACGACGGCCACGACGCTCTCCTGGTTCACCTACATGGCCATGACGCACCCGGCCGTGGCCGAGAAGCTCCGCCGCGAGCTGGCCGCCTTCGAGTCCGAGCGGCGCCGCGAGGATGGCGTCGCTCTGGTCCCCT
FIELDER_D_GENOME_1        TGATCGCCGGGCGGGACACCACGGCCACGACGCTGTCCTGGTTCACCTACATGGCCATGACGCACCCGGACGTGGCCGAGAAGCTCCGCCGCGAGCTGGCCGCCTTCGAGGCGGAGCGGCGCCGCGAGGATGGCGTCGCTCTGGTCCCCT
PLANT75_90_BPDEL          TGATCGCCGGGCGGGACACCACGGCCACGACGCTGTCCTGGTTCACCTACATGGCCATGACGCACCCGGACGTGGCCGAGAAGCTCCGCCGCGAGCTGGCCGCCTTCGAGGCGGAGCGGCGCCGCGAGGATGGCGTCGCTCTGGTCCCCT
PLANT64_96_BPDEL+2_BPINS  TGATCGCCGGGCGGGACACCACGGCCACGACGCTGTCCTGGTTCACCTACATGGCCATGACGCACCCGGACGTGGCCGAGAAGCTCCGCCGCGAGCTGGCCGCCCTTCGAGGCGGAGCGGCGCCGCGAGGATGGCGTCGCTCTGGTCCCCT


FIELDER_A_GENOME_2        GCAGCGACTCAGACGGCGACGGCTCCGACGAGGCCTTCGCCGCCCGCGTGGCCCAGTTCGCGGGGCTGCTGAGCTACGACGGGCTCGGGAAGCTGGTGTACCTCCACGCGTGCGTGACGGAGACGCTGCGCCTGTACCCGGCGGTGCCGC
PLANT14_4_BPDEL           GCAGCGACTCAGACGGCGACGGCTCCGACGAGGCCTTCGCCGCCCGCGTGGCCCAGTTCGCGGGGCTGCTGAGCTACGACGGGCTCGGGAAGCTGGTGTACCTCCACGCGTGCGTGACGGAGACGCTGCGCCTGTACCCGGCGGTGCCGC
PLANT45_C_INS            GCAGCGACTCAGACGGCGACGGCTCCGACGAGGCCTTCGCCGCCCGCGTGGCCCAGTTCGCGGGGCTGCTGAGCTACGACGGGCTCGGGAAGCTGGTGTACCTCCACGCGTGCGTGACGGAGACGCTGCGCCTGTACCCGGCGGTGCCGC
FIELDER_B_GENOME_3        GCAGCGAC     GGCGAGGGCTCCGACGAGGCCTTCGCCGCCCGCGTGGCCCAGTTCGCGGGACTCGACTACGACGGGCTCGGGAAGCTGGTGTACCTCCACGCGTGCGTGACGGAGACGCTCCGCCTGTACCCGGCGGTGCCGC
PLANT09_81_BPDEL          GCAGCGAC     GGCGAGGGCTCCGACGAGGCCTTCGCCGCCCGCGTGGCCCAGTTCGCGGGACTCGACTACGACGGGCTCGGGAAGCTGGTGTACCTCCACGCGTGCGTGACGGAGACGCTCCGCCTGTACCCGGCGGTGCCGC
PLANT30_9_BPDEL           GCAGCGAC     GGCGAGGGCTCCGACGAGGCCTTCGCCGCCCGCGTGGCCCAGTTCGCGGGACTCGACTACGACGGGCTCGGGAAGCTGGTGTACCTCCACGCGTGCGTGACGGAGACGCTCCGCCTGTACCCGGCGGTGCCGC
PLANT121_23_BPDEL         GCAGCGAC     GGCGAGGGCTCCGACGAGGCCTTCGCCGCCCGCGTGGCCCAGTTCGCGGGACTCGACTACGACGGGCTCGGGAAGCTGGTGTACCTCCACGCGTGCGTGACGGAGACGCTCCGCCTGTACCCGGCGGTGCCGC
FIELDER_D_GENOME_1        GCGGCGAC     GGCGAGGGCTCCGACGAGGCCTTCGCTGCCCGCGTGGCCCAGTTCGCGGGGTTCGACTACGACGGGCTCGGGAAGCTGGTGTACCTCCACGCGTGCGTGACGGAGACGCTGCGCCTGTACCCGGCGGTGCCGC
PLANT75_90_BPDEL          GCGGCGAC     GGCGAGGGCTCCGACGAGGCCTTCGCTGCCCGCGTGGCCCAGTTCGCGGGGTTCGACTACGACGGGCTCGGGAAGCTGGTGTACCTCCACGCGTGCGTGACGGAGACGCTGCGCCTGTAC-------------
PLANT64_96_BPDEL+2_BPINS  GCGGCGAC     GGCGAGGGCTCCGACGAGGCCTTCGCTGCCCGCGTGGCCCAGTTCGCGGGGTTCGACTACGACGGGCTCGGGAAGCTGGTGTACCTCCACGCGTGCGTGACGGAGACGCTGCGCCTGTACCCGGCGGTGCCGC


                                                                                                          GATGGTGACGTAC GTGCCCTAC
FIELDER_A_GENOME_2        AGGACCCCAAGGGCATCGCGGAGGACGACGTGCTCCCGGACACGGCACCAAGGTGCGCGCCGGCGGGATGGTGACGTAC GTGCCCTACTCCATGGGGCGGATGGAGTACAACTGGGGCCCCGACGCCGCCAGCTTCCGGCCGGAGCCGGTGG
PLANT14_4_BPDEL           AGGACCCCAAGGGCATCGCGGAGGACGACGTGCTCCCGGACACGGCACCAAGGTGCGCGCCGGCGGGATGGTGAC---- GTGCCCTACTCCATGGGGCGGATGGAGTACAACTGGGGCCCCGACGCCGCCAGCTTCCGGCCGGAGCCGGTGG
PLANT45_C_INS            AGGACCCCAAGGGCATCGCGGAGGACGACGTGCTCCCGGACACGGCACCAAGGTGCGCGCCGGCGGGATGGTGACGTAC▓GTGCCCTACTCCATGGGGCGGATGGAGTACAACTGGGGCCCCGACGCCGCCAGCTTCCGGCCGGAGCCGGTGG
FIELDER_B_GENOME_3        AGGACCCCAAGGGCATCGCGGAGGACGACGTGCTCCCGGACACGGCACCAAGGTGCGCGCCGGCGGGATGGTGACGTA--------------------------------------------------------------------------
PLANT09_81_BPDEL          AGGACCCCAAGGGCATCGCGGAGGACGACGTGCTCCCGGACACGGCACCAAGGTGCGCGCCGGCGGGATGGTGAC---------TCCATGGGGCGGATGGAGTACAACTGGGGCCCCGACGCCGCCAGCTTCCGGCCGCAGAGCGGTGG
PLANT30_9_BPDEL           AGGACCCCAAGGGCATCGCGGAGGACGACGTGCTCCCGGACACGGCACCAAGGTGCGCGCCGGCGGGATGGTGAC---------TCCATGGGGCGGATGGAGTACAACTGGGGCCCCGACGCCGCCAGCTTCCGGCCGCAGAGCGGTGG
PLANT121_23_BPDEL         AGGACCCCAAGGGCATCGCGGAGGACGACGTGCTCCCGGACACGGCACCAAGG-------------------ACGTGCCCTACTCCATGGGGCGGATGGAGTACAACTGGGGCCCCGACGCCGCCAGCTTCCGGCCGCAGAGCGGTGG
FIELDER_D_GENOME_1        AGGACCCCAAGGGCATCGCGGAGGACGACGTGCTCCCGGACACGGCACCAAGGTGCGCGCCGGCGGGATGGTGACGTAC GTGCCCTACTCCATGGGGCGGATGGAGTACAACTGGGGCCCCGACGCCGCCAGCTTCCGGCCGGAGCCGGTGG
PLANT75_90_BPDEL          -------------------------------------------------------------------------------GTGCCCTACTCCATGGGGCGGATGGAGTACAACTGGGGCCCCGACGCCGCCAGCTTCCGGCCGGAGCCGGTGG
PLANT64_96_BPDEL+2_BPINS  AGGACCCCAAGGGCATCGCGGAGGACGACGT--------------------------------------------------------------------▓▓GCTTCCGGCCGGAGCCGGTGG


FIELDER_A_GENOME_2        ATCGGCGACGACGGCGCCGTTCCGCAACGCGTCGCCGTTCAAGTTCACGGCGTTCCAGGCGGGGCCGCGGATC
PLANT14_4_BPDEL           ATCGGCGACGACGGCGCCGTTCCGCAACGCGTCGCCGTTCAAGTTCACGGCGTTCCAGGCGGGGCCGCGGATC
PLANT45_C_INS            ATCGGCGACGACGGCGCCGTTCCGCAACGCGTCGCCGTTCAAGTTCACGGCGTTCCAGGCGGGGCCGCGGATC
FIELDER_B_GENOME_3        ATCGGCGACGACGGCGCCGCCTTCCGCAACGCGTCGCCGTTCAAGTTCACGGCGTTCCAGGCGGGGCCGCGGATC
PLANT09_81_BPDEL          --------CGACGGCGCCGCCTTCCGCAACGCGTCGCCGTTCAAGTTCACGGCGTTCCAGGCGGGGCCGCGGATC
PLANT30_9_BPDEL           ATCGGCGACGACGGCGCCGCCTTCCGCAACGCGTCGCCGTTCAAGTTCACGGCGTTCCAGGCGGGGCCGCGGATC
PLANT121_23_BPDEL         ATCGGCGACGACGGCGCCGCCTTCCGCAACGCGTCGCCGTTCAAGTTCACGGCGTTCCAGGCGGGGCCGCGGATC
FIELDER_D_GENOME_1        ATCGGCGACGACGGCGCCGCCTTCCGCAACGCGTCGCCGTTCAAGTTCACGGCGTTCCAGGCGGGGCCGCGGATT
PLANT75_90_BPDEL          ATCGGCGACGACGGCGCCGCCTTCCGCAACGCGTCGCCGTTCAAGTTCACGGCGTTCCAGGCGGGGCCGCGGATT
PLANT64_96_BPDEL+2_BPINS  ATCGGCGACGACGGCGCCTTCCGCAACGCGTCGCCGTTCAAGTTCACGGCGTTCCAGGCGGGGCCGCGGATT
```

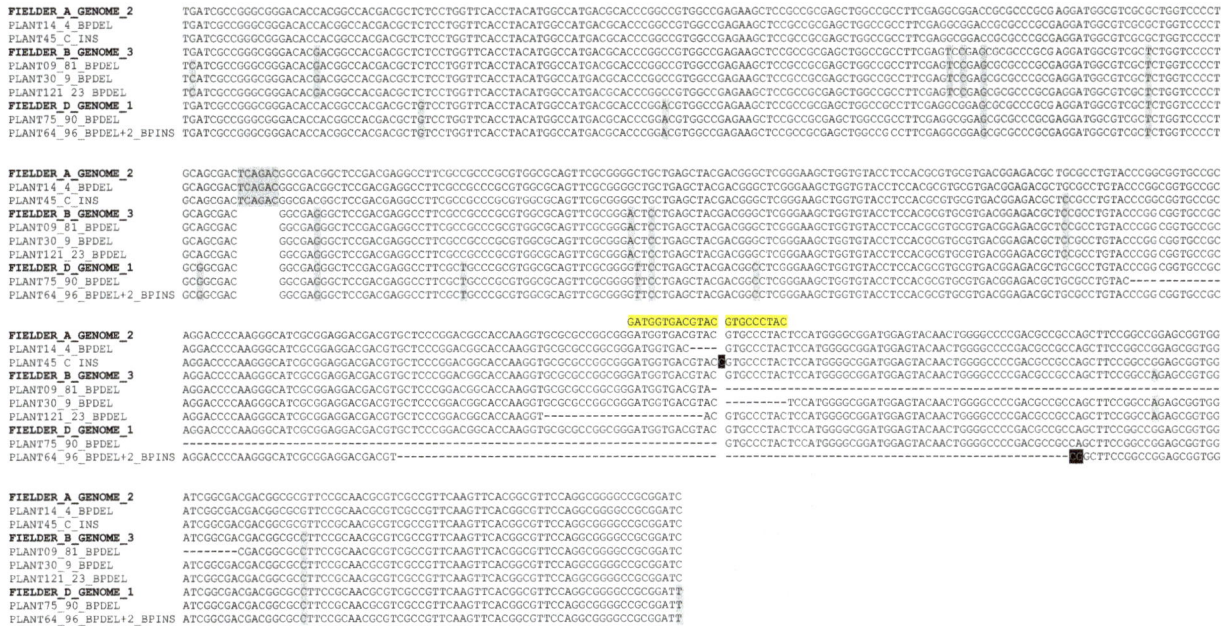

Figure 6 *TaMs26* gene mutations isolated from Fielder wheat A-, B- and D-genomes via targeted mutagenesis. Ems26+ recognition sequence is highlighted in yellow above the A-genome sequence. Deleted nucleotide sequence is represented by a hyphen ('-'). Inserted nucleotide sequence is in white font and highlighted in black. Sequence differences between the A-, B- and D-genomes are highlighted in grey. White space indicates a gap in sequence.

transformation, differences in delivery and regulation of Ems26+ were important design considerations for these targeted mutagenesis experiments. The establishment of pre-integrated and conditionally regulated endonuclease was one strategy that mitigated potential toxicity by taking advantage of a promoter capable of induction at higher temperatures and the temperature-dependent biochemical activity of this class of endonucleases (Wang *et al.*, 1997). In our studies, Ems26+ demonstrated increased nuclease activity at 37 °C both *in vitro* and *in planta*. Thus, short pulses at elevated temperatures during tissue culture or plant growth resulted in both increased transcription and nucleolytic activity of the Ems26+ gene and protein, respectively. This dual switch was an effective method for controlling DSB activity in these experiments and has been exploited as a means to rapidly generate large collections of allelic variants for gene discovery.

Pollen formation in plants is highly conserved at the biochemical and genetic level (for review see Shi *et al.*, 2015). A number of pollen- and anther-specific genes have been identified whose function is conserved across plants. The maize *Ms26* and rice *OsCYP704B2* genes encode a cytochrome P450 that plays a critical role in male fertility. In this study, conservation of sequence within these orthologous genes was used to introduce targeted mutations in rice, sorghum and wheat. Importantly, this is the first isolation and functional characterization of a male fertility gene in sorghum, a close relative to maize. As in rice, sorghum plants containing homozygous recessive mutations in the ortholog of the maize *Ms26* gene were male sterile. Moreover, a transformed copy of the maize *Ms26* gene was able to restore fertility in these sorghum mutant plants. The ability to restore fertility in these sorghum mutants using a maize *Ms26* gene further supports biochemical conservation of P450 function.

This observed functional conservation is particularly relevant to the foundational experiments described in this report where the orthologs of the maize *Ms26* gene were targeted in wheat. To date, there are only two recessive male fertility mutants that have been described in *Triticum*: *Ms1* and *Ms5* (Whitford *et al.*, 2013). Given the high sequence similarity across maize, sorghum and rice, it is likely that orthologs of *Ms26* in wheat are also required for pollen development. In contrast to these diploid species, wheat is an allohexaploid. In this study, it was observed that single-genome homozygous *TaMs26* mutations in hexaploid wheat did not confer a male sterility phenotype. These results suggest that no single *TaMs26* genomic copy from the A-, B- or D-genome is essential to fertility in wheat, as the other wild-type *TaMs26* copies present in these plants are likely to function to maintain pollen development and a male fertility phenotype. It is interesting to speculate that despite the high amino acid sequence conservation across all three wheat genome copies, functional differences may be revealed and associated with relatively minor sequence changes, in particular within the regions of amino acid diversity (Figure S5). Thus, combining these *TaMs26* gene mutations via genetic crosses will allow examination of the effects of various combinations of homozygous *TaMs26* mutations in the three genomes of hexaploid wheat (*i.e.* in one, two or all three single genomes) on male fertility.

In this report, targeted mutagenesis of the monocot male fertility gene, *Ms26*, was made possible by advances of genomics, gene editing and plant transformation that overcome two barriers in crop improvement. First, having the ability to specifically target and modify fertility genes would extend SPT systems to crops having perfect flowers enabling yield gains by heterosis not yet fully realized. Second, gene editing, when combined with advanced plant transformation methods, would allow for precision breeding of elite germplasm reducing the time needed for introgression of induced sequence diversity by backcrossing. Classical methods for introducing genetic diversity into plants include mutagenesis with chemical and physical agents or transposition. These nontargeted methods result in random and often additional and undesired changes in the genome. These approaches are both time-consuming and resource-intensive requiring multiple generations of backcrossing and selection to convert back to the original state of elite genetics for maximal crop yields. In contrast to these nontargeted approaches that slow discovery and crop improvement, adoption of the precision breeding innovations described above is essential to sustaining agriculture as population increases and as protection of natural resources becomes more challenging (Tian *et al.*, 2016; Tilman *et al.*, 2011; West *et al.*, 2014). Targeted mutagenesis of the *Ms26* gene is a good example of how advanced technologies could be deployed to rapidly and precisely generate mutant fertility alleles in elite germplasm of maize, rice, sorghum and potentially wheat. While not exclusively considered a replacement for classical breeding methods, these advances could have a positive impact and change modern agriculture. As with classical mutagenesis methods, should undesired genome modifications occur (*i.e.* random integration of plasmid DNA or off-target mutations), these changes could be easily bred away by genetic crosses to elite parental lines, which would eliminate unwanted genome alterations yet maintain targeted sequence variation. In addition, high-resolution sequencing is particularly useful in the design and testing of nucleases with unique recognition and cutting specificity, thus further improving and advancing gene editing as an accelerated breeding tool.

Experimental procedures

DNA methods

Standard DNA techniques as described in Sambrook *et al.* (1989) were used for vector construction. The Ems26+ was maize codon-optimized and designed to include an amino-terminal nuclear localization signal SV40 (MAPKKKRKV, nucleotide position +1–30) and the potato ST-LS1 intron [Accession number X04753 (nucleotides 2837–2892)] at nucleotide position +403–591 (Text S5). The chemically synthesized Ems26+, contained on a 1.280-kb NcoI-KpnI DNA fragment, was subcloned into PHP17720 (Sabelli *et al.*, 2009) digested with NcoI-KpnI that links Ems26+ to a maize constitutive ubiquitin-1 promoter, including the first intron [−899 to +1092 (Christensen *et al.*, 1992)], while transcription is terminated by the addition of the 3′ sequences from the potato proteinase inhibitor II gene (PinII) [nucleotides 2–310 (An *et al.*, 1989)] to generate UBIpro:Ems26+ vector. To generate ZmENDpro: PAT RFP fusion plasmid, the maize END2 promoter [Endosperm- and embryo-expressed promoter from maize; Chromosome 8, nt 171120325-171119383 Maize (B73) Public Genome Assembly (AGP_v3.8)] was fused to the MoPAT-DsRED (a translational fusion of the bialophos resistance gene, phosphinothricin-*N*-acetyl-transferase, and the red fluorescent protein DsRED) expression cassette previously described in Ananiev *et al.* (2009) with transcription terminated by the PinII terminator described above. To enhance expression of this expression cassette, a 485-bp fragment of the CaMV 35S promoter was inserted upstream of the ZmEnd2 promoter as described in Unger *et al.* (2001).

To generate a conditionally expressed Ems26+, a 1048-bp fragment of temperature-regulated promoter from the maize mannitol dehydrogenase gene [MDH; temperature-inducible maize promoter; Chromosome 2, nt 8783228–8784275 Maize (B73) Public Genome Assembly (AGP_v3.2)] was modified to contain an *Rca*I restriction site at the 3′ end of the sequence (Svitashev *et al.*, 2015) to accommodate translational fusion to the maize-optimized Ems26+, producing the ZmMDHpro:Ems26+

expression cassette. To construct the stably integrated Ems26+ expression cassette, a T-DNA vector containing ZmMDHpro: Ems26+ was generated and was identical to that described in Svitashev *et al.* (2015) with the exception that the ZmMDHpro: Ems26+ present on a 3.339-kb *Hind*III-*Pac*I fragment replaced the UBIpro:Cas9 portion of this vector. In addition, the duplicated DsRED fragments in this gene were separated by a 136-bp spacer that contained sequences for recognition by the Ems26+ endonuclease (Text S2). This plasmid was introduced into *Agrobacterium* strain LBA4404 (Komari *et al.*, 1996) by electroporation using a Gene Pulser II (Bio-Rad Laboratories, Inc., Hercules, CA, USA) as described by Gao *et al.* (2010) to generate *Agrobacterium* strains for plant transformation.

To generate the *ms26* fertility complementation T-DNA vector, the *Ms45* gene in PHP24490 (Wu *et al.*, 2015) was replaced by a 3.9-kb DNA fragment [Chromosome 1, nt 4509689–14506399 Maize (B73) RefGen_v2 (MGSC)], which contained the wild-type *Ms26* gene. The LTP2 promoter present in this T-DNA was also replaced by the maize END2 promoter as described above to transcriptionally regulate the RFP gene in this plasmid. The final T-DNA cassette, pG47::Bt1:ZmAA1//Ms26//35SEN-pEND2::DsRed2, was introduced into *Agrobacterium* strain LBA4404 as described above.

For PCR and DNA sequence analysis, DNA was extracted from a small amount (0.5 cm in diameter) of callus tissue or two leaf punches (four leaf punches for wheat) as described in Gao *et al.* (2010). PCR was performed using REDExtract-N-Amp™ PCR ReadyMix (Sigma-Aldrich Co., St. Louis, MO, USA) or Phusion® High-Fidelity PCR Master Mix (New England Biolabs Inc., Ipswich, MA, USA) according to the manufacturer's recommendations. Ms26 target site mutations were identified by amplification of the region by PCR using the primer pair UNIMS26 5′-2 (5′-GACGTGGTGCTCAACTTCGTGAT-3′) and UNIMS26 3′-1 (5′-GCCATGGAGAGGATGGTCATCAT-3′) and digestion of the amplified products with *Bsi*WI, the DNA restriction enzyme that recognizes the sequence 5′-CGTACG-3′, as described in Djukanovic *et al.* (2013). Products of these reactions were electrophoresed on 1% agarose gels and screened for *Bsi*WI digestion-resistant bands indicative of mutations at the Ms26 target site. PCR-amplified fragments were also subcloned for DNA sequence analysis using the pCR2.1-TOPO cloning vector (Thermo Fisher Scientific, Waltham, MA, USA).

Deep sequencing

Total genomic DNA was extracted from the temperature-treated and nontreated calli transformed with the Ems26+ homing endonuclease using a modified CTAB method (Unger *et al.*, 2001). The region surrounding the Ms26 target site was PCR-amplified with Phusion® High-Fidelity PCR Master Mix (New England Biolabs) adding on the sequences necessary for amplicon-specific barcodes and Illumina sequencing using 'tailed' primers through two rounds of PCR. The primers used in the primary PCR Ms26 reaction were Ms26F (5′-CTACACTCTTTCCC TACACGACGCTCTTCCGATCTAACCCGCGGAGGACGACGTGC TC-3′) and Ms26R (5′-CAAGCAGAAGACGGCATACGAGCT CTTCCGATCTCGTCGGGGCCCCAGTTGTAC-3′), while the primers used in the secondary PCR reaction were F2 (5′-AATGA TACGGCGACCACCGAGATCTACACTCTTTCCCTACACG-3′) and R2 (5′-CAAGCAGAAGACGGCATA-3′). Genomic DNA extracted from leaves of untransformed Fielder plants served as a negative control. The resulting PCR amplifications were concentrated using a Qiagen MinElute® PCR purification spin column (QIAGEN, Inc., Venlo, Netherlands), electrophoresed on a 2% agarose gel, and amplifications were then excised and purified with a Qiagen gel extraction spin column. The concentration of the gel-purified amplifications was measured with a Hoechst dye-based fluorometric assay, combined in an equimolar ratio, and single-read 100-nt-length deep sequencing was performed on an Illumina Genome Analyzer IIx (ELIM Biopharmaceuticals, Inc., Hayward, CA, USA) with a 30%–40% (v/v) spike of PhiX control v3 (Illumina, Inc., San Diego, CA, USA) to off-set sequence bias. Only those reads with a ≥1 nucleotide indel arising within a 6-nt window centred over the expected site of cleavage and not found in a similar level in the negative control were classified as NHEJ mutations. The total numbers of NHEJ mutations were then used to calculate the % mutant reads based on the total number of reads of an appropriate length containing a perfect match to the barcode and forward primer.

Plant methods

Rice, sorghum and wheat plant transformation and tissue culture details are outlined in Data S1.

Microscopy

Sorghum samples were dissected to remove the glumes and expose the anthers, and fixed in 2% paraformaldehyde and 4% glutaraldehyde overnight under vacuum (10 psi). After fixation, the samples were rinsed three times in 1× PBS (Phosphate Buffer Solution). Samples were then cleared in graded TDE (2, 2′-Thiodiethanol) series (10%, 25%, 50% and 97%), and the 97% was repeated to assure optimal clearing. Samples were mounted in 97% TDE and the coverslip was sealed with nail polish. Confocal images were taken with the Leica (Wetzlar, Germany) TCS SPE using the solid-state 405 nm with the DAPI setting (emission: 430–480); the 488 nm with the GFP setting (emission: 500–600); and the 561 nm with the DRED setting (emission: 650–700) laser lines and merged using the Leica LAS X software. Final image adjustments were accomplished with Adobe Systems Photoshop CS6 (San Jose, CA).

Acknowledgements

The authors thank Marc Albertsen and Peggy Ozias-Akins for reviewing the manuscript and providing comments. The authors also thank Katherine Thilges for her assistance with the sorghum microscopy and special thanks to Tracey Fisher for her assistance during manuscript preparation and for valuable comments during review. Authors also thank anonymous reviewers for insightful input.

References

An, G., Mitra, A., Choi, H.K., Costa, M.A., An, K., Thornburg, R.W. and Ryan, C.A. (1989) Functional analysis of the 3′ control region of the potato wound-inducible proteinase inhibitor II gene. *Plant Cell*, **1**, 115–122.

Ananiev, E., Wu, C., Chamberlin, M., Svitashev, S., Schwartz, C., Gordon-Kamm, W. and Tingey, S. (2009) Artificial chromosome formation in maize (*Zea mays* L.). *Chromosoma*, **118**, 157–177.

Casas, A.M., Kononowicz, A.K., Haan, T.G., Zhang, L., Tomes, D.T., Bressan, R.A. and Hasegawa, P.M. (1997) Transgenic sorghum plants obtained after microprojectile bombardment of immature inflorescences. *In Vitro Cellular & Developmental Biology-Plant*, **33**, 92–100.

Christensen, A.H., Sharrock, R.A. and Quail, P.H. (1992) Maize polyubiquitin genes: structure, thermal perturbation of expression and transcript splicing, and promoter activity following transfer to protoplasts by electroporation. *Plant Mol. Biol.* **18**, 675–689.

Cigan, A.M., Unger, E., Xu, R.-J., Kendall, T.L. and Fox, T.W. (2001) Phenotypic complementation of *ms45* maize requires tapetal expression of MS45. *Sex. Plant Reprod.* **14**, 135–142.

Cigan, A.M., Unger-Wallace, E. and Haug-Collet, K. (2005) Transcriptional gene silencing as a tool for uncovering gene function in maize. *Plant J.* **43**, 929–940.

Consortium, I.W.G.S. (2014) A chromosome-based draft sequence of the hexaploid bread wheat (*Triticum aestivum*) genome. *Science*, **345**, 1251788.

Cornu, T.I., Thibodeau-Beganny, S., Guhl, E., Alwin, S., Eichtinger, M., Joung, J.K. and Cathomen, T. (2007) DNA-binding specificity is a major determinant of the activity and toxicity of zinc-finger nucleases. *Mol. Ther.* **16**, 352–358.

Djukanovic, V., Smith, J., Lowe, K., Yang, M., Gao, H., Jones, S., Nicholson, M.G. *et al.* (2013) Male-sterile maize plants produced by targeted mutagenesis of the cytochrome P450-like gene (*MS26*) using a re-designed I-*Cre*I homing endonuclease. *Plant J.* **76**, 888–899.

Duvick, D.N. (1999) Heterosis: feeding people and protecting natural resources. In: *The Genetics and Exploitation of Heterosis in Crops* (Corrs, J.G., Pandey, S. and Gerdes, J.T. eds), pp. 19–29. Madison, WI: ASA-CSSA-SSSA.

Dwivedi, S., Perotti, E. and Ortiz, R. (2008) Towards molecular breeding of reproductive traits in cereal crops. *Plant Biotechnol. J.* **6**, 529–559.

Gao, H., Smith, J., Yang, M., Jones, S., Djukanovic, V., Nicholson, M.G., West, A. *et al.* (2010) Heritable targeted mutagenesis in maize using a designed endonuclease. *Plant J.* **61**, 176–187.

Guo, J.-X. and Liu, Y.-G. (2012) Molecular control of male reproductive development and pollen fertility in rice. *J. Integr. Plant Biol.* **54**, 967–978.

Hervé, P. and Kayano, T. (2006) Japonica rice varieties (*Oryza sativa*, Nipponbare, and others). In: *Agrobacterium Protocols* (Wang, K. ed) pp. 213–222. Totowa, NJ: Humana Press.

Hobo, T., Suwabe, K., Aya, K., Suzuki, G., Yano, K., Ishimizu, T., Fujita, M. *et al.* (2008) Various spatiotemporal expression profiles of anther-expressed genes in rice. *Plant Cell Physiol.* **49**, 1417–1428.

Komari, T., Hiei, Y., Saito, Y., Murai, N. and Kumashiro, T. (1996) Vectors carrying two separate T-DNAs for co-transformation of higher plants mediated by *Agrobacterium tumefaciens* and segregation of transformants free from selection markers. *Plant J.* **10**, 165–174.

Kumar, V. and Jain, M. (2015) The CRISPR–Cas system for plant genome editing: advances and opportunities. *J. Exp. Bot.* **66**, 47–57.

Li, J. and Yuan, L. (1999) Hybrid rice: genetics, breeding and seed production. In *Plant Breeding Reviews* (Janick, J., ed), pp. 15–158. Oxford, U.K.: John Wiley & Sons Inc.

Li, H., Pinot, F., Sauveplane, V., Werck-Reichhart, D., Diehl, P., Schreiber, L., Franke, R. *et al.* (2010) Cytochrome P450 family member CYP704B2 catalyzes the ω-hydroxylation of fatty acids and is required for anther cutin biosynthesis and pollen exine formation in rice. *The Plant Cell Online*, **22**, 173–190.

Loukides, C.A., Broadwater, A.H. and Bedinger, P.A. (1995) Two new male-sterile mutants of *Zea mays* (Poaceae) with abnormal tapetal cell morphology. *Am. J. Bot.* **82**, 1017–1023.

Melchinger, A.E. (1999) Genetic diversity and heterosis. In: *The Genetics and Exploitation of Heterosis in Crops* (Coors, J.G. and Pandy, S. eds), pp. 99–118. Madison, WI: CSSA.

Miller, F.R. (1984) Registration of RTx430 sorghum parental line. *Crop Sci.* **24**, 1224.

Neuffer, M.G., Coe, E.H. and Wessler, S.R. (1997) *Mutants of Maize*. Plainview, NY: Cold Spring Harbor Laboratory Press.

Osakabe, Y. and Osakabe, K. (2015) Genome editing with engineered nucleases in plants. *Plant Cell Physiol.* **56**, 389–400.

Parry, M.A.J. and Hawkesford, M.J. (2012) An integrated approach to crop genetic improvement. *J. Integr. Plant Biol.* **54**, 250–259.

Puchta, H. and Hohn, B. (2012) In planta somatic homologous recombination assay revisited: a successful and versatile, but delicate tool. *Plant Cell*, **24**, 4324–4331.

Ray, D.K., Mueller, N.D., West, P.C. and Foley, J.A. (2013) Yield trends are insufficient to double global crop production by 2050. *PLoS One*, **8**, e66428.

Reddy, B.V.S., Ramesh, S. and Ortiz, R. (2005) Genetic and cytoplasmic-nuclear male sterility in sorghum. In *Plant Breeding Reviews* (Janick, J., ed), pp. 139–172. Oxford, U.K.: John Wiley & Sons Inc.

Sabelli, P.A., Hoerster, G., Lizarraga, L.E., Brown, S.W., Gordon-Kamm, W.J. and Larkins, B.A. (2009) Positive regulation of minichromosome maintenance gene expression, DNA replication, and cell transformation by a plant retinoblastoma gene. *Proc. Natl Acad. Sci.* **106**, 4042–4047.

Sambrook, J., Fritsch, E.F. and Maniatis, T. (1989) *Molecular Cloning: a Laboratory Manual*. Cold Spring Harbor, NY: Cold Spring Harbor Laboratory.

Shi, J., Cui, M., Yang, L., Kim, Y.J. and Zhang, D. (2015) Genetic and biochemical mechanisms of pollen wall development. *Trends Plant Sci.* **20**, 741–753.

Southern, E.M. (1975) Detection of specific sequences among DNA fragments separated by gel electrophoresis. *J. Mol. Biol.* **98**, 503–517.

Svitashev, S., Young, J.K., Schwartz, C., Gao, H., Falco, S.C. and Cigan, A.M. (2015) Targeted mutagenesis, precise gene editing, and site-specific gene insertion in maize using Cas9 and guide RNA. *Plant Physiol.* **169**, 931–945.

Szczepek, M., Brondani, V., Buchel, J., Serrano, L., Segal, D.J. and Cathomen, T. (2007) Structure-based redesign of the dimerization interface reduces the toxicity of zinc-finger nucleases. *Nat. Biotechnol.* **25**, 786–793.

Tester, M. and Landrige, P. (2010) Breeding technologies to increase crop production in a changing world. *Science*, **327**, 818–822.

Tian, H., Lu, C., Ciais, P., Michalak, A.M., Canadell, J.G., Saikawa, E., Huntzinger, D.N. *et al.* (2016) The terrestrial biosphere as a net source of greenhouse gases to the atmosphere. *Nature*, **531**, 225–228.

Tilman, D., Balzer, C., Hill, J. and Befort, B.L. (2011) Global food demand and the sustainable intensification of agriculture. *Proc. Natl Acad. Sci. USA*, **108**, 20260–20264.

Toki, S. (1997) Rapid and efficient *Agrobacterium*-mediated transformation in rice. *Plant Mol. Biol. Rep.* **15**, 16–21.

Townsend, J.A., Wright, D.A., Winfrey, R.J., Fu, F., Maeder, M.L., Joung, J.K. and Voytas, D.F. (2009) High-frequency modification of plant genes using engineered zinc-finger nucleases. *Nature*, **459**, 442–445.

Unger, E., Betz, S., Xu, R. and Cigan, A.M. (2001) Selection and orientation of adjacent genes influences DAM-mediated male sterility in transformed maize. *Transgenic Res.* **10**, 409–422.

Voytas, D.F. and Gao, C. (2014) Precision genome engineering and agriculture: opportunities and regulatory challenges. *PLoS Biol.* **12**, e1001877.

Wang, J., Kim, H., Yuan, X. and Herrin, D. (1997) Purification, biochemical characterization and protein-DNA interactions of the I-*Cre*I endonuclease produced in *Escherichia coli*. 10.1093/nar/25.19.3767. *Nucleic Acids Res.* **25**, 3767–3776.

Werck-Reichhart, D. and Feyereisen, R. (2000) Cytochromes P450: a success story. *Genome Biol.* **1**, 3003.1–3003.9.

West, P.C., Gerber, J.S., Engstrom, P.M., Mueller, N.D., Brauman, K.A., Carlson, K.M., Cassidy, E.S. *et al.* (2014) Leverage points for improving global food security and the environment. *Science*, **345**, 325–328.

Whitford, R., Fleury, D., Reif, J.C., Garcia, M., Okada, T., Korzun, V. and Langridge, P. (2013) Hybrid breeding in wheat: technologies to improve hybrid wheat seed production. *J. Exp. Bot.* **64**, 5411–5428.

Wu, Y., Fox, T.W., Trimnell, M.R., Wang, L., Xu, R.J., Cigan, A.M., Huffman, G.A. *et al.* (2015) Development of a novel recessive genetic male sterility

system for hybrid seed production in maize and other cross-pollinating crops. *Plant Biotechnol. J.* **14**, 1046–1054.

Yamagata, Y., Doi, K., Yasui, H. and Yoshimura, A. (2007) Identification of mutants for abnormal pollen development in rice. *Breeding Science*, **57**, 331–337.

Zhao, Z.-Y., Cai, T., Tagliani, L., Miller, M., Wang, N., Pang, H., Rudert, M. *et al.* (2000) *Agrobacterium*-mediated sorghum transformation. *Plant Mol. Biol.* **44**, 789–798.

Zhu, H., Jeoung, J.M., Liang, G.H., Muthukrishnan, S., Krishnaveni, S. and Wilde, G. (1998) Biolistic transformation of sorghum using a rice chitinase gene [*Sorghum bicolor* (L.) Moench-*Oryza sativa* L.]. *Journal of Genetics & Breeding (Italy)*, **52**, 243–252.

10

CRISPR-Cas9-mediated efficient directed mutagenesis and RAD51-dependent and RAD51-independent gene targeting in the moss *Physcomitrella patens*

Cécile Collonnier[1,*], Aline Epert[1], Kostlend Mara[1], François Maclot[1], Anouchka Guyon-Debast[1], Florence Charlot[1], Charles White[2], Didier G. Schaefer[3] and Fabien Nogué[1,*]

[1]*INRA Centre de Versailles-Grignon, IJPB (UMR1318), Versailles Cedex, France*
[2]*Génétique, Reproduction et Développement, UMR CNRS 6293, Clermont Université, INSERM U1103, Université Blaise Pascal, Clermont Ferrand, France*
[3]*Laboratoire de Biologie Moléculaire et Cellulaire, Institut de Biologie, Université de Neuchâtel, Neuchâtel, Switzerland*

*Correspondence

email cecile.collonnier@versailles.inra.fr

email fabien.nogue@versailles.inra.fr

Keywords: CRISPR-Cas9, *Physcomitrella patens*, genome editing, alt-EJ, gene targeting, RAD51.

Summary

The ability to address the CRISPR-Cas9 nuclease complex to any target DNA using customizable single-guide RNAs has now permitted genome engineering in many species. Here, we report its first successful use in a nonvascular plant, the moss *Physcomitrella patens*. Single-guide RNAs (sgRNAs) were designed to target an endogenous reporter gene, *PpAPT*, whose inactivation confers resistance to 2-fluoroadenine. Transformation of moss protoplasts with these sgRNAs and the Cas9 coding sequence from *Streptococcus pyogenes* triggered mutagenesis at the *PpAPT* target in about 2% of the regenerated plants. Mainly, deletions were observed, most of them resulting from alternative end-joining (alt-EJ)-driven repair. We further demonstrate that, in the presence of a donor DNA sharing sequence homology with the *PpAPT* gene, most transgene integration events occur by homology-driven repair (HDR) at the target locus but also that Cas9-induced double-strand breaks are repaired with almost equal frequencies by mutagenic illegitimate recombination. Finally, we establish that a significant fraction of HDR-mediated gene targeting events (30%) is still possible in the absence of PpRAD51 protein, indicating that CRISPR-induced HDR is only partially mediated by the classical homologous recombination pathway.

Introduction

Over the last decades, gene editing and transgene integration have been shown to be facilitated by the generation of a DNA double-strand break (DSB) at targeted genomic locations, using homing endonucleases such as meganucleases, zinc finger nucleases (ZFNs) and TAL effector nucleases (TALENs). Recently, a new type of site-directed nucleases based on the prokaryotic type II CRISPR-Cas9 (clustered regularly interspaced short palindromic repeats/CRISPR-associated protein) immune system has been used for precise genome editing in many species with spectacular success (Lander, 2016). In this system, a Cas9 endonuclease protein from *Streptococcus pyogenes* guided by a customizable noncoding RNA introduces site-specific double-strand DNA breaks (DSBs) in the genome. Repair of these DSBs can lead to gene disruption if the break is repaired by a deleterious event resulting from a classical nonhomologous end-joining reaction (C-NHEJ), an alternative end-joining reaction (alt-EJ, also called microhomology-mediated end joining, MMEJ) or a single-strand annealing reaction (SSA; Ceccaldi et al., 2016). Alternatively, in the presence of a homologous donor DNA template, these DSBs can be repaired via a homology-directed repair (HDR) pathway, leading to accurate gene replacement (Ceccaldi et al., 2016). The feasibility of such approach has been demonstrated in several eukaryotic cells and raises great expectations for gene therapeutic approaches. However, highly efficient HDR of CRISPR-Cas9-induced genomic DSB remains so far restricted to GT-competent cells such as budding yeast (Dicarlo et al., 2013).

In plants, the CRISPR-Cas9 system has been applied with good efficiency for the induction of illegitimate recombination-mediated (IR) targeted mutagenesis—or knockout—of endogenous loci (Schaeffer and Nakata, 2015). It constitutes a revolutionary tool for functional gene analysis, but also a promising approach for the development of new traits of interest in crops (Collonnier et al., 2016; Petolino et al., 2016). By comparison, the examples of integration of a donor DNA template sharing homology with the target and leading to gene replacement through HDR-based repair, namely gene targeting (GT), are rare, hardly reaching the percentage range (Čermák et al., 2015; D'Halluin et al., 2013; Li et al., 2013; Nishizawa-Yokoi et al., 2015; Schiml et al., 2014; Shukla et al., 2009; Townsend et al., 2009). This reflects the fact that very low GT efficiency is generally observed in higher plants, reaching only 0.01% to 0.1% of the effectively transformed plants (Hanin and Paszkowski, 2003). Even if site-directed nucleases, such as the CRISPR-Cas9 system, help for the targeted modification of genes, this strategy is still challenging compared to gene knockout (Schiml and Puchta, 2016). High GT frequencies are naturally achieved only by a few species/cell lines (Schaefer, 2001), and this is thought to be associated with the fact that homologous recombination (HR) is the principal mechanism for DSBs repair, as exemplified in *Saccharomyces cerevisiae* (Pâques and Haber, 1999). The moss *Physcomitrella patens* is the only plant that naturally displays high GT efficiencies (Schaefer and Zrÿd, 1997), and recent genetic studies have shown that this feature is tightly associated with the classical RAD51-mediated HR repair pathway (Charlot et al., 2014; Kamisugi et al., 2012; Schaefer et al., 2010). Efficient GT, the availability of a completely sequenced genome and unique genetic and developmental facilities have established *P. patens* as a valuable novel model system in plant biology (Bonhomme *et al.*, 2013; Kofuji and Hasebe, 2014; Prigge and Bezanilla, 2010).

Here, we report the first successful use of the CRISPR-Cas9 system to achieve both targeted mutagenesis and gene targeting in *Physcomitrella patens*. To monitor the RNA-guided Cas9 nuclease activity, we designed sgRNAs targeting the endogenous adenine phosphoribosyltransferase (*PpAPT*) selectable marker gene whose loss of function confers resistance to 2-fluoroadenine (2-FA; Charlot *et al.*, 2014; Kamisugi *et al.*, 2012; Schaefer *et al.*, 2010; Trouiller *et al.*, 2006, 2007). We show that PEG-mediated transformation of moss protoplasts with these sgRNAs and the Cas9 coding sequence from *Streptococcus pyogenes* efficiently induces targeted mutagenesis of the *PpAPT* gene in 2% to 3% of the regenerated plants. Molecular analyses revealed that these mutations result from a diversity of deletions, insertions and/or substitutions in the *PpAPT* locus, confirming the efficiency of the CRISPR-Cas9 system for gene knockout/editing in *P. patens*.

To evaluate the impact of CRISPR/Cas9 on GT, we performed moss transformation with the sgRNA sequences, the Cas9 gene and a circular donor plasmid bearing an antibiotic resistance gene flanked by DNA fragments homologous to the genomic regions flanking the target. Our analyses reveal that HDR-mediated integration of the donor DNA in the *PpAPT* locus occurs in almost 100% of the transformed plants (i.e. proportion of 2-FA^R among antibiotic-resistant plants). Molecular analyses further indicate that the proportion of single-copy replacements is significantly increased compared to the classical approach with linearized replacement vectors. Interestingly, our data also demonstrate that approximately 40% (i.e. proportion of antibiotic resistant among 2-FA^R plants) of Cas9-induced DSBs are not repaired by HDR in this situation, indicating that a significant fraction of these DSBs are repaired by mutagenic IR or end-joining reactions.

Finally, we assessed CRISPR-Cas9-mediated GT efficiency in the *Pprad51-1-2* double mutant, as we previously established that this gene was essential to achieve GT using linearized replacement vectors (Schaefer *et al.*, 2010; Wendeler *et al.*, 2015). Unexpectedly, this analysis revealed that HDR-mediated GT was reduced but not abolished in the mutant, reaching approximately 30% of the WT level. This observation implies that other types of DNA repair pathways are involved in the integration of the donor template when Cas9 generates a DSB at the chromosomal target gene. Thus, the use of the CRISPR-Cas9 system significantly improves GT efficiency and precision in *P. patens*, expanding the range of available tools for gene function analysis in this model organism. These data also uncover novel features of CRISPR-induced HDR-mediated GT that could lead to improve the efficiency of such approach in GT noncompetent cells.

Results

Highly efficient gene knockout in *P. patens* with RNA-guided Cas9 nuclease

To evaluate the potential of the CRISPR-Cas9 system to induce targeted mutagenesis in *P. patens*, two sgRNAs matching, respectively, two target loci in exon 3 (sgRNA#2) and exon 5 (sgRNA#1) of the *PpAPT* reporter gene were designed (Figures 1 and S1). *P. patens* wild-type protoplasts were cotransformed by

PEG-mediated transformation with two plasmids, one bearing the Cas9 gene under the control of the rice actin 1 promoter (pAct-Cas9) and another bearing the sgRNA#1 or sgRNA#2, both under the control of a *P. patens* U6 promoter. Mutations leading to a loss of APT activity confer resistance to toxic adenine analogues, such as 2-fluoroadenine (2-FA; Schaff, 1994; Trouiller *et al.*, 2006). The mutation rates (expressed in percentages) were estimated by dividing the number of 2-FA-resistant plants by the number of regenerating plants observed just before the transfer on 2-FA. The mutation rates obtained using sgRNA#1 and sgRNA#2 were, respectively, 2.2% and 3.2% (Table 1) and were optimal when the sgRNAs and Cas9 plasmids were provided in a 1 : 1 ratio (Figure S2). To characterize these mutations, we amplified by PCR and sequenced the *PpAPT* gene in 34 independent clones for sgRNA#1 and 43 independent clones for sgRNA#2 (Figure 2). As expected, all the mutations were located in the vicinity of the PAM target of the Cas9-induced cleavage site (Gasiunas *et al.*, 2012) and generated loss of APT function. These mutations consisted of deletions of 1 to 588 bp, insertions of 2 to 39 bp and substitutions of 1 to 2 bp, a majority of the mutations being deletions (Table 2). Generally, the substitutions occurred inside the target sequence but they were also observed up to 22 bp downstream of the PAM. Regarding the insertions, they all occurred a few nucleotides upstream of the PAM and for sgRNA#1, 8 of 9 occurred exactly at the same position 3 bp before the PAM. Interestingly, for a large number of the deletions (12/23 simple deletion events for sgRNA#1 and 36/40 for sgRNA#2), microhomologies (of 2 to 4 bp) could be detected between the end of the deletion itself and the sequence located just upstream of the deletion (Figure 2). With sgRNA#1, 8 events could be explained by alt-EJ-mediated repair based on 3-bp-long microhomologies and 4 events on 2-bp-long microhomologies. With sgRNA#2, 34 events could be explained by alt-EJ-mediated repair based on 4-bp-long microhomologies and 2 events on 2-bp-long microhomologies (Figure 2). Thus, the CRISPR-Cas9 system is very efficient to induce targeted mutagenesis in *P. patens*, and the repair of an induced genomic DSB seems to implicate both C-NHEJ and alt-EJ mechanisms (Figure S3).

RNA-guided nuclease activity is very specific in *P. patens*

The sgRNAs used in this study were designed to minimize potential off-target cleavage in the *P. patens* genome (Phytozome 3.1) using the CRISPOR software. For both the selected targets, no perfect 20-bp matches were found but potential off-target sequences presenting 3 to 6 mismatches were identified: 9 for sgRNA#1 and 3 for sgRNA#2 (Figure S4). All these potential off-target loci were amplified with surrounding primers (Figure S5) and sequenced in 48 clones transformed with hCas9 and sgRNA#1 and 48 clones transformed with hCas9 and sgRNA#2 that were all previously identified as mutated at the *PpAPT* locus. No mutation could be detected in the potential off-target sequences for any of the tested clones.

CRISPR-Cas9 system increases gene targeting efficiency and single-copy replacement in *P. patens*

The impact of the CRISPR-Cas9 system on gene targeting efficiency was evaluated by cotransferring into protoplasts three circular plasmids, one expressing the Cas9 gene (pAct-Cas9), one expressing the sgRNA#2 and one carrying the donor cassette PpAPT-KO7 which bears a G418-resistant gene surrounded by

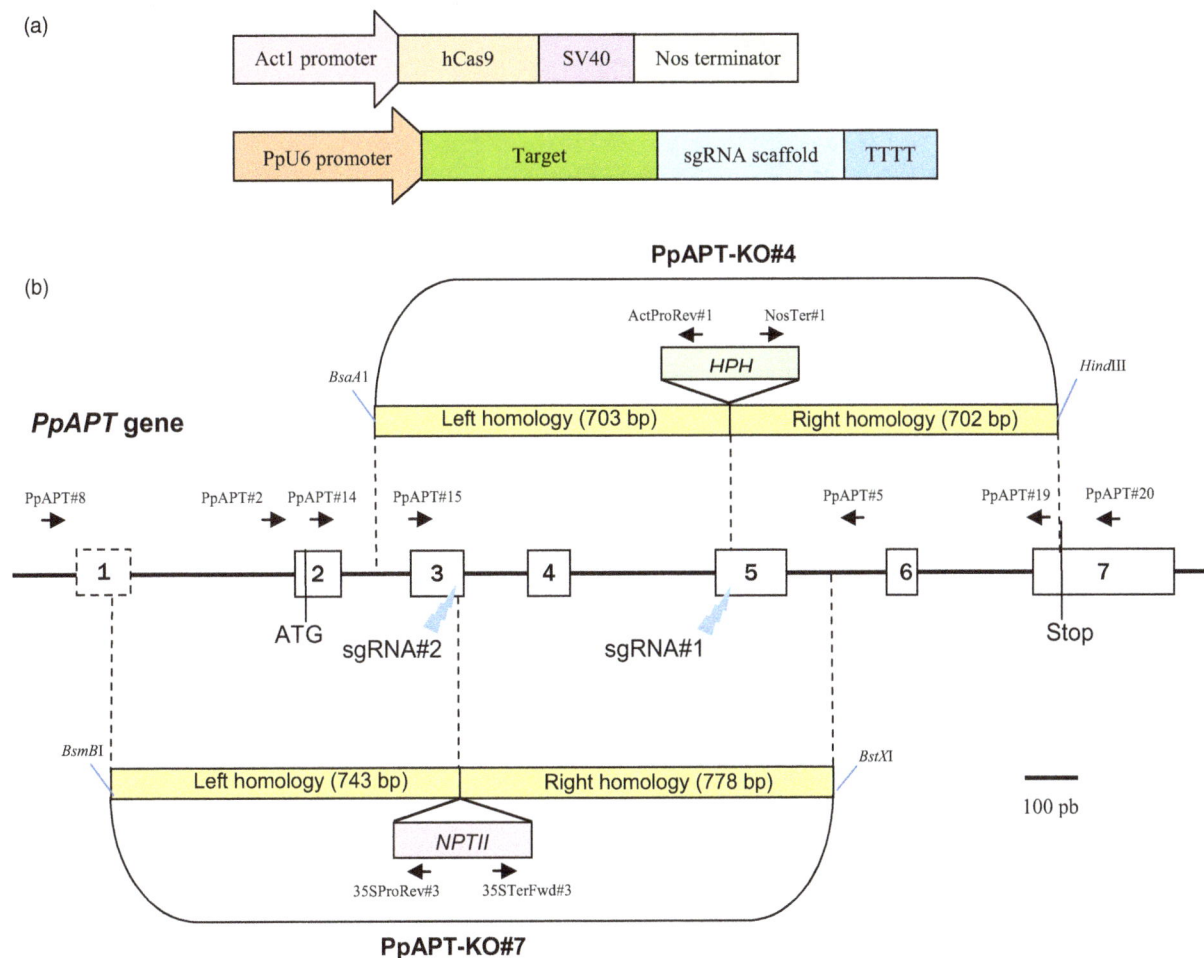

Figure 1 Schematic description of the sgRNA/Sp-hCas9 system and of the PpAPT reporter gene. (a) Design of the pAct-Cas9 and psgRNAs constructs. (b) Structure of PpAPT with the target sites and the homologies of the donor cassette (PpAPT-KO4 and PpAPT-KO7) used for gene targeting experiment (white rectangles represent exons). For classical GT experiments, donor DNA was released using restriction enzymes as indicated.

Table 1 PpAPT gene knockout efficiency using the CRISPR-Cas9 system

sgRNA used for transformation*	Regenerant clones	2-FA^R clones	Knockout efficiency (%)
No sgRNA	15 000	0	0
sgRNA#1	149 175 (49 725 ± 3857[†])	3321 (1107 ± 107[†])	2.2 ± 0.09[†]
sgRNA#2	164 650 (54 883 ± 2819[†])	5229 (1743 ± 140[†])	3.2 ± 0.16[†]

*sgRNA#1 and sgRNA#2 target exon 5 and exon 3 of the PpAPT gene respectively (see Figure 1b).

[†]Average and standard deviations were determined from three independent experiments.

PpAPT gene sequences flanking the sgRNA target (Figure 1b). In parallel, we performed a classical gene targeting experiment and transformed protoplasts with the linearized replacement cassette PpAPT-KO7 (Figure 1b). In classical GT experiments, HR-mediated integration of a replacement vector can occur in 2 distinct ways: by targeted gene replacement (TGR) or by targeted gene insertion (TGI). TGR is mediated by two HR reactions involving both homologous sequences of the vector and leads to accurate gene replacement. TGI results from a single HR involving one of the homologous sequence of the vector on one side of the target, associated with an IR event involving the other homologous

sequence on the other side (Kamisugi et al., 2006; Schaefer, 2002; Schaefer and Zrÿd, 2004). To evaluate the efficiency of gene targeting through HDR, regenerating moss plants from the 'CRISPR-Cas9' or 'classical' transformations were sequentially subcultured first on media containing the antibiotic G418 and then on 2-FA. Our analysis showed that relative transformation frequency (RTF: i.e. frequency of G-418^R plants) was much higher using the 'CRISPR-Cas9'-mediated transformation (2.1%) compared to the 'classical' transformation method (0.25%, Table 3). It further demonstrated that GT efficiencies (i.e. the percentage of 2-FA-resistant plants among G418-resistant plants) are

sgRNA#1

                                        PAM
```
AGGCCCTGTGATTAGGGaagagtatagtctagagtaTGGTACCGATTGCATTGAGATGCACGTTGG                    WT
AGGCCCTGTGATTAGGGaagagtatagtctagaCtaTGGTACCGATTGCATTGAGATGCACGTTGG                   (x1)
AGGCCCTGTGATTAGGGaagagtatagtctagaC-aTGGTACCGATTGCATTGAGATGCACGTTGG          -1    (x1)
AGGCCCTGTGATTAGGGaagagtatagtcta--gtaTGGTACCGATTGCATTGAGATGCACGTTGG          -2    (x6)
AGGCCCTGTGATTAGGGaagagtatagtctaga--aTGGTACCGATTGCATTGAGATGCACGTTGG          -2    (x1)
AGGCCCTGTGATTAGGGaagagtatag------AtaTGGTACCGATTGCATTGAGATGCACGTTGG          -6    (x1)
AGGCCCTGTGATTAGGGaagagtatag-----  -taTGGTACCGATTGCATTGAGATGCACGTTGG          -7    (x2)
AGGCCCTGTGATTAGGGaagagtatagtctag-------TACCGATTGCATTGAGATGCACGTTGG          -7    (x1)
AGGCCCTGTGATTAGGGaagagta---------   TGGTACCGATTGCATTGAGATGCACGTTGG          -12   (x8)
AGGCCCTGTGATTAGGGaagagtatagt----------  ACCGATTGCATTGAGATGCACGTTGG          -12   (x1)
AGGCCCTGTGATTAGGGaagagtatagtctaga----------  TTGCATTGAGATGCACGTTGG          -12   (x1)
--------------------------------------------GATTGCATTGAGATGCACGTTGG         -588   (x1)
AGGCCCTGTGATTAGGGaagagtatagtctagaCAAtaTGGTACCGATTGCATTGAGATGCACGTTGG         +2    (x1)
AGGCCCTGTGATTAGGGaagagtatagtctagaCCTgtaTGGTACCGATTGCATTGAGATGCACGTTGG        +3    (x1)
AGGCCCTGTGATTAGGGaagagtatagtctagaCTATAgtaTGGTACCGATTGCATTGAGATGCACGTTGG      +5    (x1)
AGGCCCTGTGATTAGGGaagagtatagtctagaCTATGTCTgtaTGGTACCGATTGCATTGAGATGCACGTTGG   +8    (x1)
AGGCCCTGTGATTAGGGaagagtatagtctagaCTATACTCAgtaTGGTACCGATTGCATTGAGATGCACGTTGG  +9    (x1)
AGGCCCTGTGATTAGGGaagagtatagtctagaTTAGGGAAGAgtaTGGTACCGATTGCATTGAGATGCACGTTGG +10   (x1)
AGGCCCTGTGATTAGGGaagagtatagtctagaTACCGGAAGAgtaTGGTACCGATTGCATTGAGATGCACGTTGG +10   (x1)
AGGCCCTGTGATTAGGGaagagtatagtctagaCTATACTAGACTATgtaTGGTACCGATTGCATTGAGATGCACGTTGG +14 (x1)
AGGCCCTGTGATTAGGGaagagtatagtctagaCTATACTATAGTATgtaTGGTACCGATTGCATTGAGATGCACGTTGG  +14 (x1)
```

sgRNA#2

                  PAM
```
TGTTgagcgttaccgggaccagaAGGTGGACGTCATTGTGGGTGCGATGCCCT                    WT
TGTTgagcgttaccggg-------  TGGACGTCATTGTGGGTGCGATGCCCT           -9   (x1)
TGTTgagcgttaccggga---------GGATGTCATTGTGGGTGCGATGCCCT           -9   (x1)
TGTTgagcgttaccgggg----------TAACGTCATTGTGGGTGCGATGCCCT          -10  (x1)
TGTTgagcgttaccgggac-------  GTCATTGTGGGTGCGATGCCCT              -12  (x32)
TGTTgagcgttaccgggac-------  GTCATTGTGGGTGCGACGCCCT              -12  (x1)
TGTTgagcgttaccgggac-------  GTTATTGTGGGTGCGATGCCCT              -12  (x1)
TGTTgagcgttaccgggac---------------ATTGTGGGTGCGATGTCCT           -15  (x1)
TGTTTg--------------------  GACGTCATTGTGGGTGCGATGCCCT            -23  (x1)
--------------------------------CATTGTGGGTGCGATGCCCT            -76  (x1)
TGTTgagcgttaccgggacGTCAACAACCTTGAagaAGGTGGACGTCATTGTGGGTGCGATGCCCT  +14  (x1)
TGTTgagcg-------gCccagGTTTAGGTTTAGGGTTTAGGTGGACGTCATTGTGGGTGCGATGCCCT -7/+16 (x1)
TGTTgagcgttaccgggacGTATAAGAGTAAGAAAAAagaAGGTGGACGTCATTGTGGGTGCGATGCCCT +17 (x1)
TGTTgagcgttaccgggaccagaGTTTAGGGTTTAGGGTTTAGGTTTAGGGTTTAGGACCAGaAGGTGGACGTCATTGTGGGTGCGATGCCCT +39 (x1)
```

Figure 2 Targeted genome editing on *PpAPT* gene in *Physcomitrella patens* protoplasts. Green letters indicate the sequences targeted by the tested sgRNAs. DNA insertions are shown in blue, point mutations in bold capital letters on a yellow background and deletions with dashes. With a grey background, the differences between two insertions of the same length. The PAM (protospacer adjacent motif) is marked in red underlined letters. In the frames, the microhomologies potentially involved in alt-EJ-mediated repair of the CRISPR-induced DSBs. For each deletion potentially due to alt-EJ, the brown frame surrounds the 5′ homology, and the black one, the position of the 3′ homology before alt-EJ occurred and produced the deletion represented by dashes (see Figure S3).

Table 2 Types of CRISPR-Cas9-induced mutations

sgRNA used for transformation*	Number of analysed clones	No. of clones with deletions (%)	No. of clones with insertions (%)	No. of clones with substitutions (%)
sgRNA#1	34	23 (67)	10 (29)	3 (9)
sgRNA#2	43	40 (93)	4 (9)	5 (9)

*sgRNA#1 and sgRNA#2 target exon 5 and exon 3 of the *PpAPT* gene respectively (see Figure 1b).

significantly increased, reaching 100% following CRISPR-mediated transformation compared to 54% with the classical strategy (Table 3, Fisher's exact test $P = 0.008$).

At the molecular level, junction analyses by PCR genotyping of the *PpAPT* locus in G418 and 2-FA-resistant plants ($n = 95$ for the 'CRISPR-Cas9'-mediated transformation and $n = 52$ for the 'classical' mediated transformation) provided evidence for the integration of the donor vector by at least one HR event in all the plants isolated from both procedures, leading to either TGR or TGI events (Figure S6). The ratio between the two types of events is not significantly different between the two methods of transformation (Table 3).

Thus, our analyses revealed that the use of CRISPR-Cas9 increases both transformation frequencies (8.4-fold) and GT

Table 3 Comparison of *PpAPT* gene targeting efficiency using the 'CRISPR-Cas9' versus 'classical' mediated transformations

Type of transformation*	RTF[†] %	AB[R] clones	2-FA[R] clones[‡]	Integration due to IR[§]	Integration due to HDR[¶]		GT[‖] %
					TGR	TGI (5'TGI + 3'TGI)	
'CRISPR-Cas9'	2.1 ± 0.2	95	95	0	84	11 (7 + 4)	100
'Classical'	0.25 ± 0.1	95	52	43	40	12 (7 + 5)	54.7

*For 'CRISPR-Cas9'-mediated transformation, wild-type protoplasts were cotransformed with pAct-Cas9, psgRNA#2 and circular PpAPT-KO7 donor DNA cassette. For 'classical' mediated transformation, wild-type protoplasts were transformed with linear PpAPT-KO7 donor DNA cassette.

[†]Relative transformation frequencies (RTF) express the frequency of stable AB[R] clones in the population of regenerated clones. A total of 2580 and 123 AB[R] clones were obtained for the 'CRISPR-Cas9' and 'classical' methods of transformation, respectively. Standard deviation was determined from 3 independent experiments.

[‡]2-FA[R] clones are the stable AB[R] clones that survived after subculture on 2-FA medium. They all result from a homology-driven recombination (HDR) event (see Figure S6).

[§]Number of AB[R] clones where the donor DNA template has been randomly inserted by illegitimate recombination (IR) and not via HDR.

[¶]Number of 2-FA[R] clones resulting from HDR (TGR or TGI)-mediated insertion of the donor DNA template at the *PpAPT* locus was determined by PCR analysis (see Figure S6). Clones resulting from TGR show 5' and 3' junction, clones resulting from 5' TGI show only a 5' junction, and clones resulting from 3' TGI show only a 3' junction.

[‖]GT efficiencies (%) express the frequency of 2-FA-resistant clones among the population of antibiotic-resistant transgenic clones.

efficiencies (1.8-fold). These results also show that when a DSB is induced at the chromosomal target site, the linearization of the donor DNA template is no longer a prerequisite for efficient GT. Finally, as observed before for classical mediated transformation, 'CRISPR-Cas9'-mediated transformation can lead to TGR or TGI of the donor DNA template.

Insertion of concatenated copies of the donor cassettes is frequent in GT experiments in *P. patens* (Kamisugi *et al.*, 2006; Schaefer and Zrÿd, 1997). We further genotyped these plants to detect TGR events in which a single copy of the donor cassette had been integrated at the target site, using genomic primers located upstream and downstream of the *PpAPT* sequences present in PpAPT-KO7 (Figure S7). This analysis showed that the number of clones carrying only one copy of the cassette was significantly higher (Fisher's exact test $P = 0.04$) with the 'CRISPR-Cas9'-mediated transformation (40.5%) than with the 'classical' method (15%; Figure 3). This significant increase in single-copy insertions at the *APT* locus using the CRISPR-Cas9 strategy has to be confirmed for others loci in *P. patens*. The use of CRISPR-Cas9

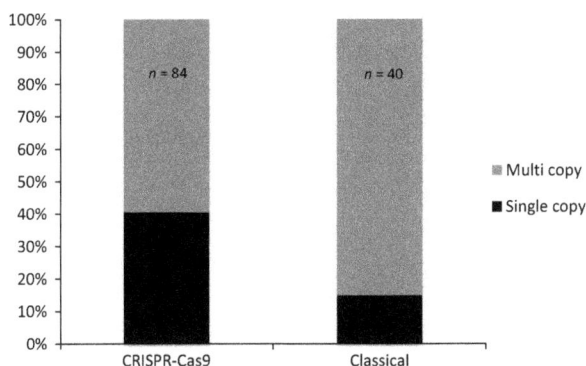

Figure 3 Ratio of single copy versus multiple copies of donor DNA template insertions using the 'CRISPR-Cas9' or the 'classical' methods of transformation in *Physcomitrella patens*. Frequency of single-copy TGR insertions was determined by genotyping the 2-FA[R] clones (84 for the 'CRISPR-Cas9' method and 40 for the 'classical' method, see Table 3) using primers located outside the sequences homologous to the gene fragments present in the PpAPT-KO7 donor DNA template (see Figure S7).

could possibly reduce the problem of insertion of concatemers at the target site, a frequent occurrence in *P. patens* transformation. Altogether, our results show that the induction of a chromosomal DSB at the target site increases the frequency, the efficiency and the accuracy of GT in *P. patens*, compared to the classical approach with linearized replacement vectors.

In the presence of a donor template, CRISPR/Cas9-induced DSBs are repaired either by HDR or by IR

As shown above, integration of a donor DNA template occurs almost exclusively by HDR when a DSB is produced by the CRISPR-Cas9 system at the chromosomal target (Table 3, 95 of 95 for sgRNA#2/PpAPT-KO7). We also showed that, in absence of a donor DNA template, Cas9-induced DSBs can be efficiently repaired via C-NHEJ or alt-EJ, leading to deletions, substitutions or insertions (Table 2). To evaluate the proportions of these two types of potentially concurrent events, we transformed moss with the Cas9 cassette, the sgRNA#2 and the donor DNA cassette PpAPT-KO7 and selected the plants initially for 2-FA resistance and then for G418 resistance among 2-FA[R] plants. In this situation, we isolated approximately 2 times more 2FA[R] plants (4%, Table 4) than upon initial selection for antibiotic resistance (2.1%, Table 3). Consistently, this experiment revealed that only 60% of CRISPR-Cas9-induced DSBs are repaired by HDR-mediated targeted integration of the donor DNA template (121/200), while the remaining 40% are repaired by IR events (C-NHEJ or alt-EJ; Table 4). This indicates clearly that both homology-driven and illegitimate recombination pathways are equally proficient to repair Cas9-induced chromosomal DSBs in *P. patens* even in the presence of a homologous donor template.

CRISPR-Cas9-induced targeted integration in *P. patens* is only partially RAD51 dependent

The RAD51 protein catalyses most homologous recombination reactions in eukaryotes and is directly involved in homology searching, homologous pairing and DNA strand transfer (Holthausen *et al.*, 2010). Previous studies have shown that gene targeting in *P. patens* was totally abolished in the absence of PpRAD51-1 and PpRAD51-2 proteins (Schaefer *et al.*, 2010). To determine whether CRISPR-induced HDR-mediated targeted integration of a donor DNA template depends on the same HR

Table 4 Proportion of HDR- versus IR-mediated repair of CRISPR-Cas9-induced DSB at the *PpAPT* locus, in the presence of a donor cassette

	Number of 2-FA^R clones analysed*	AB^R clones from 2-FA^R clones[†]	HDR-mediated DSB repair %[‡]	IR-mediated DSB repair %[§]
pAct-Cas9 + psgRNA#2 + PpAPT-KO7	200	121	60	40

*A total of 3750 2FA^R clones were obtained from 3 independent experiments.
[†]AB^R clones are the stable 2-FA^R clones that survived after subculture on G418 medium.
[‡]HDR-mediated DSB repair (%) expresses the frequency of G418-resistant clones among the population of 2-FA^R-resistant clones. HDR-mediated repair of the G418^R clones was confirmed by PCR analysis of the *PpAPT* locus (not shown, same as Figure S6).
[§]IR-mediated DSB repair (%) expresses the frequency of G418-sensitive clones among the population of 2-FA^R-resistant clones.

DNA repair pathway, CRISPR-induced GT was analysed in the double-mutant *Pprad51-1-2* (Schaefer *et al.*, 2010). We transformed wild-type and *Pprad51-1-2* protoplasts with the Cas9 cassette, the sgRNA#1 and the donor DNA cassette PpAPT-KO4, bearing a hygromycin-resistant gene surrounded by *PpAPT* gene sequences flanking the sgRNA target (Figure 1b). Regenerated plants were then sequentially selected for hygromycin resistance first, and then for resistance to 2-FA. RTFs were not significantly different between the two strains and similar to those observed in the previous CRISPR-Cas9 experiment, indicating that both genotypes are equally competent for transformation (Table 5). In the wild-type strain, 94% (47/50) of the hygromycin-resistant clones were also 2FA^R (Table 5). At the molecular level, PCR genotyping provided evidence for the presence of junctions generated by HR and for the successful disruption of the *PpAPT* locus in 46 of them, supporting HDR-mediated targeted integration (TGR or TGI) of the donor DNA template (Table 5 and Figure S8). The last plant did not show junctions corresponding to a HR event but shows an interruption of the *PpAPT* gene. This is consistent with the integration of PpAPT-KO4 through illegitimate recombination in the DSB generated by Cas9 at the *PpAPT* locus.

Thus, GT efficiency in the wild type reached 94% (47/50) using the sgRNA#1/PpAPT-KO4 donor cassette couple (Table 5), which is similar to the efficiency observed with the sgRNA#2/ PpAPT-KO7 couple (Table 3) and confirms that the 'CRISPR-Cas9'-mediated GT is remarkably efficient in moss.

For the *Pprad51-1-2* mutant, the percentage of 2-FA^R plants among hygromycin-resistant plants reached 91% (93/102), which was similar to what was observed in the wild type (Table 5). However, PCR genotyping of the *PpAPT* locus and of the recombined junctions identified 3 distinct classes of transformed plants (Table 5 and Figure S8). Disruption of the *PpAPT* locus associated with the generation of at least one of the junctions *via* HR was observed in 29 of them, providing evidence for HDR-mediated targeted integration of the vector at the *PpAPT* locus. In 30 of them, the *PpAPT* locus appeared intact and no junctions could be detected, which is consistent with a random integration of the donor DNA template in the genome accompanied by the repair of the CRISPR-Cas9-induced DSB in the *PpAPT* gene through C-NHEJ or alt-EJ pathways. This was further confirmed by sequence analysis of the PCR product (data not shown). In the last 34 plants, PCR data indicated that the *PpAPT* locus was disrupted without the generation of predicted recombined junctions. Such pattern is consistent with the integration of the donor vector by IR in the DSBs generated by Cas9 at the *PpAPT* locus as previously observed in one WT plant described above. Finally, junction analysis of HDR-mediated events also revealed a significant change of the TGR/TGI ratio: TGR events occurred in 85% (39/46 HDR events) of the WT, as previously observed with sgRNA#2 (88%, Table 3), but only in 10.3% (3/29 HDR events) of the *Pprad51-1-2*-transformed plants (Figure 4, Fisher's exact test $P = 1.8 \times 10^{-4}$).

These data show that in the *Pprad51-1-2* double-mutant, HDR-mediated targeted integration of the donor DNA template reached 28.4% (29/102) and is decreased by threefold compared to the wild type [92% (46/50) to 28.4%], while integration by illegitimate recombination at random loci or at Cas9-generated DSB is dramatically increased [2% (1/50) in the WT *versus* 63% (64/102) in the mutant]. Furthermore, the residual HDR-mediated targeted integrations found in the mutant consist mostly of TGI events resulting from a single HR reaction within one of the homologous sequence of the donor DNA template (confirmed by sequencing analysis, data not shown), accompanied by an apparent NHEJ reaction at the other end of the cassette. The RAD51 function being essential for DNA repair *via* homologous

Table 5 Comparison of 'CRISPR-Cas9'-mediated *PpAPT* gene targeting efficiency in wild-type and the double-mutant *Pprad51-1-2*

	AB^R clones*	2-FA^R clones[†]	Integration due to HDR[‡]		Integration due to IR		GT[¶] %
			TGR	TGI (5'TGI + 3'TGI)	PpAPT locus[‡]	Random[§]	
Wild type	50	47	39	7 (3 + 4)	1	0	94
Pprad51-1-2	102	93	3	26 (7 + 19)	34	30	91

*From 2 independent experiments, a total of 1680 and 1800 AB^R clones were obtained for the wild-type and the *Pprad51-1-2* mutant, respectively.
[†]2-FA^R clones are the stable AB^R clones that survived after subculture on 2-FA medium.
[‡]Number of 2-FA^R clones resulting from HDR- (TGR or TGI) or IR-mediated insertion of the donor DNA template at the *PpAPT* locus was determined by PCR analysis (see Figure S8). Clones resulting from TGR show 5' and 3' junction, clones resulting from 5' TGI show only a 5' junction, and clones resulting from 3' TGI show only a 3' junction.
[§]Number of 2-FA^R clones where the donor DNA template has been randomly inserted by IR was determined by PCR analysis (see Figure S8).
[¶]GT efficiencies (%) express here the frequency of 2-FA^R clones resulting from HDR-mediated targeted insertion among the population of antibiotic-resistant transgenic clones.

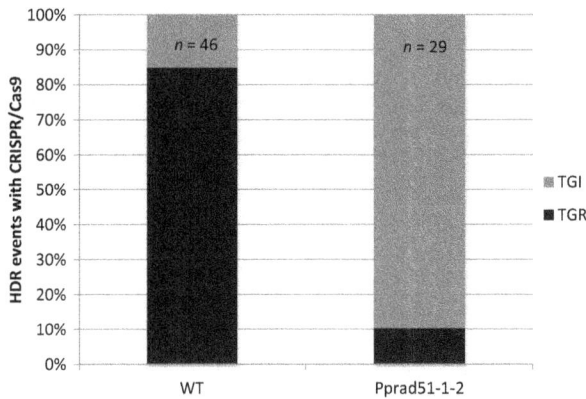

Figure 4 Ratio of TGR versus TGI insertions of the donor DNA template in the wild-type and *Pprad51-1-2* mutant. Frequency of TGR and TGI insertions was determined by genotyping the 2-FA^R clones resulting from HDR (46 for the wild-type and 29 for the *Pprad51-1-2* mutant, see Table 4) using primers specific to the PpAPT-KO4 cassette and primers located on the *PpAPT* gene but outside of the genomic fragments present on the donor DNA cassette (see Figure S8).

recombination, the residual HDR events found in the *Pprad51-1-2* mutant context are probably due to other homology-driven DNA repair mechanisms.

Discussion

In the present study, we set up and demonstrate for the first time in a bryophyte, *Physcomitrella patens,* the potential of the CRISPR-Cas9 system to induce targeted mutagenesis in the genome. We obtained fully developed 2-FA^R plants at very high frequencies (2.2% to 3.2%) and harbouring a large variety of mutations including deletions, substitutions and insertions at the *PpAPT* target gene. The deletions we observed could result from two repair pathways, either from C-NHEJ or from alt-EJ, this last pathway relying on 2- to 4-bp microhomology regions (Figure S3). One way to check this hypothesis will be to perform the same experiments in mutants impacted in these DNA repair pathways. The types of mutations we obtained are consistent with those reported in different plant species where the CRISPR-Cas9 system has been used so far, and that describes mainly small deletions (usually <20 bp), small insertions of a few bp and rare single-nucleotide substitutions.

Off-targeting is a concern for CRISPR-Cas9-induced mutagenesis in human cells (Fu *et al.*, 2013), and new Cas9 proteins with enhanced specificity have been engineered recently (Kleinstiver *et al.*, 2015; Slaymaker *et al.*, 2015). In mice and zebrafish, lower off-target activities have been described compared to human cells (Hruscha *et al.*, 2013; Wang *et al.*, 2013). In plants, no or very low off-targeting activity has been reported (Feng *et al.*, 2014; Xie and Yang, 2013). Potential off-targeting of the sgRNAs used in this study was evaluated by searching the *P. patens* genome for other genomic occurrences of the selected target sequences with a tolerance of a few mismatches. After sequencing of those targets, no off-targeting could be detected. A whole-genome deep sequencing could be more extensive.

We then compared GT efficiencies and monitored targeted integration of a donor plasmid following CRISPR-Cas9-mediated generation of a DSB in the *PpAPT* locus *versus* classical transfection with a linear *PpAPT* replacement vector. We consider that the

major difference between these two situations resides in the fact that the cells have to deal with the repair of a single DSB in the former one, and with a massive signal of DSB damage in the latter. Our data first show that the integration of a transgene is 10-fold more frequent when a single chromosomal DSB is generated by the CRISPR-Cas9 system compared to the delivery of linearized vectors (cf. relative transformation frequencies in Table 3). They also indicate that the cellular competency for transgene integration through HDR is significantly increased when the moss cell has to repair a single chromosomal break, as previously reported in many other systems (Puchta, 2005). Our data further demonstrate that GT efficiency (i.e. % of 2-FA^R in AB^R plants in Table 3) reaches almost 100% following CRISPR transformation, which slightly but significantly (Fisher's exact test: $P = 0.008$) improves the naturally high GT efficiencies of *P. patens* (Tables 3 and 5). They also show that the pattern of GT is not significantly different and that Cas9-induced HDR leads to either TGR or TGI events as observed in classical transformation. This indicates that transgene integration occurs almost always in a Cas9-generated DSB and that the cells use a similar combination of HR and IR reactions to repair this DSB. Our analysis also revealed a significant decrease in the frequency of tandem repeat integration of the donor DNA after HDR-mediated GT (Figure 3), which is probably associated with the fact that the donor DNA is provided as a circular molecule. Finally, we show that only 60% of Cas9-induced DSBs are repaired by HDR when selection is performed for loss of PpAPT function prior to donor DNA integration (Table 4). This clearly demonstrates that Cas9-induced DSB are repaired with almost equal efficiencies by HR or IR in moss and is in sharp contrast with the general idea that efficient GT correlates with the dominance of the HR pathway in DSB repair.

Altogether, this analysis shows that the use of CRISPR/Cas9 significantly improves the naturally high capacities to perform accurate modifications of the moss genome, with GT efficiencies reaching 100%. Such efficiencies have only been reported so far for *S. cerevisiae* (Dicarlo *et al.*, 2013) and clearly demonstrate that the competency for GT is essential to achieve efficient CRISPR-Cas9 HDR-mediated gene replacement, meaning that in nonproficient GT organisms, such as vascular plants for example, if CRISPR-Cas9 can help HDR-mediated gene replacement, it is not sufficient to use this strategy in routine. Our results show that the competency for GT found in *P. patens* is not due to the Cas9-induced DSB being essentially repaired through the HR pathway as both IR and HR equally contribute to these events. Further work will be needed to decipher the mechanism controlling the choice for DSB repair pathways in *P. patens*, and to elucidate the respective contributions of C-NHEJ and alt-EJ in the IR repair pathway.

GT in *P. patens* using linearized replacement vectors was shown to be strictly dependent on the core protein of the HR pathway, RAD51 (Schaefer *et al.*, 2010; Wendeler *et al.*, 2015). Therefore, we assessed its implication in CRISPR-Cas9-induced GT. This experiment showed that GT efficiencies in the absence of PpRAD51 were still possible but were reduced to ca. 30% of those observed in the WT (Table 5). A strong rise of the contribution of IR reactions in transgene integration is consistent with our previous characterisation of *Pprad51* mutants (Schaefer *et al.*, 2010). Yet, HDR-mediated GT in the absence of PpRAD51 is a surprising result and uncovers an alternative HDR-mediated GT pathway that seems to be only active following CRISPR-Cas9-induced GT. Further experiments are needed to identify the mechanisms leading to homology-mediated integration of a donor template without RAD51 proteins. One candidate pathway could be the well-described RAD52-dependant SSA pathway (Ceccaldi *et al.*, 2016; Symington, 2002), and we are currently investigating this possibility. Indeed, RAD52 is essential for HDR-mediated targeted integrations and loss of RAD51 function hardly affects GT efficiency in *S. cerevisiae* (Schiestl *et al.*, 1994).

In this report, we have shown that the use of CRISPR-Cas9 allows efficient targeted mutagenesis and significantly improves GT efficiency and precision in *P. patens*, expanding the range of available tools for gene function analysis in this model organism and facilitating the production of moss-made pharmaceutical, a very promising new area of biotechnology (Reski *et al.*, 2015). Finally, our work also uncovers novel features of CRISPR-induced HDR-mediated GT that could lead to improve the efficiency of such approach in GT noncompetent cells.

Experimental procedures

Protoplast isolation and transformation

Tissues from the Gransden wild-type strain of *Physcomitrella patens* (Ashton and Cove, 1977) and from the P*prad51-1-2* double mutant (Schaefer *et al.*, 2010) were propagated and protoplasts isolated and transformed as previously described (Schaefer and Zrÿd, 1997). Plasmid DNA was extracted with the Nucleobond XA kit (Macherey-Nagel, France). Protoplasts were transformed with 10 to 25 µg circular DNA and then spread on a regeneration medium composed of PpNO3 medium (Ashton *et al.*, 1979), supplemented with 2.7 mm NH4-tartrate (PpNH4 medium) and 0.33 m mannitol for a week before selection.

Molecular cloning

The pAct-Cas9 plasmid used in this study contains a Cas9 expression cassette containing the rice actin 1 promoter and a codon-optimized version of Cas9 from *Streptococcus pyogenes* fused to a SV40 nuclear localization (Mali *et al.*, 2013). The pAct-Cas9 plasmid was constructed as follows: the hCas9 plasmid (plasmid#41815 from AddGene) was digested by *Nco*I and *Pme*I and the hCas9 gene was ligated to the pCOR104-CaMVter plasmid (Proust *et al.*, 2011) previously digested by *Nco*I and *Sma*I.

Two sgRNA expression cassettes were designed, each containing a U6 promoter from *P. patens*, the 5′-G-N$_{(19)}$-3′ sequences targeting *PpAPT* and the tracrRNA scaffold (Mali *et al.*, 2013; Figures 1 and S1). *P. patens* genomic sequence for the U6 gene (coordinates 5050300–5050958 on chromosome 1) was identified by Basic Local Alignment Search Tool (http://www.phytozome.net/physcomitrella_er.php) using the Arabidopsis U6-26 snRNA sequence (X52528; Li *et al.*, 2007) as query. U6 promoter sequence coordinates used for gRNA expression are 5050300–5050621 on chromosome 1. For the design of CRISPR-Cas targets in the *PpAPT* gene, both strands of the *P. patens* adenine phosphoribosyltransferase gene (PpAPT, Phytozome # Pp3c8_16590) were searched using the CRISPOR, free software (http://tefor.net/crispor/crispor.cgi), for sequences of the form 5′-G-N(18 or 19)NGG-3′ with respect to the U6 promoter and Cas9 specificity conditions. Two target loci were selected, one in exon 5 (sgRNA#1) and one in exon 3 (sgRNA#2) of the *PpAPT* gene (Figure 1). The sgRNA1 and sgRNA2 cassettes were synthesized as gBlocks® by IDT (www.idtdna.com), PCR-amplified and introduced into pCR®II-TOPO®

TA-cloning vectors (www.lifetechnologies.com) to give the plasmids psgRNA#1 and psgRNA#2.

Two donor DNA cassettes were used for gene targeting experiments. The PpAPT-KO4 knockout cassette used for gene targeting experiments bears a 715-bp 5′ targeting fragment (coordinate 772–1486 on Pp3c8_16590 in Phytozome) and a 702-bp 3′ targeting fragment (coordinate 1487–2188 on Pp3c8_16590 in Phytozome) of the *PpAPT* gene, flanking a pAct :: hygroR cassette from the pActHygR plasmid. The pActHygR carries a *HPH* gene for resistance to hygromycin (Bilang *et al.*, 1991) in fusion with the rice actin 1 promoter from pCOR104 (McElroy *et al.*, 1991) and before a NOS terminator. The 5′ and 3′ sequences of the *PpAPT* gene present in the PpAPT-KO4 cassette are flanking the predicted CRISPR-mediated DSB for target#1 sequence (coordinate 1468–1487 on Pp3c8_16590 in Phytozome). The PpAPT-KO7 knockout cassette bears a 743-bp 5′ targeting fragment (coordinate 156–898 on Pp3c8_16590 in Phytozome) and a 778-bp 3′ targeting fragment (coordinate 917–1694 on Pp3c8_16590 in Phytozome) of the *PpAPT* gene, flanking a 35S :: neoR cassette from pBNRF for resistance to G418 (Schaefer *et al.*, 2010) cloned in a pCR®II-TOPO® TA-cloning vector. The 5′ and 3′ sequences of the *PpAPT* gene present in the PpAPT-KO7 cassette are flanking the predicted CRISPR-mediated DSB for the target#2 sequence (coordinate 890–909 on Pp3c8_16590 in Phytozome).

Gene knockout assays

Moss protoplasts (4.8×10^5) were cotransformed with the pAct-Cas9 and psgRNA#1 or psgRNA#2 plasmids. Non-sense mutations in the *PpAPT* gene confer resistance to the toxic adenine analogue 2-fluoroadenine (2-FA). Regenerating protoplasts were selected on PpNH4 supplemented with 10 µM 2-FA (Fluorochem) to detect clones which had been disrupted at the *PpAPT* locus. Experiments were repeated three times.

For the characterisation of the mutations triggered in the *PpAPT* gene by the sgRNA/Cas9 system, two sets of primers surrounding the target sites were designed. For target#1, we used the primers PpAPT#15 and PpAPT#19. For target#2, we used the primers PpAPT#5 and PpAPT#14. To check the specificity of our two sgRNAs, we designed primers surrounding all the potential off-target sites identified with the CRISPOR software (Figure S4).

Gene targeting assays

For the CRISPR-Cas9-mediated gene targeting experiment, moss protoplasts (4.8×10^5) were cotransformed with the pAct-Cas9, psgRNA#1 and PpAPT-KO4 plasmids or with the pAct-Cas9, psgRNA#2 and PpAPT-KO7 plasmids. For the 'classical' gene targeting experiment, moss protoplasts (4.8×10^5) were transformed with the PpAPT-KO4 plasmid digested with *Bsa*AI and *Hind*III or with the PpAPT-KO7 plasmid digested with *Bsm*BI and *Bst*XI. Targeted integration of the PpAPT-KO4 or PpAPT-KO7 cassettes at the *PpAPT* gene confers resistance to 2-FA. We selected primary transformants (unstable + stable) with 50 mg/L G418 (Duchefa). Integrative transformants were isolated following a second round of selection on G418. Small pieces of protonema tissue from these transformants were then transferred onto PpNH4 medium containing 10 µM 2-FA to detect *PpAPT* gene targeting events. Experiments were repeated three times. GT efficiencies were determined as the frequency of 2-FA-resistant plants among antibiotic-resistant transformants (targeted + random insertion of the donor DNA template).

For analysis of the nature of the HDR event, that is targeted gene replacement (TGR) versus targeted gene insertion (TGI; Kamisugi et al., 2006), the antibiotic-resistant clones that were also 2-FA resistant were genotyped. The molecular analysis of the left and right junctions of the insertions was performed using primers specific to the PpAPT-KO4 or PpAPT-KO7 cassettes and primers located in the PpAPT gene but outside of the genomic fragments present on the donor DNA cassettes. The 5' junction was detected using the primers PpAPT#8 and 35SProRev#3, and the 3' junction using PpAPT#5 and 35STerFwd#3. The number of inserted copies of the donor DNA template at the target site was estimated using a set of primers located outside of the genomic fragments present in the knockout cassette, PpAPT#5 and PpAPT#8, which amplified fragments of 1700 bp for wild-type clones and fragments of 3700 bp for monocopy insertions. Multiple insertions of the cassette led to no amplification in our conditions.

DNA DSB repair pathway choice assays

For the characterisation of the nature of DNA DSB repair after CRISPR-Cas9-mediated DSB in presence of a donor DNA template, moss protoplasts (4.8×10^5) were cotransformed with the pAct-Cas9, psgRNA#2 and PpAPT-KO7 plasmids. Transformants where the PpAPT gene has been disrupted due to non-sense mutations (via IR) or to targeted integration (via HDR) were selected on PpNH4 supplemented with 10 μM 2-FA for 2 weeks. Small pieces of protonema tissue from these transformants were then transferred onto PpNH4 medium containing 50 mg/L G418 (Duchefa) to detect PpAPT gene targeting events. Experiments were repeated three times. Proportion of the DSBs that were repaired via HDR was determined as the rate of G418-resistant plants among 2-FA-resistant transformants.

PCR analysis of the transformants

For PCR analysis, genomic DNA was extracted from 50 mg of fresh tissue using a genomic DNA quick preparation procedure previously described (Trouiller et al., 2006). For the sequences of the PCR primers used in this study, see Figures S5–S7. The quality of the DNA samples was controlled using primers targeting the SGS1 gene from P. patens: sgs1-Fwd#7 and sgs1-Rev#8.

Acknowledgements

This work was supported by the French government through the programme 'Investissements d'Avenir' and by the French National Research Agency, in the frame of the research project called 'GENIUS' (Genome ENgineering Improvement for Useful plants of a Sustainable agriculture), ref # ANR11-BTBR-0001-GENIUS.

References

Ashton, N.W. and Cove, D.J. (1977) The isolation and preliminary characterisation of auxotrophic and analogue resistant mutants of the moss, Physcomitrella patens. Mol. Gen. Genet. **154**, 87–95.

Ashton, N.W., Grimsley, N.H. and Cove, D.J. (1979) Analysis of gametophytic development in the moss, Physcomitrella patens, using auxin and cytokinin resistant mutants. Planta, **144**, 427–435.

Bilang, R., Iida, S., Peterhans, A., Potrykus, I. and Paszkowski, J. (1991) The 3'-terminal region of the hygromycin-B-resistance gene is important for its activity in Escherichia coli and Nicotiana tabacum. Gene, **100**, 247–250.

Bonhomme, S., Nogué, F., Rameau, C. and Schaefer, D.G. (2013) Usefulness of Physcomitrella patens for studying plant organogenesis. Methods Mol. Biol. **959**, 21–43.

Ceccaldi, R., Rondinelli, B. and Andrea, A.D.D. (2016) Repair pathway choices and consequences at the double-strand break. Trends Cell Biol. **26**, 52–64.

Čermák, T., Baltes, N.J., Čegan, R., Zhang, Y. and Voytas, D.F. (2015) High-frequency, precise modification of the tomato genome. Genome Biol. **16**, 232.

Charlot, F., Chelysheva, L., Kamisugi, Y., Vrielynck, N., Guyon, A., Epert, A., Le Guin, S., et al. (2014) RAD51B plays an essentialrole during somatic and meiotic recombination in Physcomitrella. NucleicAcids Res. **42**, 11965–11978.

Collonnier, C., Nogué, F. and Casacuberta, J.M. (2016) Targeted genetic modification in crops using site-directed nucleases. In Genetically Modified Organisms in Food (Watson, R.R. and Preedy, V.R., eds), pp. 133–145. London: Elsevier.

D'Halluin, K., Vanderstraeten, C., Van Hulle, J. et al. (2013) Targeted molecular trait stacking in cotton through targeted double-strand break induction. Plant Biotechnol. J. **11**, 933–941.

Dicarlo, J.E., Norville, J.E., Mali, P., Rios, X., Aach, J. and Church, G.M. (2013) Genome engineering in Saccharomyces cerevisiae using CRISPR-Cas systems. Nucleic Acids Res. **41**, 4336–4343.

Feng, Z., Mao, Y., Xu, N. et al. (2014) Multigeneration analysis reveals the inheritance, specificity, and patterns of CRISPR/Cas-induced gene modifications in Arabidopsis. Proc. Natl Acad. Sci. USA, **111**, 4632–4637.

Fu, Y., Foden, J.A., Khayter, C., Maeder, M.L., Reyon, D., Joung, J.K. and Sander, J.D. (2013) High-frequency off-target mutagenesis induced by CRISPR-Cas nucleases in human cells. Nat. Biotechnol. **31**, 822–826.

Gasiunas, G., Barrangou, R., Horvath, P. and Siksnys, V. (2012) Cas9-crRNA ribonucleoprotein complex mediates specific DNA cleavage for adaptive immunity in bacteria. Proc. Natl Acad. Sci. USA, **109**, E2579–E2586.

Hanin, M. and Paszkowski, J. (2003) Plant genome modification by homologous recombination. Curr. Opin. Plant Biol. **6**, 157–162.

Holthausen, J.T., Wyman, C. and Kanaar, R. (2010) Regulation of DNA strand exchange in homologous recombination. DNA Repair (Amst.) **9**, 1264–1272.

Hruscha, A., Krawitz, P., Rechenberg, A., Heinrich, V., Hecht, J., Haass, C. and Schmid, B. (2013) Efficient CRISPR/Cas9 genome editing with low off-target effects in zebrafish. Development, **140**, 4982–4987.

Kamisugi, Y., Schlink, K., Rensing, S.A., Schween, G., von Stackelberg, M., Cuming, A.C., Reski, R. et al. (2006) The mechanism of gene targeting in Physcomitrella patens: homologous recombination, concatenation and multiple integration. Nucleic Acids Res. **34**, 6205–6214.

Kamisugi, Y., Schaefer, D.G., Kozak, J., Charlot, F., Vrielynck, N., Holá, M., Angelis, K.J. et al. (2012) MRE11 and RAD50, but not NBS1, are essential for gene targeting in the moss Physcomitrella patens. Nucleic Acids Res. **40**, 3496–3510.

Kleinstiver, B.P., Prew, M.S., Tsai, S.Q. et al. (2015) Engineered CRISPR-Cas9 nucleases with altered PAM specificities. Nature, **523**, 481–485.

Kofuji, R. and Hasebe, M. (2014) Eight types of stem cells in the life cycle of the moss Physcomitrella patens. Curr. Opin. Plant Biol. **17**, 13–21.

Lander, E.S. (2016) The Heroes of CRISPR. Cell, **164**, 18–28.

Li, X., Jiang, D.H., Yong, K. and Zhang, D.B. (2007) Varied transcriptional efficiencies of multiple Arabidopsis U6 small nuclear RNA genes. J. Integr. Plant Biol. **49**, 222–229.

Li, J.-F., Norville, J.E., Aach, J., McCormack, M., Zhang, D., Bush, J., Church, G.M. et al. (2013) Multiplex and homologous recombination-mediated genome editing in Arabidopsis and Nicotiana benthamiana using guide RNA and Cas9. Nat. Biotechnol. **31**, 688–691.

Mali, P., Esvelt, K.M. and Church, G.M. (2013) Cas9 as a versatile tool for engineering biology. Nat. Methods, **10**, 957–963.

McElroy, D., Blowers, A.D., Jenes, B. and Wu, R. (1991) Construction of expression vectors based on the rice actin 1 (Act1) 5' region for use in monocot transformation. Mol. Gen. Genet. **231**, 150–160.

Nishizawa-Yokoi, A., Endo, M., Ohtsuki, N., Saika, H. and Toki, S. (2015) Precision genome editing in plants via gene targeting and piggyBac-mediated marker excision. Plant J. **81**, 160–168.

Pàques, F. and Haber, J.E. (1999) Multiple pathways of recombination induced by double-strand breaks in Saccharomyces cerevisiae. Microbiol. Mol. Biol. Rev. **63**, 349–404.

Petolino, J.F., Srivastava, V. and Daniell, H. (2016) Editing Plant Genomes: a new era of crop improvement. Plant Biotechnol. J. **14**, 435–436.

Prigge, M.J. and Bezanilla, M. (2010) Evolutionary crossroads in developmental biology: *Physcomitrella patens*. *Development*, **137**, 3535–3543.

Proust, H., Hoffmann, B., Xie, X., Yoneyama, K., Schaefer, D.G., Yoneyama, K., Nogué, F. *et al.* (2011) Strigolactones regulate protonema branching and act as a quorum sensing-like signal in the moss *Physcomitrella patens*. *Development*, **138**, 1531–1539.

Puchta, H. (2005) The repair of double-strand breaks in plants: mechanisms and consequences for genome evolution. *J. Exp. Bot.* **56**, 1–14.

Reski, R., Parsons, J. and Decker, E.L. (2015) Moss-made pharmaceuticals: from bench to bedside. *Plant Biotechnol. J.* **13**, 1191–1198.

Schaefer, D.G. (2001) Gene targeting in *Physcomitrella patens*. *Curr. Opin. Plant Biol.* **4**, 143–150.

Schaefer, D.G. (2002) A new moss genetics: targeted mutagenesis in *Physcomitrella patens*. *Annu. Rev. Plant Biol.* **53**, 477–501.

Schaefer, D.G. and Zrÿd, J.P. (1997) Efficient gene targeting in the moss *Physcomitrella patens*. *Plant J.* **11**, 1195–1206.

Schaefer, D. and Zrÿd, J.-P. (2004) Principles of targeted mutagenesis in the moss *Physcomitrella patens*. In *New Frontiers in Bryology* (Wood, A.J., Oliver, M.J. and Cove, D.J., eds), pp. 37–49. Dordrecht: Springer Netherlands.

Schaefer, D.G., Delacote, F., Charlot, F., Vrielynck, N., Guyon-Debast, A., Le Guin, S., Neuhaus, J.M. *et al.* (2010) RAD51 loss of function abolishes gene targeting and de-represses illegitimate integration in the moss *Physcomitrella patens*. *DNA Repair (Amst.)* **9**, 526–533.

Schaeffer, S.M. and Nakata, P.A. (2015) CRISPR/Cas9-mediated genome editing and gene replacement in plants: transitioning from lab to field. *Plant Sci.* **240**, 130–142.

Schaff, D.A. (1994) The adenine phosphoribosyltransferase (APRT) selectable marker system. *Plant Sci.* **101**, 3–9.

Schiestl, R.H., Zhu, J. and Petes, T.D. (1994) Effect of mutations in genes affecting homologous recombination on restriction enzyme-mediated and illegitimate recombination in *Saccharomyces cerevisiae*. *Mol. Cell. Biol.* **14**, 4493–4500.

Schiml, S. and Puchta, H. (2016) Revolutionizing plant biology: multiple ways of genome engineering by CRISPR/Cas. *Plant Methods*, **12**, 8.

Schiml, S., Fauser, F. and Puchta, H. (2014) The CRISPR/Cas system can be used as nuclease for in planta gene targeting and as paired nickases for directed mutagenesis in *Arabidopsis* resulting in heritable progeny. *Plant J.* **9**, 1139–1150.

Shukla, V.K., Doyon, Y., Miller, J.C. *et al.* (2009) Precise genome modification in the crop species *Zea mays* using zinc-finger nucleases. *Nature*, **459**, 437–441.

Slaymaker, I.M., Gao, L., Zetsche, B., Scott, D.A., Yan, W.X. and Zhang, F. (2015) Rationally engineered Cas9 nucleases with improved specificity. *Science*, **351**, 84–88.

Symington, L.S. (2002) Role of RAD52 epistasis group genes in homologous recombination and double-strand break repair. *Microbiol. Mol. Biol. Rev.* **66**, 630–670.

Townsend, J.A., Wright, D.A., Winfrey, R.J., Fu, F., Maeder, M.L., Joung, J.K. and Voytas, D.F. (2009) High-frequency modification of plant genes using engineered zinc-finger nucleases. *Nature*, **459**, 442–445.

Trouiller, B., Schaefer, D.G., Charlot, F. and Nogué, F. (2006) MSH2 is essential for the preservation of genome integrity and prevents homeologous recombination in the moss *Physcomitrella patens*. *Nucleic Acids Res.* **34**, 232–242.

Trouiller, B., Charlot, F., Choinard, S., Schaefer, D.G. and Nogué, F. (2007) Comparison of gene targeting efficiencies in two mosses suggests that it is a conserved feature of Bryophyte transformation. *Biotechnol. Lett.* **29**, 1591–1598.

Wang, H., Yang, H., Shivalila, C.S., Dawlaty, M.M., Cheng, A.W., Zhang, F. and Jaenisch, R. (2013) One-step generation of mice carrying mutations in multiple genes by CRISPR/Cas-mediated genome engineering. *Cell*, **153**, 910–918.

Wendeler, E., Zobell, O., Chrost, B. and Reiss, B. (2015) Recombination products suggest the frequent occurrence of aberrant gene replacement in the moss *Physcomitrella patens*. *Plant J.* **81**, 548–558.

Xie, K. and Yang, Y. (2013) RNA-guided genome editing in plants using a CRISPR-Cas system. *Mol. Plant*, **6**, 1975–1983.

11

Approaches for enhancement of N$_2$ fixation efficiency of chickpea (*Cicer arietinum* L.) under limiting nitrogen conditions

Maryam Nasr Esfahani[1], Saad Sulieman[2,3], Joachim Schulze[4], Kazuko Yamaguchi-Shinozaki[5], Kazuo Shinozaki[6] and Lam-Son Phan Tran[2,*]

[1]*Department of Biology, Faculty of Sciences, Lorestan University, Khorramabad, Iran*
[2]*Signaling Pathway Research Unit, RIKEN Center for Sustainable Resource Science (CSRS), Suehiro-cho, Tsurumi, Yokohama, Japan*
[3]*Department of Agronomy, Faculty of Agriculture, University of Khartoum, Shambat, Khartoum North, Sudan*
[4]*Section of Plant Nutrition, Department of Crop Sciences, Georg-August-University of Göttingen, Göttingen, Germany*
[5]*Laboratory of Plant Molecular Physiology, Graduate School of Agricultural and Life Sciences, University of Tokyo, Bunkyo-ku, Tokyo, Japan*
[6]*Gene Discovery Research Group, RIKEN Center for Sustainable Resource Science (CSRS), Tsurumi, Yokohama, Japan*

*Correspondence

email son.tran@riken.jp

Summary

Chickpea (*Cicer arietinum*) is an important pulse crop in many countries in the world. The symbioses between chickpea and Mesorhizobia, which fix N$_2$ inside the root nodules, are of particular importance for chickpea's productivity. With the aim of enhancing symbiotic efficiency in chickpea, we compared the symbiotic efficiency of C-15, Ch-191 and CP-36 strains of *Mesorhizobium ciceri* in association with the local elite chickpea cultivar 'Bivanij' as well as studied the mechanism underlying the improvement of N$_2$ fixation efficiency. Our data revealed that C-15 strain manifested the most efficient N$_2$ fixation in comparison with Ch-191 or CP-36. This finding was supported by higher plant productivity and expression levels of the *nifHDK* genes in C-15 nodules. Nodule specific activity was significantly higher in C-15 combination, partially as a result of higher electron allocation to N$_2$ *versus* H$^+$. Interestingly, a striking difference in nodule carbon and nitrogen composition was observed. Sucrose cleavage enzymes displayed comparatively lower activity in nodules established by either Ch-191 or CP-36. Organic acid formation, particularly that of malate, was remarkably higher in nodules induced by C-15 strain. As a result, the best symbiotic efficiency observed with C-15-induced nodules was reflected in a higher concentration of the total and several major amino metabolites, namely asparagine, glutamine, glutamate and aspartate. Collectively, our findings demonstrated that the improved efficiency in chickpea symbiotic system, established with C-15, was associated with the enhanced capacity of organic acid formation and the activities of the key enzymes connected to the nodule carbon and nitrogen metabolism.

Keywords: *Cicer arietinum*, C-metabolism, gene expression, *Mesorhizobium ciceri*, N-metabolism, symbiotic efficiency.

Introduction

Chickpea (*Cicer arietinum* L.), a staple basic food crop in many tropical and subtropical Afro–Asian countries, is one of the world's major pulse crops which is traditionally cultivated in marginal areas and semi-arid regions (Mhadhbi *et al.*, 2008). Globally, it is ranked as the third most important pulse crop (Flowers *et al.*, 2010), with an average yield of 913 kg/ha and a total production of 10.9 million metric tonnes in 2010 (Gaur *et al.*, 2012). Currently, chickpea is grown in over 50 countries across the Indian subcontinent, North Africa, the Middle East, southern Europe, the Americas and Australia. Traditionally, chickpea plays an essential role by providing protein, oil, vitamins and micronutrients that complement cereals and starch in the diets of humankind. Being a leguminous crop, chickpea exhibits an important characteristic of fixing atmospheric N$_2$ through symbiotic association with compatible *Mesorhizobium* soil bacteria, the common chickpea-specific rhizobial species. Symbiotic N$_2$ fixation is the major route for providing a large nitrogen proportion for human consumption and animal feed and contributes to agriculture sustainability (Bhattacharyya and Jha, 2012; Geurts *et al.*, 2012; Sulieman and Tran, 2013; Tran and Nguyen, 2009). Indeed, N$_2$-fixing leguminous plants form the most significant natural pathway to introduce nitrogen into the biosphere along with their enormous contribution to soil fertility (Drevon *et al.*, 2001; Mwanamwenge *et al.*, 1998). Recently, there has been a growing level of interest in environmentally friendly sustainable agricultural practices and organic farming systems (Bhattacharyya and Jha, 2012; Mafongoya *et al.*, 2009; Peoples, 2009). Through its symbiosis with rhizobia, nodulating chickpea could be cultivated in many nitrogen-deficient soils in many countries.

As a large part of nitrogen requirements for chickpea is provided by symbiosis, association with an efficient symbiotic microsymbiont is critical to support growth and productivity of chickpea. The association of symbioses between legume hosts and their specific rhizobia has been examined in a number of studies that suggested that the variation in symbiotic performance is probably due to the differences in the rhizobial symbiotic efficiency and/or the degree of compatibility with the host plant (Ben Romdhane *et al.*, 2007; Icgen *et al.*, 2002; Kyei-Boahen *et al.*, 2002; Laranjo *et al.*, 2002). In addition, the role of rhizobial strains on the symbiotic efficiency has also been investigated under unfavourable conditions, such as osmotic stress (Mhadhbi et al., 2008; Nasr Esfahani and Mostajeran, 2011), combined nitrogen application (Sulieman et al., 2010) and phosphorus deficiency (Sulieman et al., 2013). According to these studies, the symbiotic performance in terms of biomass production and nitrogen-fixing capacity under stressful conditions are correlated with their behaviour under normal conditions.

The variability in symbiotic efficiency of various strains provides an impression that the N_2-fixing ability of symbiotic bacteria could be improved by strain selection and/or genetic manipulation. Increasing evidence has indicated that the selection of the most efficient symbiotic strain is a good strategy to enhance the growth and production of various legume crops through the improvement of nodulation and N_2 fixation capacity (Bhattacharyya and Jha, 2012; Purcell, 2009). On the other hand, the precise understanding of factors affecting legume–bacterial association will also enable us to improve the symbiotic N_2 fixation by modern genetic engineering. Having realized these potentials, one of the aims of the current study was to utilize biochemical and molecular approaches to characterize the symbiotic efficiency of the Mesorhizobium C-15, Ch-191 and CP-36 strains in association with chickpea to identify its superior symbiotic partner for agricultural applications. The assessment was performed based on the association with the widely used high-yielding local elite cultivar 'Bivanij'. To this end, we examined several key carbon and nitrogen metabolic markers and their relating enzymes in the nodule machinery to understand the mechanisms affecting N_2 fixation. Additionally, we compared the gene expression levels of nifH, nifD and nifK in nodules of the three studied Mesorhizobium–chickpea symbioses to evaluate their relationship with N_2-fixing capacity. The data obtained indicate a high metabolic and genetic diversity among the tested Mesorhizobium bacteria and quantify their potential as microsymbiont partners. Our results reported in the present study also provide knowledge that could be applied for improvement of Mesorhizobium–chickpea symbioses using biotechnological approaches.

Results

Influence of various M. ciceri strains on nodulation and plant biomass production

As a means to compare the symbiosis efficiencies established between C. arietinum L. cultivar 'Bivanij' and each of the three selected M. ciceri C-15, Ch-191 and CP-36 strains, first we measured the nodulation and dry matter formation of each association. Unlike the root fraction, the dry matter of shoot, nodule and total biomass varied significantly according to the type of M. ciceri inoculated strains (Table 1). The accumulated shoot, nodule and total dry matters of C-15-inoculated plants were significantly higher than those established with either Ch-191 or CP-36. Although individual nodule dry matter was not statistically significant among the studied strains, the nodule number of C-15-inoculated plants was 38% and 32% higher than that of plants inoculated with CP-36 and Ch-191, respectively. Overall, C-15-inoculated plants exhibited the best performance in terms of nodulation and dry matter accumulation, while no significant differences were noticed between Ch-191- and CP-36-inoculated plants (Table 1).

Table 1 Dry matter production and nodulation characteristics in chickpea cv. 'Bivanij' inoculated with either Mesorhizobium ciceri C-15, Ch-191 or CP-36 strain and grown over a period of 28 days after sowing in fine quartz sand. Data presented are the means ± SE of three replicates

DM and nodule characteristics	Mesorhizobium ciceri		
	C-15	Ch-191	CP-36
DM [g/plant]			
Shoot	$2.56^a \pm 0.40$	$1.50^b \pm 0.12$	$1.47^b \pm 0.13$
Root	$0.87^a \pm 0.03$	$0.86^a \pm 0.11$	$0.88^a \pm 0.07$
Nodules	$0.30^a \pm 0.04$	$0.24^b \pm 0.03$	$0.21^b \pm 0.03$
Total	$3.72^a \pm 0.40$	$2.61^b \pm 0.15$	$2.54^b \pm 0.30$
Individual nodule DM (mg)	$1.53^a \pm 0.18$	$1.65^a \pm 0.14$	$1.76^a \pm 0.05$
Nodule number/plant	$196^a \pm 23$	$132^b \pm 12$	$121^b \pm 14$

DM, dry matter. Data with different superscript letters are significantly different as measured by Duncan's multiple-range test (DMRT) ($P \leq 0.05$).

Influence of various M. ciceri strains on N_2 fixation activities

In the next line of our study, we performed the gas exchange measurements in nodules induced with the three strains to compare the N_2 fixation activities of chickpea nodules formed from the three chickpea–M. ciceri associations. In general, the three M. ciceri strains used in symbiosis had different effect on N_2 fixation activity that was estimated by H_2 evolution. In comparison with Ch-191 and CP-36 strains, N_2 fixation was the best in plants inoculated with C-15 strain as evidenced by the significantly higher values of apparent nitrogenase activity (ANA) and total nitrogenase activity (TNA) (Table 2). Moreover, the highest N_2 fixation activity detected in nodules of plants inoculated with C-15 strain was, at least in part, resulted from the improved specific efficiency of electron allocation. This result was further supported by the higher specific activity of nodules (fixed nitrogen per unit nodule biomass) of plants inoculated with C-15, with approximately 186% and 163% increase over that of Ch-191- and CP-36-inoculated nodules, respectively. When per plant H_2 evolution was translated into N_2 fixation rate (NFR), the value for combination with C-15 was about 218% and 195% of that obtained for combination with Ch-191 and CP-36, respectively (Table 2). In addition, the CO_2 evolution measurements for the root/nodule respiration also showed a higher tendency for C-15-inoculated plants. The value recorded for inoculation with C-15 was approximately 21% and 27% higher when compared with that resulted from treatment with Ch-191 and CP-36 strains, respectively (Table 2), further indicating that C-15 is the best symbiotic partner for the high-yielding 'Bivanij' cultivar.

Correlation between nitrogenase activity and expression levels of nifHDK genes

Next, we were interested in whether the highest nitrogenase activity observed with C-15 combination in comparison with Ch-191 and CP-36 was resulted from the increase in expression levels of the nifHDK genes that encode the subunits of the nitrogenase (Rubio and Ludden, 2008). Thus, we analysed the expression of these genes in nodules induced by the three studied strains (Figure 1). Reverse transcription polymerase chain reaction (RT-PCR) assays were performed on cDNAs derived from nodule total RNA, and the transcript levels of nifH (encoding component

Table 2 Nitrogenase activity, electron allocation, N$_2$ fixation and CO$_2$ evolution in chickpea cv. 'Bivanij' inoculated with either *Mesorhizobium ciceri* C-15, Ch-191 or CP-36 strain and grown over a period of 28 days after sowing in fine quartz sand. Data presented are the means ± SE of three replicates

Nitrogenase activity and CO$_2$ evolution		*Mesorhizobium ciceri*		
Parameter	Unit	C-15	Ch-191	CP-36
ANA	μM H$_2$/h/Plant	3.21^a ± 0.13	1.78^b ± 0.14	2.15^b ± 0.02
TNA	μM H$_2$/h/Plant	7.02^a ± 0.18	3.52^b ± 0.28	4.11^b ± 0.10
EAC		0.54^a ± 0.02	0.49^b ± 0.02	0.47^b ± 0.01
NFR	μM N$_{fixed}$/h/plant	1.27^a ± 0.07	0.58^b ± 0.06	0.65^b ± 0.02
Specific N$_2$ fixation	μg N$_{fixed}$/h/mg nodule	0.13^a ± 0.02	0.07^b ± 0.00	0.08^b ± 0.01
CO$_2$ evolution	μM CO$_2$/h/plant	139^a ± 2.3	115^b ± 2.7	110^b ± 5.3

ANA, apparent nitrogenase activity; EAC, electron allocation coefficient; NFR, N$_2$ fixation rate; N$_{fixed}$ per unit nodule biomass, specific N$_2$ fixation; TNA, total nitrogenase activity. Data with different superscript letters are significantly different as measured by Duncan's multiple-range test (DMRT) ($P \leq 0.05$).

II of nitrogenase, or the Fe protein) and *nifD* and *nifK* (encoding component I of nitrogenase, or Mo–Fe protein) were evaluated. A striking difference in the expression levels of nitrogenase structural genes was observed in nodules induced by the tested strains (Figure 1). The transcript abundance of *nifH*, *nifD* and *nifK* genes was significantly higher in C-15-induced nodules when compared to their respective ones detected in Ch-191- and CP-36-induced nodules (Figure 1). This result demonstrated that the highest nitrogenase activity or in other words the best N$_2$ fixation activity conferred by C-15 inoculation is positively correlated with the expression levels of the *nifHDK* genes.

Total soluble protein in nodules and enzyme activities

The concentration of total soluble protein was higher in nodules of C-15-inoculated plants by 82% and 73% when compared to those obtained in the nodules formed by Ch-191 and CP-36, respectively (Figure 2). Measurements of the activities of the key enzymes involved in nodule carbon and nitrogen metabolisms revealed clear differences among the studied strains (Figure 3). While the specific activity of sucrose synthase (SS, EC 2.4.1.13) was significantly increased in C-15-induced nodules and that of alkaline invertase (AI, EC 3.2.1.26) was remarkably enhanced in nodules of CP-36 inoculated plants (Figure 4a). AI is known to be the alternative enzyme for the cleavage of sucrose in nodules (Figure 3) (Gordon and James, 1997). The specific activities of phosphoenolpyruvate carboxylase (PEPC, EC 4.1.1.31) (Figure 4a) and malate dehydrogenase (MDH, EC 1.1.1.37) (Figure 4b) were significantly higher in nodules of C-15-inoculated plants relative to those observed in nodules of either Ch-191- or CP-36-infected plants. No significant difference was detected for the activities of glutamine synthetase (GS, EC 6.3.1.2) (Figure 5a) and glutamate synthase (NADH-GOGAT, EC 1.4.1.14) (Figure 5b) among the

strains examined. Interestingly, aspartate aminotransferase (AAT, EC 2.6.1.1) activity was significantly enhanced in nodules of C-15-inoculated plants, reaching approximately 49% higher activity compared to that detected in nodules formed by inoculation with either Ch-191 or CP-36 strain (Figure 4c).

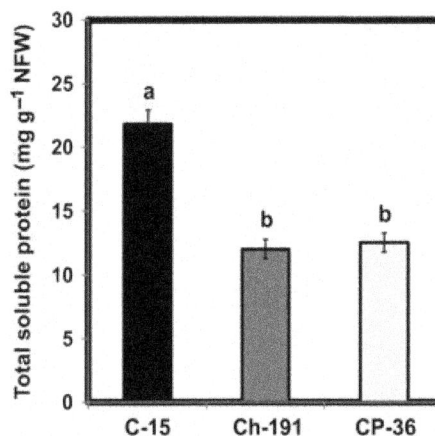

Figure 2 Total soluble protein extracted from nodules of chickpea cv. 'Bivanij' inoculated with either *Mesorhizobium ciceri* C-15, Ch-191 or CP-36 strain and grown over a period of 28 days after sowing in fine quartz sand. Data are means of three replicates. Error bars represent standard errors. Data with different letters are significantly different as measured by Duncan's multiple-range test (DMRT) ($P \leq 0.05$). NFW, nodule fresh weight.

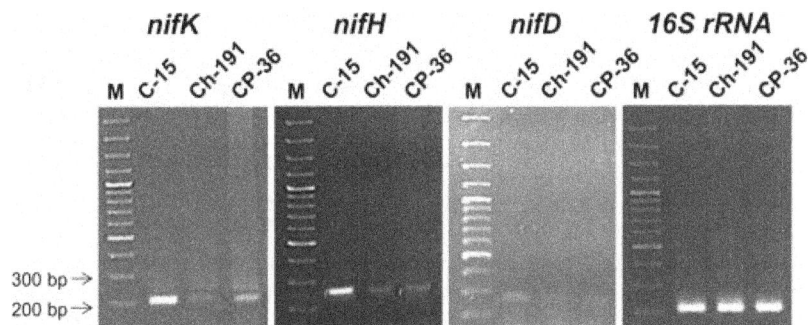

Figure 1 Expression of *nifH*, *nifD* and *nifK* genes in nodules of chickpea cv. 'Bivanij' inoculated with either *Mesorhizobium ciceri* C-15, Ch-191 or CP-36 strain and grown over a period of 28 days after sowing in fine quartz sand. The *16S rRNA* was used as reference gene. M, DNA marker.

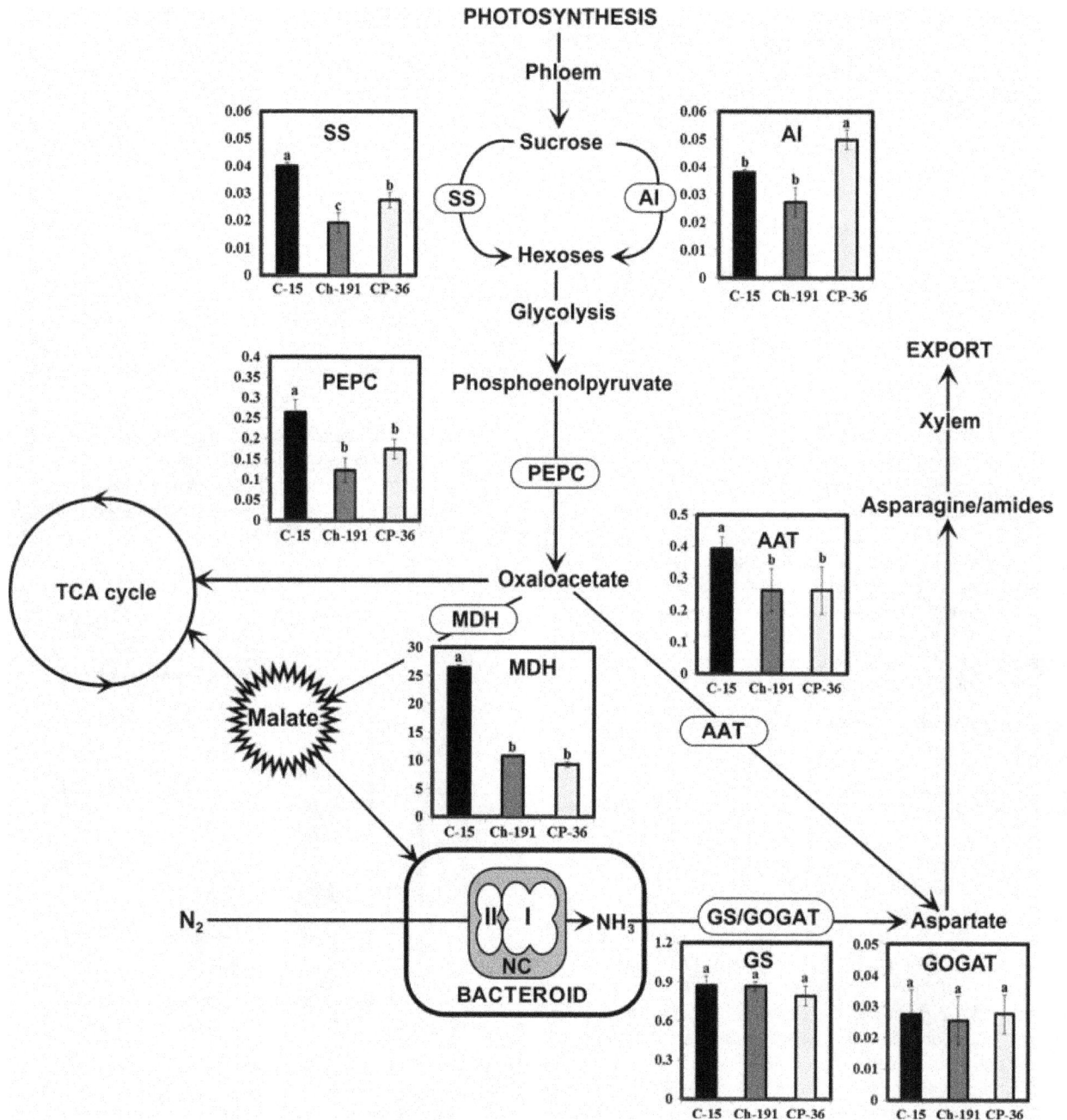

Figure 3 A model of carbon and nitrogen metabolisms representing quantification of the key enzymes involved in nodule of chickpea cv. 'Bivanij' inoculated with either *Mesorhizobium ciceri* C-15, Ch-191 or CP-36 strain and grown over a period of 28 days after sowing in fine quartz sand. AAT, aspartate aminotransferase; AI, alkaline invertase; GOGAT, glutamate synthase; GS, glutamine synthetase; MDH, malate dehydrogenase; NC, nitrogenase enzyme complex; PEPC, phosphoenolpyruvate carboxylase; SS, sucrose synthase.

Nodule sucrose and organic acid concentrations

Nodule sucrose concentration was not markedly affected by the type of studied strains (Figure 6a). Sucrose has been established as the primary source of energy, preferentially imported via the phloem, and is used to sustain nodule function in various leguminous species (Figure 3) (Vance, 2008; White *et al.*, 2007). In contrast with similarity found in free sucrose concentrations of nodules induced with the three strains, the patterns in organic acids were quite different (Figure 6b). Total organic acid concentration was significantly higher in nodules formed by C-15 strain compared with that established by either Ch-191 or CP-

36 (Figure 6b). Of the organic acids detected in the nodules, malate found in nodules of C-15-inoculated plants was about sixfold and eightfold higher in concentration when compared with that obtained in nodules formed by Ch-191 and CP-36 inoculations, respectively (Figure 6b). Malate is the organic acid of pivotal importance to legume nodule function (Figure 3) (Fischinger and Schulze, 2010). Unlike malate, tartrate concentration was significantly increased in nodules formed by Ch-191 strain, while the concentrations of the other detected organic acids, namely the succinate, lactate and fumarate, were not significantly altered by the type of strains used for inoculation (Figure 6b).

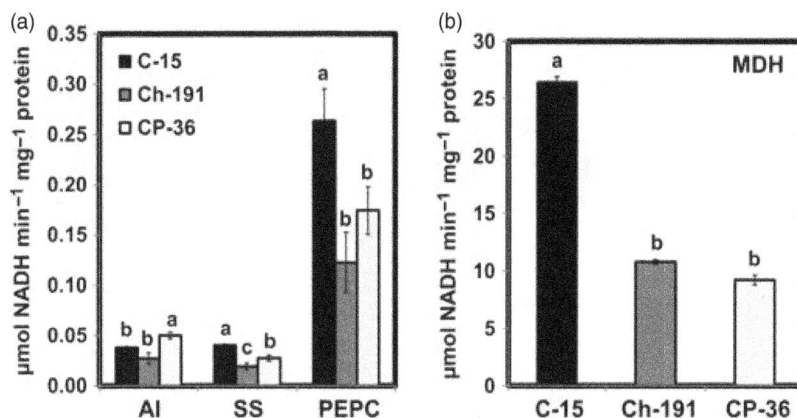

Figure 4 *In vitro* carbon–enzyme activities (a) alkaline invertase (AI), sucrose synthase (SS), phosphoenolpyruvate carboxylase (PEPC) and (b) malate dehydrogenase (MDH) in nodules of chickpea cv. 'Bivanij' inoculated with either *Mesorhizobium ciceri* C-15, Ch-191 or CP-36 strain and grown over a period of 28 days after sowing in fine quartz sand. Data are means of three replicates. Error bars represent standard errors. Data with different letters are significantly different as measured by Duncan's multiple-range test (DMRT) ($P \leq 0.05$).

Nodule amino acid concentration

The higher per plant and specific (per mg dry matter nodule) N_2 fixation activities in nodules formed by C-15 strain compared to those established by Ch-191 and CP-36 strains were also reflected in the higher nodule total free amino acid concentration (Figure 7a). The total amino acid concentration in nodules induced by C-15 strain was about 138% and 147% of those fractions in nodules induced by Ch-191 and CP-36, respectively. This increase was largely attributed to the increase in asparagine content (Figure 7a), which has also been reported to be the predominant amino acid in nodules of many other leguminous plants (Figure 3) (Lea *et al.*, 2007; Sulieman and Tran, 2013; Sulieman *et al.*, 2010). The asparagine constituted approximately 59%, 49% and 48% of the total amino acid content in the nodules formed by C-15, Ch-191 and CP-36, respectively (Figure 7b). Apart from asparagine, the concentrations of other major amino acid constituents, namely aspartate, glutamate and glutamine, were also more significantly increased in nodules induced by C-15 compared to those established by Ch-191 and CP-36 strains (Figure 7a).

Influence of various *M. ciceri* strains on yield and yield components

The plant productivity was markedly affected by the type of studied strains or nitrogen source used for supporting the growth of the plants (Table 3). The plant yield for the exposed treatments was evaluated on the basis of the numbers of pods and seeds per each plant, 100 seed weight and total seed and hay yields per plant. The present results are in agreement and strongly support the tendency of the growth and performance of the three studied *M. ciceri* strain characterized in the controlled environment (Tables 1 and 2). The yield and yield components for C-15-inoculated plants were significantly higher when compared with those associated with Ch-191 or CP-36. Although not reaching the significant level, the yield production of C-15-inoculated plants was relatively better than that of plants that were solely dependent on external combined nitrogen supply. With exception to 100 seed yield, the productivity of either Ch-191- or CP-36-infected plants was significantly higher compared to those grown without any source for nitrogen supply (N treatment) (Table 3).

Discussion

Of all the essential nutrients, nitrogen is known as the most limiting factor for crop production, which is required in large quantities. The deficiency of this element is a universal phenomenon that severely affects crop growth and productivity in most extensive agricultural systems (Vance, 2001). Although it is the most abundant element in the earth, most of the arable lands are classified as nitrogen deficient, particularly in tropical areas. The implementation of sustainable farming systems in tropical lands is greatly affected by the reduced nitrogen-use efficiency by various crops (Graham and Vance, 2003). This obstacle could be resolved using an efficient symbiotic strain that would be used to enhance legume production and simultaneously reduce the use of inorganic nitrogen fertilizers. Apparently, application of rhizobia remains one of the effective and cheapest strategies that could improve the production of legumes in various tropical and subtropical countries. Several reports have showed that more N_2 could be fixed by chickpea when it was inoculated with a *M. ciceri* symbiotic bacterium (Mhadhbi *et al.*, 2004, 2008; Nasr Esfahani and Mostajeran, 2011), suggesting that screening for an efficient symbiotic strain and research towards understanding the mechanism underlying the N_2 fixation efficiency between chickpea and its symbiont are important tasks for improvement of chickpea productivity at low nitrogen fertilizer usage.

In the present study, we examined the interaction of chickpea and three *M. ciceri* strains to establish the best symbiotic combination conferring the highest N_2 fixing capacity. Our results indicated that among the three *M. ciceri* strains examined, chickpea cv. 'Bivanij' had significantly higher symbiotic efficiency in combination with C-15 than with either Ch-191 or CP-36 strain. This finding was reflected in the remarkable increase in dry matter accumulation (Table 1) and H_2 and CO_2 evolutions (Table 2) in C-15-nodulated plants in comparison with those found in plants inoculated with either Ch-191 or CP-36. These results were also supported by the highest NFR observed in the nodules induced by C-15 relative to those formed by Ch-191 or CP-36 (Table 2). More and above, further solid evidence was documented when the plant productivity was quantified and compared for the three symbiotic combinations in question (Table 3). As expected, the higher N_2 capacity for C-15-induced plants was significantly reflected in the final yield and yield components when compared with those established with either Ch-191 or CP-36. When the reasons behind the lower symbiotic efficiency of Ch-191 and CP-36 associations were examined, it was evident that at least the lower relative efficiency in electron allocation to N_2 *versus* H^+, which resulted in the lower specific activity of nodules (fixed nitrogen per unit nodule biomass), contributed to this suboptimal efficiency (Table 2). In agreement with our finding, other studies investigating the suboptimal

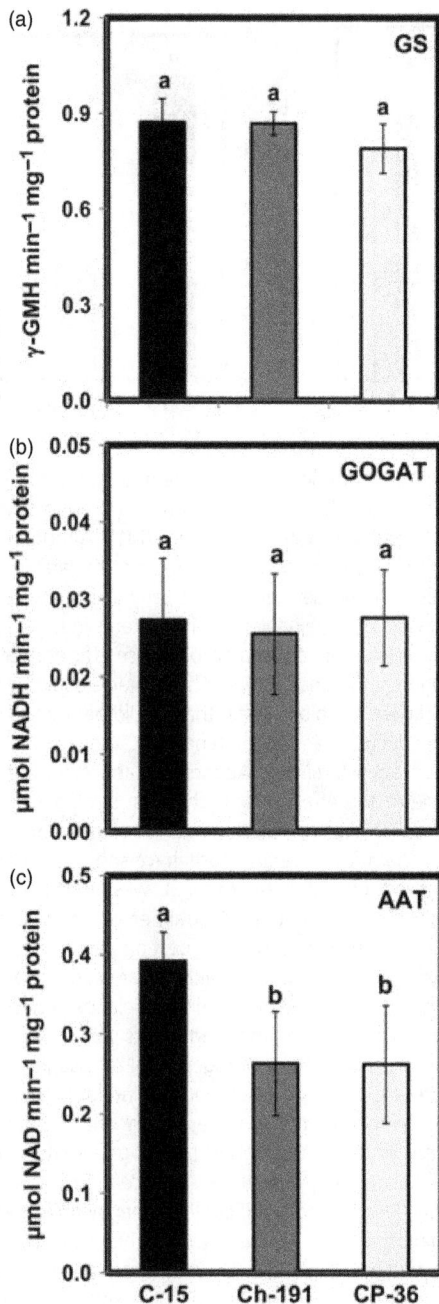

Figure 5 *In vitro* nitrogen–enzyme activities (a) glutamine synthetase (GS), (b) glutamate synthase (GOGAT) and (c) aspartate aminotransferase (AAT) in nodules of chickpea cv. 'Bivanij' inoculated with either *Mesorhizobium ciceri* C-15, Ch-191 or CP-36 strain and grown over a period of 28 days after sowing in fine quartz sand. Data are means of three replicates. Error bars represent standard errors. Data with different letters are significantly different as measured by Duncan's multiple-range test (DMRT) ($P \leq 0.05$).

symbiotic efficiency of *M. truncatula* cv. Jemalong A17 in combining with *Sinorhizobium meliloti* strain 2011 also reported lower electron allocation coefficient (EAC) values for this model symbiotic combination (Sulieman and Schulze, 2010a; Sulieman et al., 2013).

As suspected, the much higher per plant and specific nitrogenase activity observed in nodules of plants inoculated with C-15

as compared to that in nodules induced with either Ch-191 or CP-36 was also validated by a remarkable increase in expression levels of the *nifHDK* genes encoding for nitrogenase (Figure 1). The transcript levels of all three *nifH*, *nifD* and *nifK* genes were found to be more significantly abundant in C-15 nodules compared to Ch-191 and CP-36 nodules (Figure 1). Previous studies have identified a close association between nitrogenase efficiency and *nifH* expression level (Burbano et al., 2011; Deslippe and Egger, 2006). Our data clearly demonstrated that the enhanced nitrogenase activity in C-15–chickpea association was well correlated with an increase in expression levels of *nifH*, *nifD* and *nifK* genes (Figure 1). In other words, this result strongly suggested that the enhanced nitrogenase activity in nodules of chickpea plants inoculated with C-15 was due to the up-regulation of the *nifH*, *nifD* and *nifK* genes.

Recently, it has been revealed that there is a close association between the whole plant symbiotic efficiency and nodule carbon metabolism in *Medicago* species (Fischinger et al., 2010; Lopez et al., 2008; Sulieman and Schulze, 2010b; Sulieman et al., 2013). The present analyses demonstrated critical differences in nodule carbon metabolism between the three symbiotic combinations (Figure 3). In this regard, several aspects showing that the significant variations in the symbiotic efficiency of the cultivar 'Bivanij' in association with C-15, Ch-191 and CP-36 strains are closely related to the capacity of organic acid formation in nodules and thus the carbon provision for bacteroid metabolism and amino acid biosynthesis (Vance, 2008). No significant difference was detected in nodule sucrose availability among the three symbiotic associations (Figure 6a), suggesting that carbon supply from the host plant was not behind the comparatively lower symbiotic efficiency observed with either Ch-191 or CP-36 associations. It might be possible that the symbiotic efficiencies of Ch-191 and CP-36 were limited due to the decreased ability of their induced nodule to hydrolyse sucrose. Indeed, the content of malate, which was the main organic acid hydrolysed from sucrose to provide carbon supply to bacteroid respiration, enabling efficient atmospheric N_2 fixing by nitrogenase, was much lower in Ch-191- and CP-36-induced nodules relative to that found in nodules formed by C-15. Malate is known to play a pivotal role in determining the symbiotic efficiency and the level of nitrogenase activity in various annual and perennial pasture legumes (Fischinger et al., 2010; Sulieman et al., 2013). In this sequence of events, sucrose breakdown is the initial critical step in nitrogenase carbon provision which is performed with the action of either SS or AI (Figure 3) (Gordon et al., 1999). While the specific activity of SS was significantly higher in nodules induced with C-15, AI was only greatly enhanced in CP-36 symbiotic fraction (Figure 4a), suggesting that the increase in SS activity was responsible for the higher symbiotic efficiency observed with C-15 strain. SS is known to be the main enzyme involved in nodule sucrose metabolism, while AI represents the alternative route for sucrose breakdown (White et al., 2007). These data collectively indicated that the lack of SS activity could not be compensated by AI (Figure 3). This might be related to the less ATP cost required for the function of SS than for that of AI (Gordon and James, 1997; Gordon et al., 1999). In agreement with our finding, SS has been described as the principle enzyme responsible for sucrose degradation in N_2-fixing nodules (Baier et al., 2007; Ben Salah et al., 2010). Modifications of SS expression and activity through mutations (Craig et al., 1999; Horst et al., 2007) or antisense techniques (Baier et al., 2007) have shown that SS activity is essential for efficient nodule

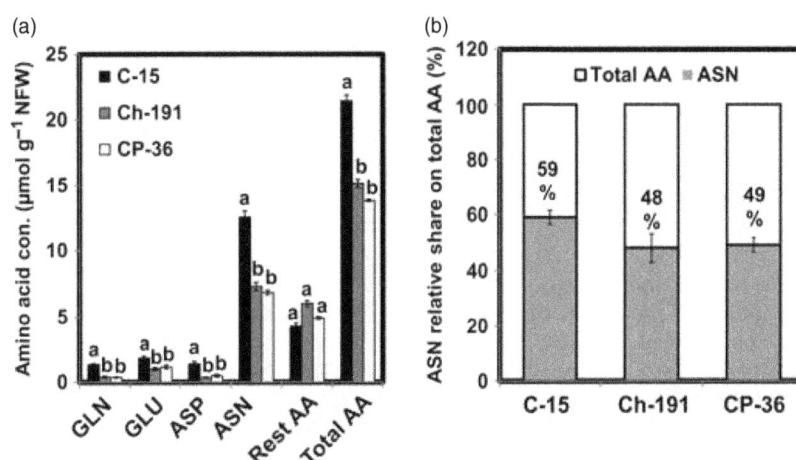

Figure 6 Nodule (a) sucrose and (b) organic acid concentrations of chickpea cv. 'Bivanij' inoculated with either *Mesorhizobium ciceri* C-15, Ch-191 or CP-36 strain and grown over a period of 28 days after sowing in fine quartz sand. Data are means of three replicates. Error bars represent standard errors. Data with different letters are significantly different as measured by Duncan's multiple-range test (DMRT) ($P \leq 0.05$). FW, fresh weight; NFW, nodule fresh weight; OA, organic acid.

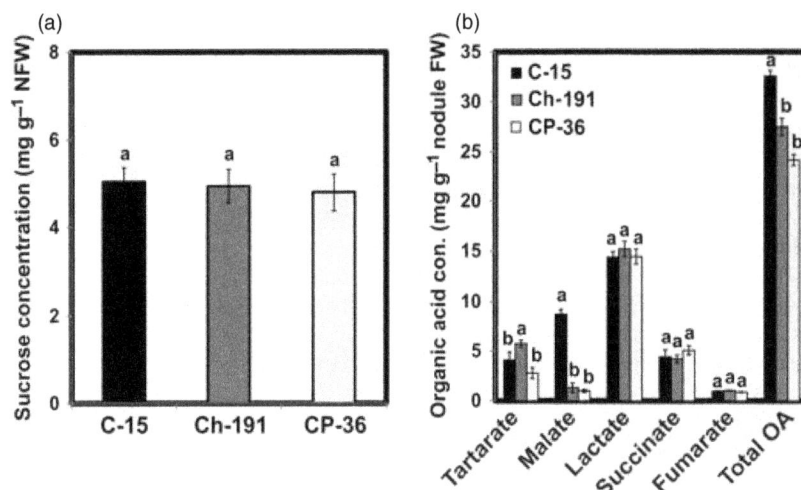

Figure 7 (a) Nodule amino composition and (b) asparagine (ASN) proportion on total amino acid concentration in nodules of chickpea cv. 'Bivanij' inoculated with either *Mesorhizobium ciceri* C-15, Ch-191 or CP-36 strain and grown over a period of 28 days after sowing in fine quartz sand. Data are means of three replicates. Error bars represent standard errors. Data with different letters are significantly different as measured by Duncan's multiple-range test (DMRT) ($P \leq 0.05$). AA, amino acids; ASN, asparagine; ASP, aspartate; GLN, glutamine; GLU, glutamate; NFW, nodule fresh weight.

Table 3 Yield and yield components of chickpea cv. 'Bivanij' inoculated with either *Mesorhizobium ciceri* C-15, CP-36 or Ch-191 strains or noninoculated and grown over a period of 82 days after sowing in fine quartz sand and kept in natural field conditions. Noninoculated plants were supplied with either full nitrogen (+ N) or nitrogen-free solution (- N). Data presented are the means ± SE of five replicates

	Nodulation			Non-nodulation	
	C-15	Ch-191	CP-36	+ N	- N
Pod number per plant	23.14 ± 0.30[a]	13.96 ± 0.46[b]	12.94 ± 0.47[b]	23.4 ± 0.52[a]	4.38 ± 0.10[c]
Seed number per pod	1.14 ± 0.24[a]	0.92 ± 0.06[b]	0.90 ± 0.03[b]	1.18 ± 0.05[a]	0.52 ± 0.04[c]
Total seed yield (g/plant)	7.82 ± 0.18[a]	3.98 ± 0.17[b]	3.78 ± 0.14[b]	7.32 ± 0.21[a]	0.69 ± 0.06[c]
100 seed weight (g)	29.8 ± 0.96[a]	26.6 ± 0.67[b]	26.8 ± 0.73[b]	28.6 ± 1.25[a]	24.8 ± 0.58[b]
Hay yield (g plant)	3.61 ± 0.11[a]	2.02 ± 0.05[b]	1.97 ± 0.04[b]	3.53 ± 0.12[a]	0.82 ± 0.02[c]

Data with different superscript letters are significantly different as measured by Duncan's multiple-range test (DMRT) ($P \leq 0.05$).

functioning. Moreover, the inhibition of SS activity in response to a wide range of environmental constraints was highly correlated with the decline in N_2 fixation under the same conditions (Arrese-Igor et al., 1999; Gonzalez et al., 2001). In addition to SS, two other critical enzymes are known to be involved in nodule organic acid biosynthesis, namely PEPC and MDH (Figure 3) (Fedorova et al., 1999; Schulze, 2004; Schulze et al., 1998). Both are integral enzymes in carbon provision in the form of organic acids that are supplied to bacteroids to fuel N_2 fixation (Vance, 2008). The significantly higher activities of PEPC and MDH in nodules of

C-15-inoculated plants as compared to those detected in nodules of Ch-191- and CP-36-inoculated plants might result in greater carbon efficiency to effectively supply bacteroids with malate (Figure 4a, b). Consistent with our findings, previous studies reported a remarkable decrease in N_2 fixation following PEPC down-regulation (Nomura et al., 2006), while MDH overexpression markedly enhanced nodule specific activity (Schulze et al., 2002). Moreover, both enzymes have been shown to be implicated in the physiological modulation of the oxygen diffusion barrier (Bargaz et al., 2012; Miller et al., 1998). Change in oxygen diffusion barrier remains one of the main mechanisms that have been used to explain the regulation of nitrogenase activity under various environmental conditions (Minchin et al., 2008; Schulze, 2004). The shift towards a significantly more effective SS-PEPC-MDH pathway in nodules induced by C-15 was reflected in the higher capacity to produce organic acids, particularly malate (Figure 3) (Schulze, 2004). This C_4 dicarboxylic acid is the main respiratory substrate for nitrogenase activity and carbon skeletons for ammonia assimilation (Fischinger et al., 2010; Schulze et al., 2002). The constitutive content of this metabolite was higher in C-15 nodules, approximately 6.5-fold and 8.7-fold as compared to Ch-191 and CP-36 nodules, respectively (Figure 6b). These data are consistent with the highest nitrogenase activity observed in the nodules of C-15 association as reflected by the highest values of H_2 and CO_2 evolutions that were measured (Table 2).

As nitrogen assimilation is inseparably linked to carbon metabolism, the differential capacity of organic acid formation may lead to the differential amino acid biosynthesis activity in nodules of the three symbiotic combinations (Figure 3). The significant abundance of total organic acid in nodules induced by C-15 strain was accompanied by the significantly higher content of total amino acids, particularly in the content of asparagine, when compared with those in nodules formed by either Ch-191 or CP-36 (Figure 7a, B). In accordance with these measurements, total soluble protein was significantly higher in C-15-induced nodules compared to those obtained in nodules induced by Ch-191 or CP-36 (Figure 2). In agreement with our result, remarkable reduction was also reported in most amino acids following the application of ineffective bacteroids (Barsch et al., 2006). In addition, although the GS/GOGAT cycle revealed no significant difference among the studied symbiotic systems, glutamine and glutamate pool sizes were significantly lower in Ch-191- and CP-36-induced nodules (Figures 5a, b, 7a), verifying that GS and NADH-GOGAT form the major ammonia assimilation pathway in almost all higher plant tissues as reported earlier (Figure 3) (Groat and Vance, 1981). This might be due to the relatively lower nitrogen turnover observed in the nodules formed by inoculation with either Ch-191 or CP-36. Furthermore, the smaller pool sizes of aspartate in Ch-191- and CP-36-formed nodules, when compared to that found in C-15-induced nodules, were reflected in a significant lower AAT activity (Figures 5c, 7a). Aspartate remains the prerequisite for asparagine biosynthesis in nodule nitrogen biochemical pathway (Figure 3) (Vance, 2008). This amide has already been described as the primary assimilation product from N_2 fixation in legumes and the predominant nitrogen-transported substrate in various leguminous species (Figure 3) (Lea et al., 2007; Sulieman and Tran, 2013).

In summary, the present study provides strong lines of evidence that C-15 strain possesses the highest symbiotic efficiency in combination with chickpea among the three examined M. ciceri strains, suggesting that C-15 could be useful for applications in agriculture. Detailed comparative analyses of carbon and nitrogen metabolisms and their relating enzymes in the nodule machinery indicated that in comparison with Ch-191 and CP-36, C15 formed the best association with chickpea due to its ability to induce nodules that exhibited enhanced malate formation. This finding in turn suggests that an improvement of nodule CO_2 fixation and malate formation in nodules might be a promising genetic engineering strategy for improving N_2 fixation efficiency in chickpea and perhaps other leguminous crops as well.

Experimental procedures

Growth chamber experiment

Plant and bacterial materials and growth conditions

To enhance chickpea N_2 fixation efficiency, the potential symbiotic capability of the famous local Iranian high-yielding cultivar namely 'Bivanij' was initially examined in combination with eight M. ciceri strains originating from different sources (Nasr Esfahani, 2011). This cultivar is a high-yielding elite variety of Kabuli type with relatively large seeds (~ 400 mg) and semi-erect growth habit. Five of the examined symbiotic companions namely C-15, C-22, SWRI7, SWRI4 and SWRI9 were native Iranian M. ciceri strains [courtesy from the Soil and Water Research Institute (SWRI), Iran], while IC-59 was introduced from ICRISAT (India), and CP-36 and Ch-191 were kindly provided from the International Center for Agricultural Research in the Dry Areas (ICARDA), Syria. Among the examined strains, C-15 scored the highest performance, while Ch-191 and CP-36 revealed the lowest growth and nodulation (Nasr Esfahani, 2011). Based on the prescreening investigations, the present study was further aimed to (i) confirm whether the contrasting (superior/inferior) symbiotic efficiencies of the M. ciceri strains will have significant reflection in the final growth and productivity of the elite cultivar 'Bivanij' (ii) and characterize the potential biochemical and molecular reasons behind the observed significant variations in the symbiotic capabilities for the highest/lowest chickpea–M. ciceri combinations. Such approach remains a vital prerequisite which ultimately paves the road for a better understanding of the mechanisms controlling the symbiotic efficiency, thereby improving the symbiotic N_2 fixation by modern genetic engineering for chickpea as well as other leguminous plants.

The M. ciceri strains were grown under standard conditions in yeast extract mannitol (YEM) as described by Nasr Esfahani and Mostajeran (2011) and subsequently used for inoculation. C. arietinum L. cv. 'Bivanij' seeds [provided by Dryland Agriculture Research Institute (DARI), Iran] were sterilized with 95% (v/v) ethanol for 10 s, 0.2% $HgCl_2$ for 2 min, followed by rinsing several times with sterile water. Following surface sterilization, seeds were placed in sterilized water for 4 h. Subsequently, seeds were sown on sterilized fine quartz sand (Ø = 0.1–0.5 mm, Quarzsandwerke Weferlingen, Germany) in PVC tubes. Plants were maintained in a controlled growth chamber in conditions as described by Nasr Esfahani and Mostajeran (2011). For inoculation, one mL of bacterial suspension was added for each plant at sowing and at days 3 and 6 after emergence. Plants were watered daily with nitrogen-free nutrient solution (Vadez et al., 2000) to 80% of the maximum water-holding capacity. 28 days after sowing, plants were harvested and divided into shoot, root and nodule fractions. These plant samples were dried to a constant weight at 60 °C for dry matter determination. Separate nodule sample sets were kept frozen at −80 °C until further biochemical and molecular analyses.

H_2 and CO_2 evolution measurements

For the H_2 and CO_2 evolution measurements, nodulated root systems of intact plants were sealed in the PVC tubes and connected to an open-flow gas exchange measurement system under N_2/O_2 (80/20, v/v) using an electrochemical H_2 and CO_2 analyzers (Qubit Systems, Canada), as described previously (Sulieman and Schulze, 2010b). When stable H_2 outflow from the root/nodule compartment was reached, this value was taken as ANA. Subsequently, the inflow air composition was switched to Ar:O$_2$ (80:20 v/v) that enabled TNA measurement. NFR was calculated as (TNA-ANA)/3, while EAC was estimated as 1-(ANA/TNA) (Curtis et al., 2004).

Determination of expression levels of the nifHDK genes

For the determination of the expression levels of the *nifH*, *nifD* and *nifK* genes in nodules, total nodule RNA was isolated using RNX[TM] (-Plus) kit (CinnaGen, Iran) according to the manufacturer's instructions. Genomic DNA was removed from the RNA samples by incubating 2 µg of RNA with RNase-free DNase I, (Fermentas, Thermo Fisher Scientific, Inc., MA) at 37 °C for 30 min according to the manufacturer's protocol. First-strand cDNA synthesis was performed using 1 µg of DNase I-treated total RNA with the 2-steps RT-PCR kit (Vivantis Technologies, Malaysia) according to the manufacturer's protocol.

Gene-specific primers for *nifH* (AY318755) (F: 5'-CATCCTCAAATATGCCCATTC-3' and R: 5-'GTGGATCTTCTCGGCCAGAG-3'), *nifD* (AY626911) (F: 5'-GCATACTGCTTGAGGAGATAG-3' and R: 5'-TTGGCAATGACCTTTTCGGTC-3') and *nifK* (AY873924) (F: 5'-AGTCATGTCGACGGCTATGAC-3' and R: 5-'ATCGAACTGGTCAGAGGCATC-3') were designed using Oligo analyzer 3.1 (http://www.idtdna.com/analyzer/Applications/OligoAnalyzer). The *16S rRNA* (NR_025953) (F: 5'-TACTGACGCTGAGGTGCGAAAG-3' and R: 5'-AAACCACATGCTCCACCGCTTG-3') was used as a reference gene in expression analysis. RT-PCR was performed with 2 µL of prepared cDNAs. PCR conditions were optimized for each primer pair as follows: initial denaturation of 2 min at 94 °C; 33 cycles of 94 °C, 55 °C (*nifH*), 57 °C (*nifD*), 59 °C (*nifK*) and 63 °C (*16S rRNA*) for 30S; 72 °C for 30 s and final extension of 7 min at 72 °C.

Preparation of nodule extracts and enzyme activity assays

For the determination of *in vitro* nodule enzyme activity, 300 mg of freshly detached nodules were homogenized in a mortar and pestle in extraction buffer (50 M/m^3 MOPS, 10 M/m^3 2-mercaptoethanol, 2% PVPP, 1 M/m^3 EDTA, 5 M/m^3 MgCl$_2$, 10% glycerol and 1 M/m^3 PMSF, pH 7 at 0–4 °C), and the homogenized samples were subsequently centrifuged for 30 min at 15000 g at 4 °C to obtain the protein fraction. Protein content was measured using Bradford's reagent (Bradford, 1976). SS was measured according to Gordon (1991). The activities of AI and GS were measured as previously described (Gonzalez et al., 1995). The activities of PEPC, MDH and AAT were measured in a direct or coupled assay, respectively, monitoring the disappearance of NADH at A$_{340}$ and 20 °C using the method described previously (Egli et al., 1989). NADH-GOGAT activity was assayed by monitoring the reduction of NAD$^+$ or oxidation of NADH according to the protocol described by Groat and Vance (1981).

Nodule metabolite analyses

For the detection of free amino acids and organic acids, frozen nodules were ground to a fine powder in liquid N$_2$ using a mortar and pestle. The samples (approximately 0.5 g fresh weight) were extracted with 3 mL of 50% ethanol (v/v) for 20 min at 45°C in a water bath. The solution from each sample was subsequently centrifuged (8000 rpm, 30 min, 10 °C), and the supernatant was immediately applied to an HPLC system (Waters, Milford, MA) after filtration (0.45 µm). Free amino acid contents were measured with a fluorescence detector after precolumn derivatization by orthophthaldialdehyde (Chen et al., 1979). Organic acids were determined by a photodiode array detector (Keutgen

and Pawelzik, 2008). Sucrose content was measured spectrophotometrically according to the protocol described previously (Gonzalez et al., 1995).

Field experiment

For yield quantification, a second experiment was carried out in outdoor pot culture under natural daylight in the Experimental Farm at Lorestan University, Khorramabad (latitude 33°29'16" N and longitude 48°21'21"E). The average temperature was between 35 and 39 °C by day and 15–22 °C by night with 10% relative humidity during the growing period (June–September 2013). Plants were exposed to natural light intensity, temperature and day length conditions throughout the entire course of the experiment. Seeds of chickpea cv. 'Bivanij' were treated as mentioned above and transferred to plastic pots (diameter of 37 cm, height of 22 cm) containing nitrogen-free autoclaved silica sand (Ø = 1.0–1.5 mm). Nodulation plants were inoculated with C-15, CP-36 or Ch-191 of *M. ciceri* strains and irrigated as described by Nasr Esfahani and Mostajeran (2011). To buffer the proton excretion of N$_2$-fixing nodules and roots, 0.13 g CaCO$_3$ was added per each kilogram sand. For controls, non-nodulation plants were grown using the same pot medium and flushed with either the half-strength nitrogen Hoagland solution (Hoagland and Arnon, 1950) (referred as + N) or a solution containing the same chemical composition but lacking combined nitrogen (referred as - N). Each treatment was carried out in five replicates. Plants were watered daily with respective nutrient solution to 80% of the maximum water-holding capacity. 82 days after sowing, plants were harvested and fractioned into shoot, root and nodule for quantification of the yield and yield components.

Statistical analysis

Data were subjected to an analysis of variance (one-way ANOVA) procedure using the SPSS software package 16.0. In the case of homogenous sample variances, mean separation procedures were carried out using the Duncan's multiple-range test (DMRT) ($P \leq 0.05$).

Acknowledgements

We would like to thank Dr. A. Asgharzadeh from Soil and Water Research Institute (SWRI), Iran, and Mr. F. Afandi from the International Center for Agricultural Research in the Dry Areas (ICARDA) for providing *M. ciceri* strains C-15 and Ch-191 and CP-36, respectively. S. Sulieman is supported by a postdoc fellowship from Japan Society for the Promotion of Science (JSPS).

References

Arrese-Igor, C., Gonzalez, E.M., Gordon, A.J., Minchin, F.R., Galvez, L., Royuela, M., Cabrerizo, P.M. and Aparicio-Tejo, P.M. (1999) Sucrose synthase and nodule nitrogen fixation under drought and other environmental stresses. *Symbiosis*, **27**, 189–212.

Baier, M.C., Barsch, A., Kuster, H. and Hohnjec, N. (2007) Antisense repression of the *Medicago truncatula* nodule-enhanced sucrose synthase leads to a handicapped nitrogen fixation mirrored by specific alterations in the symbiotic transcriptome and metabolome. *Plant Physiol.* **145**, 1600–1618.

Bargaz, A., Ghoulam, C., Amenc, L., Lazali, M., Faghire, M., Abadie, J. and Drevon, J. (2012) A phosphoenol pyruvate phosphatase transcript is induced in the root nodule cortex of *Phaseolus vulgaris* under conditions of phosphorus deficiency. *J. Exp. Bot.* **63**, 4723–4730.

Barsch, A., Tellstrom, V., Patschkowski, T., Kuster, H. and Niehaus, K. (2006) Metabolite profiles of nodulated alfalfa plants indicate that distinct stages of nodule organogenesis are accompanied by global physiological adaptations. *Mol. Plant Microbe Interact.* **19**, 998–1013.

Ben Romdhane, S., Tajini, F., Trabelsi, M., Aouani, M.E. and Mhamdi, R. (2007) Competition for nodule formation between introduced strains of *Mesorhizobium ciceri* and the native populations of rhizobia nodulating chickpea (*Cicer arietinum*) in Tunisia. *World J. Microbiol. Biotechnol.* **23**, 1195–1201.

Ben Salah, I., Slatni, T., Albacete, A., Gandour, M., Andujar, C.M., Houmani, H., Ben Hamed, K., Martinez, V., Perez-Alfocea, F. and Abdelly, C. (2010) Salt tolerance of nitrogen fixation in *Medicago ciliaris* is related to nodule sucrose metabolism performance rather than antioxidant system. *Symbiosis*, **51**, 187–195.

Bhattacharyya, P.N. and Jha, D.K. (2012) Plant growth-promoting rhizobacteria (PGPR): emergence in agriculture. *World J. Microbiol. Biotechnol.* **28**, 1327–1350.

Bradford, M.M. (1976) Rapid and sensitive method for quantitation of microgram quantities of protein utilizing principle of protein-dye binding. *Anal. Biochem.* **72**, 248–254.

Burbano, C.S., Liu, Y., Rosner, K.L., Reis, V.M., Caballero-Mellado, J., Reinhold-Hurek, B. and Hurek, T. (2011) Predominant *nifH* transcript phylotypes related to Rhizobium rosettiformans in field-grown sugarcane plants and in Norway spruce. *Env. Microbiol. Rep.* **3**, 383–389.

Chen, R.F., Scott, C. and Trepman, E. (1979) Fluorescence properties of Ortho-Phthaldialdehyde derivatives of amino-acids. *Biochim. Biophys. Acta* **576**, 440–455.

Craig, J., Barratt, P., Tatge, H., Dejardin, A., Handley, L., Gardner, C.D., Barber, L., Wang, T., Hedley, C., Martin, C. and Smith, A.M. (1999) Mutations at the *rug4* locus alter the carbon and nitrogen metabolism of pea plants through an effect on sucrose synthase. *Plant J.* **17**, 353–362.

Curtis, J., Shearer, G. and Kohl, D.H. (2004) Bacteroid proline catabolism affects N_2 fixation rate of drought-stressed soybeans. *Plant Physiol.* **136**, 3313–3318.

Deslippe, J.R. and Egger, K.N. (2006) Molecular diversity of *nifH* genes from bacteria associated with high arctic dwarf shrubs. *Microb. Ecol.* **51**, 516–525.

Drevon, J., Abdelly, C., Amarger, N., Aouani, M., Aurag, J., Gherbi, H., Jebara, M., Lluch, C., Payre, H., Schump, O., Soussi, M., Sifi, B. and Trabelsi, M. (2001) An interdisciplinary research strategy to improve symbiotic nitrogen fixation and yield of common bean (*Phaseolus vulgaris*) in salinised areas of the Mediterranean basin. *J. Biotechnol.* **91**, 257–268.

Egli, M.A., Griffith, S.M., Miller, S.S., Anderson, M.P. and Vance, C.P. (1989) Nitrogen assimilating enzyme-activities and enzyme protein during development and senescence of effective and plant gene-controlled ineffective alfalfa nodules. *Plant Physiol.* **91**, 898–904.

Fedorova, M., Tikhonovich, I.A. and Vance, C.P. (1999) Expression of C-assimilating enzymes in pea (*Pisum sativum* L.) root nodules. *In situ* localization in effective nodules. *Plant, Cell Environ.* **22**, 1249–1262.

Fischinger, S.A. and Schulze, J. (2010) The importance of nodule CO_2 fixation for the efficiency of symbiotic nitrogen fixation in pea at vegetative growth and during pod formation. *J. Exp. Bot.* **61**, 2281–2291.

Fischinger, S.A., Hristozkova, M., Mainassara, Z.A. and Schulze, J. (2010) Elevated CO_2 concentration around alfalfa nodules increases N_2 fixation. *J. Exp. Bot.* **61**, 121–130.

Flowers, T.J., Gaur, P.M., Gowda, C.L.L., Krishnamurthy, L., Samineni, S., Siddique, K.H.M., Turner, N.C., Vadez, V., Varshney, R.K. and Colmer, T.D. (2010) Salt sensitivity in chickpea. *Plant, Cell and Environ.* **33**, 490–509.

Gaur, P.M., Jukantil, A.K. and Varshney, R.K. (2012) Impact of genomic technologies on chickpea breeding strategies. *Agronomy* **2**, 199–221.

Geurts, R., Lillo, A. and Bisseling, T. (2012) Exploiting an ancient signalling machinery to enjoy a nitrogen fixing symbiosis. *Curr. Opin. Plant Biol.* **15**, 438–443.

Gonzalez, E.M., Gordon, A.J., James, C.L. and Arreseigor, C. (1995) The role of sucrose synthase in the response of soybean nodules to drought. *J. Exp. Bot.* **46**, 1515–1523.

Gonzalez, E.M., Galvez, L., Royuela, M., Aparicio-Tejo, P.M. and Arrese-Igor, C. (2001) Insights into the regulation of nitrogen fixation in pea nodules: lessons from drought, abscisic acid and increased photoassimilate availability. *Agronomie*, **21**, 607–613.

Gordon, A.J. (1991) Enzyme distribution between the cortex and the infected region of soybean nodules. *J. Exp. Bot.* **42**, 961–967.

Gordon, A.J. and James, C.L. (1997) Enzymes of carbohydrate and amino acid metabolism in developing and mature nodules of white clover. *J. Exp. Bot.* **48**, 895–903.

Gordon, A.J., Minchin, F.R., James, C.L. and Komina, O. (1999) Sucrose synthase in legume nodules is essential for nitrogen fixation. *Plant Physiol.* **120**, 867–877.

Graham, P.H. and Vance, C.P. (2003) Legumes: Importance and constraints to greater use. *Plant Physiol.* **131**, 872–877.

Groat, R.G. and Vance, C.P. (1981) Root nodule enzymes of ammonia assimilation in alfalfa (*Medicago Sativa* L) - developmental patterns and response to applied nitrogen. *Plant Physiol.* **67**, 1198–1203.

Hoagland, D.R. and Arnon, D.I. (1950) A water culture method for growing plants without soil. Circular 347, pp. 25–32. Berkley: University of California Agricultural Experiment Station.

Horst, I., Welham, T., Kelly, S., Kaneko, T., Sato, S., Tabata, S., Parniske, M. and Wang, T.L. (2007) TILLING mutants of *Lotus japonicus* reveal that nitrogen assimilation and fixation can occur in the absence of nodule-enhanced sucrose synthase. *Plant Physiol.* **144**, 806–820.

Icgen, B., Ozcengiz, G. and Alaeddinoglu, N.G. (2002) Evaluation of symbiotic effectiveness of various *Rhizobium cicer* strains. *Res. Microbiol.* **153**, 369–372.

Keutgen, A.J. and Pawelzik, E. (2008) Contribution of amino acids to strawberry fruit quality and their relevance as stress indicators under NaCl salinity. *Food Chem.* **111**, 642–647.

Kyei-Boahen, S., Slinkard, A.E. and Walley, F.L. (2002) Evaluation of rhizobial inoculation methods for chickpea. *Agron J.* **94**, 851–859.

Laranjo, M., Branco, C., Soares, R., Alho, L., Carvalho, M.D.E. and Oliveira, S. (2002) Comparison of chickpea rhizobia isolates from diverse Portuguese natural populations based on symbiotic effectiveness and DNA fingerprint. *J. Appl. Microbiol.* **92**, 1043–1050.

Lea, P.J., Sodek, L., Parry, M.A.J., Shewry, R. and Halford, N.G. (2007) Asparagine in plants. *Ann. Appl. Biol.* **150**, 1–26.

Lopez, M., Herrera-Cervera, J.A., Iribarne, C., Tejera, N.A. and Lluch, C. (2008) Growth and nitrogen fixation in *Lotus japonicus* and *Medicago truncatula* under NaCl stress: nodule carbon metabolism. *J. Plant Physiol.* **165**, 641–650.

Mafongoya, P.L., Mpepereki, S. and Mudyazhezha, S. (2009) The importance of biological nitrogen fixation in cropping systems in nonindustrialized nations. In *Nitrogen Fixation in Crop Production* (Emerich, D.W. and Krishnan, H.B., eds), pp. 329–348. Madison, WI, USA: American Society of Agronomy, Crop Science Society of America, Soil Science Society of America.

Mhadhbi, H., Jebara, M., Limam, F. and Aouani, M.E. (2004) Rhizobial strain involvement in plant growth, nodule protein composition and antioxidant enzyme activities of chickpea-rhizobia symbioses: modulation by salt stress. *Plant Physiol. Biochem.* **42**, 717–722.

Mhadhbi, H., Jebara, M., Zitoun, A., Limam, F. and Aouani, M.E. (2008) Symbiotic effectiveness and response to mannitol-mediated osmotic stress of various chickpea-rhizobia associations. *World J. Microb. Biot.* **24**, 1027–1035.

Miller, S.S., Driscoll, B.T., Gregerson, R.G., Gantt, J.S. and Vance, C.P. (1998) Alfalfa malate dehydrogenase (MDH): molecular cloning and characterization of five different forms reveals a unique nodule-enhanced MDH. *Plant J.* **15**, 173–184.

Minchin, F.R., James, E.K. and Becana, M. (2008) Oxygen diffusion, production of reactive oxygen and nitrogen species, and antioxidants in legume nodules. In *Nitrogen-fixing Leguminous Symbioses* (Dilworth, M.J., James, E.K., Sprent, J.I. and Newton, W.E., eds), pp. 321–362. Dordrecht, The Netherlands: Springer.

Mwanamwenge, J., Loss, S.P., Siddique, K.H.M. and Cocks, P.S. (1998) Growth, seed yield and water use of faba bean (*Vicia faba* L) in a short-season Mediterranean-type environment. *Aust. J. Exp. Agr.* **38**, 171–180.

Nasr Esfahani, M. (2011) The effect of drought stress on the nitrogenase activity in sensitive and tolerance symbioses of chickpea (*Cicer arietinum* L.)-*Mesorhizobium ciceri*. PhD thesis, University of Isfahan, Iran.

Nasr Esfahani, M. and Mostajeran, A. (2011) Rhizobial strain involvement in symbiosis efficiency of chickpea-rhizobia under drought stress: plant growth, nitrogen fixation and antioxidant enzyme activities. *Acta Physiol. Plant* **33**, 1075–1083.

Nomura, M., Mai, H.T., Fujii, M., Hata, S., Izui, K. and Tajima, S. (2006) Phosphoenolpyruvate carboxylase plays a crucial role in limiting nitrogen fixation in *Lotus japonicus* nodules. *Plant Cell Physiol.* **47**, 613–621.

Peoples, M.B. (2009) The potential environmental benefits and risks derived from legumes in rotations. In *Nitrogen Fixation in Crop Production* (Emerich, D.W. and Krishnan, H.B., eds), pp. 349–385. Madison, WI, USA: American Society of Agronomy, Crop Science Society of America, Soil Science Society of America.

Purcell, L.C. (2009) Physiological responses of N_2 fixation to drought and selecting genotypes for improved N_2 fixation. In *Nitrogen Fixation in Crop Production* (Emerich, D.W. and Krishnan, H.B., eds), pp. 211–238. Madison, WI, USA: American Society of Agronomy, Crop Science Society of America, Soil Science Society of America.

Rubio, L.M. and Ludden, P.W. (2008) Biosynthesis of the iron-molybdenum cofactor of nitrogenase. *Annu. Rev. Microbiol.* **62**, 93–111.

Schulze, J. (2004) How are nitrogen fixation rates regulated in legumes? *J. Plant Nutr. Soil Sci.* **167**, 125–137.

Schulze, J., Shi, L.F., Blumenthal, J., Samac, D.A., Gantt, J.S. and Vance, C.P. (1998) Inhibition of alfalfa root nodule phosphoenolpyruvate carboxylase through an antisense strategy impacts nitrogen fixation and plant growth. *Phytochemistry*, **49**, 341–346.

Schulze, J., Tesfaye, M., Litjens, R.H.M.G., Bucciarelli, B., Trepp, G., Miller, S., Samac, D., Allan, D. and Vance, C.P. (2002) Malate plays a central role in plant nutrition. *Plant Soil*, **247**, 133–139.

Sulieman, S. and Schulze, J. (2010a) The efficiency of nitrogen fixation of the model legume *Medicago truncatula* (Jemalong A17) is low compared to *Medicago sativa*. *J. Plant Physiol.* **167**, 683–692.

Sulieman, S. and Schulze, J. (2010b) Phloem-derived gamma-aminobutyric acid (GABA) is involved in upregulating nodule N_2 fixation efficiency in the model legume *Medicago truncatula*. *Plant, Cell Environ.* **33**, 2162–2172.

Sulieman, S. and Tran, L.S. (2013) Asparagine: an amide of particular distinction in the regulation of symbiotic nitrogen fixation of legumes. *Crit. Rev. Biotechnol.* **33**, 309–327.

Sulieman, S., Fischinger, S.A., Gresshoff, P.M. and Schulze, J. (2010) Asparagine as a major factor in the N-feedback regulation of N_2 fixation in *Medicago truncatula*. *Physiol. Plant.* **140**, 21–31.

Sulieman, S., Schulze, J. and Tran, L.S. (2013) Comparative Analysis of the Symbiotic Efficiency of *Medicago truncatula* and *Medicago sativa* under Phosphorus Deficiency. *Int. J. Mol. Sci.* **14**, 5198–5213.

Tran, L.S. and Nguyen, H.T. (2009) Future biotechnology of legumes. In *Nitrogen Fixation in Crop Production* (Emerich, D.W. and Krishnan, H.B., eds), pp. 265–308. Madison, WI, USA: American Society of Agronomy, Crop Science Society of America, Soil Science Society of America.

Vadez, V., Sinclair, T.R. and Serraj, R. (2000) Asparagine and ureide accumulation in nodules and shoots as feedback inhibitors of N_2 fixation in soybean. *Physiol. Plant.* **110**, 215–223.

Vance, C.P. (2001) Symbiotic nitrogen fixation and phosphorus acquisition. Plant nutrition in a world of declining renewable resources. *Plant Physiol.* **127**, 390–397.

Vance, C.P. (2008) Carbon and nitrogen metabolism in legume nodules. In *Nitrogen-fixing Leguminous Symbioses* (Dilworth, M.J., James, E.K., Sprent, J.I. and Newton, W.E., eds), pp. 293–320. Dordrecht, The Netherlands: Springer.

White, J., Prell, J., James, E.K. and Poole, P. (2007) Nutrient sharing between symbionts. *Plant Physiol.* **144**, 604–614.

12

Expression of the MYB transcription factor gene *BplMYB46* affects abiotic stress tolerance and secondary cell wall deposition in *Betula platyphylla*

Huiyan Guo[1,2], Yucheng Wang[1], Liuqiang Wang[1], Ping Hu[1], Yanmin Wang[1,3], Yuanyuan Jia[1], Chunrui Zhang[1], Yu Zhang[1], Yiming Zhang[1], Chao Wang[1,*] and Chuanping Yang[1,*]

[1]State Key Laboratory of Tree Genetics and Breeding, Northeast Forestry University, Harbin, China
[2]Department of Life Science and Technology, Mudanjiang Normal College, Mudanjiang, China
[3]Key Laboratory of Fast-Growing Tree Cultivating of Heilongjiang Province, Forestry Science Research Institute of Heilongjiang Province, Harbin, China

*Correspondence
email yangcp@nefu.edu.cn
email wangchao@nefu.edu.cn

Keywords: *Betula platyphylla*, *BplMYB46*, abiotic stress, secondary wall deposition.

Summary

Plant MYB transcription factors control diverse biological processes, such as differentiation, development and abiotic stress responses. In this study, we characterized *BplMYB46*, an *MYB* gene from *Betula platyphylla* (birch) that is involved in both abiotic stress tolerance and secondary wall biosynthesis. BplMYB46 can act as a transcriptional activator in yeast and tobacco. We generated transgenic birch plants with overexpressing or silencing of *BplMYB46* and subjected them to gain- or loss-of-function analysis. The results suggest that BplMYB46 improves salt and osmotic tolerance by affecting the expression of genes including *SOD*, *POD* and *P5CS* to increase both reactive oxygen species scavenging and proline levels. In addition, BplMYB46 appears to be involved in controlling stomatal aperture to reduce water loss. Overexpression of BplMYB46 increases lignin deposition, secondary cell wall thickness and the expression of genes in secondary cell wall formation. Further analysis indicated that BplMYB46 binds to MYBCORE and AC-box motifs and may directly activate the expression of genes involved in abiotic stress responses and secondary cell wall biosynthesis whose promoters contain these motifs. The transgenic *BplMYB46*-overexpressing birch plants, which have improved salt and osmotic stress tolerance, higher lignin and cellulose content and lower hemicellulose content than the control, have potential applications in the forestry industry.

Introduction

Plant growth and development are strongly influenced by various stresses, such as salinity, drought and extreme temperatures (Su *et al.*, 2014). A large number of transcription factors (TFs) mediate stress responses in plants, including MYB (Oh *et al.*, 2011), NAC (Mao *et al.*, 2012), bZIP (Uno *et al.*, 2000) and WRKY (Mare *et al.*, 2004) family members. The MYB family is one of the largest families of TFs. *Arabidopsis* contains more than 198 *MYB* genes (Yanhui *et al.*, 2006), cotton and *Populus* contain approximately 200 (Cedroni *et al.*, 2003), maize contains 157 (Du *et al.*, 2012a) and soybean contains 252 (Du *et al.*, 2012b). The MYB family is divided into four classes, including 1R-, R2R3-, 3R- and 4R-MYB proteins, according to the number of MYB domains (Dubos *et al.*, 2010). R2R3-MYB appears to be specific to plants and has been widely investigated (Kim *et al.*, 2015; Li *et al.*, 2006; Prouse and Campbell, 2012). MYBs bind to several *cis*-acting motifs, including the following: MBSI (T/C)AAC (G/T)G(A/C/T)(A/C/T), which is involved in cell cycle control and resistance to low temperatures (Ma *et al.*, 2009; Prouse and Campbell, 2012); MBSII (A/G)(G/T)T(A/T)GGT(A/G), which is involved in regulating secondary cell wall biosynthesis (Kim *et al.*, 2012); MBSIIG, ACC(A/T)ACC(A/C/T), which is related to flavonoid biosynthesis (Grotewold *et al.*, 1994); MYBCORE, CAGTTA and CTGTTG, which are associated with drought tolerance (Ithal and Reddy, 2004); and AC-box, ACC(A/T)A(A/C) (T/C), which is related to secondary cell wall deposition (Zhong *et al.*, 2013).

Plant MYBs regulate cell differentiation, organ formation, leaf morphogenesis, secondary metabolism and abiotic stress responses (Ambawat *et al.*, 2013; Qi *et al.*, 2015; Sun *et al.*, 2015; Xu *et al.*, 2015). For instance, *OsMYB4* overexpression enhances freezing tolerance in rice, and it also improves acclimation to cold and drought in transgenic apple (Pasquali *et al.*, 2008; Vannini *et al.*, 2004). TaMYB73 in wheat induces the expression of stress signalling genes and increases salinity tolerance in transgenic *Arabidopsis* (He *et al.*, 2012). Overexpression of OsMYB48-1 induces the expression of stress-response genes and improves salinity and drought stress tolerance in rice (Xiong *et al.*, 2014). The functions of MYBs in secondary wall biosynthesis have also been investigated. For example, *Arabidopsis* AtMYB46 and AtMYB83 function as master switches in a transcriptional network promoting secondary wall deposition that is regulated by the secondary wall-associated NAC domain protein 1 (SND1) (McCarthy *et al.*, 2009; Zhong *et al.*, 2007). PtrMYB2, 3, 20 and 21 positively regulate secondary wall biosynthesis during wood formation in poplar trees (Zhong *et al.*, 2013). Gain- and loss-of-function analysis showed that AtMYB61 regulates both stomatal aperture and the expression of genes involved in lignin deposition (Liang *et al.*, 2005; Newman *et al.*, 2004; Romano *et al.*, 2012). *AtMYB52* is a target gene regulated by AtMYB46 during secondary wall biosynthesis. Overexpression of *AtMYB52* also improves drought tolerance by regulating the ABA signal transduction pathway, suggesting a possible connection between secondary wall deposition and ABA responses (Ko *et al.*, 2009; Park *et al.*, 2011). Although MYBs have been investigated in diverse biological systems, their functional roles have not been fully elucidated. In addition, many members of the large MYB transcription factor family have not yet been characterized.

In this study, we cloned and functionally characterized an MYB transcription factor, BplMYB46, from birch (*Betula platyphylla*), a pioneer tree species widely distributed from Europe to Asia that has important applications for the paper, building and furniture industries (Zhang *et al.*, 2012). Our results show that overexpression of BplMYB46 improves salt and osmotic stress tolerance and mediates secondary cell wall deposition in transgenic birch. This study increases our knowledge of the crosstalk between the abiotic stress and secondary cell wall biosynthesis pathways and provides insights into the functions of MYB proteins.

Results

BplMYB46 is an R2R3-type MYB protein

The MYB transcription factor gene *BplMYB46* (GenBank accession number: KP711284) was isolated from *B. platyphylla*. Multiple sequence alignments (Figure S1) and phylogenetic analysis (Figure S2) indicate that BplMYB46 belongs to the R2R3-type MYB subfamily, with a highly conserved domain in the N-terminus containing R2 and R3 functional regions. Members of this subfamily, including VvMYB46, PtMYB2, PtMYB21 and AtMYB46, function in secondary cell wall biosynthesis.

Transactivating BplMYB46 localizes to the nucleus

We transformed 35S : BplMYB46-GFP into onion epidermal cells by particle bombardment, using 35S : GFP as the control. Green fluorescent signals from the 35S: BplMYB46-GFP transformed cells were detected in the nuclei, which were stained using DAPI. By contrast, 35S : GFP signals were uniformly distributed throughout the cell (Figure 1a), indicating that BplMYB46 is a nuclear protein.

To investigate whether BplMYB46 activated transcription and to identify the activation domain, a series of deletions of the BplMYB46 CDS were fused with the GAL4 DNA-binding domain sequence in pGBKT7 (Clontech). The resulting constructs were, respectively, transformed into yeast cells for transcriptional activation analysis using the yeast two-hybrid system (Y2H). Yeast cells harbouring the full CDS of BplMYB46 grew normally on SD/-Trp/-His/-Ade/X-α-Gal medium, and x-α-gal was activated (Figure 1b), suggesting that BplMYB46 is a transcriptional activator. Furthermore, analysis of deletions with truncated CDS of BplMYB46 suggested that the transcriptional activation domain is located in a region from amino acid 233 to 304 in BplMYB46 (Figure 1b).

Analysis of the expression of BplMYB46

To gain insight into the biological role of BplMYB46, we analysed the expression patterns of *BplMYB46* in response to NaCl, ABA and mannitol treatment using real-time PCR. The results show that *BplMYB46* expression was induced by NaCl, ABA and mannitol after 6–24 h of treatment (Figure 2a). In addition, the expression patterns of *BplMYB46* in response to NaCl and ABA treatment were quite similar, that is induction after 6 to 24 h of treatment, reaching a peak at 24 h. Mannitol treatment increased *BplMYB46* expression at 6 to 24 h, with a peak at 12 h. We further examined *BplMYB46* expression in roots, leaves and various stem internodes of 6-month-old birch (Figure 2b). The expression of *BplMYB46* was higher in the stem internodes

than in roots or leaves, with 60-fold higher expression in the 18th internodes compared with leaves, whereas no significant difference in expression was observed between roots and leaves (Figure 2c). *BplMYB46* expression was lowest in the 1st internodes, with an increasing gradient from the stem tip to base. The highest expression level was observed in the 18th internodes of stems, with levels approximately 18-fold those of the 1st internodes (Figure 2c). Therefore, *BplMYB46* is predominantly expressed in stems and is more highly expressed in mature tissue (base of stem) than in juvenile tissue (tip of stem).

BpMYB46 binds to the MYBCORE and AC-box motifs

Previous studies have shown that MYB proteins bind to the MYBCORE and AC-box motifs (Ithal and Reddy, 2004; Zhong *et al.*, 2013). To investigate the binding of BplMYB46 to these motifs, three tandem copies of these motif sequences (MYB-CORE: CAGTTA; AC-box: ACCACCT) were, respectively, cloned into pHIS2 and their interactions with BplMYB46 were determined using Y1H analysis (Figure 3a). The results indicate that yeast cells cotransformed with BplMYB46-effector and different reporters grew on TDO/3-AT medium, demonstrating that BplMYB46 also binds to the MYBCORE and AC-box motifs (Figure 3b).

To further verify the above interactions identified by Y1H, the pROK2-*BplMYB46* construct was used as an effector, and three tandem copies of MYBCORE and AC-box, together with their mutants, were, respectively, fused with the minimal 35S promoter (−46 to +1) to drive the *GUS* reporter gene (Figure 3c). GUS activity was detected in tobacco leaves following cotransformation of lines harbouring pROK2-*BplMYB46* with the MYBCORE and AC-box motifs. However, very low GUS activity was detected in cotransformed lines harbouring pROK2-*BplMYB46* and the respective mutant sequences (Figure 3d, e). MYBCORE is involved in drought stress responses (Ithal and Reddy, 2004), and the AC-box plays a role in secondary wall biosynthesis (Zhong *et al.*, 2013).

Generation of transgenic birch plants with overexpression and knock-down of *BplMYB46*

We generated 16 *BplMYB46*-overexpressing (OE) and 15 RNAi-silenced *BplMYB46* (SE) transgenic birch lines and examined the expression of *BplMYB46* using real-time RT-PCR. The transgenic and wild-type (WT, nontransgenic) plants were generated from a single birch clone, which indicates that they have the same genetic background. In addition, the expression levels of endogenous *BplMYB46* in the transgenic and WT plants were similar. Therefore, the expression levels of the transgene *BplMYB46* and the functions of this gene could be investigated in the OE, SE and WT plants. Our results indicate that the expression of *BplMYB46* was significantly increased in the OE lines, with levels 3-fold–38-fold higher than those of WT, but were significantly reduced in the SE lines, with a 50%–91% decrease relative to WT (Figure S3). *BplMYB46*-overexpressing transgenic lines 9 and 10 (termed OE9 and OE10, respectively), which showed moderate and high *BplMYB46* expression, respectively, were selected for further study. Two RNAi-silenced *BplMYB46* lines, lines 3 and 15 (termed SE3 and SE15), which exhibited a high degree of silencing of *BplMYB46*, were also employed for further study.

BplMYB46 confers salt and osmotic stress tolerance

Soil-grown transgenic birch plants, including OE, WT and SE lines, were exposed to salt or mannitol to evaluate their stress

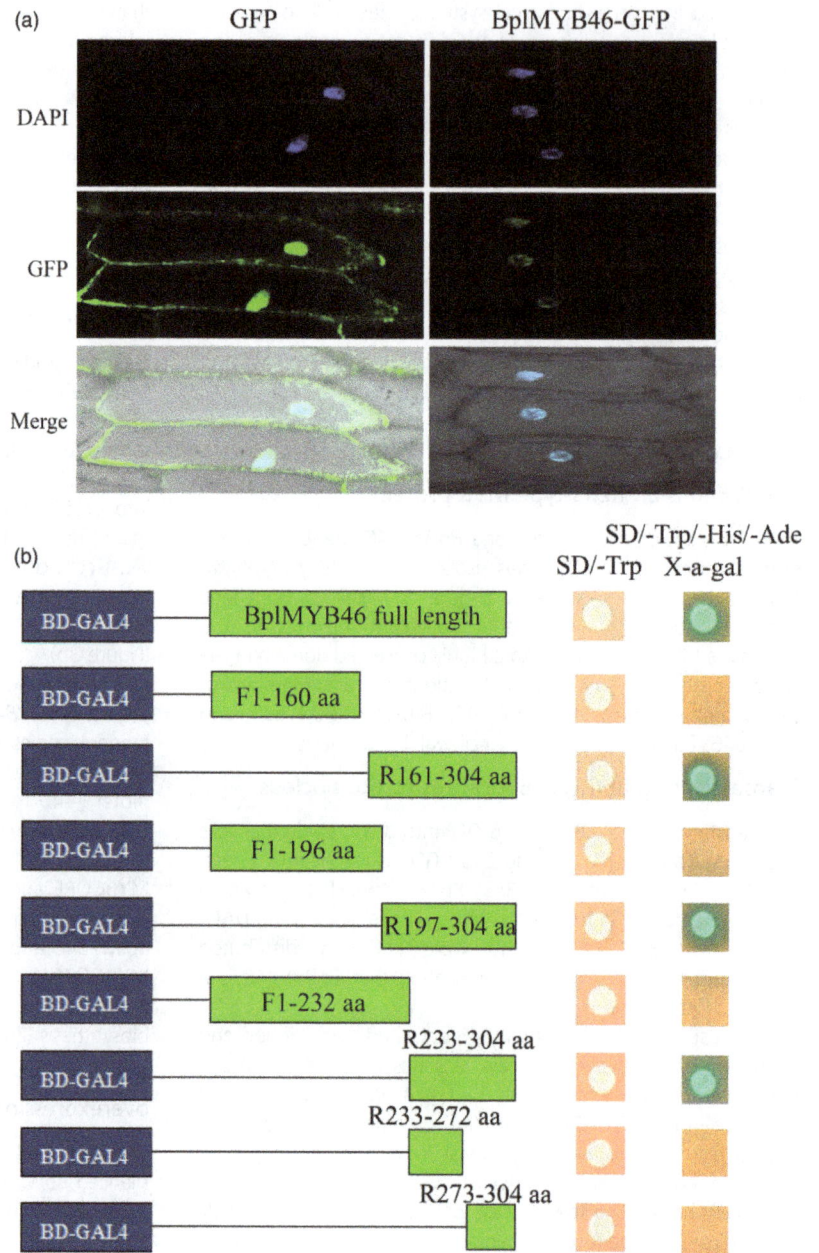

Figure 1 Subcellular localization and transactivation assay of BplMYB46. (a) The 35S : *BplMYB46-GFP* fusion gene and 35S:*GFP* control plasmid were transformed into onion epidermal cells using particle bombardment. The transformed cells were imaged by confocal microscopy after transformation for 24 h. DAPI: DAPI staining of nucleus; GFP: GFP fluorescence detection; merge: the merged images of bright-field, GFP and DAPI staining. (b) Various truncated sequences of the CDS of *BplMYB46* were fused in-frame, respectively, to the GAL4 DNA-binding domain in pGBKT7 and transformed into AH109 yeast cells. The transformed cells were plated onto SD/-Trp (growth control) or SD/-Trp/-His/-Ade/X-a-Gal medium. The pGBKT7 empty vector was used as a negative control.

tolerance. There was no substantial difference in phenotype, growth rate, fresh weight or root length among OE, WT and SE lines under control conditions (Figure 4a), suggesting that *BplMYB46* does not affect the growth phenotype or growth rate of the plants. Under NaCl or mannitol treatment, compared with WT plants, both OE9 and OE10 displayed significantly higher growth rates, were greener and exhibited less wilting, in addition to having significantly higher fresh weights and root lengths. By contrast, lines SE3 and SE15 exhibited more severe leaf rolling and wilting, significantly reduced fresh weights and root lengths and the loss of green coloration compared with WT (Figure 4b, c). In addition, the chlorophyll contents were similar among OE, WT and SE lines under control conditions. However, under salt and osmotic stress conditions, compared with WT plants, both OE9 and OE10 had significantly higher chlorophyll levels, while SE3 and SE15 had significantly lower chlorophyll

levels (Figure 4d). Overall, these results suggest that *BplMYB46* overexpression significantly improves abiotic stress tolerance in birch.

BplMYB46 affects reactive oxygen species (ROS) scavenging

In view of the key role of ROS in abiotic stress tolerance, we investigated whether *BpMYB46* affects ROS scavenging. We evaluated the levels of H_2O_2, a main ROS species, using DAB *in situ* staining. No obvious difference in DAB staining was observed among OE, WT and SE lines under control conditions. However, under NaCl or mannitol treatment conditions, compared with WT plants, OE9 and OE10 exhibited reduced DAB staining, whereas the two SE lines displayed strongly increased DAB staining, indicating that *BplMYB46* overexpression reduces H_2O_2 accumulation in plants (Figure 5a). We further measured

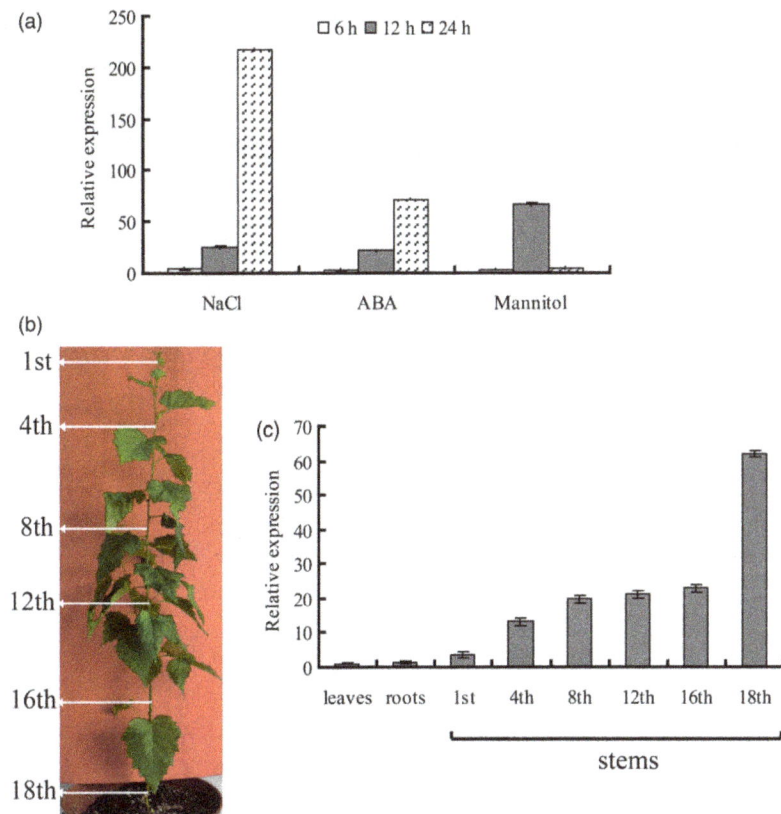

Figure 2 Expression patterns of *BplMYB46*. (a) Expression of *BplMYB46* in response to NaCl, ABA and mannitol treatment. Well-watered plants were used as controls to normalize the expression level at each time point. (b) The positions of the birch stem internodes analysed for gene expression. Arrows indicate the different stem internodes used in the analysis. (c) *BplMYB46* mRNA levels in the roots, leaves and 1st, 4th, 8th, 12th, 16th and 18th stem internodes from 6-month-old birch plants. The *BplMYB46* mRNA level in birch leaves was set to 1 to normalize its transcription level to that of roots and stem internodes. The error bars indicate the standard deviation (SD) from three biological replicates.

H_2O_2 levels. Consistent with the DAB staining results, there was no difference among OE, WT and SE lines under control conditions. However, under salt and osmotic stress conditions, the H_2O_2 levels were highest in SE3 and SE15, followed by WT, whereas OE9 and OE10 had the lowest H_2O_2 levels (Figure 5b).

We also examined ROS levels in guard cells using DCFH-DA fluorescence staining (Figure 5c). When exposed to NaCl or mannitol, ROS strongly accumulated in guard cells in the two SE lines, while both OE lines had substantially reduced ROS levels compared with WT plants. As ROS levels were significantly different in the OE, WT and SE lines when exposed to NaCl or mannitol, we also examined the activities of superoxide dismutase (SOD) and peroxidase (POD), the two major ROS scavenging enzymes in plants. The OE lines had significantly higher SOD and POD activity than the other lines, followed by WT, whereas the SE lines had significantly lower SOD and POD activity than the other lines (Figure 5d, E).

As the SOD and POD activities in these lines were altered in response to NaCl and mannitol treatment, we examined the expression of six *SOD* and eight *POD* genes in these lines. Under NaCl or mannitol treatment, the expression of all *SOD* and *POD* genes was highest in the OE lines, followed by WT and the SE lines (Figure 5f). These results indicate that BpMYB46 enhances ROS scavenging by affecting the expression of *SOD* and *POD*.

BplMYB46 affects proline biosynthesis

The proline levels were similar among OE, WT and SE lines under control conditions. However, under NaCl or mannitol treatment, the OE lines had significantly higher proline levels, followed by the WT and SE lines (Figure 5g). We further analysed the expression

of genes related to proline biosynthesis (including *P5CS1* and *P5CS2)* and proline degradation (including *P5CDH* and *ProDH*). Compared with WT plants, the two *P5CS* genes were significantly more highly expressed in the OE lines but had reduced expression in the SE lines. Conversely, the proline degradation genes, including two *P5CDH* genes and one *ProDH* gene, were significantly down-regulated in the OE lines but significantly up-regulated in the SE lines compared with WT plants (Figure 5h).

Overexpression of BplMYB46 reduces cell death

Propidium iodide (PI) staining reflects cell membrane damage based on fluorescence levels. We used PI staining to investigate cell death under NaCl or mannitol stress conditions. Compared with WT plants, the OE lines displayed relatively weak fluorescence, indicating reduced cell death, while the SE lines exhibited stronger fluorescence, suggesting severe cell death (Figure 6a). Consistently, the OE lines had the lowest electrolyte leakage rates, but the SE lines had the highest rates, when exposed to NaCl or mannitol (Figure 6b). Together, these results suggest that *BplMYB46* overexpression reduces cell death under salt and osmotic stress conditions.

BplMYB46 overexpression decreases water loss and reduces stomatal apertures

The OE lines showed significantly reduced water loss rates. The SE lines had elevated water loss rates compared with WT, with significantly higher water loss detected in detached leaves exposed to air (Figure 7a), indicating that *BplMYB46* overexpression reduces the transpiration rate in plants. Under control conditions, the stomatal apertures were smaller in the OE lines

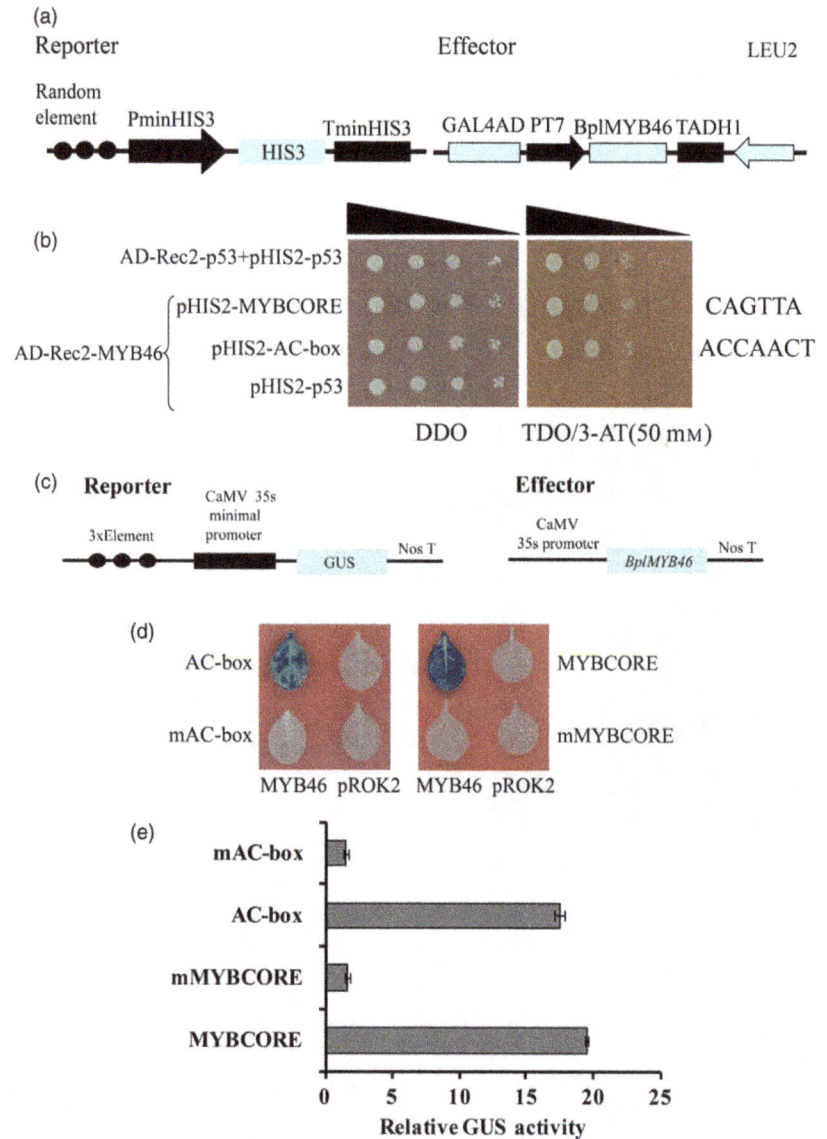

Figure 3 Analyses of BplMYB46 binding motifs. (a) Schematic diagram of the effector and reporter constructs used in Y1H analysis. (b) Analysis of binding of BplMYB46 to MYBCORE and AC-box using Y1H. (c) Schematic diagram of the effector and reporter constructs used for coexpression in tobacco plants. (d, e) GUS staining (d) and GUS activity (e) Assays of the binding of BplMYB46 to the AC-box and MYBCORE motifs in tobacco plants. The error bars indicate the standard deviation (SD) from three biological replicates.

than in WT plants, whereas the stomatal apertures in the SE lines were slight larger than those of WT. Under salt and osmotic stress, the stomates in the OE lines were almost closed, in contrast to the stomates in the SE lines, which remained open (Figure 7b). In addition, the width/length ratios of stomatal apertures in the OE lines were significantly smaller than those of the WT and SE lines. However, under NaCl or mannitol stress, the width/length ratios of the stomatal apertures of the SE plants were significantly larger than those of the OE and WT plants (Figure 7c).

We also analysed the expression of *BpMYB61* (GenBank number: KT344120), a gene homologous to *Arabidopsis* *AtMYB61*, which encodes a protein that regulates stomatal aperture (Liang *et al.*, 2005). Under control conditions, the expression of *BpMYB61* was significantly induced in the OE lines compared with WT plants, but no significant differences in *BpMYB61* expression were detected between the WT and SE lines. Under salt or osmotic stress conditions, the expression of *BpMYB61* was significantly induced in the OE lines. However, its expression was significantly reduced in the SE lines compared with WT plants (Figure 7d).

BplMYB46 affects vessel dimension and secondary wall thickening in fibres

We stained stem sections with phloroglucinol-HCl and toluidine blue, finding that the number of vessels was higher in the OE lines than in WT plants, whereas the dimensions of the vessels in the OE lines were smaller than those of WT. Conversely, the number of vessels in the SE lines was reduced, and the vessel dimension was larger, compared to WT (Figure 8a–c, g–i). We measured the ratios of the vessel area to total area in anatomical sections of plants. Compared with WT plants, the ratio of vessel area to total area was lower in the OE lines but higher in the SE lines (Figure 8n).

Additionally, toluidine blue staining showed no substantial difference in secondary wall thickening in vessels among OE, WT and SE lines (Figure 8g–i). However, the secondary wall thickening of xylem fibres in WT plants was lower than that in OE lines but higher than that in SE lines (Figure 8j–l). Measurements using Image J software further indicated that the secondary wall thickness of xylem fibres in the OE lines was approximately 2 µm (±0.07), approximately 1.45 µm (±0.1) in WT and almost 0.9 µm

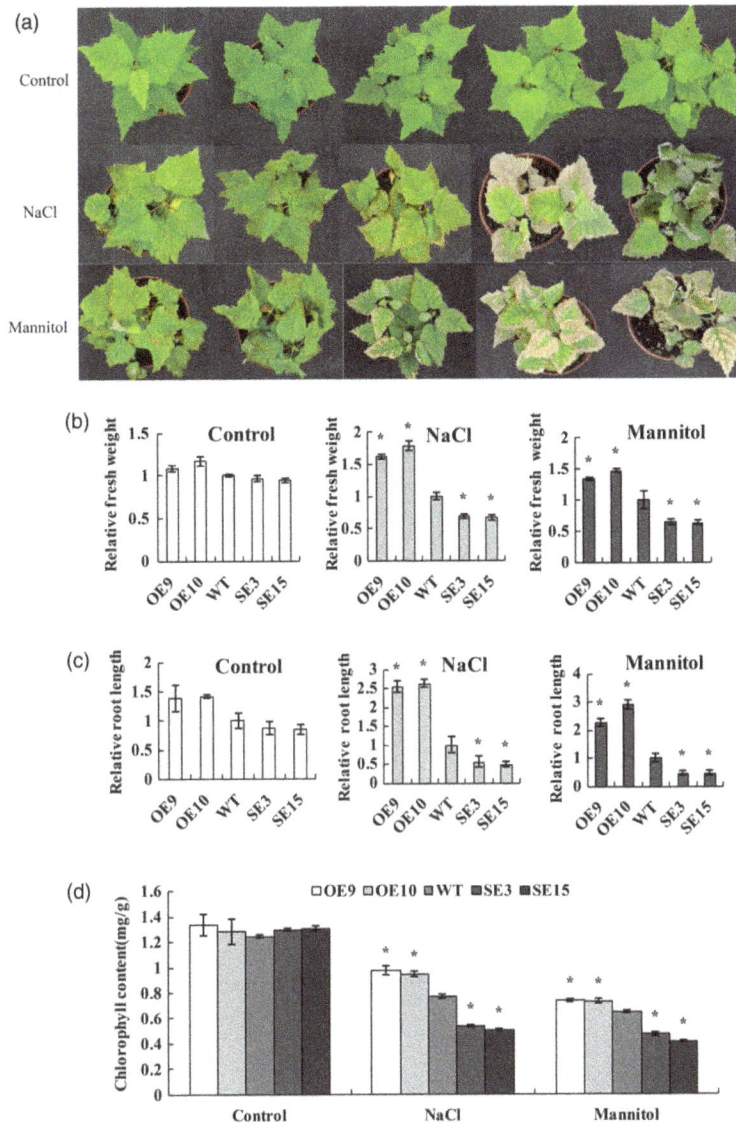

Figure 4 Salt and osmotic stress tolerance. (a) Comparison of growth phenotypes among OE, WT and SE birch plants under control, salt and osmotic stress conditions. (b–d) analysis of relative fresh weight (b), root length (c) and chlorophyll contents (d). Asterisk indicates $P < 0.05$. The error bars indicate the standard deviation (SD) from three biological replicates. ANOVA was used to determine statistically significant differences between results.

($\pm$0.1) in the SE lines (Figure 8m). These results suggest that BplMYB46 controls secondary cell wall thickness in fibres.

In the sections stained with phloroglucinol-HCl, which stains lignin, the red staining was more intense in the OE lines but less intense in the SE lines compared with WT (Figure 8a–c). These results indicate that higher lignin deposition occurred in the OE lines than in WT, but less lignin was deposited in the SE lines than in WT. Detection of lignin autofluorescence (Figure 8d–f) indicated a stronger fluorescent signal in OE (Figure 8d) than in WT (Figure 8e) but a weaker signal in the SE lines (Figure 8f) than in WT, suggesting that BplMYB46 promotes lignin deposition. We also found significantly increased lignin content in the OE lines and significantly reduced content in the SE lines compared with WT, as determined by chemical analysis (Table 1), which is consistent with the results of lignin autofluorescence analysis (Figure 8d–f). The chemical analysis also suggested that the cellulose content in the OE lines was higher than that of WT. However, the hemicellulose content in the OE lines was dramatically reduced compared with WT (Table 1). These results indicate that BplMYB46 has a positive effect on

lignin and cellulose content but a negative effect on hemicellulose content.

BplMYB46 affects the expression of secondary wall biosynthesis genes

We investigated the expression of the birch genes *phenylalanine ammonia lyase* (*PAL*), *caffeoyl-CoA O-methyltransferase* (*CCoAOMT*), *4-coumarate-coa ligase* (*4CL*), *POD*, *Laccase* (*LAC*), *cinnamoyl-CoA reductase* (*CCR*), *cellulose synthase* (*CESA*), *fragile fibre* (*FRA*) and *irregular xylem* (*IRX*), which are homologous to lignin, cellulose and hemicellulose biosynthesis genes, via real-time RT-PCR. We also investigate the expression of the birch genes *BplMYB1*, *BplMYB2* and *BplMYB3*, which are homologous to *AtMYB42*, *AtMYB103* and *AtMYB52*, respectively; these genes are involved in secondary cell wall formation in *Arabidopsis*. The expression of genes related to lignin and cellulose biosynthesis was significantly induced in the OE lines but significantly down-regulated in the SE lines. However, *FRA* and *IRX*, which are related to hemicellulose biosynthesis, were significantly down-regulated in the OE lines but significantly up-regulated in

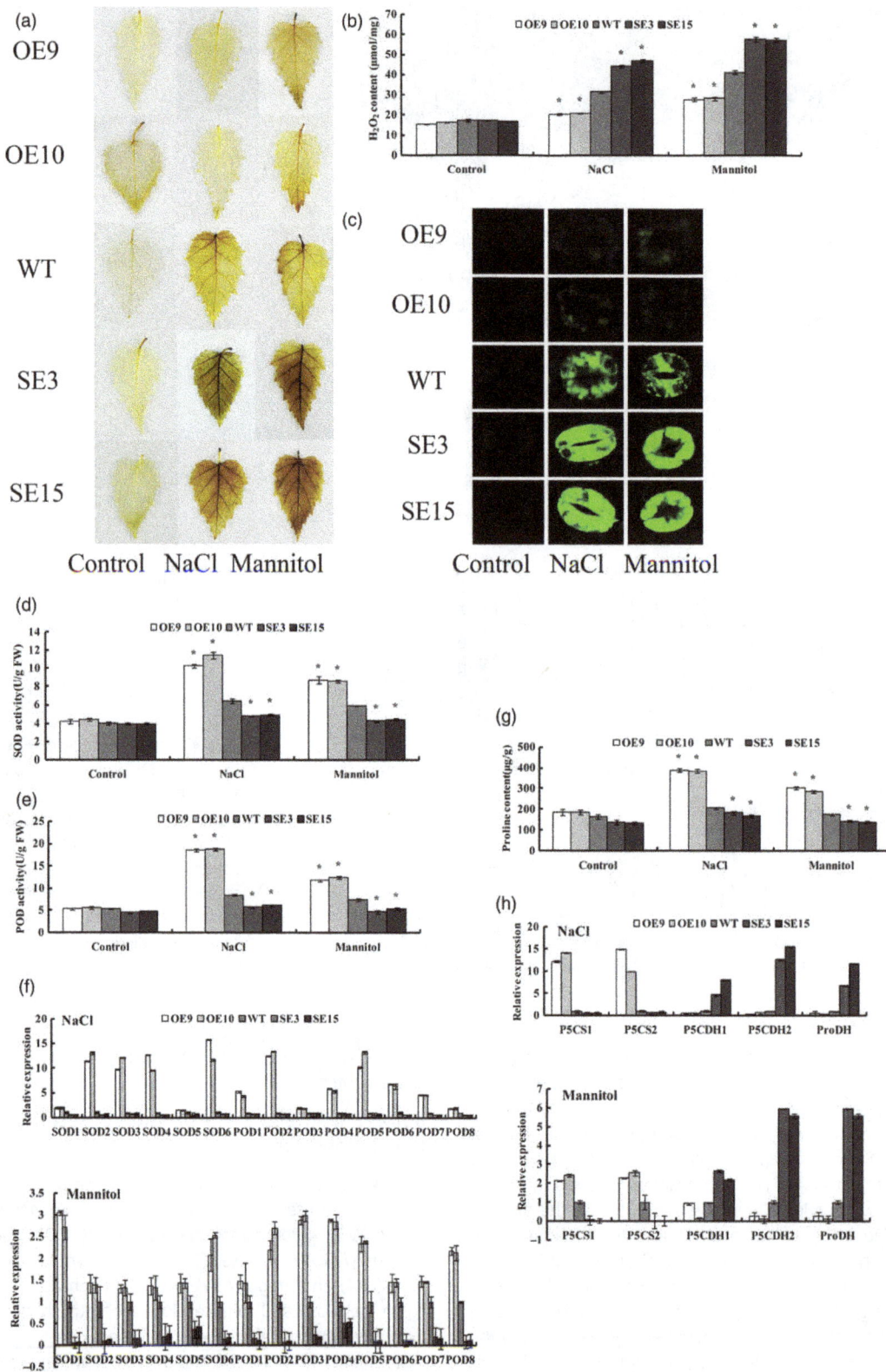

Figure 5 Detection of ROS scavenging and analysis of proline biosynthesis. (a) Detection of ROS using DAB *in situ* staining. (b) Assay of H_2O_2 levels. (c) Analyses of ROS production in intact guard cells by H2DCF-DA staining. (d) SOD activity assay. (e) POD activity assay. (f) Expression analysis of *POD* and *SOD* genes under NaCl and mannitol treatment using qRT-PCR. The expressions of the genes in WT plants were used as calculators to normalize their expressions in OE and SE lines. (g) Analysis of proline levels in OE, WT and SE lines. (h) Expression analysis of proline biosynthesis and degradation-related genes following NaCl and mannitol treatment using qRT-PCR. The expressions of the genes in WT plants were used to normalize their expressions in OE and SE lines. Asterisk indicates *P* < 0.05. The error bars indicate the standard deviation (SD) from three biological replicates. ANOVA was used to determine statistically significant differences between results.

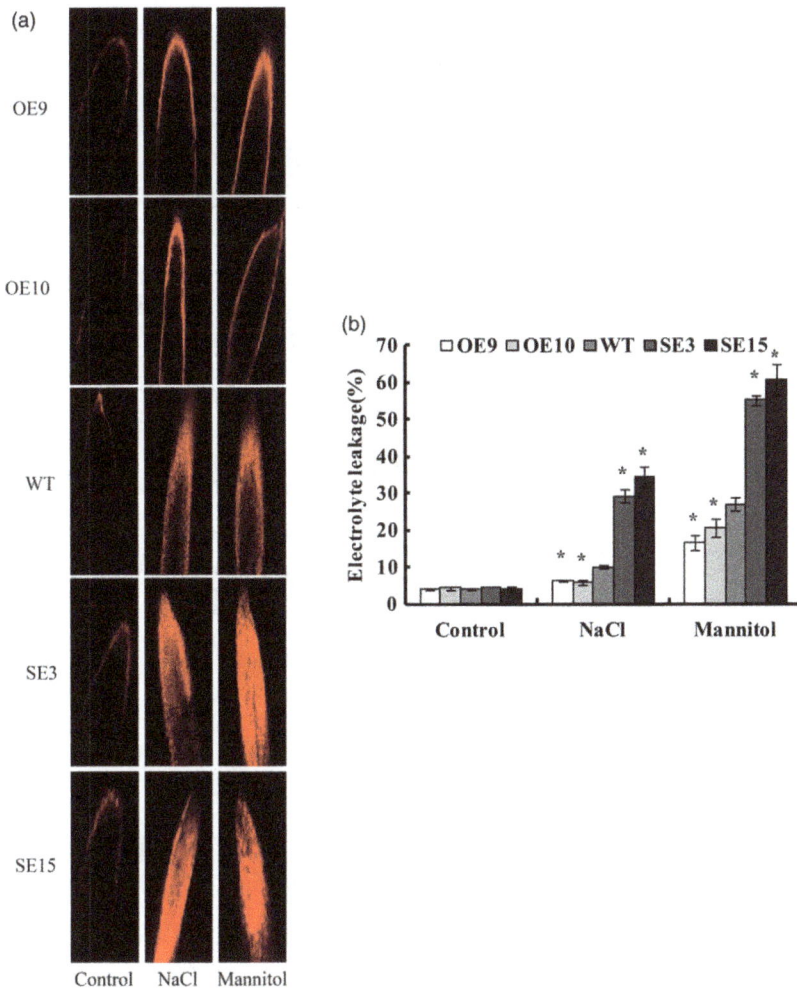

Figure 6 Analysis of cell death. (a) Assay of cell death among OE, WT and SE lines under NaCl and mannitol treatment using propidium iodide (PI) fluorescence staining. (b) Electrolyte leakage assay. Asterisk indicates P < 0.05. The error bars indicate the standard deviation (SD) from three biological replicates. ANOVA was used to determine statistically significant differences between results.

the SE lines (Figure 7o). These results suggest that BplMYB46 affects the expression of genes related to lignin, cellulose and hemicellulose biosynthesis in birch.

ChIP and promoter–reporter analyses indicate BplMYB46 binding

We performed ChIP analysis to determine whether BplMYB46 directly binds to the promoters of genes related to (1) abiotic stress (oxidative and osmotic stress), including *POD*, *SOD* and *P5CS* and (2) secondary wall deposition, including *PAL*, *CcoAOMT*, *4CL*, *POD*, *LAC*, *CESA* and *MYB*. As BplMYB46 binds to MYBCORE and AC-box motifs in genes related to abiotic stress and secondary wall deposition (Figure 9), and these two motifs are present in the promoters of target genes of BplMYB46 (Table S1), the primers for ChIP were designed to amplify the regions containing the MYBCORE and/or AC-box motif. Quantitative ChIP-PCR showed that the promoters of these putative target genes were significantly enriched (>threefold; Figure 9a) in the chromatin immunoprecipitated with GFP antibody (ChIP+) when compared with those in Mock sample (ChIP-). These results indicate that BplMYB46 preferentially binds to the promoters of these genes, suggesting that BplMYB46 may directly regulate a set of genes that mediate abiotic stress responses and secondary wall biosynthesis.

To further verify that BplMYB46 binds to the promoters of this gene set as determined by ChIP, we cotransformed the pROK2-BplMYB46 and the truncated promoters containing the

MYBCORE or AC-box motif into tobacco leaves, and cotransformation with 35S : LUC was used to normalize the transformation efficiency values (Figure 9b). GUS activity measurement suggested that BplMYB46 can bind to the promoters of all genes investigated (Figure 9c).

Discussion

Mette *et al.* (2000) showed that double-stranded promoter RNA hairpins cause *trans*-silencing of target genes triggered by methylation, which silences target genes more specifically than RNAi based on coding region. In the present study, as the genome of birch was not available early in our study, we employed RNAi-based silencing of the promoter sequence to specifically knock down the expression of *BplMYB46*. The results show that the expression of *BplMYB46* was significantly reduced in the transgenic plants, with a minimum reduction of 90.77% (Figure S3), indicating that the RNAi-based promoter used in our study was also functional in birch plants.

Although determining the copy number of a transgenic cassette is important for characterizing transgenic plants, the aim of the present study was to generate transgenic birch lines suitable for gain- and loss-of-function analysis of *BplMYB46*. Therefore, we directly determined the expression levels of *BplMYB46* in these plants, but we did not investigate the copy numbers of the cassette in OE and RNAi-silenced lines. The results show that the expression of *BplMYB46* was significantly elevated

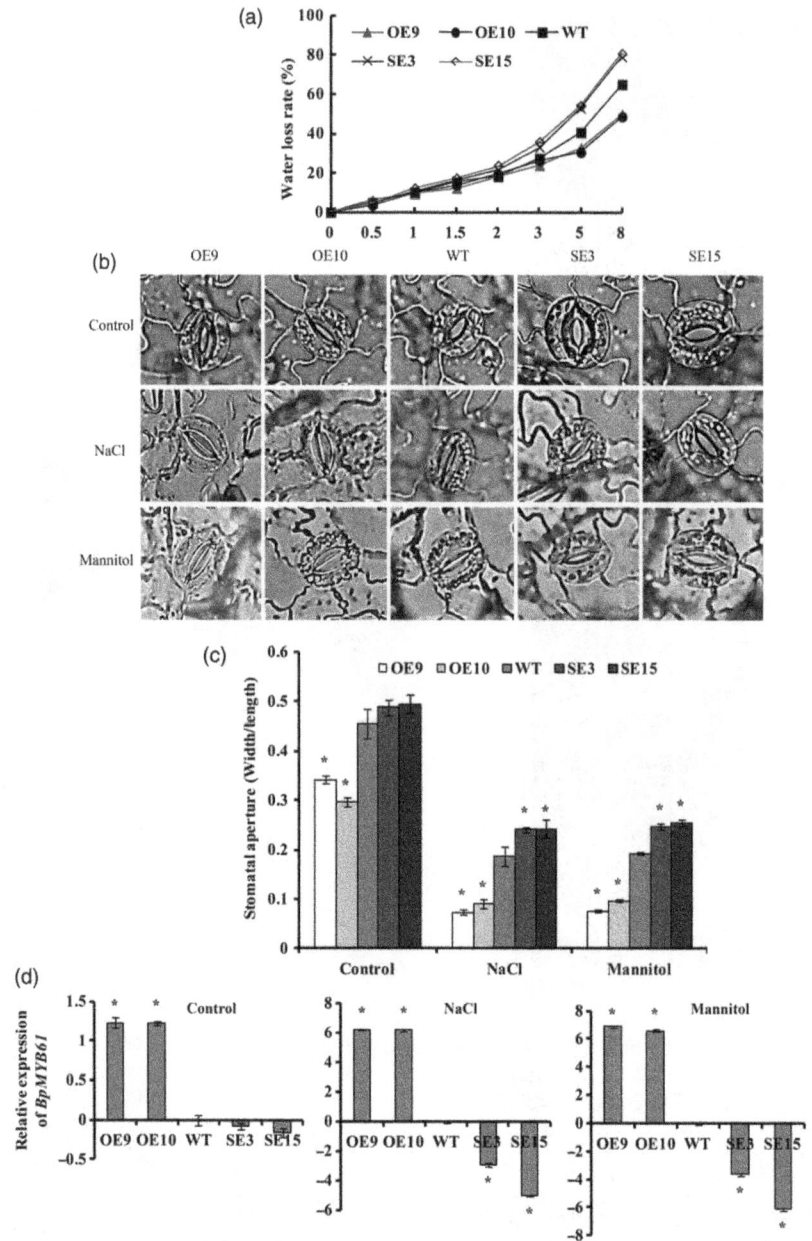

Figure 7 Analysis of water loss rate, stomatal aperture and *BpMYB61* expression in OE, WT and SE *BplMYB46* lines. (a) Water loss rates and (b, c) measurement of stomatal aperture under control conditions, 50 mM NaCl and 50 mM mannitol. (d) Expression of *BpMYB61* (GenBank accession number: KT344120) in OE, WT and SE *BplMYB46* lines under control conditions, 50 mM NaCl and 50 mM mannitol. Asterisk indicates *P* < 0.05. The error bars indicate the standard deviation (SD) from three biological replicates. ANOVA was used to determine statistically significant differences between results.

in the OE lines and significantly reduced in the SE lines, indicating that the OE and SE lines were suitable for gain- and loss-of-function analysis, respectively.

In the present study, we used 200 mM NaCl or 300 mM mannitol solution for birch plant treatments. Sudden application of NaCl and mannitol in high concentration to plants would first cause osmotic shock rather than osmotic/salt stress, and followed by the imposition of an ionic shock (Shavrukov, 2013). Therefore, there is an overlap of changes in gene expression relating to ionic and osmotic responses, and these genes mainly involve in signal transduction, osmotic regulation, water loss and ionic component of salt stress response (Shavrukov, 2013).

In plants, Δ1-pyrroline-5-carboxylate synthetase (P5CS) catalyses the rate-limiting step in proline biosynthesis. *P5CS* overexpression greatly increases proline levels, while reducing *P5CS* expression abrogates the ability of the plant to accumulate proline (Deuschle *et al.*, 2004). Two mitochondrial enzymes, pro-

dehydrogenase (ProDH) and P5C-dehydrogenase (P5CDH), play sequential roles in catalysing proline degradation (Deuschle *et al.*, 2004). In the current study, *BplMYB46* expression was positively correlated with the expression of *P5CS* but negatively correlated with the expression of *P5CDH* and *ProDH* (Figure 5h). Therefore, proline levels were positively correlated with *P5CS* expression but negatively correlated with *P5CDH* and *ProDH* expression. These results suggest that BplMYB46 induces proline biosynthetic genes and inhibits the expression of proline degradation genes, resulting in elevated proline levels and improved abiotic stress tolerance.

The transpiration rate in a plant is closely related to water stress tolerance. Our results indicate that *BplMYB46* expression reduced the transpiration rates of transgenic plants by controlling stomatal aperture to reduce water loss (Figure 7b, c). Indeed, overexpression of *AtMYB61* confers resistance to drought in transgenic *Arabidopsis* by reducing stomatal aperture (Liang *et al.*, 2005). Therefore, to investigate whether BplMYB46 controls stomatal

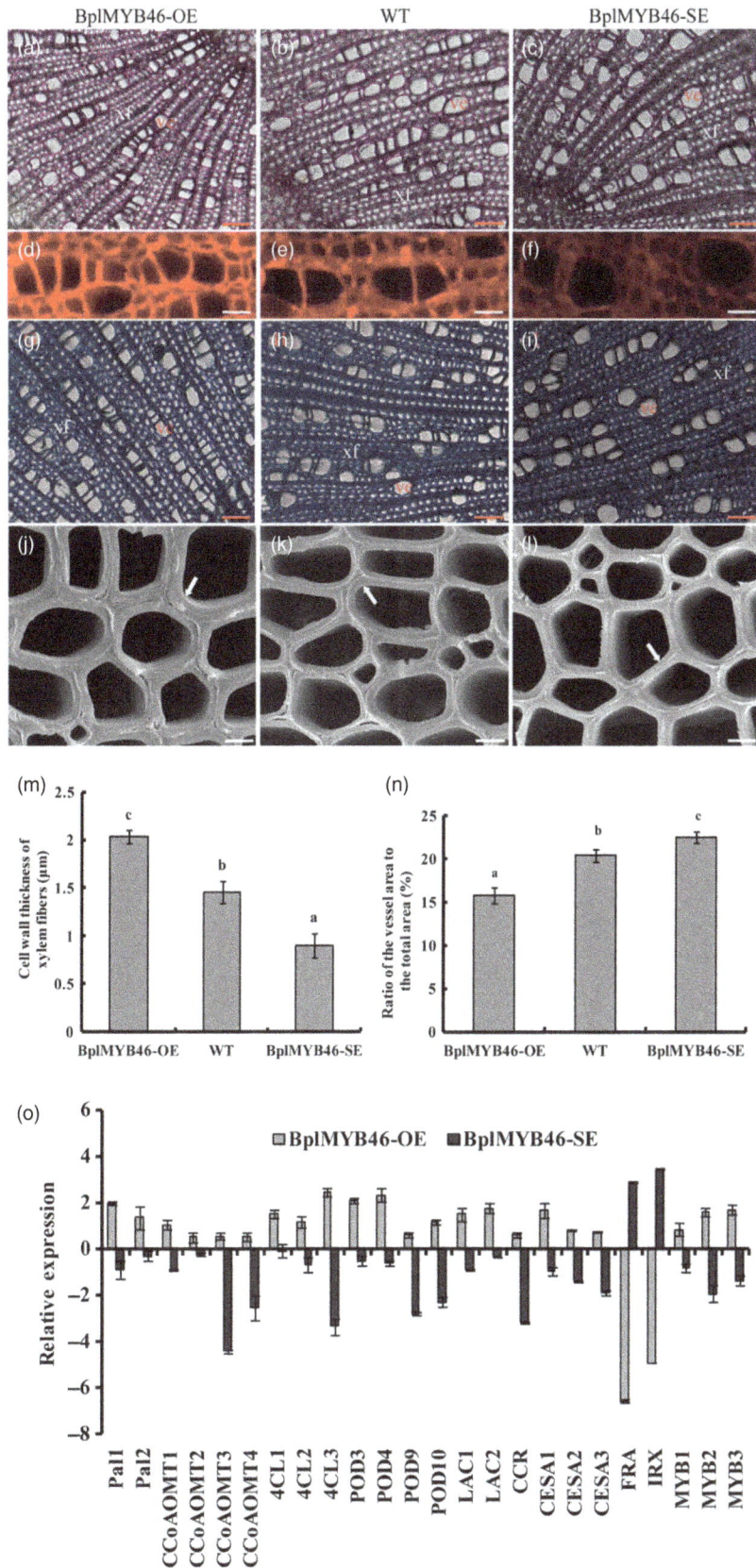

Figure 8 Microstructure of xylem vessels and fibres, and the expression of genes involved in secondary wall deposition. (a–c) Sections stained with phloroglucinol-HCl (20×) in OE (a), WT (b) and SE plants (c); ve, vessel; xf, xylem fibre; bar = 50 μm. (d–f) Lignin autofluorescence in sections observed by microscopy (63×) in OE (d), WT (e) and SE (f), bar = 15 μm. (g–i) Sections stained with toluidine blue (20×) in OE (g), WT (h) and SE (i); ve, vessel; xf, xylem fibre, bar = 50 μm. (j–l) Sections analysed by scanning electron microscopy (3000×) in OE (j), WT (k) and SE (l); the arrow refers to the wall thickness of the xylem fibre, bar = 5 μm. (m) Secondary wall thickness of xylem fibres in OE, WT and SE. (n) Ratio of the vessel area to the total area (%) in OE, WT and SE. (o) Relative expression of genes related to secondary wall deposition in transgenic lines overexpressing *BplMYB46* (OE) and RNAi-silenced *BplMYB46* (SE) lines. The sections were, respectively, analysed from the same stem positions of different plants of the same transgenic line. The error bars indicate the standard deviation (SD) from three biological replicates. ANOVA was used to determine statistically significant differences between results. a, b and c represent significant difference ($P < 0.05$) in OE, SE and WT birch plants. GenBank accession numbers: *Pal1,2*: KP711309 and KP711310; *CCoAOMT1-4*: KP711311, KP711312, KT223488 and KT223489; *4CL1-3*: KP711313, KP711314 and KP711315; *POD3,4,9,10*: KP711298, KP711299, KP711304 and KP711305; *LAC1,2*: KP711316 and KT223490; *CCR*: JQ783349; *CESA1-3*: KP711317, KP711318 and KP711319; FRA: KU168419; *IRX*: KU168420; *MYB1-3*: KP711285, KP711286 and KP711287.

aperture by regulating the expression of genes homologous to *AtMYB61* in birch, we compared the sequences of BpMYB61 and AtMYB61, finding that they were highly homologous. We investigated the expression of *BpMYB61* in the OE, WT and SE lines. Our results suggest that BplMYB46 positively regulates the expression of *BpMYB61* (Figure 7d). These results suggest that

Table 1 Secondary cell wall composition of transgenic birch and WT plants

Line	Lignin (%)	Cellulose (%)	Hemicellulose (%)
BplMYB46-OE	25.80 ± 0. 11	44.34 ± 0. 89	21.46 ± 1.3
Wild type	24.56 ± 0. 16	42.17 ± 0. 97	25.09 ± 1.0
BplMYB46-SE	22.52 ± 0. 31	40.49 ± 0. 37	29.10 ± 2.8

Measurements were conducted on transgenic birch plants overexpressing *BplMYB46*, wild-type (WT) plants and RNAi-silenced *BplMYB46* plants. Values represent mean and standard deviation. All transgenic lines displayed significantly different values compared to WT (*P* < 0.05%). *n* = 11 (12 biological replicates).

BplMYB46 reduces water loss by positively regulating *BpMYB61*, which reduces the stomatal aperture to prevent water loss.

Overexpression of *PtrMYB3* and *PtrMYB20* in poplar increases the ectopic deposition of cellulose, xylan and lignin and regulates the expression of genes such as *CCoAOMT1*, *4CL*, *FRA8*, *IRX8*, *IRX9*, *CesA4*, *CesA7* and *CesA8*, which are related to cellulose, xylan and lignin biosynthesis (McCarthy *et al.*, 2010), suggesting that MYB transcription factors play an important role in secondary wall biosynthesis. In the current study, *BplMYB46* was more highly expressed in stems than in other tissues (Figure 2). Furthermore, overexpression of *BplMYB46* increased secondary wall thickening (Figure 8m). Knock-down of *BplMYB46* reduced the secondary wall thickness in xylem fibres and decreased the lignin content (Figure 8m, Table 1). RT-PCR analysis showed that

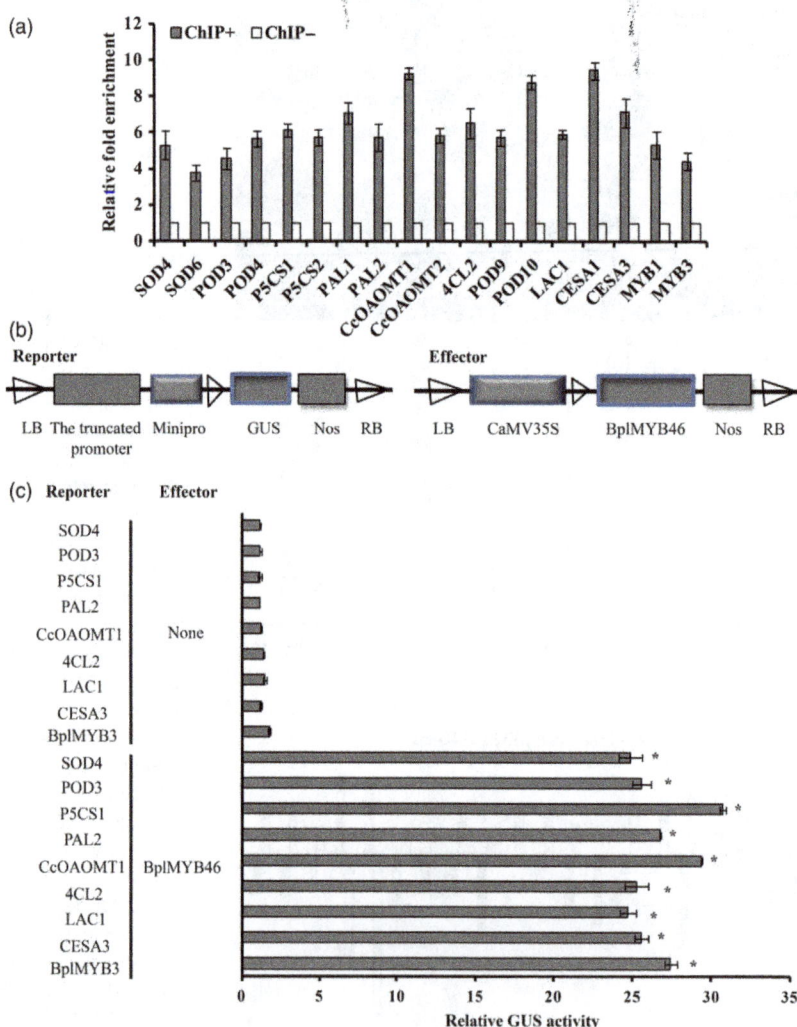

Figure 9 The regulation of target genes by BplMYB46, as determined by ChIP and GUS activity analyses. (a) Real-time quantitative PCR analysis showing the enrichment of the promoter sequences of genes after chromatin immunoprecipitation. ChIP+: The sonicated chromatin was immunoprecipitated with GFP antibody; mock: the sonicated chromatin was immunoprecipitated with anti-hemagglutinin (HA); input: the sonicated chromatin used as positive control. Three biological replicates were performed. Chromatin from whole seedlings was isolated from pROK2-35S: BplMYB46-GFP birch plants produced by *Agrobacterium tumefaciens*-mediated transient transformation. The *tubulin* sequence was used as an internal control. After normalization against *tubulin*, the values of the enrichment of the promoter sequences of target genes in ChIP- were set to 1. The error bars indicate the standard deviation (SD) from three biological replicates. (b) Schematic diagram of the reporter and effector constructs used for coexpression in tobacco plants. (c) Relative GUS activity of the truncated promoters of genes. Asterisk indicates P < 0.05. The error bars indicate the standard deviation (SD) from three biological replicates. ANOVA was used to determine statistically significant differences between results. The GenBank accession numbers of promoters used in (a) and (c): KX373440–KX373458.

overexpression and silencing BplMYB46 alter the expression of lignin, cellulose and hemicellulose biosynthesis-related genes, including *PAL*, *CCoAOMT*, *4CL*, *POD*, *CCR*, *LAC*, *CESA*, *FRA* and *IRX* (Figure 8o). In addition, ChIP-qPCR shows BplMYB46 preferentially binds to these same promoters (Figure 9). These results strongly suggest that BplMYB46 may regulate a set of lignin, cellulose and hemicellulose biosynthesis-related genes.

In the present study, we studied the bindings of BplMYB46 to promoters using ChIP method (Figure 9). However, the evidence of the bindings between the BplMYB46 and promoters was from plant lines with overexpression of the BplMYB46-GFP fusion protein and therefore could be an artefact of ectopic expression. Additionally, a set of promoters were selected and used for ChIP-PCR to investigate their bindings to BplMYB46, which does not provide a full picture of the impact of the native BplMYB46 on global steady state transcripts like RNA-seq would. Therefore, to study the native BplMYB46 on global steady state transcripts in the future, ChIP-Seq will be performed using BplMYB46 antibody to immunoprecipitate the chromatin bound by endogenous BplMYB46 in WT birch plants.

Birch plants are self-sterile, meaning that it is impossible to generate homozygote plants by selfing. Additionally, birch trees do not reach the reproductive stage to produce seeds until they are more than 12 years old. Therefore, as it is impossible to produce homozygous transgenic plants by selfing, hemizygous transgenic plants were used in this study. In the future, when the transgenic plants reach the flowering stage, we plan to perform anther culture to generate haploid plants and to generate transgenic homozygous plants by performing chromosome doubling of the haploid plants.

Experimental procedures

Cloning and subcellular localization of BplMYB46

The cDNA sequence of *BplMYB46* was obtained from the birch transcriptome (Wang et al., 2014). The CDS of *BplMYB46* without the stop codon fused in-frame to the N-terminus of green fluorescent protein (GFP) was transformed into the pROK2 vector under the control of the CaMV 35S promoter (35S : MYB-GFP; primer sequences are shown in Table S2). The GFP protein under the control of the 35S promoter was used as a control (35S : GFP). The 35S : MYB-GFP and 35S : GFP constructs were introduced into onion epidermal cells by particle bombardment (Bio-Rad laboratories, Inc. Hercules, California, USA). After incubation for 48 h, the transformed onion epidermal cells were stained with DAPI (100 ng/mL) and visualized under an LSM700 confocal laser microscope (Zeiss, Jena, Germany).

Transactivation assay

The complete and various truncated versions of the CDS of *BplMYB46* were PCR amplified (using the primers listed in Table S3) and fused in-frame to the GAL4 DNA-binding domain in the pGBKT7 vector to generate the pGBKT7-*BplMYB46* construct (Clontech laboratories, Inc. Mountain View, California, USA). The pGBKT7-*BplMYB46* construct was transformed into AH109 yeast cells, which were incubated on SD/-Trp or SD/-Trp/-His/-Ade/X-α-Gal medium at 30 °C for 3–5 days.

Plant materials and stress treatments

Birch seeds were planted in pots with a 12-cm diameter containing a mixture of perlite/vermiculite/soil (1 : 1 : 4) in a greenhouse under controlled conditions (16/8 h light/dark, 25 °C

and 70%–75% relative humidity). Each pot, which contained four seedlings, was thoroughly watered with deionized water every day. After 2 months, the plants were watered with 200 mM NaCl, 100 μM ABA or 300 mM mannitol solution, which was applied to the top of the soil. All seedlings were collected at the same time after treatment for 6, 12 or 24 h, and seedlings watered with deionized water were used as a control. Three independent biological replicates were performed, and each replicate included four seedlings. The roots, leaves and internodes of stems from 6-month-old birch plants were harvested, including the 1st, 4th, 8th, 12th, 16th and 18th stem internodes.

Real-time PCR

Total RNA was isolated from birch using the CTAB method (Chang et al., 1993) and treated with DNase I to remove DNA contamination. Total RNA was reverse transcribed into cDNA using a PrimeScript™ RT reagent Kit (Takara Bio Inc. Kusatsu, Shiga, Japan) for RT-PCR analysis. Real-time RT-PCR was performed with a TransStart Top Green qPCR SuperMix kit (TransGen Biotech, Beijing, China) using the primer sequences listed in Table S4. The amplification procedure was conducted using the following parameters: 94 °C for 30 s; 45 cycles at 94 °C for 12 s, 58 °C for 30 s and 72 °C for 45 s; and 79 °C for 1 s for plate reading. Three independent experiments were performed in triplicate. The tubulin (GenBank accession number: FG067376) and ubiquitin (GenBank accession number: FG065618) genes were used as the internal controls. The relative expression level of each gene was calculated using the delta–delta CT method (Livak and Schmittgen, 2001).

Examining the binding of BplMYB46 to the MYBCORE and AC-box motifs using Y1H

Three tandem copies of MYBCORE (CAGTTA) and AC-box (ACCAACT) were inserted into pHIS2 (Clontech) upstream of the reporter gene *HIS3*, respectively. The CDS of *BplMYB46* was cloned into pGADT7-Rec2 (Clontech) as the effector (pGADT7-*BplMYB46*). The constructs were cotransformed into Y187 cells, which were plated onto SD/-Trp/-His/ (DDO) and SD/-Trp/-His/-Leu/ (TDO) medium supplemented with 50 mM 3-AT (3-amino-1, 2, 4-triazole) and incubated at 30 °C for 3–5 days. All primers used are listed in Table S5.

Binding of BplMYB46 to motifs and promoters

Three tandem copies of MYBCORE, AC-box, their mutated sequences and the truncated promoter were, respectively, fused with the 35S CaMV minimal promoter (−46 bp to +1) to drive the *GUS* gene in a modified pCAMBIA1301 vector (in which the 35S:hygromycin region was deleted) using the primers listed in Tables S6 and S7. The full CDS of *BplMYB46* was cloned into pROK2 under the control of the 35S promoter (35S : *BplMYB46*) as the effector. The effector was co-transformed with each reporter into tobacco leaves by *Agrobacterium tumefaciens*-mediated transient transformation (Ji et al., 2014). The firefly luciferase (*LUC*) gene driven by the 35S promoter was co-transformed into tobacco leaves as a control for normalization of transformation efficiency. GUS and luciferase activities were determined as previously described (Gampala et al., 2001).

Generation of plants with overexpression and silencing of *BplMYB46*

To silence *BplMYB46*, a truncated promoter of *BplMYB46* (223 bp in length) with sense and antisense sequences was inserted into pFGC5941 in forward and reverse direction to form an inverted repeat truncated promoter (pFGC5941-*BplMYB46*; the primers were listed in Table S8). Transgenic birch lines were constructed using *Agrobacterium tumefaciens*-mediated transformation to generate birch plants with overexpression (OE) or silencing (SE) of *BplMYB46*. Briefly, birch leaves were cut into small pieces, soaked in *Agrobacterium* suspension culture (OD_{600} = 0.5) for 5 min, washed three times with sterile water and incubated on woody plant medium (WPM + 2% [w/v] sucrose, pH 5.8) for 48 h. The plants were grown in selection medium containing WPM + 1.0 mg/L 6-BA + 2% (w/v) sucrose + 50 mg/L kanamycin (for 35S : *BplMYB46* transformation) or 2 mg/L glufosinate (for pFGC5941-*BplMYB46* transformation) + 600 mg/L carbenicillin, pH 5.8 to induce resistant callus. After the antibiotic-resistant calli were regenerated, they were transferred to growth medium (WPM + 1.0 mg/L 6-BA + 50 mg/L kanamycin or 2 mg/L glufosinate) for bud differentiation. Adventitious buds with 3–5 cm shoots were cut and transferred to root generation medium (WPM + 0.2 mg/L NAA + 50 mg/L kanamycin or 2 mg/L glufosinate).

Analysis of stress tolerance phenotype

The transgenic OE, SE and WT birch plants were grown in pots containing a mixture of perlite/vermiculite/soil (1 : 1 : 4) under a 16-/8-h light/dark cycle at 25 °C and relative humidity of 75%. Each pot contained four plants. The seedlings were thoroughly watered with deionized water. After 2 months, the soil was directly watered with 200 mM NaCl or 300 mM mannitol every day for 10 d, and well-watered plants served as controls. There were three replicates and four seedlings per replicates. The relative fresh weight, root length and chlorophyll content of each line were measured as described previously (Lichtenthaler, 1987).

Histochemical analysis of stress response

Leaves detached at the third stem nodes of birch plants cultured for 2 months were incubated in a solution of 150 mM NaCl or 200 mM mannitol (or water for the control) for 6 h, stained with 3,3′-diaminobenzidine (DAB, 1.0 mg/mL) (Zhang *et al.*, 2011) and photographed. To observe ROS accumulation in intact guard cells, detached epidermal peels of leaves were incubated in deionized water (control) or a solution of 30 mM KCl and 10 mM MES-KOH (pH 6.15) supplemented with 150 mM NaCl or 200 mM mannitol for 2 h. ROS were detected using 2,7-dichlorofluorescin diacetate (H2DCF-DA) as described previously (Zhang *et al.*, 2011). For PI staining, the birch plants were incubated in a solution of deionized water (as a control), 150 mM NaCl or 200 mM mannitol for 24 h and 50 μg/mL of PI for 30 min, followed by three rinses with sterile water. The root tips were visualized under an LSM700 confocal laser microscope.

Measurement of physiological parameters related to stress tolerance

OE, WT and SE *BplMYB46* plants grown in soil for 2 months were watered with a solution of 200 mM NaCl or 300 mM mannitol for 24 h. Well-watered plants were used as controls. Superoxide dismutase (SOD) activity, peroxidase (POD) activity and electrolyte leakage were measured (Wang *et al.*, 2012). Proline content was determined as previously described (Bates *et al.*, 1973). H_2O_2 levels were measured as described previously (Sergiev *et al.*, 1997). Each sample included at least three plants, and all experiments were conducted in triplicate.

Water loss, stomatal aperture measurements and *BpMYB61* expression

The fresh weights (FW) of detached leaves were determined, and the leaves were then desiccated under normal atmospheric conditions. The leaves were weighed (desiccated weight) after exposure to air for 0.5, 1, 1.5, 2, 3, 5 and 8 h, dried overnight at 80 °C and their dry weights (DW) determined. The water loss rates (WLR) were calculated using the formula: WLR (%) = [(FW − desiccated weight)/(FW − DW)] × 100.

Epidermal peels were stripped from the leaves of 2-month soil-grown OE, SE and WT plants and floated in a solution of 30 mM KCl and 10 mM MES-KOH (pH 6.15), followed by incubation for 2 h in the light at 22 °C to induce stomatal opening (Cheng *et al.*, 2013). Then, 50 mM NaCl or 50 mM mannitol was added to the buffer solution. The samples were incubated for an additional 2 h. Stomatal apertures were photographed using light microscopy (Olympus BX43, Olympus Corporation, Shinjuku-ku, Tokyo, Japan). The ratios of the widths and lengths of stomatal apertures under different treatments were calculated.

The expression of *BpMYB61*, a gene homologous to *AtMYB61*, encoding a protein that regulates stomatal aperture in *Arabidopsis*, was investigated in the OE, WT and SE *BplMYB46* lines under control conditions, 50 mM NaCl and 50 mM mannitol stress using the primers listed in Table S9.

Histological analysis

Stems of 6-month soil-grown OE, WT and SE *BplMYB46* birch plants were fixed in FAA solution (70% ethanol: glacial acetic acid: formaldehyde; 90: 5: 5, v/v) and embedded in frozen sectioning medium (OCT; Thermo Scientific, Waltham, MA) to obtain 25-μm-thick stem base sections using a Microtome Cryostat (Thermo Scientific HM560). The stem sections were stained with phloroglucinol-HCl and toluidine blue and examined by light microscopy (Zhong *et al.*, 2006). Lignin autofluorescence was observed under a confocal laser microscope (Zeiss, Jena, Germany). For scanning electron microscopy (SEM), 0.2-cm-thick sections of 12-month soil-grown OE, WT and SE *BplMYB46* birch plants were obtained manually and observed under a scanning electron microscope (S-4800, HITACHI, Tokyo, Japan). The ratio of vessel area to total area was measured from 12 anatomical sections representing each genotype after staining with phloroglucinol-HCl. The secondary wall thickness of xylem fibres in the SEM micrographs was quantified in 45 cells using Image J software (http://rsbweb.nih.gov/ij/).

Determination of secondary wall composition

The lignin, cellulose and hemicellulose levels of stems from the OE, WT and SE lines grown in soil for 12 months were determined according to the Klason procedure (Whiting *et al.*, 1981) using an automatic fibre analyser (Ankom 2000i; Ankom, Macedon, NY). Twelve biological replicates were performed in this experiment.

Expression analysis of *BplMYB46* target genes

For stress tolerance-related gene expression analysis, birch plants grown for 2 months in soil were treated with 200 mM NaCl or 300 mM mannitol for 24 h; plants watered with deionized water only were harvested at the same time and used as controls. For wood formation-related gene analysis, stems of the OE, WT and SE *BplMYB46* birch plants grown for 6 months in soil were harvested. All primers used are listed in Table S10. Three independent experiments were performed, with three biological replicates.

ChIP assay

The 35S : BplMYB46-GFP construct was transformed into birch plants for the ChIP assay. The ChIP assay was performed according to the published method (Haring et al., 2007). Briefly, after protein and chromatin were cross-linked, the chromatin was sheared into 0.2–0.8 kb fragments by sonication, and 1/10 (volume) of each sample was used for the input control. Sonicated chromatin was immunoprecipitated with GFP antibody (Abmart) (ChIP+), and chromatin immunoprecipitated with antihemagglutinin (HA) antibody was used as a negative control (Mock). The antibody-bound complex was precipitated with protein A + G agarose beads. The immunoprecipitated DNA was purified by chloroform extraction. The enrichment of the truncated promoters in immunoprecipitated samples were determined by real-time PCR, and the tubulin sequence was used as an internal control. Three biological replications were performed. The primer sequences used for ChIP amplification are listed in Table S11.

Statistical analysis

Histological indices involving stress responses and measurement of wood characters related to secondary wall deposition were analysed using ANOVA. All statistical analyses were performed using SPSS software (IBM. Chicago, IL, USA), version 18.0. Differential analysis of wood characters was performed using the DUNCAN method.

Acknowledgements

This work was supported by Innovation Project of State Key Laboratory of Tree Genetics and Breeding (Northeast Forestry University, 2013A05) and National Natural Science Foundation of China (No. 31270703 and No. 31470671).

References

Ambawat, S., Sharma, P., Yadav, N.R. and Yadav, R.C. (2013) MYB transcription factor genes as regulators for plant responses: an overview. Physiol. Mol. Biol. Plants, 19, 307–321.

Bates, L.S., Waldren, R.P. and Teare, J.D. (1973) Rapid determination of free proline for water stress studies. Plant Soil, 39, 205–207.

Cedroni, M.L., Cronn, R.C., Adams, K.L., Wilkins, T.A. and Wendel, J.F. (2003) Evolution and expression of MYB genes in diploid and polyploid cotton. Plant Mol. Biol. 51, 313–325.

Chang, S., Puryear, J. and Cairney, J. (1993) A simple and efficient method for isolating RNA from pine trees. Plant Mol. Biol. Rep. 11, 113–116.

Cheng, M.C., Liao, P.M., Kuo, W.W. and Lin, T.P. (2013) The Arabidopsis ETHYLENE RESPONSE FACTOR1 regulates abiotic stress-responsive gene expression by binding to different cis-acting elements in response to different stress signals. Plant Physiol. 162, 1566–1582.

Deuschle, K., Funck, D., Forlani, G., Stransky, H., Biehl, A., Leister, D., van der Graaff, E. et al. (2004) The role of [Delta]1-pyrroline-5-carboxylate dehydrogenase in proline degradation. Plant Cell, 16, 3413–3425.

Du, H., Feng, B.R., Yang, S.S., Huang, Y.B. and Tang, Y.X. (2012a) The R2R3-MYB transcription factor gene family in maize. PLoS ONE, 7, e37463.

Du, H., Yang, S.S., Liang, Z., Feng, B.R., Liu, L., Huang, Y.B. and Tang, Y.X. (2012b) Genome-wide analysis of the MYB transcription factor superfamily in soybean. BMC Plant Biol. 12, 106.

Dubos, C., Stracke, R., Grotewold, E., Weisshaar, B., Martin, C. and Lepiniec, L. (2010) MYB transcription factors in Arabidopsis. Trends Plant Sci. 15, 573–581.

Gampala, S.S., Hagenbeek, D. and Rock, C.D. (2001) Functional interactions of lanthanum and phospholipase D with the abscisic acid signaling effectors VP1 and ABI1-1 in rice protoplasts. J. Biol. Chem. 276, 9855–9860.

Grotewold, E., Drummond, B.J., Bowen, B. and Peterson, T. (1994) The MYB-homologous P gene controls phlobaphene pigmentation in maize floral organs by directly activating a flavonoid biosynthetic gene subset. Cell, 76, 543–553.

Haring, M., Offermann, S., Danker, T., Horst, I., Peterhansel, C. and Stam, M. (2007) Chromatin immunoprecipitation: optimization, quantitative analysis and data normalization. Plant Methods, 3, 11.

He, Y., Li, W., Lv, J., Jia, Y., Wang, M. and Xia, G. (2012) Ectopic expression of a wheat MYB transcription factor gene, TaMYB73, improves salinity stress tolerance in Arabidopsis thaliana. J. Exp. Bot. 63, 1511–1522.

Ithal, N. and Reddy, A.R. (2004) Rice flavonoid pathway genes, OsDfr and OsAns, are induced by dehydration, high salt and ABA, and contain stress responsive promoter elements that interact with the transcription activator, OsC1-MYB. Plant Sci. 166, 1505–1513.

Ji, X., Zheng, L., Liu, Y., Nie, X., Liu, S. and Wang, Y. (2014) A transient transformation system for the functional characterization of genes involved in stress response. Plant Mol. Biol. Rep. 32, 732–739.

Kim, W.C., Ko, J.H. and Han, K.H. (2012) Identification of a cis-acting regulatory motif recognized by MYB46, a master transcriptional regulator of secondary wall biosynthesis. Plant Mol. Biol. 78, 489–501.

Kim, J.H., Hyun, W.Y., Nguyen, H.N., Jeong, C.Y., Xiong, L., Hong, S.W. and Lee, H. (2015) AtMyb7, a subgroup 4 R2R3 Myb, negatively regulates ABA-induced inhibition of seed germination by blocking the expression of the bZIP transcription factor ABI5. Plant, Cell Environ. 38, 559–571.

Ko, J.H., Kim, W.C. and Han, K.H. (2009) Ectopic expression of MYB46 identifies transcriptional regulatory genes involved in secondary wall biosynthesis in Arabidopsis. Plant J. 60, 649–665.

Li, J., Li, X., Guo, L., Lu, F., Feng, X., He, K., Wei, L. et al. (2006) A subgroup of MYB transcription factor genes undergoes highly conserved alternative splicing in Arabidopsis and rice. J. Exp. Bot. 57, 1263–1273.

Liang, Y.K., Dubos, C., Dodd, I.C., Holroyd, G.H., Hetherington, A.M. and Campbell, M.M. (2005) AtMYB61, an R2R3-MYB transcription factor controlling stomatal aperture in Arabidopsis thaliana. Curr. Biol. 15, 1201–1206.

Lichtenthaler, H.K. (1987) Chlorophylls and carotenoids: pigments of photosynthetic biomembranes. Methods Enzymol. 148, 350–382.

Livak, K.J. and Schmittgen, T.D. (2001) Analysis of relative gene expression data using real-time quantitative PCR and the 2(-Delta Delta C(T)) method. Methods, 25, 402–408.

Ma, Q., Dai, X., Xu, Y., Guo, J., Liu, Y., Chen, N., Xiao, J. et al. (2009) Enhanced tolerance to chilling stress in OsMYB3R-2 transgenic rice is mediated by alteration in cell cycle and ectopic expression of stress genes. Plant Physiol. 150, 244–256.

Mao, X., Zhang, H., Qian, X., Li, A., Zhao, G. and Jing, R. (2012) TaNAC2, a NAC-type wheat transcription factor conferring enhanced multiple abiotic stress tolerances in Arabidopsis. J. Exp. Bot. 63, 2933–2946.

Mare, C., Mazzucotelli, E., Crosatti, C., Francia, E., Stanca, A.M. and Cattivelli, L. (2004) Hv-WRKY38: a new transcription factor involved in cold- and drought-response in barley. Plant Mol. Biol. 55, 399–416.

McCarthy, R.L., Zhong, R. and Ye, Z.H. (2009) MYB83 is a direct target of SND1 and acts redundantly with MYB46 in the regulation of secondary cell wall biosynthesis in Arabidopsis. Plant Cell Physiol. 50, 1950–1964.

McCarthy, R.L., Zhong, R., Fowler, S., Lyskowski, D., Piyasena, H., Carleton, K., Spicer, C. et al. (2010) The poplar MYB transcription factors, PtrMYB3 and PtrMYB20, are involved in the regulation of secondary wall biosynthesis. Plant Cell Physiol. 51, 1084–1090.

Mette, M.F., Aufsatz, W., van der Winden, J., Matzke, M.A. and Matzke, A.J. (2000) Transcriptional silencing and promoter methylation triggered by double-stranded RNA. The EMBO journal, 19, 5194–5201.

Newman, L.J., Perazza, D.E., Juda, L. and Campbell, M.M. (2004) Involvement of the R2R3-MYB, AtMYB61, in the ectopic lignification and dark-photomorphogenic components of the det3 mutant phenotype. Plant J. 37, 239–250.

Oh, J.E., Kwon, Y., Kim, J.H., Noh, H., Hong, S.W. and Lee, H. (2011) A dual role for MYB60 in stomatal regulation and root growth of Arabidopsis thaliana under drought stress. Plant Mol. Biol. 77, 91–103.

Park, M.Y., Kang, J.Y. and Kim, S.Y. (2011) Overexpression of AtMYB52 confers ABA hypersensitivity and drought tolerance. Mol. Cells, 31, 447–454.

Pasquali, G., Biricolti, S., Locatelli, F., Baldoni, E. and Mattana, M. (2008) Osmyb4 expression improves adaptive responses to drought and cold stress in transgenic apples. *Plant Cell Rep.* **27**, 1677–1686.

Prouse, M.B. and Campbell, M.M. (2012) The interaction between MYB proteins and their target DNA binding sites. *Biochim. Biophys. Acta*, **1819**, 67–77.

Qi, T., Huang, H., Song, S. and Xie, D. (2015) Regulation of jasmonate-mediated stamen development and seed production by a bHLH-MYB complex in *Arabidopsis*. *Plant Cell*, **27**, 1620–1633.

Romano, J.M., Dubos, C., Prouse, M.B., Wilkins, O., Hong, H., Poole, M., Kang, K.Y. *et al.* (2012) AtMYB61, an R2R3-MYB transcription factor, functions as a pleiotropic regulator via a small gene network. *New Phytolog.* **195**, 774–786.

Sergiev, I., Alexieva, V. and Karanov, E. (1997) Effect of spermine, atrazine and combination between them on some endogenous protective systems and stress markers in plants. *Compt. Rend. Acad. Bulg. Sci.* **51**, 121–124.

Shavrukov, Y. (2013) Salt stress or salt shock: which genes are we studying? *J. Exp. Bot.* **64**, 119–127.

Su, L.T., Li, J.W., Liu, D.Q., Zhai, Y., Zhang, H.J., Li, X.W., Zhang, Q.L. *et al.* (2014) A novel MYB transcription factor, GmMYBJ1, from soybean confers drought and cold tolerance in *Arabidopsis thaliana*. *Gene*, **538**, 46–55.

Sun, X., Gong, S.Y., Nie, X.Y., Li, Y., Li, W., Huang, G.Q. and Li, X.B. (2015) A R2R3-MYB transcription factor that is specifically expressed in cotton (*Gossypium hirsutum*) fibers affects secondary cell wall biosynthesis and deposition in transgenic *Arabidopsis*. *Physiol. Plant.* **154**, 420–432.

Uno, Y., Furihata, T., Abe, H., Yoshida, R., Shinozaki, K. and Yamaguchi-Shinozaki, K. (2000) *Arabidopsis* basic leucine zipper transcription factors involved in an abscisic acid-dependent signal transduction pathway under drought and high-salinity conditions. *Proc. Natl Acad. Sci. USA*, **97**, 11632–11637.

Vannini, C., Locatelli, F., Bracale, M., Magnani, E., Marsoni, M., Osnato, M., Mattana, M. *et al.* (2004) Overexpression of the rice Osmyb4 gene increases chilling and freezing tolerance of *Arabidopsis thaliana* plants. *Plant J.* **37**, 115–127.

Wang, L., Xu, C., Wang, C. and Wang, Y. (2012) Characterization of a eukaryotic translation initiation factor 5A homolog from *Tamarix androssowii* involved in plant abiotic stress tolerance. *BMC Plant Biol.* **12**, 118.

Wang, C., Zhang, N., Gao, C., Cui, Z., Sun, D., Yang, C. and Wang, Y. (2014) Comprehensive transcriptome analysis of developing xylem responding to artificial bending and gravitational stimuli in *Betula platyphylla*. *PLoS ONE*, **9**, e87566.

Whiting, P., Favis, B., St-Germain, F. and Goring, D. (1981) Fractional separation of middle lamella and secondary wall tissue from spruce wood. *J. Wood Chem. Technol.* **1**, 29–42.

Xiong, H., Li, J., Liu, P., Duan, J., Zhao, Y., Guo, X., Li, Y. *et al.* (2014) Overexpression of OsMYB48-1, a novel MYB-related transcription factor, enhances drought and salinity tolerance in rice. *PLoS ONE*, **9**, e92913.

Xu, R., Wang, Y., Zheng, H., Lu, W., Wu, C., Huang, J., Yan, K. *et al.* (2015) Salt-induced transcription factor MYB74 is regulated by the RNA-directed DNA methylation pathway in Arabidopsis. *J. Exp. Bot.* **66**, 5997–6008.

Yanhui, C., Xiaoyuan, Y., Kun, H., Meihua, L., Jigang, L., Zhaofeng, G., Zhiqiang, L. *et al.* (2006) The MYB transcription factor superfamily of *Arabidopsis*: expression analysis and phylogenetic comparison with the rice MYB family. *Plant Mol. Biol.* **60**, 107–124.

Zhang, X., Wang, L., Meng, H., Wen, H., Fan, Y. and Zhao, J. (2011) Maize ABP9 enhances tolerance to multiple stresses in transgenic *Arabidopsis* by modulating ABA signaling and cellular levels of reactive oxygen species. *Plant Mol. Biol.* **75**, 365–378.

Zhang, R., Yang, C., Wang, C., Wei, Z., Xia, D., Wang, Y., Liu, G. *et al.* (2012) Time-course analyses of abscisic acid level and the expression of genes involved in abscisic acid biosynthesis in the leaves of *Betula platyphylla*. *Mol. Biol. Rep.* **39**, 2505–2513.

Zhong, R., Demura, T. and Ye, Z.H. (2006) SND1, a NAC domain transcription factor, is a key regulator of secondary wall synthesis in fibers of *Arabidopsis*. *Plant Cell*, **18**, 3158–3170.

Zhong, R., Richardson, E.A. and Ye, Z.H. (2007) The MYB46 transcription factor is a direct target of SND1 and regulates secondary wall biosynthesis in *Arabidopsis*. *Plant Cell*, **19**, 2776–2792.

Zhong, R., McCarthy, R.L., Haghighat, M. and Ye, Z.H. (2013) The poplar MYB master switches bind to the SMRE site and activate the secondary wall biosynthetic program during wood formation. *PLoS ONE*, **8**, e69219.

13

Application of nontargeted metabolite profiling to discover novel markers of quality traits in an advanced population of malting barley

Adam L. Heuberger[1], Corey D. Broeckling[1], Kaylyn R. Kirkpatrick[1] and Jessica E. Prenni[1,2,]*

[1]Proteomics and Metabolomics Facility, Colorado State University, Fort Collins, CO, USA
[2]Department of Biochemistry and Molecular Biology, Colorado State University, Fort Collins, CO, USA

*Correspondence
email jessica.
prenni@colostate.edu

Keywords: metabolomics, *Hordeum vulgare*, malting barley, breeding, UPLC-MS, association study.

Summary

The process of breeding superior varieties for the agricultural industry is lengthy and expensive. Plant metabolites may act as markers of quality traits, potentially expediting the appraisal of experimental lines during breeding. Here, we evaluated the utility of metabolites as markers by assessing metabolic variation influenced by genetic and environmental factors in an advanced breeding setting and in relation to the phenotypic distribution of 20 quality traits. Nontargeted liquid chromatography–mass spectrometry metabolite profiling was performed on barley (*Hordeum vulgare* L.) grain and malt from 72 advanced malting barley lines grown at two distinct but climatically similar locations, with 2-row and 6-row barley as the main genetic factors. 27 420 molecular features were detected, and the metabolite and quality trait profiles were similarly influenced by genotype and environment; however, malt was more influenced by genotype compared with barley. An O2PLS model characterized molecular features and quality traits that covaried, and 1319 features associated with at least one of 20 quality traits. An indiscriminant MS/MS acquisition and novel data analysis method facilitated the identification of metabolites. The analysis described 216 primary and secondary metabolites that correlated with multiple quality traits and included amines, amino acids, alkaloids, polyphenolics and lipids. The mechanisms governing quality trait–metabolite associations were interpreted based on colocalization to genetic markers and their gene annotations. The results of this study support the hypothesis that metabolism and quality traits are co-influenced by relatively narrow genetic and environmental factors and illustrate the utility of grain metabolites as functional markers of quality traits.

Introduction

Plant metabolites are potential biomarkers for variation in agronomically important phenotypes (Meyer *et al.*, 2007; Riedelsheimer *et al.*, 2012; Schauer *et al.*, 2006; Steinfath *et al.*, 2010). Metabolites can be defined as small molecules (<1200 Da) that act as both substrates and end products of metabolism and include organic acids, amines, amino acids, lipids, carbohydrates, alkaloids and others. A metabolite profile provides a snapshot of metabolism at a given time. The term 'metabolomics' describes metabolite profiling of many samples and can be used to explain the biochemistry that underlies complex plant phenotypes by measuring the inherent metabolic variation in a given system.

The identification of major genetic mechanisms related to crop metabolism is a new area of research (Keurentjes, 2009). Recent studies have shown that metabolic variation is associated with genetic diversity within a plant species (Heuberger *et al.*, 2010; Mensack *et al.*, 2010; Mochida *et al.*, 2009). Furthermore, metabolites have been mapped as quantitative traits in plants, including *Arabidopsis* and *O. sativa* (Keurentjes *et al.*, 2006; Matsuda *et al.*, 2012), and metabolite variation due to growing environment and genotype–environment interactions has been documented in wheat and maize (Asiago *et al.*, 2012; Beleggia *et al.*, 2013; Skogerson *et al.*, 2010). The co-influence of

genotypic and environmental effects on crop metabolism and quality traits has been described in wheat and potato (Carreno-Quintero *et al.*, 2012; Shewry *et al.*, 2010; Steinfath *et al.*, 2010). While such investigations show that factors influencing crop metabolism (e.g. genotype, environment) similarly affect commercially important traits (e.g. food quality), most experimental designs inflate the genetic and/or environmental diversity to levels not normally observed in an advanced breeding trial. The extent of metabolic variation in advanced breeding population grown at locations ideal for commercial production has not been reported, and therefore the true utility of metabolic markers for varietal selection remains unknown.

Malting barley (*Hordeum vulgare* L., and herein referred to as barley) is an ideal system to evaluate metabolic variation given the chemical composition of the grain is a major contributor to malting quality. For brewing, barley grain is converted to malt by first storing the grain, adding moisture to instigate germination (steeping), and then drying the germinated grain at high temperature (kilning). This process, referred to as malting, activates select enzymes and proteins and begins the breakdown of complex oligosaccharides. Thus, malt represents processed barley grain tissue whose metabolites directly contribute energy required for fermentation, as well as flavour, aroma and appearance characteristics. An example malting quality trait is 'fine extract', a crude measure of soluble malt material that includes levels of fermentable carbohydrates. Barley metabolites may also be indirectly related to malting quality (or agronomic quality) traits as a result of coregulation at the genomic, transcriptional, translational or metabolic level. Previous studies have shown barley metabolites to be associated with qualitative traits such as abiotic and biotic stress (Bollina et al., 2010; Kogel et al., 2010; Roessner et al., 2006; Sicher et al., 2012; Widodo et al., 2009), and the malting process itself has been shown to contribute to metabolite variation (Frank et al., 2011; Gorzolka et al., 2012). A recent study showed that diversity in genetic factors related to quality traits is maintained in an elite barley breeding programme (Munoz-Amatriain et al., 2010). Therefore the genetic variation in an advanced population, while relatively small, may nevertheless translate to diversity in whole-plant metabolism that influences the chemical composition of the barley grain and malt, and therefore malting quality.

Nontargeted metabolite profiling by ultra-performance liquid chromatography coupled with mass spectrometry (UPLC-MS) is an analytical approach well suited for large-scale investigations of metabolism in biological systems. Nontargeted metabolite profiling is a global approach not limited to a predetermined panel of known metabolites and thus allows for novel discoveries related to plant metabolism. Previous investigations of crop metabolic variation used gas chromatography (GC) to detect metabolites involved in primary metabolism (Asiago et al., 2012; Beleggia et al., 2013; Carreno-Quintero et al., 2012; Steinfath et al., 2010). However, while GC methods are optimal for the analysis of small polar metabolites, they are limited by the need to chemically modify each molecule for increased volatility and therefore fail to detect most of the larger, secondary plant metabolites such as complex alkaloids (e.g. barley hordatines) and glycerophospholipids.

Nontargeted UPLC-MS metabolomics is a complementary analytical approach to GC-MS, however the primary challenge is the confident annotation of the molecular signals observed in the mass spectrometer to the biological metabolites from which they are derived. Furthermore, each metabolite in a sample results in multiple signals in the mass spectrometer which can complicate the annotation process (reviewed in Dettmer et al., 2007). We recently developed a novel data analysis workflow to group molecular signals into a reconstructed mass spectrum (Broeckling et al., 2013). This data reconstruction is coupled with the use of indiscriminant data acquisition (i.e. no precursor ion selection) to enable generation of both low-energy in-source (idMS) and high-energy fragmentation (idMS/MS) spectra for each detected signal. Using this approach, the reconstructed spectra can be searched against spectral libraries to generate highly confident metabolite annotation of the mass spectrometry data.

Here, UPLC-MS was utilized to characterize metabolic variation as related to phenotypic distribution of multiple malting and agronomic quality traits. To our knowledge, this is the first description of metabolic variation in an advanced breeding population, and the first to use nontargeted UPLC-MS metabolomics for marker discovery related to breeding. The results of this study support that genotype and growing environment contribute to variation in both barley and malt metabolite profiles and quality traits, and therefore metabolites may be novel markers of downstream quality traits. A comparison of the barley and malt metabolome supports that the influence of barley genetic variation on brewing quality is exacerbated upon the conversion of barley to malt. Metabolite–trait associations were further investigated by mapping important metabolites to genetic loci. These data described herein can serve as a foundation for metabolite-assisted breeding in malting barley and also highlight the utility of our UPLC-MS nontargeted metabolomics workflow for other plant systems, specifically when multiple metabolites and quality traits are under investigation.

Results and discussion

Experimental design and influence of genetics and environment on quality traits

The experiment was designed to evaluate the effect of genetic and environmental factors on both barley and malt metabolites and quality traits. For barley, the major varietal group is defined by the number of florets per barley rachis node (2-row or 6-row type). The 2-row and 6-row phenotypes have been described as due to a single genetic factor (Komatsuda et al., 2007; Ramsay et al., 2011). Furthermore, the two groups are largely maintained in separate breeding pools (Hayes and Szucs, 2006), and thus 'row type' defines a major unit of genetic diversity in malting barley. The relationship between large plant genetic diversity and metabolite variation has been well documented (Heuberger et al., 2010; Mensack et al., 2010; Mochida et al., 2009). The present study expands on this concept by only assessing varieties that range between 'good'-to-'great' agronomic and malting quality performance.

The genetic relatedness for the barley population displayed a structure associated with 2-row and 6-row differences, but was less structured within each varietal group (Figure 1a). A comparable design was established by growing the narrow pool of 2-row and 6-row genotypes in two similar environments: Fairfield, MT and Idaho Falls, ID (USA). The environments are at approximately the same latitudes, grow similar varieties for commercial production (American Malting Barley Association, 2010) and rely on similar crop production methods (e.g. irrigation, pesticide application). In the 2010 growing season, the major measurable difference was that Idaho Falls consistently experienced higher monthly mean temperatures during the growing season (Figure 1b). Given that temperature has been shown to affect regulation of barley seed development (Mangelsen et al., 2011), it was hypothesized that the seemingly minute differences between the two environments would nevertheless induce variation in quality traits, as well as in the barley grain and malt metabolite profiles.

Twenty malting and agronomic traits were chosen based on their importance to selection during barley breeding for brewing (Table 1). Malting quality is defined by protein, carbohydrate and enzyme content, fermentation parameters and agronomic quality such as grain size, growth habit and yield. The 20 traits were assessed for 72 barley varieties grown at Fairfield and Idaho Falls, and all traits exhibited a normal, quantitative distribution consistent with variation normally observed in advanced breeding trials (Blake et al., 2012). While limited in genetic and environmental diversity, the quality trait distribution was sufficient to discover novel associations between metabolism and quality traits.

Principal component analysis (PCA) was conducted to assess the contribution of genetic (2-row and 6-row types) and environmental (Fairfield and Idaho Falls) factors to quality trait

(a)

(b)

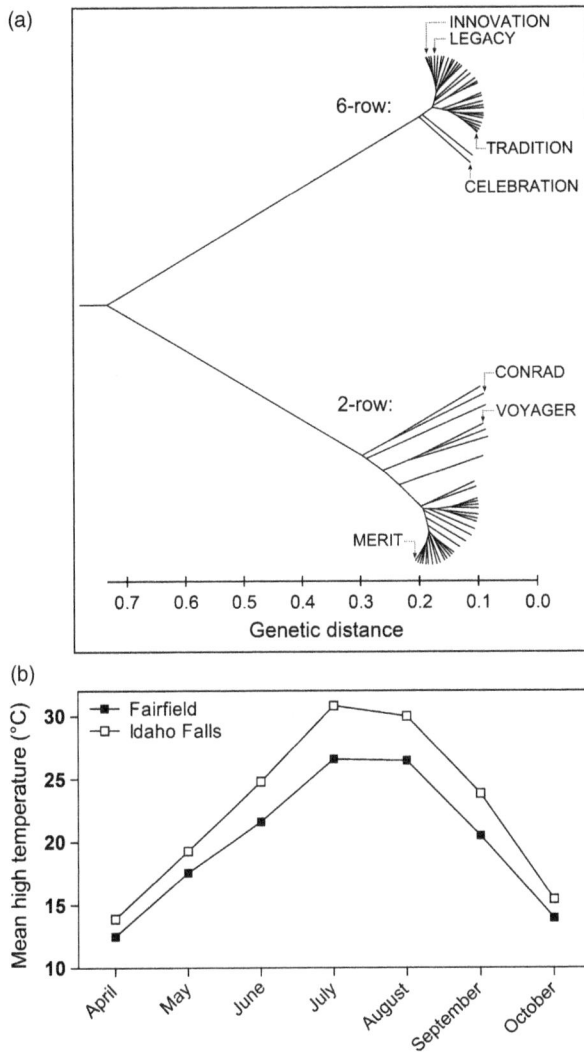

Figure 1 Genotypes and environments assessed for influence on metabolite variation in barley and malt. (a) Genetic relatedness of barley (n = 69 varieties) via unweighted paired group method with arithmetic mean on 1961 single nucleotide polymorphisms. Released malting barley varieties are denoted by name. (b) Mean monthly high temperature at Fairfield, MT and Idaho Falls, ID (USA) as reported by the National Oceanic and Atmospheric Administration weather stations.

variation. The quality trait PCA (Figure 2a) explained 52.2% of the variation between PCs 1 and 2. PC2 largely explained separation of genotypes (2-row and 6-row types), and PC1 largely explained separation by environment (Fairfield, Idaho Falls). The PCA loadings show several correlations among the 20 traits (Figure 2b). For example, a negative correlation is observed between fine extract content (FEC) (simple carbohydrates) and malt protein (corresponding to PC1). A positive correlation is observed between oligosaccharide content (beta-glucan) and viscosity that corresponded to PC2, a relationship described in many cereals (Boros et al., 1993; Grosjean et al., 1999; Saulnier et al., 1995). For a single variety grown at two environments, the Euclidian distance between PC1 and PC2 is representative of quality trait stability. A larger Euclidian distance indicates larger variation associated with the environmental effect. The calculated environment-induced Euclidian distances showed a normal

distribution for quality trait stability (Figure 2c). Based on this calculation, 6-row varieties were found to be more stable than 2-row varieties (Figure 2c, Student's t-test, $P < 0.0001$), largely attributed to separation in PC1, which was mostly associated with carbohydrate and protein related traits (observed as loadings in Figure 2b). This is consistent with a winter barley study that characterized 6-row varieties as more stable than 2-row varieties for grain yield across environmental conditions (del Moral et al., 2003).

Taken together, these data show that the selected barley population contains major genetic variation (2-row and 6-row types), as well as allelic diversity within each row type. There was sufficient diversity within each row type to establish quality trait phenotypes that were quantitative and normally distributed. The quality trait data were affected by both genotype and environment, providing the opportunity to identify covarying metabolites.

Influence of genetics and environment on barley and malt metabolite content

Metabolites of barley grain and malt were assessed by UPLC-MS nontargeted profiling to determine whether, like quality traits, genetic and/or environmental diversity contribute to metabolite variation. MS metabolomics is based on the analysis of molecular features, which is defined as a detected mass (m/z) at a given retention time, and multiple features may correspond to a single metabolite. 27 420 molecular features were detected among the barley and malt extracts.

Principal component analysis was conducted on the 27 420 molecular features of the barley and malt data sets to determine whether variation was explained by row type and/or environmental differences (Table 2). In the barley metabolite profiling data, PC1 and PC2 were independently associated with row type and environment effects, respectively. In the quality trait and malt metabolite profiling data sets, both PC1 and PC2 were associated with row type-environment interactions, as neither PC1 nor PC2 independently explained a single factor. The PC scores plot for barley metabolites (Figure 3a) shows separation between Fairfield and Idaho Falls, corresponding to 26.5% of the variation explained by PC1. PC2 (5.3% of the variation) explains differences by row type (ANOVA on PC2 scores, $P < 0.0001$). The malt metabolite profiles also show separation for both row type and environment (Figure 3b). PC1 largely explains differences in environment (28.5% of the variation), and PC2 explains metabolite differences associated 2-row and 6-row varieties (18.3% of the variation). Compared with barley, the malt metabolite PC score clusters are more defined for PC2, indicating that the metabolite variation associated with genotype is stronger in the malt metabolite data set than for barley metabolites. Furthermore, the four clusters in the barley metabolite PCA scores plot separate along the PC1 and PC2 axes. Conversely, for the malt metabolite PCA, the combined effect of PC1 and PC2 was necessary to delineate the four clusters.

For barley and malt, PC1 and PC2 explained variation attributed to genetic and environmental factors, and therefore the Euclidian distance between PC scores for a single variety from two locations represents metabolomic stability, akin to the genotype–environment interaction term. A large Euclidian distance indicates the environment induced a large effect on metabolite variation. The varietal stability metrics exhibited a normal, quantitative distribution for both barley and malt

Table 1 Quality traits for malting barley

Abbreviation	Class	Subclass	Measure (units)
Pro_nir	Malting quality	Protein content	Total malt protein (%)
Pro_wga	Malting quality	Protein content	Barley wheat germ agglutinin (ppm)
Pro_soluble	Malting quality	Protein content	Wort soluble protein (%)
Pro_fan	Malting quality	Protein content	Free amino nitrogen (ppm)
Pro_st	Malting quality	Protein content	Soluble/malt protein (Kolbach Index%)
Enz_aa	Malting quality	Enzymatic activity	α-amylase (20°du)
Enz_dp	Malting quality	Enzymatic activity	Diastatic power (°asbc)
Carb_bg	Malting quality	Carbohydrate content	β-glucan (ppm)
Carb_fe	Malting quality	Carbohydrate content	Fine extract (%)
Carb_ce	Malting quality	Carbohydrate content	Coarse extract (%)
Carb_fce	Malting quality	Carbohydrate content	Fine/coarse extract (%)
Ferm_rdf	Malting quality	Fermentation	Real degree fermentation (%)
Ferm_turb	Malting quality	Fermentation	Turbidity (FTU)
Ferm_visc	Malting quality	Fermentation	Viscosity (cP)
Agr_plump	Agronomic quality	Grain type	Plump grain (%)
Agr_medium	Agronomic quality	Grain type	Medium grain (%)
Agr_thin	Agronomic quality	Grain type	Thin grain (%)
Agr_heading	Agronomic quality	Growing habit	Heading date (days)
Agr_worth	Agronomic quality	Yield	Plot worth (AU)
Agr_yield	Agronomic quality	Yield	Yield (kg/ha)

(Figure 3c,d). No significant difference in stability between 2-row and 6-row barley (Student's t-test, $P = 0.10$) was observed. In the malt metabolite analysis, 2-row type are more stable than 6-row type (Student's t-test, $P = 0.03$). Although the malt metabolite stability (2-row type more stable than 6-row type) was in contrast to quality trait stability (6-row type more stable than 2-row type, Figure 2c), select varieties were consistent across the quality trait, barley and malt metabolite data sets, such as Celebration (stable) and Merit (unstable). Additionally, the differences in stability metrics for barley and malt suggest that environmental factors impact the malting process itself. This may include factors that control the protein or carbohydrate modifications that occur during malting.

The effects of genotype and environment (and the interaction) were confirmed using ANOVA conducted on the entire metabolite profiling data sets for both barley and malt (Figure 3e). For barley metabolites, 11.0, 56.8 and 0.9% of the features were influenced by row type, environment and the interaction term, respectively (Benjamini–Hochberg adjusted $P < 0.01$). For malt metabolites, 61.3, 52.1 and 14.7% of the features were influenced by row type, environment and the interaction term, respectively. For molecular features that covaried between the barley and malt data sets, the P-values from the malt data set were lower in 83.6% (row type), 61.2% (environment) and 91.1% (interaction) of the data set, suggesting the malt metabolite data are more responsive to the experimental design than the barley metabolite data. The ANOVA results agree with the PCA in showing that malt metabolites were more influenced by genotype and genotype–environment interactions than barley metabolites, however both barley and malt metabolites were equally affected by environment (Figure 3f). Given the greater effect of row type on malt metabolites than barley, these data support that the malting process itself exacerbates the genetic variation between 2-row and 6-row barley, which are known to produce different qualities of brews (Goldammer, 1999).

Metabolomic association of barley and malt metabolites to malting quality traits

Both the unbiased (PCA) and biased (ANOVA) analyses showed the greater responsiveness of the malt metabolite data compared with barley, and thus we chose to focus on malt to interrogate the relationship between metabolites and quality traits. The malt metabolite PC scores plot (Figure 3b) is similar to the quality traits PC scores plot (Figure 2a) in that genotype was a major contributor to variation. The PC that explained genetic factors for quality traits and malt metabolites correlated at both environments (Figure 4a,b, malt $r_s = 0.84$ at Fairfield, $r_s = 0.72$ at Idaho Falls, $P < 0.0001$), indicating that the factors contributing to malting and agronomic quality trait are related to the factors influencing malt metabolite variation.

Given each PC is defined by variation of a set of metabolites or quality traits, a multivariate covariance method was determined to be more appropriate than a standard correlation analysis between individual features and quality traits. O2PLS, a multivariate regression method that builds on orthogonal projection to latent structure (O-PLS) by incorporating multiple y variables (Trygg, 2002), was applied to identify malt metabolites that correlated with quality traits. The O2PLS method was recently described as a robust algorithm to identify covariation across two multivariate data sets (Bylesjo et al., 2007), including applications in transcriptomics and metabolomics. Covariation between the quality trait data set and the malt metabolite profiling data set was determined by treating the quality traits as O2PLS y variables and molecular features as x variables. The O2PLS model characterized four components in the malt metabolite data set (Figure 5a, Table S1). Twelve of the 20 quality traits have cumulative prediction rates (Q^2Y) of greater than 50%, indicating which quality traits could be estimated based on malt metabolite data.

Multiple trends were noted based on the distribution of components in the O2PLS model. For example, FEC (an estimate of total soluble material from the malt, including carbohydrates

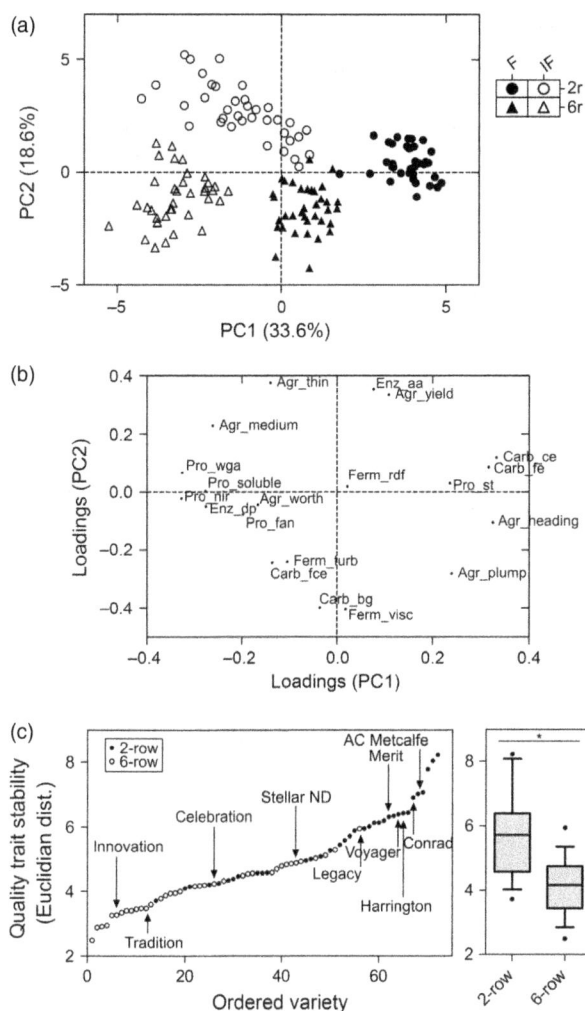

Figure 2 Variation in quality traits attributed to row type and environment. (a) Principal component (PC) scores plot of 72 barley varieties grown at two environments shows both row type and environment contribute to quality trait variation. (b) PC loadings plot of 20 quality traits (Table 1) show correlations between multiple traits. (c) Euclidian distance for each variety based on PC1 and PC2 scores from panel (a) indicates quality trait stability among environments. Arrows indicate released varieties. Box plot indicates the mean stability metric for 2-row and 6-row barley, and variation is denoted by an asterisk (Student's t-test, $P < 0.0001$).

and protein) was mostly explained by molecular features that contribute to component 1, whereas beta-glucan was mostly associated with component 2. Component 1 also explained many of the protein traits, which agrees with the negative correlation identified between protein and extract (as in Figure 2b). Fermentation-related quality traits are explained by a combination of components, agreeing with the lack of covariance among these traits (Figure 2b). The agronomic quality traits are mostly explained by a combination of components 1 and 2.

The annotation of molecular features that contribute to each component is discussed in the section below. The annotation process characterized 216 metabolites that associated with quality traits (Table S2). These included alkaloids, amines, amino acids, lipids (glycerophospholipids, lipids amides and oxylipins), organic acids, peptides, polyphenolics, purines, saccharides and many unknowns. The correlation between the 216 metabolites and each quality trait is reported in Table S2. Hierarchical clustering was used on O2PLS loadings to define metabolite patterns across the four O2PLS components. The amine/amino acid group showed little variation among metabolites (Figure 5b). Alternatively, the glycerolipids formed two major clusters (Figure 5c). For 18-carbon length lyso-phosphatidylcholines and ethanolamines, the sn2 and sn1 lipids positively and negatively contributed to component 1, respectively. The distinction between some isomers (sn1/sn2) may be explained by specificity of enzymes involved in lipid metabolism, although some lipid isomers covaried (e.g. LysoPC(16:0)-sn1 and LysoPC(16:0)-sn2). The alkaloids resulted in two major clusters based on components 1 and 2 (Figure 5d). Alkaloids have been previously shown to be affected by genetic and environmental factors in wild barley (Batchu *et al.*, 2006), and our study expands on this by highlighting variation within the alkaloid class, including between alkaloid isomers (e.g. hordatine B diglycoside). Polyphenolic metabolites showed only minor variation in the data set (Figure 5e).

In summary, this report defines a portion of the barley metabolome that covaries with multiple quality traits. The trends across and within metabolite classes (both primary and secondary) indicate a complex regulation of barley metabolism across genetic and environmental factors. Therefore, in advanced breeding populations, there are likely functional links between regulation of metabolism during the growing season, the final metabolite content of the barley grain and malt, and the subsequent variation in agronomic and malting quality traits.

Annotation of the metabolomics data set facilitated by idMS and idMS/MS spectra

Metabolite annotation is a significant challenge in nontargeted metabolite profiling experiments (Wishart, 2011), and annotation

Table 2 Row type and environment effects determined by ANOVA on PC scores

Data set	PC	% variation*	Row type[†]	Environment[†]	Interaction[†]
Quality traits	PC1	33.6	1.76×10^{-11}	1.46×10^{-11}	1.70×10^{-4}
	PC2	18.6	1.79×10^{-11}	2.45×10^{-11}	2.45×10^{-6}
Barley metabolites	PC1	26.5	0.62	1.52×10^{-11}	0.22
	PC2	5.3	2.21×10^{-11}	0.12	0.84
Malt metabolites	PC1	20.7	2.07×10^{-11}	0.02	2.24×10^{-11}
	PC2	12.1	1.65×10^{-11}	0.524	1.75×10^{-11}

PCA, Principal component analysis.

*Determined by PCA.

[†]P-value by ANOVA.

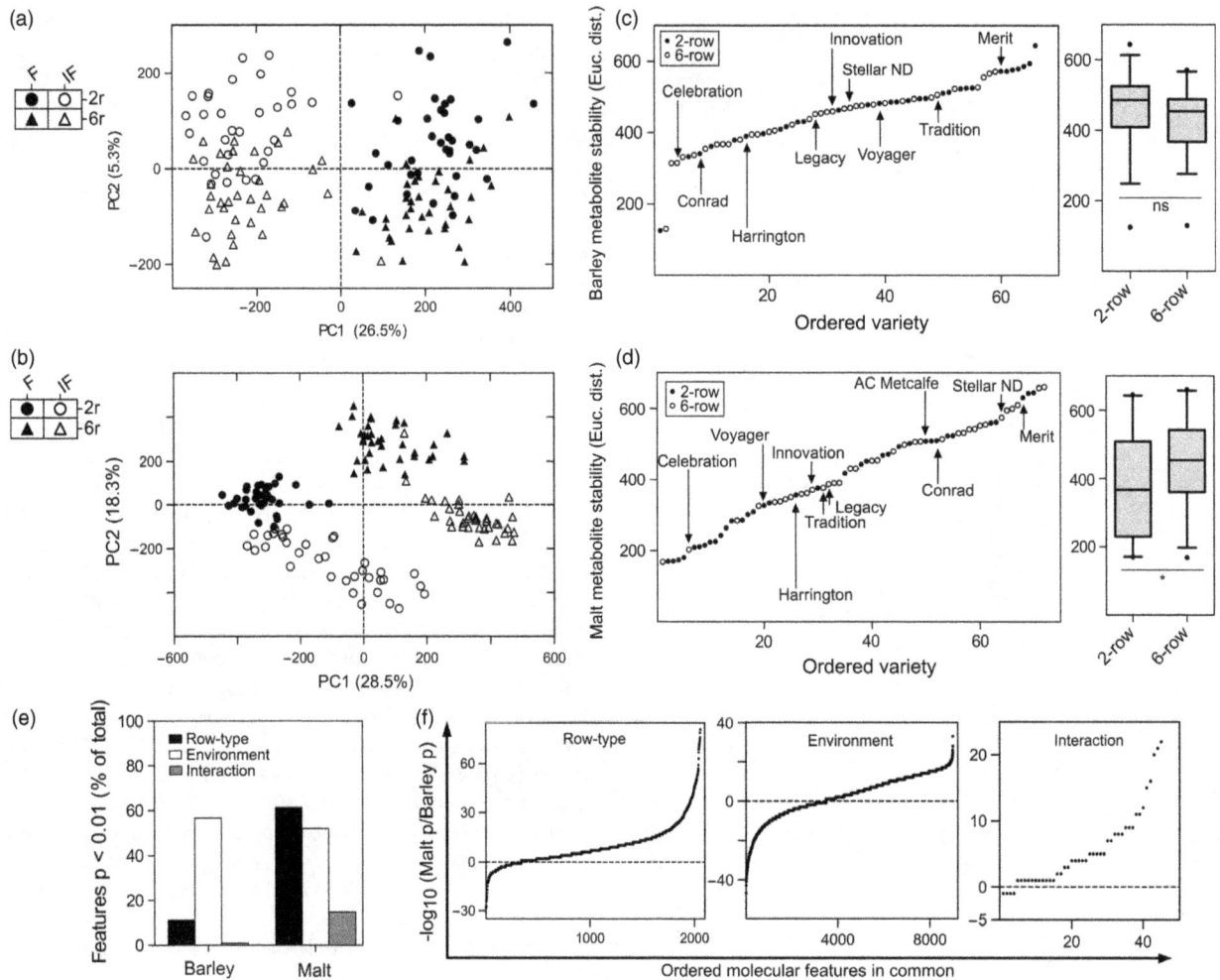

Figure 3 Metabolomic variation malt and barley extracts detected by ultra-performance liquid chromatography coupled with mass spectrometry (UPLC-MS) nontargeted metabolite profiling. Principal component analysis (PCA) scores plot for (a) barley metabolites from 2-row (2r) and 6-row (6r) varieties grown at two locations: Fairfield (F) and Idaho Falls (IF). (b) Metabolite stability scores for each variety as determined by Euclidian distances for each variety based on PC1 and PC2 scores from panel (a). Arrows indicate released barley varieties. The box plot indicates the mean and distribution of the stability metric for the barley metabolite data set (Student's t-test, asterisk = $P < 0.05$, ns = not significant). Malt metabolites were assessed for (c) PC scores and (d) stability metrics. (e) Percentage of the 27 420 molecular features that vary among row type, environment and row type/environment interactions in the barley and malt data sets (ANOVA, Benjamini–Hochberg adjusted $P < 0.01$). (f) Distribution of P-value ratios in barley and malt metabolite data sets for molecular features that varied due to row type, environment or the interaction term. Values greater than zero indicate increased statistical confidence observed in the malt metabolite data set, values less than zero increased confidence in the barley metabolite data set.

of all 27 420 detected molecular features is not feasible. Of the 27 420 molecular features, 1319 were associated with at least one quality trait based on the O2PLS model (z-transformation described in the Experimental procedures section). The molecular features of interest were reconstructed into mass spectra based on the covariation across the full data set within a restricted time window (Broeckling et al., 2013). These spectra consisted of both idMS-level (low collision energy) and idMS/MS-level (high collision energy, increased fragmentation) data. These idMS-level data were used for quantitative analysis, and a combination of the idMS and the idMS/MS data was used to facilitate annotations.

The incorporation of indiscriminant data acquisition in our workflow and the subsequent generation of idMS and idMS/MS spectra were critical to annotating the 1319 significant molecular features in the malt metabolite data set. An example of the metabolite annotation process is provided in Figure 6. Here, two O2PLS-significant molecular features, 357.185 and 579.150 m/z, co-eluted at 200 s (Figure 6a). Traditional workflows lack the

ability to differentiate the two masses (m/z values) and recreate individual spectra. However, because the two masses did not covary (Figure 6b), two unique spectra were recreated (idMS in Figure 6c,d and idMS/MS in Figure 6e,f). The idMS/MS spectra for each feature were searched against the most current version of the NIST LC-MS/MS spectral library. The idMS/MS spectrum for 579.150 m/z displayed a strong match to a library spectrum for procyanidin B2, a barley tannin (Dvorakova et al., 2008; Figure 6e). The idMS/MS spectrum for the feature at 357.185 m/z (Figure 6f) showed a partial match to the library spectrum for agmatine, a substructural component of barley alkaloids, and to hexose fragments. Further manual annotation assigned 276.158 m/z as M + 2H of the barley alkaloid hordatine A based on a description in the literature (von Ropenack et al., 1998). The molecular feature at 357.185 m/z also exhibited an isotope pattern indicating a doubly charged ion, corresponding to a singly charged ion of 713.362 m/z. The difference between the inferred peak at 713.362 m/z and the expected M+H for

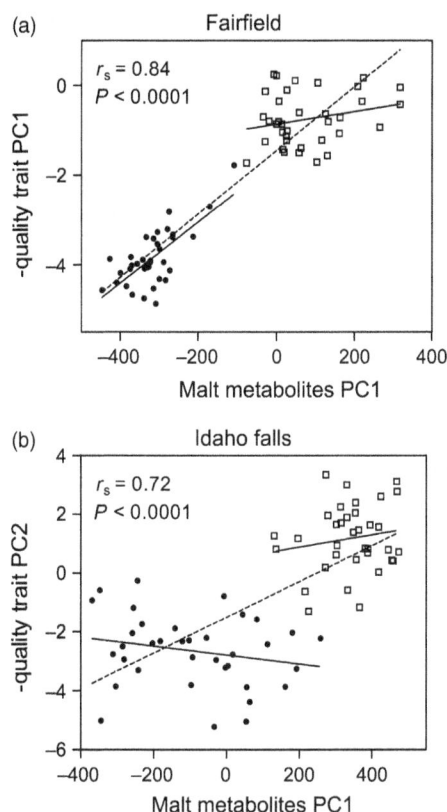

Figure 4 Correlation between genetic principal components (PCs) for quality traits and malt metabolites. Spearman's rank correlation for PCs associated with genetic effects for malt from (a) Fairfield and (b) Idaho Falls. Genotypes are denoted by 2-row (black circles) or 6-row type (white squares). Linear regression was performed to display relationships within row type (solid lines) or with row types combined (dashed lines).

hordatine A (551.308 m/z) is 162.054, indicating the addition of hexopyranose and resulting in annotation of the metabolite as a hordatine A glycoside. This compound was only recently described in the literature although no structural MS/MS data were reported (Kohyama and Ono, 2013). These two examples illustrate the power of our workflow for the identification of a well-described compound (procyanidin B2) and a compound previously described in the literature, but that has not been structurally characterized by mass spectrometry (hordatine A glycoside), is not represented in metabolite databases, and for which no commercial standard is available.

The idMS and idMS/MS spectra generated in this study are included in this report (Data S1). The spectra are provided as a .txt file and formatted for .msp, allowing for direct import into publicly available mass spectrometry viewing software and subsequent searching against other experimental data. It is important to note that these spectra differ from what is present for standard compounds in the major metabolite databases, which are typically collected using targeted fragmentation approaches employing precursor ion isolation.

Genetic colocalization as a tool to describe putative functional associations between metabolites and quality traits

While malting barley metabolites were observed to correlate with quality traits, the cause of these associations remains unclear. Metabolites that act as 'perfect' or 'functional' markers, whereby the metabolite directly affects the quality trait, would provide greater confidence in predicting trait values across additional varieties and environments. Combining genetic analysis with metabolite profiling to discover mechanisms of metabolite regulation has been demonstrated (Matsuda et al., 2012) and can be an important component to improve the genome annotation process (May et al., 2008; Shen et al., 2013).

Here, the relationship between metabolites and quality traits was investigated using a genetic colocalization approach. Quality traits and the 216 metabolites of interest were individually mapped to the barley genome based on the variation of 1961 single nucleotide polymorphisms (SNPs). The mapping was performed using a linear mixed model and included a kinship term to account for false positives related to population structure (Figure S1). Three quantitative variables were explored for use in genetic mapping: data from the Fairfield environment, the Idaho Falls environment and the mean values of phenotypes across environments. The genetic localization of quality traits and metabolites are reported in Data S2.

Example metabolites that correlated with FEC

Fine extract content, a measure of total water soluble material from the malt for brewing, was chosen for further investigation for its importance to the brewing industry and the large number of correlating barley metabolites (Figure 5, Table S2). The ability to predict FEC in an experimental variety using molecular markers is ideal, as the full measurement procedure is time-consuming and costly and requires a large quantity of malted barley.

Two hypothesized functional metabolite markers for FEC were glycine betaine (an amine) and LysoPC(18:2)-sn2 (a glycerophospholipid), which both correlated with FEC in barley and malt (Figure 7a,b). Both betaines and lipids have been widely documented to vary with heat stress (Allakhverdiev et al., 2008; Sakamoto and Murata, 2002), an important environmental parameter in the present experiment (Figure 1b). Regulation of betaine and regulation of membrane lipids are both thought to protect photosynthesis against high temperature (Allakhverdiev et al., 2007; Gombos et al., 1994). Moreover, both betaine and phospholipid contents have been shown to vary with genotype (Nayyar and Walia, 2004; Su et al., 2009) and genotype–environment interactions in maize (Harrigan et al., 2007).

Seven genetic markers were identified for FEC (Benjamini–Hochberg adjusted P < 0.001; Figure 7c). Malt-derived betaine and LysoPC(18:2)-sn2 were also mapped and resulted in 11 additional SNP markers of interest (Benjamini–Hochberg adjusted P < 0.001; Figure 7d,e) The SNPs of interest were assigned to genes based on synteny with the rice genome (Table 3). Colocalization of genetic markers for fine extract, betaine and LysoPC(18:2) phenotypes was determined based on relative proximities (chromosome, arm and map unit).

Colocalization of FEC and betaine

FEC and betaine colocalized to markers at 70.4 and 72.64 cM on the short arm of chromosome 7H (notation f and m). The SNP most strongly associated with FEC is within an HVA22-protein-encoding gene (Table 3 notation f). HVA22 protein regulates the mobilization of seed nutrients during germination (Guo and Ho, 2008). Additionally, HVA22 expression is induced by abscisic acid (ABA), a hormone known to regulate plant response to environmental parameters including heat stress (Larkindale and Knight, 2002) and specifically for seed growth (Cheikh and Jones, 1994).

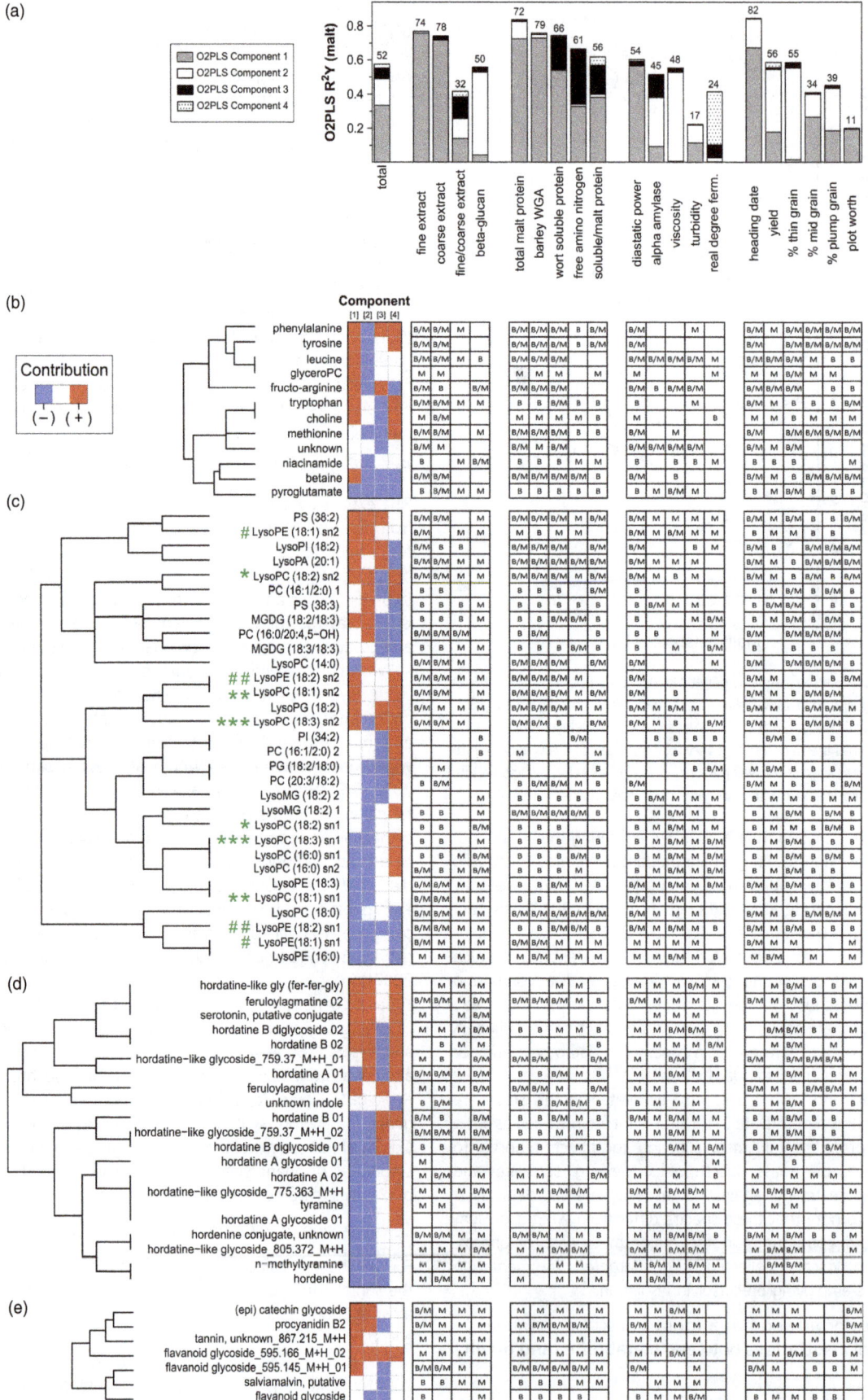

Figure 5 Metabolites that correlate to quality traits in malting barley. (a) Overview of O2PLS regression of the quality trait malt data set against the metabolite data set, including the total variation explained by each component (R^2Y). Cumulative predicted explained variation (Q^2Y) is denoted as numerical values above each bar. Loading scores for each of the four components were z-transformed, metabolites were separated by class, and hierarchical clustering was performed on the O2PLS loadings z values for (b) amines (c) glycerophospholipids (d) alkaloids and (e) polyphenolics. Green symbols denote sn1/sn2 positional isomer pairs for 18-carbon LysoPC and LysoPE. Metabolites with identical spectra at multiple retention times were denoted as '01, 02, etc.' to indicate uncharacterized isomers. Heat map colours indicate a positive (red) or negative (blue) contribution to O2PLS components 1–4, respectively. Metabolites were assessed for association to each trait using Spearman's rank correlation, and B denotes $P < 0.01$ for the barley data set, M denotes $P < 0.01$ for the malt data set, and B/M denotes $P < 0.01$ for both data sets. The B, M and B/M notation correspond to the metabolite to the left and the trait above. Metabolite abbreviations and nomenclature, O2PLS loading z values and Spearman correlation r_s and P-values are described in Table S2.

Directly adjacent is this locus is sucrose synthase, which was associated with betaine (Table 3 notation m), implicating a relationship between betaine synthesis and saccharide content in the grain. The betaine–sucrose synthase association was additionally observed when evaluating the two environments independently (Data S2, Benjamini–Hochberg adjusted $P < 0.01$ for both Fairfield and Idaho Falls). The two may be linked through ABA, which has been shown to increase betaine synthesis in many plants (Gao *et al.*, 2004; Jagendorf and Takabe, 2001; McDonnell *et al.*, 1983; Nakamura *et al.*, 2001; Nayyar and Walia, 2004; Saneoka *et al.*, 2001) and observed to decrease grain saccharide content in wheat grain (Ahmadi and Baker, 1999).

Colocalization of FEC and LysoPC(18:2)

The direct link between FEC and malt metabolites is also indicated by the colocalization with LysoPC(18:2) on the long arm of chromosome 1 (Table 3 notation a, q). The thionin and indole-3-glycerol phosphate lyase genes are known to contribute to plant responses to biotic stress (Andresen *et al.*, 1992; Bohlmann *et al.*, 1988; Frey *et al.*, 1997), implying a second set of environmentally

responsive genes that associated with both quality trait and metabolite variation.

The relationship between betaine and LysoPC(18:2)-sn2 could be explained by marker 4988-858 (Table 3 notation j), which corresponds to a patatin T5 precursor gene. Patatin is a potato storage protein with lipolytic activity (Shewry, 2003), and patatin-domain containing lipases have been shown to be central to lipid catabolism during seed germination (Eastmond, 2006), a process integral to malting. Patatin lipases have also been linked to oxylipin synthesis (Yang *et al.*, 2012), for which many molecular signals were found to covary with quality traits over genetic and environmental factors (Table S2). The patatin lipases have also shown to prefer sn1 lysophospholipids (Anderson *et al.*, 2002), potentially explaining the sn1/sn2 variation observed in 18 carbon lysophospholipids in our study (Figure 5c).

Summary of genetic colocalization results

Based on the colocalization of the FEC and metabolite pheno-types to environmentally responsive genes, our data support both charged amines and glycerophospholipids as functional grain markers of quality traits in an advanced breeding population. In

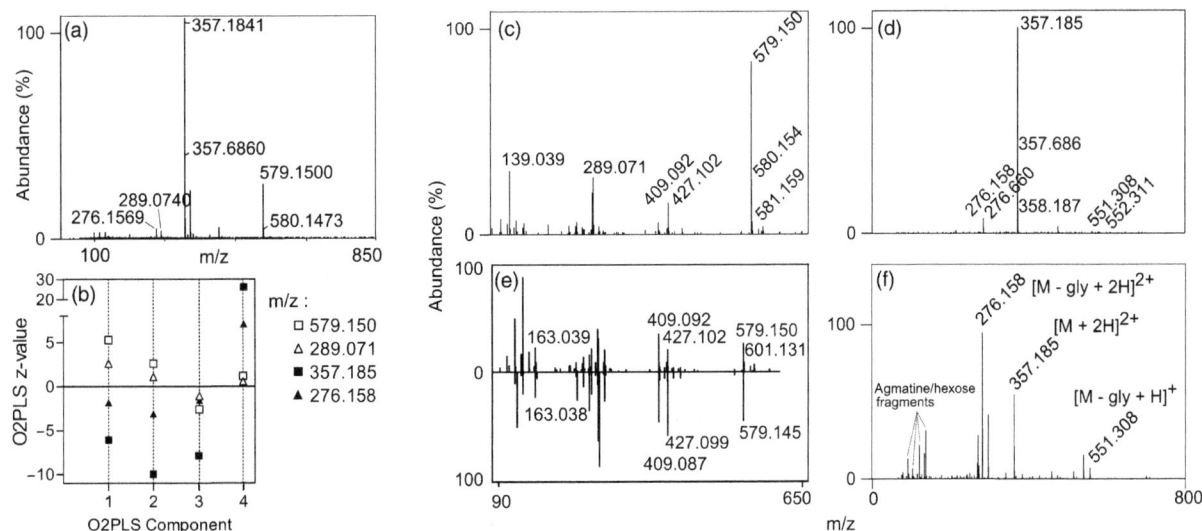

Figure 6 Example annotation for two co-eluting plant compounds. (a) The low collision energy spectrum for a malt metabolite extract at 200 s shows multiple m/z values to be annotated as metabolites. (b) At 200 s, four select m/z values contributed to the O2PLS model but show varying trends across the four O2PLS components (white or black shades). idMS spectra were constructed based on m/z values that correlate with (c) 579.150 m/z and/or (d) 357.185 m/z between 199 and 201 s using Pearson's r threshold of 0.75. (e) 579.150 m/z idMS/MS spectrum for high collision energy scans result in high fragmentation. The idMS/MS spectrum (top) was screened in the NIST metabolite database and matched to procyanidin B2 (bottom), a plant tannin. (f) idMS/MS spectrum for 357.185 m/z partially matched to agmatine and hexose in the NIST database (not shown). Other m/z values were manually interpreted and accurate mass, protonation (H) and charge state, and loss of hexose molecules (gly) indicate 357.185 m/z corresponds to hordatine A glycoside, a barley metabolite not described in available MS databases. Differences in m/z values between panel a and b–f are due to m/z averaging that occurs during peak detection (see Experimental Procedures).

this study, the higher temperatures at Idaho Falls may have affected the mobilization of metabolites to the barley grain and therefore altered downstream malting quality. Barley lines grown at Idaho Falls had reduced FEC, betaine and LysoPC(18:2). However, FEC and metabolites also varied by genotype, indicating a genetic component potentially mediated through SNP variation in the genes described in Table 3, including genes related to abiotic and biotic stress. Thus, grain metabolites have the potential to act as functional markers representative of variation induced by growing location and genotype. This could provide a novel method to predict environmental effects on quality traits based on metabolite abundances and also to evaluate experimental lines for performance in new environments.

Summary

It was found that barley and malt metabolites correlated with multiple quality traits in an advanced barley population. Both metabolites (primary and secondary) and quality traits covaried based on genetic and environmental parameters. While variation in quality traits related to genotypic differences (2-row and 6-row barley) can be predicted based on genetic markers alone, metabolites allow for the ability to assess the effect of growing environment and genotype–environment interactions. Therefore, metabolite profiling represents a novel method to assess a variety's stability across growing locations. We hypothesize that metabolites and quality traits are coregulated by a plant's response to environmental stimuli, such as heat stress. Thus, metabolite markers may enable prediction of quality trait stability for a variety grown in multiple locations, potentially expediting the evaluation of advanced or experimental lines in a breeding population. Furthermore, it was found that two diverse metabolites covaried with the fine extract quality trait and both mapped to genetic locations related to biotic and abiotic stress. In summary, the results presented here illustrate the application of nontargeted UPLC-MS metabolite profiling for the detection of metabolites (some previously undefined) that covaried with quality traits in malting barley and provides a foundation for expanding these types of studies to other plant systems.

Experimental procedures

Genetic relatedness of the barley population

The genetic relatedness of the barley population was determined using an unweighted pair group method with arithmetic mean performed in MEGA, v5 (Tamura et al., 2011). The analysis was conducted based on 1961 SNPs provided by the BarleyCAP initiative for 69 of the 72 barley accessions.

Plant material, quality traits and metabolite extraction

Barley grain and malt were provided by Busch Agricultural Resources for plants grown at Fairfield, MT or Idaho Falls, ID

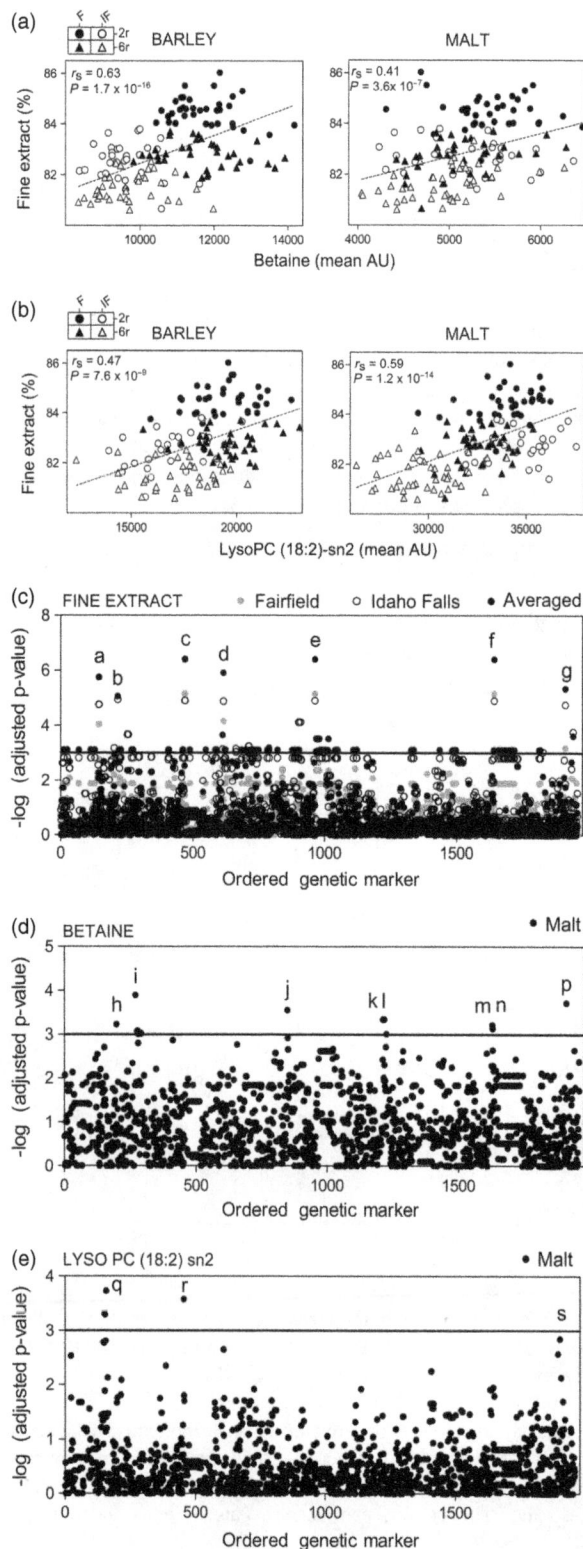

Figure 7 Phenotypic and genetic covariation among fine extract content (FEC), betaine and LysoPC(18:2)-sn2 for barley and malt. Pairwise associations between FEC and the metabolites (a) glycine betaine and (b) LysoPC(18:2)-sn2 for barley and malt. Sample points indicate row type (2-row and 6-row type) and environment (Fairfield, Idaho Falls), and metabolite quantities are reported in peak area units (AU). Dashed lines represent linear regression performed on the data set. Spearman's correlation r_s and P-values are described in Table S2. (c) Manhattan plot for fine extract based on data from Fairfield and Idaho Falls, and averaged values were calculated based on the mean of the two environments. Manhattan plots for malt were generated for (d) betaine and (e) LysoPC (18:2)-sn2. Genetic mapping was determined using a linear mixed model association via likelihood ratio tests with maximum likelihood estimates (Benjamini–Hochberg adjusted $P < 0.001$). Letters correspond to genetic markers described in Table 3.

Table 3 Genetic markers associated to fine extract, betaine and LysoPC(18:2)-sn2

Barley marker	Locus	Phenotype	Figure 7 notation	Rice locus	Rice annotation
U32_185_829	1H / L / 126.48	Fine extract	a	LOC_Os06g32550	Leaf-specific thionin BTH6 precursor, putative, expressed
U32_3959_548	1H / L / 135.56	LysoPC(18:2)-sn2	q	LOC_Os07g08430	Protein indole-3-glycerol phosphate lyase, chloroplast precursor, putative, expressed
U32_6062_476	2H / S / 15.15	Betaine	h	LOC_Os04g13210	Multidrug resistance-associated protein 4, putative, expressed
7144-973	2H / S / 42.7	Fine extract	b	LOC_Os07g48890	Poly (A) polymerase, putative, expressed
3806-486	2H / L / 55.63	Betaine	i	LOC_Os04g39210	Agmatine deiminase, putative, expressed
ABC22737_1_28	3H / S / 28.44	LysoPC(18:2)-sn2	r	LOC_Os11g33110	Protein germin-like protein subfamily 3 member 4 precursor, putative, expressed
U32_5641_239	3H / S / 48.63	Fine extract	c	LOC_Os03g48850	Valyl-tRNA synthetase, putative, expressed
ABC19616_1_756	3H / L / 124.84	Fine extract	d	LOC_Os01g66860	Ankyrin-kinase, putative, expressed
4988-858	4H / L / 135.52	Betaine	j	LOC_Os08g37180	Patatin T5 precursor, putative, expressed
U35_21417_466	5H / S / 46.23	Fine extract	e	LOC_Os03g39650	Cytochrome P450 71D8, putative, expressed
U32_6342_1490	5H / L / 161.58	Betaine	l	LOC_Os01g25386	Multidrug resistance-associated protein 4, putative, expressed
3931-1434	5H / L / 220.18	Betaine	k	LOC_Os03g59760	RING finger 126, putative, expressed
2669-1012	7H / unk / 73.31	Betaine	n	LOC_Os06g09679	Chaperonin, chloroplast precursor, putative, expressed
U35_4735_535	7H / S / 70.4	Fine extract	f	LOC_Os06g12220	HVA22-like a, putative, expressed
943-3107	7H / S / 72.64	Betaine	m	LOC_Os06g09450	Sucrose synthase 1, putative, expressed
U32_9521_149	unk	Fine extract	g	LOC_Os03g49770	Oxysterol-binding protein 1, putative, expressed
u32_711_642	unk	Betaine	p	LOC_Os01g16030	ADP-ribosylation factor, putative, expressed
U32_259_1252	unk	LysoPC(18:2)-sn2	s	LOC_Os05g49880	Protein malate dehydrogenase, mitochondrial precursor, putative, expressed

(USA) from the 2010 season. Planting, harvest and grain storage were executed under normal breeding conditions. Grain was pooled per variety and malting parameters (a 4-day method that includes germination, kilning and removing rootlets), and quality assessment was executed as previously described (Lapitan *et al.*, 2009).

For metabolite profiling, barley grain and malt from the same pool used for quality assessment were ground to a fine powder, and metabolites were extracted by adding 1.5 mL of methanol/water (70 : 30, v : v) to 120 mg of tissue. Samples were shaken on a vortex mixer for 2 h at room temperature and centrifuged at 13 500 *g* for 10 min, and the supernatant was collected and stored at −80 °C until further analysis.

Nontargeted metabolite profiling by ultra-performance liquid chromatography–mass spectrometry

One microlitre of metabolite extract was injected twice ($n = 2$ replicates) onto a Waters Acquity UPLC system in discrete, randomized blocks and separated using a Waters Acquity UPLC T3 column (1.8 μm, 1.0 × 100 mm), using a gradient from solvent A (water, 0.1% formic acid) to solvent B (acetonitrile, 0.1% formic acid). Injections were made in 100% A, held at 100% A for 1 min and followed a linear gradient to 95% B in 12 min, held at 95% B for 3 min, returned to starting conditions over 0.05 min and allowed to re-equilibrate for 3.95 min, with a 200 μL/min flow rate. The column and samples were held at 50 °C and 5 °C, respectively. The column eluent was infused into a Waters Xevo G2 Q-TOF-MS with an electrospray source in positive mode, scanning 50–1200 m/z at 0.2 s per scan, alternating between MS (6 V collision energy) and MS^E mode (15–30 V ramp). Calibration was performed using sodium formate with 1 ppm mass accuracy. The capillary voltage was

held at 2200 V, source temp at 150 °C and nitrogen desolvation temp at 350 °C with a flow rate of 800 L/h.

MS files were converted to netCDF format and processed with XCMS (Smith *et al.*, 2006) for peak detection, retention time alignment and feature grouping on both low and high collision energy channels. The data sets were separated following alignment, normalized to total ion signal, averaged via the mean injection replicates, and idMS and idMS/MS spectra were generated as previously described (Broeckling *et al.*, 2013). Spectra were searched against in-house, NISTv12, Metlin and Massbank databases for annotations. Annotation confidence levels are as recommended by the Metabolites Standards Initiative (Sumner *et al.*, 2007). Metabolite isomers were inferred by identical idMS and idMS/MS spectra at multiple retention times, and glycerolipid positional isomers were classified as previously described (Dong *et al.*, 2010).

Data preprocessing and statistics

The MS-level data set was used for all statistics applications. Seven samples were determined to contain poor injections and removed from the data set for all subsequent analysis. PCA was conducted using SIMCA P+ (v12.0, Umetrics) on molecular feature abundance values that were mean-centred and Pareto-scaled. Differences in metabolite stability for 2-row versus 6-row barley and malt were determined using Student's *t*-test on the Euclidian distance of PC1 and PC2 scores with a *P*-value threshold of 0.05. Two-way analysis of variance (ANOVA) was conducted with row type, environment and genotype–environment interactions as factors for PC1 and PC2 using the aov function in R. In a separate analysis, two-way ANOVA was conducted on each molecular feature with row type, environment and genotype–environment interactions as factors using the aov function in R, and *P*-values were adjusted for false positives using the p.adjust function in R (Benjamini and Hochberg, 1995). For O2PLS (SIMCA P+ v12.0, Umetrics), quality traits values were scaled to unit variance and molecular features were Pareto-scaled to reduce the influence of noise. The O2PLS model was a regression of quality traits (*y* variables) onto molecular features (*x* variables). A molecular feature's contribution to O2PLS was determined by *z*-transformation of a component loading based on the mean and standard deviation of each component loading. Hierarchical clustering and Spearman's rank correlation were performed in the R v2.14.2 using the hclust, cor and cor.test functions (R Core Team, 2012).

Genetic mapping of quality traits and metabolites

Associations were determined using the EMMA package (Kang *et al.*, 2008) in R, v2.9.2., via a linear mixed model via likelihood ratio tests with maximum likelihood estimates, and *P*-values were adjusted via Benjamini–Hochberg. The genetic location and rice gene annotation for each marker were as previously described (Blake *et al.*, 2012; Munoz-Amatriain *et al.*, 2011).

Acknowledgements

This research was supported by the American Malting Barley Association, Inc. and Anheuser-Busch, Inc. We would like to thank Gary Hanning and his team for contributions and guidance critical to this investigation and manuscript.

References

Ahmadi, A. and Baker, D.A. (1999) Effects of abscisic acid (ABA) on grain filling processes in wheat. *Plant Growth Regul.* **28**, 187–197.

Allakhverdiev, S.I., Los, D.A., Mohanty, P., Nishiyama, Y. and Murata, N. (2007) Glycinebetaine alleviates the inhibitory repair of photosystem II effect of moderate heat stress on the during photoinhibition. *BBA-Bioenergetics*, **1767**, 1363–1371.

Allakhverdiev, S.I., Kreslavski, V.D., Klimov, V.V., Los, D.A., Carpentier, R. and Mohanty, P. (2008) Heat stress: an overview of molecular responses in photosynthesis. *Photosynth. Res.* **98**, 541–550.

Anderson, C., Pinsirodom, P. and Parkin, K.L. (2002) Hydrolytic selectivity of patatin (lipid acyl hydrolase) from potato (*Solanum tuberosum* L.) tubers toward various lipids. *J. Food Biochem.* **26**, 63–74.

Andresen, I., Becker, W., Schluter, K., Burges, J., Parthier, B. and Apel, K. (1992) The identification of leaf thionin as one of the main jasmonate-induced proteins of barley (*Hordeum vulgare*). *Plant Mol. Biol.* **19**, 193–204.

Asiago, V.M., Hazebroek, J., Harp, T. and Zhong, C. (2012) Effects of genetics and environment on the metabolome of commercial maize hybrids: a multisite study. *J. Agric. Food Chem.* **60**, 11498–11508.

American Malting Barley Association, Inc. (2010) *Barley Variety Survey*. Milwaukee, WI. Available at: http:\\ambainc.org.

Batchu, A.K., Zimmermann, D., Schulze-Lefert, P. and Koprek, T. (2006) Correlation between hordatine accumulation, environmental factors and genetic diversity in wild barley (*Hordeum spontaneum* C. Koch) accessions from the Near East Fertile Crescent. *Genetica*, **127**, 87–99.

Beleggia, R., Platani, C., Nigro, F., De Vita, P., Cattivelli, L. and Papa, R. (2013) Effect of genotype, environment and genotype-by-environment interaction on metabolite profiling in durum wheat (*Triticum durum* Desf.) grain. *J. Cereal Sci.* **57**, 183–192.

Benjamini, Y. and Hochberg, Y. (1995) Controlling the false discovery rate – a practical and powerful approach to multiple testing. *J. R. Stat. Soc. Series B Stat. Methodol.* **57**, 289–300.

Blake, V.C., Kling, J.G., Hayes, P.M., Jannink, J.L., Jillella, S.R., Lee, J., Matthews, D.E., Chao, S., Close, T.J., Muehlbauer, G.J., Smith, K.P., Wise,

R.P. and Dickerson, J.A. (2012) The hordeum toolbox: the Barley coordinated agricultural project genotype and phenotype resource. *Plant Genome*, **5**, 81–91.

Bohlmann, H., Clausen, S., Behnke, S., Giese, H., Hiller, C., Reimannphilipp, U., Schrader, G., Barkholt, V. and Apel, K. (1988) Leaf-specific thionins of barley – a novel class of cell-wall proteins toxic to plant-pathogenic fungi and possibly involved in the defense-mechanism of plants. *EMBO J.* **7**, 1559–1565.

Bollina, V., Kumaraswamy, G.K., Kushalappa, A.C., Choo, T.M., Dion, Y., Rioux, S., Faubert, D. and Hamzehzarghani, H. (2010) Mass spectrometry-based metabolomics application to identify quantitative resistance-related metabolites in barley against Fusarium head blight. *Mol. Plant. Pathol.* **11**, 769–782.

Boros, D., Marquardt, R.R., Slominski, B.A. and Guenter, W. (1993) Extract viscosity as an indirect assay for water-soluble pentosan content in rye. *Cereal Chem.* **70**, 575–580.

Broeckling, C.D., Heuberger, A.L., Prince, J.A., Ingelsson, E. and Prenni, J.E. (2013) Assigning precursor–product ion relationships in indiscriminant MS/MS data from non-targeted metabolite profiling studies. *Metabolomics*, **9**, 33–43.

Bylesjo, M., Eriksson, D., Kusano, M., Moritz, T. and Trygg, J. (2007) Data integration in plant biology: the O2PLS method for combined modeling of transcript and metabolite data. *Plant J.* **52**, 1181–1191.

Carreno-Quintero, N., Acharjee, A., Maliepaard, C., Bachem, C.W.B., Mumm, R., Bouwmeester, H., Visser, R.G.F. and Keurentjes, J.J.B. (2012) Untargeted metabolic quantitative trait loci analyses reveal a relationship between primary metabolism and potato tuber quality. *Plant Physiol.* **158**, 1306–1318.

Cheikh, N. and Jones, R.J. (1994) Disruption of maize kernel growth and development by heat-stress - role of cytokinin abscisic-acid balance. *Plant Physiol.* **106**, 45–51.

Dettmer, K., Aronov, P.A. and Hammock, B.D. (2007) Mass spectrometry-based metabolomics. *Mass Spectrom. Rev.* **26**, 51–78.

Dong, J., Cai, X.M., Zhao, L.L., Xue, X.Y., Zou, L.J., Zhang, X.L. and Liang, X.M. (2010) Lysophosphatidylcholine profiling of plasma: discrimination of isomers and discovery of lung cancer biomarkers. *Metabolomics*, **6**, 478–488.

Dvorakova, M., Moreira, M.M., Dostalek, P., Skulilova, Z., Guido, L.F. and Barros, A.A. (2008) Characterization of monomeric and oligomeric flavan-3-ols from barley and malt by liquid chromatography–ultraviolet detection–electrospray ionization mass spectrometry. *J. Chromatogr. A*, **1189**, 398–405.

Eastmond, P.J. (2006) SUGAR-DEPENDENT1 encodes a patatin domain triacylglycerol lipase that initiates storage oil breakdown in germinating Arabidopsis seeds. *Plant Cell*, **18**, 665–675.

Frank, T., Scholz, B., Peter, S. and Engel, K.H. (2011) Metabolite profiling of barley: influence of the malting process. *Food Chem.* **124**, 948–957.

Frey, M., Chomet, P., Glawischnig, E., Stettner, C., Grun, S., Winklmair, A., Eisenreich, W., Bacher, A., Meeley, R.B., Briggs, S.P., Simcox, K. and Gierl, A. (1997) Analysis of a chemical plant defense mechanism in grasses. *Science*, **277**, 696–699.

Gao, X.P., Pan, Q.H., Li, M.J., Zhang, L.Y., Wang, X.F., Shen, Y.Y., Lu, Y.F., Chen, S.W., Liang, Z. and Zhang, D.P. (2004) Abscisic acid is involved in the water stress-induced betaine accumulation in pear leaves. *Plant Cell Physiol.* **45**, 742–750.

Goldammer, T. (1999) *The Brewers' Handbook*. Clifton, VA: Apex Publishers.

Gombos, Z., Wada, H., Hideg, E. and Murata, N. (1994) The unsaturation of membrane-lipids stabilizes photosynthesis against heat-stress. *Plant Physiol.* **104**, 563–567.

Gorzolka, K., Lissel, M., Kessler, N., Loch-Ahring, S. and Niehaus, K. (2012) Metabolite fingerprinting of barley whole seeds, endosperms, and embryos during industrial malting. *J. Biotechnol.* **159**, 177–187.

Grosjean, F., Saulnier, L., Maupetit, P., Beaux, M.F., Flatres, M.C., Magnin, M., Le Pavec, P. and Victoire, C. (1999) Variability of wheat and other cereal water extract viscosity. 1 – improvements in measuring viscosity. *J. Sci. Food Agric.* **79**, 116–122.

Guo, W.J. and Ho, T.H.D. (2008) An abscisic acid-induced protein, HVA22, inhibits gibberellin-mediated programmed cell death in cereal aleurone cells (vol 147, pg 1710–1722, 2008). *Plant Physiol.* **148**, 1182.

Harrigan, G.G., Stork, L.G., Riordan, S.G., Reynolds, T.L., Ridley, W.P., Masucci, J.D., MacIsaac, S., Halls, S.C., Orth, R., Smith, R.G., Wen, L., Brown, W.E., Welsch, M., Riley, R., Mcfarland, D., Pandravada, A. and Glenn, K.C. (2007) Impact of genetics and environment on nutritional and metabolite components of maize grain. *J. Agric. Food Chem.* **55**, 6177–6185.

Hayes, P. and Szucs, P. (2006) Disequilibrium and association in barley: thinking outside the glass. *Proc. Natl Acad. Sci. USA*, **103**, 18385–18386.

Heuberger, A.L., Lewis, M.R., Chen, M-H., Brick, M.A., Leach, J.E. and Ryan, E.P. (2010) Metabolomic and functional genomic analyses reveal varietal differences in bioactive compounds of cooked rice. *PLoS ONE*, **5**, e12915.

Jagendorf, A.T. and Takabe, T. (2001) Inducers of glycinebetaine synthesis in barley. *Plant Physiol.* **127**, 1827–1835.

Kang, H.M., Zaitlen, N.A., Wade, C.M., Kirby, A., Heckerman, D., Daly, M.J. and Eskin, E. (2008) Efficient control of population structure in model organism association mapping. *Genetics*, **178**, 1709–1723.

Keurentjes, J.J.B. (2009) Genetical metabolomics: closing in on phenotypes. *Curr. Opin. Plant Biol.* **12**, 223–230.

Keurentjes, J.J.B., Fu, J.Y., de Vos, C.H.R., Lommen, A., Hall, R.D., Bino, R.J., van der Plas, L.H.W., Jansen, R.C., Vreugdenhil, D. and Koornneef, M. (2006) The genetics of plant metabolism. *Nat. Genet.* **38**, 842–849.

Kogel, K-H., Voll, L.M., Schäfer, P., Jansen, C., Wu, Y., Langen, G., Imani, J., Hofmann, J., Schmiedl, A., Sonnewald, S., von Wettstein, D., Cook, R.J. and Sonnewald, U. (2010) Transcriptome and metabolome profiling of field-grown transgenic barley lack induced differences but show cultivar-specific variances. *Proc. Natl Acad. Sci. USA*, **104**, 6198–6203.

Kohyama, N. and Ono, H. (2013) Hordatine A ß-D-glucopyranoside from ungerminated barley grains. *J. Agric. Food Chem.* **61**, 1112–1116.

Komatsuda, T., Pourkheirandish, M., He, C.F., Azhaguvel, P., Kanamori, H., Perovic, D., Stein, N., Graner, A., Wicker, T., Tagiri, A., Lundqvist, U., Fujimura, T., Matsuoka, M., Matsumoto, T. and Yano, M. (2007) Six-rowed barley originated from a mutation in a homeodomain-leucine zipper I-class homeobox gene. *Proc. Natl Acad. Sci. USA*, **104**, 1424–1429.

Lapitan, N.L.V., Hess, A., Cooper, B., Botha, A.M., Badillo, D., Iyer, H., Menert, J., Close, T., Wright, L., Hanning, G., Tahir, M. and Lawrence, C. (2009) Differentially expressed genes during malting and correlation with malting quality phenotypes in barley (*Hordeum vulgare* L.). *Theor. Appl. Genet.* **118**, 937–952.

Larkindale, J. and Knight, M.R. (2002) Protection against heat stress-induced oxidative damage in arabidopsis involves calcium, abscisic acid, ethylene, and salicylic acid. *Plant Physiol.* **128**, 682–695.

Mangelsen, E., Kilian, J., Harter, K., Jansson, C., Wanke, D. and Sundberg, E. (2011) Transcriptome analysis of high-temperature stress in developing barley caryopses: early stress responses and effects on storage compound biosynthesis. *Mol. Plant.* **4**, 97–115.

Matsuda, F., Okazaki, Y., Oikawa, A., Kusano, M., Nakabayashi, R., Kikuchi, J., Yonemaru, J-I., Ebana, K., Yano, M. and Saito, K. (2012) Dissection of genotype-phenotype associations in rice grains using metabolome quantitative trait loci analysis. *Plant J.* **70**, 624–636.

May, P., Wienkoop, S., Kempa, S., Usadel, B., Christian, N., Rupprecht, J., Weiss, J., Recuenco-Munoz, L., Ebenhoh, O., Weckwerth, W. and Walther, D. (2008) Metabolomics- and proteomics-assisted genome annotation and analysis of the draft metabolic network of *Chlamydomonas reinhardtii*. *Genetics*, **179**, 157–166.

McDonnell, E.M., Coughlan, S.J. and Jones, R.G.W. (1983) Differential-effects of abscisic-acid on glycinebetaine and proline accumulation in 3 plant-species. *Z. Pflanzenphysiol.* **109**, 207–213.

Mensack, M.M., Fitzgerald, V.K., Ryan, E.P., Lewis, M.R., Thompson, H.J. and Brick, M.A. (2010) Evaluation of diversity among common beans (*Phaseolus vulgaris* L.) from two centers of domestication using 'omics' technologies. *BMC Genomics*, **11**, 1–11.

Meyer, R.C., Steinfath, M., Lisec, J., Becher, M., Witucka-Wall, H., Torjek, O., Fiehn, O., Eckardt, A., Willmitzer, L., Selbig, J. and Altmann, T. (2007) The metabolic signature related to high plant growth rate in *Arabidopsis thaliana*. *Proc. Natl Acad. Sci. USA*, **104**, 4759–4764.

Mochida, K., Furuta, T., Ebana, K., Shinozaki, K. and Kikuchi, J. (2009) Correlation exploration of metabolic and genomic diversity in rice. *BMC Genomics*, **10**, 568. doi: 10.1186/1471-2164-10-568.

del Moral, L.F.G., del Moral, M.B.G., Molina-Cano, J.L. and Slafer, G.A. (2003) Yield stability and development in two- and six-rowed winter barleys under Mediterranean conditions. *Field Crop Res.* **81**, 109–119.

Munoz-Amatriain, M., Xiong, Y.W., Schmitt, M.R., Bilgic, H., Budde, A.D., Chao, S.A.M., Smith, K.P. and Muehlbauer, G.J. (2010) Transcriptome analysis of a barley breeding program examines gene expression diversity and reveals target genes for malting quality improvement. *BMC Genomics*, **11**, 653. doi: 10.1186/1471-2164-11-653.

Munoz-Amatriain, M., Moscou, M.J., Bhat, P.R., Svensson, J.T., Bartos, J., Suchankova, P., Simkova, H., Endo, T.R., Fenton, R.D., Lonardi, S., Castillo, A.M., Chao, S.M., Cistue, L., Cuesta-Marcos, A., Forrest, K.L., Hayden, M.J., Hayes, P.M., Horsley, R.D., Makoto, K., Moody, D., Sato, K., Valles, M.P., Wulff, B.B.H., Muehlbauer, G.J., Dolezel, J. and Close, T.J. (2011) An improved consensus linkage map of barley based on flow-sorted chromosomes and single nucleotide polymorphism markers. *Plant Genome*, **4**, 238–249.

Nakamura, T., Nomura, M., Mori, H., Jagendorf, A.T., Ueda, A. and Takabe, T. (2001) An isozyme of betaine aldehyde dehydrogenase in barley. *Plant Cell Physiol.* **42**, 1088–1092.

Nayyar, H. and Walia, D.P. (2004) Genotypic variation in wheat in response to water stress and abscisic acid-induced accumulation of osmolytes in developing grains. *J. Agron. Crop Sci.* **190**, 39–45.

R Core Team (2013) *R: A language and environment for statistical computing* Vienna, Austria: R Foundation for Statistical Computing. Available at: http://www.R-project.org/.

Ramsay, L., Comadran, J., Druka, A., Marshall, D.F., Thomas, W.T.B., Macaulay, M., MacKenzie, K., Simpson, C., Fuller, J., Bonar, N., Hayes, P.M., Lundqvist, U., Franckowiak, J.D., Close, T.J., Muehlbauer, G.J. and Waugh, R. (2011) INTERMEDIUM-C, a modifier of lateral spikelet fertility in barley, is an ortholog of the maize domestication gene TEOSINTE BRANCHED 1. *Nat. Genet.* **43**, 169–172.

Riedelsheimer, C., Czedik-Eysenberg, A., Grieder, C., Lisec, J., Technow, F., Sulpice, R., Altmann, T., Stitt, M., Willmitzer, L. and Melchinger, A.E. (2012) Genomic and metabolic prediction of complex heterotic traits in hybrid maize. *Nat. Genet.* **44**, 217–220.

Roessner, U., Patterson, J.H., Forbes, M.G., Fincher, G.B., Langridge, P. and Bacic, A. (2006) An investigation of boron toxicity in barley using metabolomics. *Plant Physiol.* **142**, 1087–1101.

von Ropenack, E., Parr, A. and Schulze-Lefert, P. (1998) Structural analyses and dynamics of soluble and cell wall-bound phenolics in a broad spectrum resistance to the powdery mildew fungus in barley. *J. Biol. Chem.* **273**, 9013–9022.

Sakamoto, A. and Murata, N. (2002) The role of glycine betaine in the protection of plants from stress: clues from transgenic plants. *Plant, Cell Environ.* **25**, 163–171.

Saneoka, H., Ishiguro, S. and Moghaieb, R.E.A. (2001) Effect of salinity and abscisic acid on accumulation of glycinebetaine and betaine aldehyde dehydrogenase mRNA in Sorghum leaves (*Sorghum bicolor*). *J. Plant Physiol.* **158**, 853–859.

Saulnier, L., Peneau, N. and Thibault, J.F. (1995) Variability in grain extract viscosity and water-soluble arabinoxylan content in wheat. *J. Cereal Sci.* **22**, 259–264.

Schauer, N., Semel, Y., Roessner, U., Gur, A., Balbo, I., Carrari, F., Pleban, T., Perez-Melis, A., Bruedigam, C., Kopka, J., Willmitzer, L., Zamir, D. and Fernie, A.R. (2006) Comprehensive metabolic profiling and phenotyping of interspecific introgression lines for tomato improvement. *Nat. Biotechnol.* **24**, 447–454.

Shen, M., Broeckling, C.D., Yiyi Chu, E., Ziegler, G., Baxter, I.R., Prenni, J.E. and Hoekenga, O.A. (2013) Leveraging non-targeted metabolite profiling via statistical genomics. *PLoS ONE*, **8**, e57667.

Shewry, P.R. (2003) Tuber storage proteins. *Ann. Bot.* **91**, 755–769.

Shewry, P.R., Piironen, V., Lampi, A.M., Edelmann, M., Kariluoto, S., Nurmi, T., Fernandez-Orozco, R., Ravel, C., Charmet, G., Andersson, A.A.M., Aman, P., Boros, D., Gebruers, K., Dornez, E., Courtin, C.M., Delcour, J.A., Rakszegi, M., Bedo, Z. and Ward, J.L. (2010) The HEALTHGRAIN wheat diversity screen: effects of genotype and environment on phytochemicals and dietary fiber components. *J. Agric. Food Chem.* **58**, 9291–9298.

Sicher, R.C., Timlin, D. and Bailey, B. (2012) Responses of growth and primary metabolism of water-stressed barley roots to rehydration. *J. Plant Physiol.* **169**, 686–695.

Skogerson, K., Harrigan, G.G., Reynolds, T.L., Halls, S.C., Ruebelt, M., Iandolino, A., Pandravada, A., Glenn, K.C. and Fiehn, O. (2010) Impact of genetics and environment on the metabolite composition of maize grain. *J. Agric. Food Chem.* **58**, 3600–3610.

Smith, C.A., Want, E.J., O'Maille, G., Abagyan, R. and Siuzdak, G. (2006) XCMS: processing mass spectrometry data for metabolite profiling using nonlinear peak alignment, matching, and identification. *Anal. Chem.* **78**, 779–787.

Steinfath, M., Strehmel, N., Peters, R., Schauer, N., Groth, D., Hummel, J., Steup, M., Selbig, J., Kopka, J., Geigenberger, P. and Van Dongen, J.T. (2010) Discovering plant metabolic biomarkers for phenotype prediction using an untargeted approach. *Plant Biotechnol. J.* **8**, 900–911.

Su, K.M., Bremer, D.J., Jeannotte, R., Welti, R. and Yang, C. (2009) Membrane lipid composition and heat tolerance in cool-season turfgrasses, including a hybrid bluegrass. *J. Am. Soc. Hortic. Sci.* **134**, 511–520.

Sumner, L., Amberg, A., Barrett, D., Beale, M., Beger, R., Daykin, C., Fan, T., Fiehn, O., Goodacre, R., Griffin, J., Hankemeier, T., Hardy, N., Harnly, J., Higashi, R., Kopka, J., Lane, A., Lindon, J., Marriott, P., Nicholls, A., Reily, M., Thaden, J. and Viant, M. (2007) Proposed minimum reporting standards for chemical analysis. *Metabolomics*, **3**, 211–221.

Tamura, K., Peterson, D., Peterson, N., Stecher, G., Nei, M. and Kumar, S. (2011) MEGA5: molecular evolutionary genetics analysis using maximum likelihood, evolutionary distance, and maximum parsimony methods. *Mol. Biol. Evol.* **28**, 2731–2739.

Trygg, J. (2002) O2-PLS for qualitative and quantitative analysis in multivariate calibration. *J. Chemometr.* **16**, 283–293.

Widodo, J.H.P., Patterson, J.H., Newbigin, E., Tester, M., Bacic, A. and Roessner, U. (2009) Metabolic responses to salt stress of barley (*Hordeum vulgare* L.) cultivars, Sahara and Clipper, which differ in salinity tolerance. *J. Exp. Bot.* **60**, 4089–4103.

Wishart, D.S. (2011) Advances in metabolite identification. *Bioanalysis*, **3**, 1769–1782.

Yang, W.Y., Zheng, Y., Bahn, S.C., Pan, X.Q., Li, M.Y., Vu, H.S., Roth, M.R., Scheu, B., Welti, R., Hong, Y.Y. and Wang, X.M. (2012) The patatin-containing phospholipase A pPLAII alpha modulates oxylipin formation and water loss in *Arabidopsis thaliana*. *Mol. Plant.* **5**, 452–460.

14

Genetic enhancement of palmitic acid accumulation in cotton seed oil through RNAi down-regulation of *ghKAS2* encoding β-ketoacyl-ACP synthase II (KASII)

Qing Liu[1]*, Man Wu[1,2], Baolong Zhang[1,3], Pushkar Shrestha[1], James Petrie[1], Allan G. Green[1] and Surinder P. Singh[1]

[1]*CSIRO Agriculture & Food, Canberra, ACT, Australia*

[2]*State Key Laboratory of Cotton Biology, Cotton Research Institute, Chinese Academy of Agricultural Sciences, Anyang, China*

[3]*Jiangsu Provincial Key Laboratory of Agrobiology, Jiangsu Academy of Agricultural Sciences, Nanjing, China*

*Correspondence
email qing.liu@csiro.au

Summary

Palmitic acid (C16:0) already makes up approximately 25% of the total fatty acids in the conventional cotton seed oil. However, further enhancements in palmitic acid content at the expense of the predominant unsaturated fatty acids would provide increased oxidative stability of cotton seed oil and also impart the high melting point required for making margarine, shortening and confectionary products free of *trans* fatty acids. Seed-specific RNAi-mediated down-regulation of β-ketoacyl-ACP synthase II (KASII) catalysing the elongation of palmitoyl-ACP to stearoyl-ACP has succeeded in dramatically increasing the C16 fatty acid content of cotton seed oil to well beyond its natural limits, reaching up to 65% of total fatty acids. The elevated C16 levels were comprised of predominantly palmitic acid (C16:0, 51%) and to a lesser extent palmitoleic acid (C16:1, 11%) and hexadecadienoic acid (C16:2, 3%), and were stably inherited. Despite of the dramatic alteration of fatty acid composition and a slight yet significant reduction in oil content in these high-palmitic (HP) lines, seed germination remained unaffected. Regiochemical analysis of triacylglycerols (TAG) showed that the increased levels of palmitic acid mainly occurred at the outer positions, while C16:1 and C16:2 were predominantly found in the sn-2 position in both TAG and phosphatidylcholine. Crossing the HP line with previously created high-oleic (HO) and high-stearic (HS) genotypes demonstrated that HP and HO traits could be achieved simultaneously; however, elevation of stearic acid was hindered in the presence of high level of palmitic acid.

Keywords: palmitic acid, KASII, cotton seed oil, RNAi, fatty acids, TAG.

Introduction

Cottonseed oil is a highly polyunsaturated vegetable oil because more than half of its total fatty acid content is linoleic acid (C18:2$^{\Delta9,12}$) that is oxidatively unstable and makes it unsuitable for direct food applications (Jones and King, 1993; O'Brien, 2002). Partial hydrogenation that converts much of linoleic acid into monounsaturated and saturated fatty acids has often been used for whole-food applications, such as in providing hard stocks with high melting point for margarine and shortening production. However, *trans* fatty acids are commonly produced as a by-product of the partial hydrogenation process and have been increasingly recognised to have significant cholesterol-raising properties and increase the risk of cardiovascular disease based on evidence derived from epidemiologic and clinical studies (Mozaffarian *et al.*, 2006; Oomen *et al.*, 2001; de Souza *et al.*, 2015). A recent systematic review and meta-analysis of observational studies in the past 10 years, including 41 studies of the association between saturated fat intake and health outcomes, covering more than 300 000 people, and 20 studies of *trans* fat intake and health outcomes that covered more than 200 000 people, have found positive associations between consumption of *trans* fat and total coronary heart disease (CHD) and fatal CHD, but not between consumption of saturated fat and CHD, cardiovascular disease (CVD), stroke and type 2 diabetes (de Souza *et al.*, 2015). Although further research on saturated fatty acids is required, there is no doubt that the decades trading saturated fats for *trans* fat have had an enormous influence on

the rising incidences of heart disease and many of the so-called metabolic syndromes. Alternative cotton seed oils with similar functionality to partially hydrogenated oil could therefore be nutritionally desirable.

Cotton seed oil rich in oxidatively stable fatty acids, such as oleic acid (C18:1$^{\Delta9}$), stearic acid (C18:0) and palmitic acid (C16:0), would meet this purpose. Palmitic acid makes up approximately 25% of the total fatty acids in the conventional cotton seed oil, and its further enhancement is anticipated to not only increase the oxidative stability of cotton seed oil by offsetting the instability of linoleic acid but also impart the high melting point required for making such products as margarine, shortening and confectionary products free of *trans* fatty acids (Neff and List, 1999; Nzikou *et al.*, 2009).

Tropical oils such as palm oil contain about 50% of palmitic acid, which is increasingly being used as an alternative to partially hydrogenated vegetable oils in baking and processed food applications (Hayes and Pronczuk, 2010; L'Abbe *et al.*, 2009). The mid-fraction of a palm oil (PMF) rich in symmetrical palmitate triglycerides (POP) has also been most commonly used to formulate cocoa butter substitute by interesterification with oils derived from stearate-rich tropical oils (L'Abbe *et al.*, 2009; Zaliha *et al.*, 2004). Oil palm plantings have continued to expand in recent years, in response to rapidly increasing demand for food oil, as well as for biofuel. As the conversion of virgin agricultural land to palm plantation in tropical countries has become an ever more sensitive issue, development of cotton seed oil with similar fatty acid composition and functionality to palm oil may offer an attractive alternative because of its continuous availability that is driven by the continuing demand for cotton fibre.

In higher plants, the plastid is the major site of *de novo* fatty acid biosynthesis pathway, generating fatty acyl precursors for fatty acids of different chain lengths and saturation levels. The *de novo* fatty acid biosynthesis is performed by a complex of soluble proteins known as fatty acid synthases (FAS), with the β-ketoacyl-ACP synthase (KAS) enzyme family catalysing the elongation of malonyl-acyl carrier protein (malonyl-ACP) by reiteratively adding C2 units to a growing fatty acyl chain through Claisen condensation (Ohlrogge and Jaworski, 1997). KASIII catalyses the condensation of C2-CoA to C4, while KASI prefers C4- to C14-ACP substrates leading to the production of palmitoyl-ACP that is then elongated to stearoyl-ACP by KASII and subsequently desaturated to form oleoyl-ACP by Δ9 stearoyl-ACP desaturase (SAD). The saturated fatty acids, mostly palmitic acid in cotton seed, can be cleaved from palmitoyl-ACP by the action of the palmitoyl-ACP thioesterase (FatB), allowing for transportation of free palmitic acid into cytoplasm where it becomes available for further desaturation and triacylglycerol (TAG) assembly.

Because palmitoyl-ACP is the substrate for two major activities, KASII and FatB, it represents a key branch point in fatty acid biosynthesis (Cahoon and Shanklin, 2000). Most fatty acids in the seed oil pass through a C16 form during biosynthesis, and the content of palmitic acid remaining in the final cotton seed oil is therefore determined, to a large extent, by the competing action of FatB and KASII (Cahoon and Shanklin, 2000; Martz *et al.*, 2006). These enzymes therefore represent possible targets for genetic manipulation of palmitic acid levels in cotton seed oil (Aslan *et al.*, 2015; Pidkowich *et al.*, 2007; Sun *et al.*, 2014).

Genetic engineering of *FatB* has previously been successful in raising or lowering palmitic acid levels in seeds of other plants. Overexpression of the Arabidopsis *FatB1* in seeds resulted in a nearly four fold increase in seed palmitic acid content (Dormann *et al.*, 2000). Transgenic expression of a *FatB* gene derived from *Cuphea hookerinana* resulted in raised palmitic acid from 6% to 34% in rapeseed oil (Dehesh *et al.*, 1996). An even more effective approach for raising palmitic acid was demonstrated by RNAi down-regulation of *KAS2* that encodes KASII in Arabidopsis where palmitic acid level was raised to as high as 53% beyond which level the seeds were aborted (Pidkowich *et al.*, 2007). It is

assumed that the significantly lowered KASII activity resulting from the attenuated *KAS2* transcription reduces the flow of metabolites from palmitoyl-ACP to stearoyl-ACP, causing a significant accumulation of the palmitoyl-ACP and consequently enriched palmitic acid content in TAG.

We have employed a similar approach to raise palmitic acid in cotton seed oil. In this report, we describe the characterisation of two different *KAS2* genes from developing cotton embryos and their RNAi down-regulation that led to substantial enhancement of palmitic acid accumulation. Attempts have also been made to combine the high-palmitic (HP) trait by crossing with either or both high-oleic (HO) and high-stearic (HS) traits that have been previously generated using RNAi-mediated gene down-regulation of *ghFAD2-1* or *ghSAD-1*, respectively (Liu *et al.*, 2002).

Results

Isolation of two different *KAS2* cDNAs from developing cotton seed

From a cDNA library derived from developing embryos of upland cotton (*Gossypium hirsutum* L.) cv DP-16, two distinct cDNAs were isolated and designated as *ghKAS2-1* (GenBank accession No. KF611921) and *ghKAS2-2* (GenBank accession No. KF611922). Full length sequences of these cDNAs were completed by 5' race using RNAs derived from developing embryos as templates. The complete *ghKAS2-1* is 2268 bp long, encoding 577 amino acids. The *ghKAS2-2* cDNA is 2192 bp long, encoding 526 amino acids. The putative polypeptides encoded by the coding region of *ghKAS2-1* and *ghKAS2-2* showed 78.7% identity to each other. The phylogenetic relationship of these two genes and their orthologs in a selected number of plant species is illustrated in Fig. 1. It appears that *ghKAS2-1* and *ghKAS2-2* clustered together with their ortholog derived from *Theobroma cacao* that is taxonomically closely related to cotton. By searching the *G. raimondii* genome database, *ghKAS2-1* was mapped onto chromosome 13 and its coding region was interrupted by 12 introns and the 13th intron was found in the 3' UTR. *ghKAS2-2* was mapped onto chromosome 4, and its coding region was interrupted by 12 introns. The position and length of introns were relatively conserved between *ghKAS2-1* and *ghKAS2-2*. It is noteworthy that about half of the *ghKAS2-1*

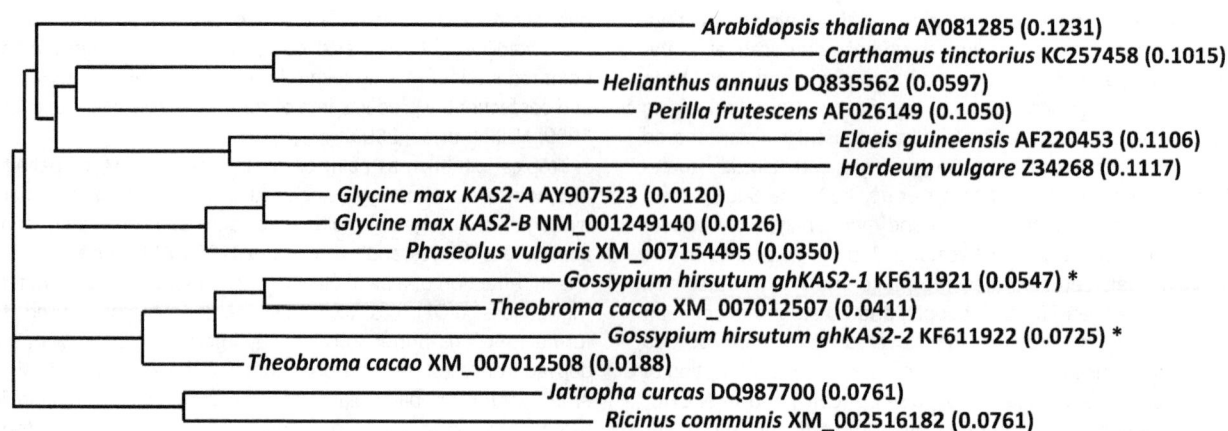

Figure 1 Phylogenetic comparison of cotton *ghKAS2* gene family and orthologous *KAS2s* from some selected plant species. The phylogenetic tree was constructed from multiple sequence alignment of KASII protein from the GenBank generated by the AlignX module of Vector NTI suite using neighbour-joining clustering algorithm in Clustal X2 program. The GenBank accession numbers and the sources of *KAS2* genes and bootstrap values of the clusters are indicated in phylogram. The sequences of *ghKASII* from this study are marked with an asterisk.

cDNA clones isolated from the cotton developing embryo cDNA library contained intron 12 that was not properly spliced. It is not known whether this constitutes one level of post-transcriptional regulation and is responsible for the higher level of palmitic acid in cotton seed relative to many other temperate oilseed crops.

RNAi construct designing and selection of primary transgenic cotton plants

As detailed in the Experimental procedures, a chimeric DNA sequence consisting of an inverted repeat structure of *ghKAS2-1* was used as the trigger DNA sequence in an RNAi cassette. Instead of a typical intron sequence, the upstream adjacent sequence of the 400-bp sense arm of the inverted repeat was used as a spacer separating the head to head inverted repeat. Although the trigger sequence was designed based on the *ghKAS2-1*, it was chosen as the most conserved region of the gene. The trigger sequence has 87.5% homology with *ghKAS2-2*, with several stretches of continuous homologous regions longer than 21 bp, enabling it to achieve a level of cross-silencing of *ghKAS2-2*. A seed-specific promoter derived from soya bean lectin gene *LEC1* (Cho *et al.*, 1995) was used to drive the transgene. Six independent cotton lines transformed with the *RNAi-ghKAS2-1* construct were regenerated and allowed to grow to maturity. Four of the six lines were male sterile, a much higher frequency than seen in other typical cotton transformations in our laboratory, suggesting a possible impact of severe *KAS2* down-regulation rather than a side effect of tissue culture. The other two fertile lines, KIR-1 and KIR-10, were carried through for further molecular and biochemical analyses. PCR was first performed to confirm the transgene presence, followed by Southern blot analysis with multiple restriction enzymes on the T_1 primary transgenic plants using the promoter region of the transgene as a probe. Single restriction bands were observed in both KIR-1 and KIR-10, indicating single transgene insertions (data not shown). Under glasshouse conditions, neither of these two lines exhibited any growth abnormalities nor was there any apparent penalty on plant vigour or seed yield.

Analysis of fatty acid composition was carried out along with the selection and establishment of homozygous transgenic plants. In comparison with wild type (WT) and null segregants, fatty acid profiles of the transgenic seeds derived from KIR-1 and KIR-10 have altered significantly, featuring enhanced C16 fatty acids with concomitant reduction in C18 fatty acids. Fig. 2 shows the content of palmitic acid and C16 unsaturated fatty acids in the cotyledons of 65 and 51 randomly selected individual T_2 seeds in KIR-1 and KIR-10, respectively. Among the T_2 seeds, the segregation of null seeds is evident as they have similar fatty acid composition to WT plants, while all other seeds showed raised palmitic acid at various levels, with the majority exceeding 50% of total fatty acids in both transgenic lines. The ratios of transgenic seeds to null segregants for KIR-1 (54 : 11) and KIR-10 (49 : 9) fitted a 3 : 1 ratio expected for the segregation of a single dominant gene ($x^2 = 0.19$ and 0.10, respectively), consistent with the molecular assessment of single copy transgenes as confirmed by Southern blot analysis. In KIR-1, a single seed with the highest accumulation of total C16 fatty acids reached 85% of total fatty acids. Palmitoleic acid was raised to as high as 11% of total fatty acids. C16:2 that is normally undetectable in cotton seed accounted for up to 3.3% of total fatty acids. Compared to KIR-1, the extent of increase in palmitic acid and unsaturated C16 fatty acid levels in KIR-10 seeds is relatively consistent among the transgenic T_2 seeds. Most of the transgenic seeds showing

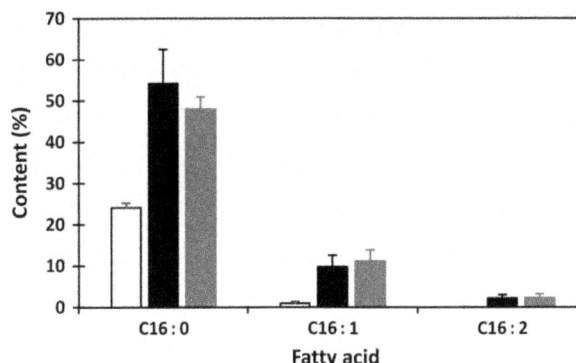

Figure 2 Accumulation of C16 fatty acids in the primary transgenic plants of transgenic lines KIR-1 and KIR-10. WT (open box); KIR-1 (closed box) and KIR-10 (shaded box).

palmitic acid level between 45% and 50%. Similar to KIR-1, the sum of palmitoleic acid and C16:2 was also increased to about 15% of total fatty acids in individual KIR-10 seeds. As the result, the contents of C18 fatty acids were all reduced, except for *cis*-vaccenic acid (C18:1$^{\Delta 11}$) that is the elongation product of palmitoleic acid. There is an approximate doubling of *cis*-vaccenic acid content in both KIR-1 and KIR-10 lines compared to untransformed control.

The T_2 seeds with significantly raised palmitic acid were selected and germinated for plant establishment and assessment of the inheritance stability of the gained HP trait in the progenies of both KIR-1 and KIR-10. Fifteen individual mature T_3 seeds borne on each of the established T_2 plants were subjected to analysis of fatty acid composition. There were clear segregation of null seeds with WT-like fatty acid profile among the T_3 seeds, and the homozygous HP lines were able to be established at T_4 generation for both KIR-1 and KIR-10.

The fatty acid profiles of T_2 (null excluded), T_3 and T_4 seeds were tabulated in Table 1. In all the three generations, the increased contents of palmitic acid and its derivative C16 unsaturated fatty acids were clearly a prominent feature and consistent among generations. The increase of C16 fatty acids was at the expense of all three major C18 fatty acids, including stearic, oleic and linoleic acids. In KIR-1, distinct from the high variability in primary transgenic seeds, the palmitic acid accumulation in T_3–T_4 generations was more consistent among individual seeds without significant variation.

Oil content analysis and seed germination test

Assessment of oil content was carried out in mature homozygous seeds harvested from the T_4 and T_5 HP lines and WT plants that were grown alongside each other under controlled greenhouse condition. Across the two generations, there was a slight yet significant decline of oil content in both HP lines ($P < 0.05$), while there is no significant variation between KIR-1 and KIR-10 (Table 2). Two-way analysis of variance (ANOVA) revealed significant variation in oil content neither between generations nor interaction between genotype and generation (Table S1).

Germination response to temperature of homozygous T_4 seeds of KIR-1 and KIR-10 lines was evaluated at cool (18 °C) and warm (28 °C) temperature regimes. All the seeds derived from both WT and HP lines showed significantly higher germination rate at 28 °C compared to 18 °C, but there was no significant variation among cotton lines ($P < 0.05$; Table 2; Table S2).

Table 1 Fatty acid composition of transgenic cotton expressing RNAi-ghKAS2-1 in three successive progeny generations

Generation	Genotype	C14:0	C16:0	C16:1	C16:2	C18:0	C18:1$^{\Delta 9}$	18:1$^{\Delta 11}$	C18:2
T$_2$	Null	0.2 ± 0.1	24.1 ± 1.1	0.9 ± 0.4	0.0 ± 0.0	2.1 ± 0.6	13.3 ± 2.9	0.8 ± 0.6	58.1 ± 3.1
	KIR-1	0.2 ± 0.1	54.1 ± 8.4	9.7 ± 2.7	2.2 ± 0.8	1.2 ± 0.5	3.5 ± 2.0	1.5 ± 0.4	26.8 ± 7.2
	KIR-10	0.2 ± 0.1	48.0 ± 2.9	11.2 ± 2.7	2.2 ± 0.9	1.0 ± 0.1	3.9 ± 1.4	1.7 ± 0.2	31.3 ± 4.8
T$_3$	KIR-1	0.2 ± 0.1	51.0 ± 2.0	11.2 ± 2.1	2.8 ± 0.5	1.0 ± 0.0	2.3 ± 0.1	1.5 ± 0.0	29.2 ± 3.4
	KIR-10	0.3 ± 0.1	49.5 ± 1.2	10.7 ± 2.3	2.1 ± 0.7	1.0 ± 0.1	3.7 ± 1.3	1.6 ± 0.1	30.4 ± 2.1
T$_4$	KIR-1	0.2 ± 0.1	50.2 ± 1.7	10.9 ± 2.0	2.2 ± 0.7	1.1 ± 0.1	4.7 ± 1.8	1.8 ± 0.3	28.4 ± 2.8
	KIR-10	0.2 ± 0.1	48.9 ± 1.8	11.8 ± 2.2	2.61 ± 0.7	1.1 ± 0.1	4.0 ± 1.6	1.7 ± 0.2	29.0 ± 2.5

Mean ± Std ($n = 3$).

Table 2 Oil content analysis of T$_4$ and T$_5$ cotton seeds and the germination rate of T$_4$ cotton seeds at two different temperatures (18 °C and 28 °C)

Genotype	Oil content (%)		Germination rate (%)	
	T$_4$	T$_5$	18 °C	28 °C
Coker 315	22.8 ± 0.7$_a$	23.6 ± 0.5$_a$	81.7 ± 4.3$_a$	97.5 ± 3.2$_b$
KIR-1	21.3 ± 1.1$_b$	22.0 ± 0.1$_b$	80.0 ± 4.7$_a$	96.7 ± 2.7$_b$
KIR-10	21.3 ± 0.4$_b$	21.5 ± 0.4$_b$	83.3 ± 6.1$_a$	95.8 ± 3.2$_b$

Mean ± Std (oil content analysis, $n = 3$; germination test, $n = 4$). Columns with different letters represent significant differences (LSD) at $p < 0.05$.

Real-time RT-PCR gene expression

To confirm the molecular basis of the alteration in fatty acid composition in KIR-1 and KIR-10, real-time quantitative PCR (RT-qPCR) was performed to evaluate the transcription level of the target *KAS2* gene using template RNAs derived from mid-maturity developing embryos as previous studies have established that this is the most active period of fatty acid biosynthesis and oil accumulation (Liu *et al.*, 1999). As shown in Fig. 3, the reduction in the expression of both *ghKAS2-1* and *ghKAS2-2* genes in the T$_4$ homozygous transgenic lines was evident compared to WT control. It is also noteworthy that expression of *ghKAS2-1* appeared to be lower than *ghKAS2-2* in developing cotton seeds. This result is consistent with our observation that cDNA library screening yields 3 times more clones of *ghKAS2-2* than *ghKAS2-1*. Taken together, we may assume that these two genes do not have equal contributions to the conversion of palmitic acid to stearic acid. Although the *ghKAS2-1* sequence alone was used as the RNAi trigger sequence, the expression of both *ghKAS2-1* and *ghKAS2-2* has been effectively attenuated.

Lipid analysis of HP cotton seed oil

Palmitic acid content in the total lipids extracted from mature whole cotton seeds of KIR-1 and KIR-10 was measured as 52.6% and 50.1%, respectively, compared to 22.2% in WT (Fig. 4). In the TAG fraction, palmitic acid content in KIR-1 and KIR-10 seeds was 52.8% and 50.3% of total fatty acids, respectively. Similarly, palmitic acid content in the polar lipid (PL) fraction of KIR-1 and KIR-10 seed oil was 34.7% and 35.3%, respectively, compared to 21.6% in WT; palmitic acid content in the phosphatidylcholine (PC) fraction of KIR-1 and KIR-10 was 30.5% and 29.7%, respectively, in contrast to 19.5% in WT (Fig. 4). The increase in unsaturated C16 fatty acids was consistent among TAG, total PL and PC. Linoleic acid decreased with a concomitant increase in

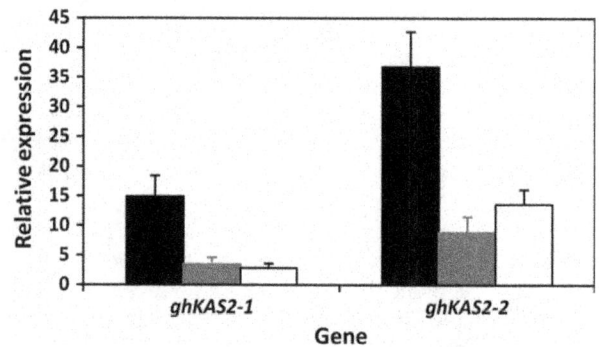

Figure 3 Real-time RT-qPCR analysis of *ghKAS2-1* and *ghKAS2-2* in immature cotton embryos at mid-developmental stage (30 DAA). Coker315 (closed box); KIR-1 (shaded box) and KIR-10 (open box).

palmitic acid and other C16 fatty acids, more evident in TAG compared to total PL and PC. The relatively smaller alteration of fatty acid composition in total PL and PC might explain why there was little impact on seed germination and plant growth by RNAi down-regulation of *KAS2* expression in cotton seed.

Positional analysis of palmitic acid

To determine the regiospecificity of palmitic acid and its two C16 derivative fatty acids, palmitoleic acid and C16:2, in KIR-1 and KIR-10, the TAG fraction was isolated from the homozygous T$_4$ seeds and digested with appropriate lipase that preferentially cleaves fatty acid from the outer positions (*sn*-1 and *sn*-3), releasing two free fatty acids and a 2-monoacylglycerol (2-MAG) molecule retaining the *sn*-2 acyl chain. In KIR-1 and KIR-10, consistent with WT, the outer positions of TAG were predominantly occupied by palmitic acid, while the *sn*-2 position was mainly occupied by unsaturated fatty acids, mostly linoleic acid. Palmitoleic acid and C16:2 were found in both *sn*-2 and outer positions, with some degree of preference for *sn*-2 (Fig. 5). This was consistent with the general observation that, in plant seed oil, saturated fatty acids preferentially occupy the outer positions in TAG, whereas unsaturated fatty acids are predominantly found in *sn*-2 position.

For both HP lines, we also analysed the positional distribution of fatty acids on PC, an important intermediate in TAG synthesis and membrane constituents. The isolated PC fraction was digested with phospholipase A2, an enzyme that preferentially cleaves fatty acids from the *sn*-2 of PC, releasing a free fatty acid and a lyso-PC molecule retaining the *sn*-1 acyl chain. This positional analysis showed that consistent with WT, the *sn*-1 position was predominantly occupied by palmitic acid, while *sn*-2

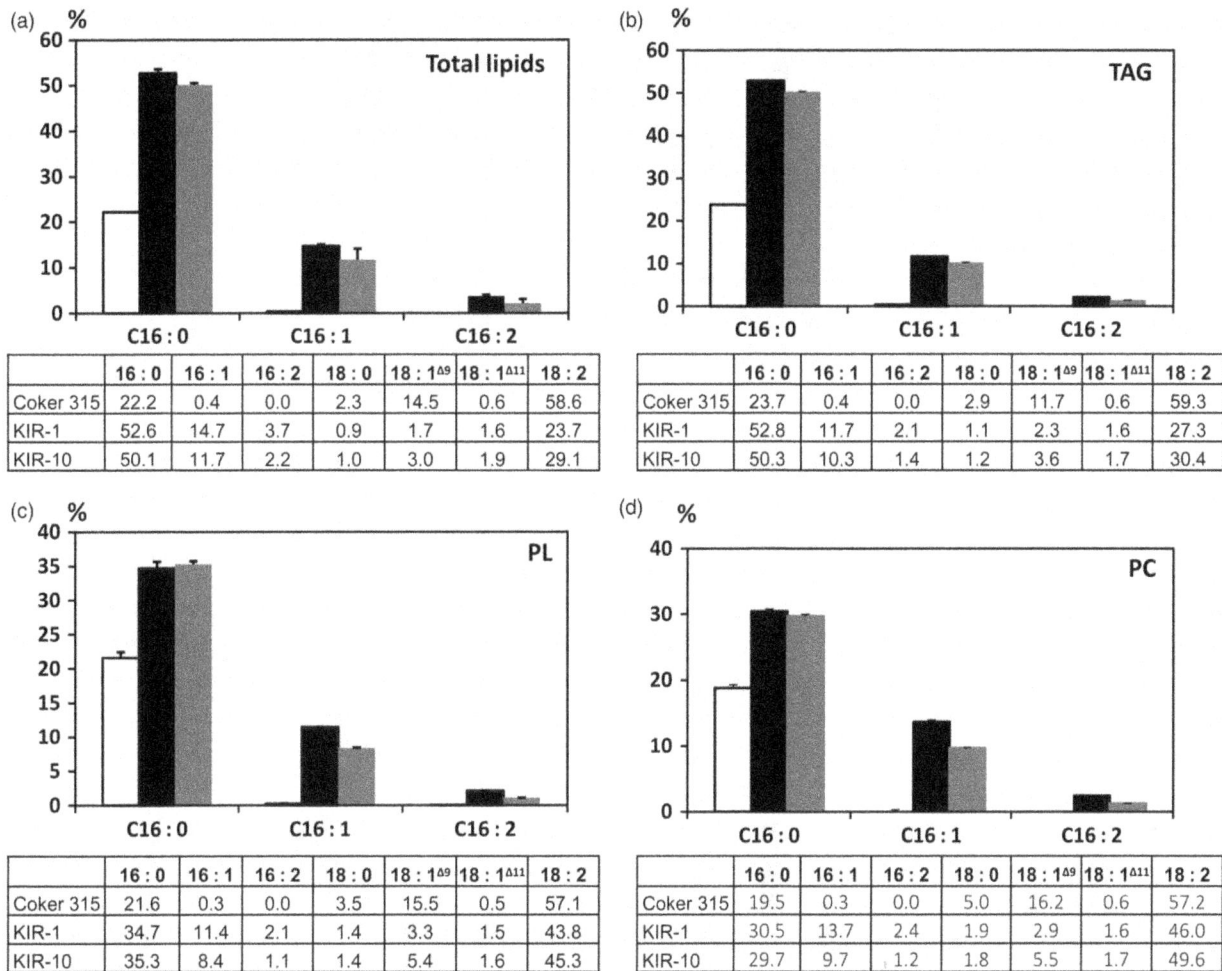

(a) Total lipids

	16:0	16:1	16:2	18:0	18:1$^{\Delta 9}$	18:1$^{\Delta 11}$	18:2
Coker 315	22.2	0.4	0.0	2.3	14.5	0.6	58.6
KIR-1	52.6	14.7	3.7	0.9	1.7	1.6	23.7
KIR-10	50.1	11.7	2.2	1.0	3.0	1.9	29.1

(b) TAG

	16:0	16:1	16:2	18:0	18:1$^{\Delta 9}$	18:1$^{\Delta 11}$	18:2
Coker 315	23.7	0.4	0.0	2.9	11.7	0.6	59.3
KIR-1	52.8	11.7	2.1	1.1	2.3	1.6	27.3
KIR-10	50.3	10.3	1.4	1.2	3.6	1.7	30.4

(c) PL

	16:0	16:1	16:2	18:0	18:1$^{\Delta 9}$	18:1$^{\Delta 11}$	18:2
Coker 315	21.6	0.3	0.0	3.5	15.5	0.5	57.1
KIR-1	34.7	11.4	2.1	1.4	3.3	1.5	43.8
KIR-10	35.3	8.4	1.1	1.4	5.4	1.6	45.3

(d) PC

	16:0	16:1	16:2	18:0	18:1$^{\Delta 9}$	18:1$^{\Delta 11}$	18:2
Coker 315	19.5	0.3	0.0	5.0	16.2	0.6	57.2
KIR-1	30.5	13.7	2.4	1.9	2.9	1.6	46.0
KIR-10	29.7	9.7	1.2	1.8	5.5	1.7	49.6

Figure 4 Fatty acid composition (%) of total lipids (a), TAG (b), PL (c) and PC (d). HP cotton seed oil derived from KIR-1 (closed box) and KIR-10 (shaded box) were compared to WT (Open box). Lipids were extracted from homozygous T$_4$ seeds and fractionated by TLC to generate neutral and polar lipid fractions. Data are presented as mean ± SD (n = 3). The full fatty acid composition of cotton seed oil is tabulated below the bar diagram.

position was mainly occupied by unsaturated fatty acids (Fig. 6). Palmitoleic acid and C16:2 were found on both sn-1 and sn-2 positions, with a clear preference for sn-2. It was also clear that palmitic acid in KIR-1 and KIR-10 was mainly found on sn-1 position, while its increase on sn-2 position was relatively minor.

Incorporation of HP trait into HS, HO and HSO cotton genotypes

To incorporate the HP trait into some other genotypes with genetically altered fatty acid profiles, KIR-10 was crossed with previously generated HO, HS lines and their homozygous hybrid (HSO) (Liu et al., 2002). The fatty acid profiles of the total lipids, TAG, PL and PC fractions in the F$_1$ hybrid seeds derived from a cross between KIR-10 and a HS line, HS-35, are shown in Fig. 7. Palmitic acid in the F$_1$ seeds was increased to a similar level as in its HP parent in the total lipids and TAG fraction, but its increase in PL and PC fractions was not as prominent. Correspondingly, the reduction in linoleic acid was also significantly more prominent in total lipids and TAG fraction than in polar lipids. In contrast to the high level accumulation of palmitic acid, the stearic acid level in all the four lipid pools in the HP x HS F$_1$ seeds remains low, about 4%, which is significantly higher than WT, but much lower than its HS-35 parent.

The F$_1$ hybrid seeds derived from a cross between KIR-10 with a HO cotton line, HO-30, showed the simultaneous increase of both palmitic and oleic acids, mainly at the expense of linoleic acid (Fig. 7). The average level of palmitic acid in the total seed lipids of the HP × HO cross was 41.7%, compared to 50% in the HP parent. The level of oleic acid in the HP × HO cross averaged 38.2% of total seed fatty acids, which in absolute terms is only half of the 78% oleic in the HO parent. However, in the context of the reduced pool of C18 fatty acids in the presence of the HP trait, the 38.2% of oleic acid does represent approx. 85% of the total C18 fatty acids, which is approaching the very high preponderance of oleic acid in the C18 fatty acid pool of the HO parent (94%). Reflecting the effect of FAD2 silencing, the linoleic acid content in the HP × HO hybrid appeared to be at a similar level to that in its HO parent, maintaining below 5%, while its HP parent contained 31.2% linoleic acid. Such a profile was consistent in the TAG, total PL and PC fractions. It is also worthy of noting that the ratio of palmitic/oleic acids was substantially lower in total PL and PC fraction, compared to total lipids and TAG fraction. Palmitoleic acid level remained at a similar level among the four lipid pools examined.

The hybrid (HPSO) was obtained by crossing KIR-10 with a homozygous HO/HS cotton plant (HO/HS-9) that was generated

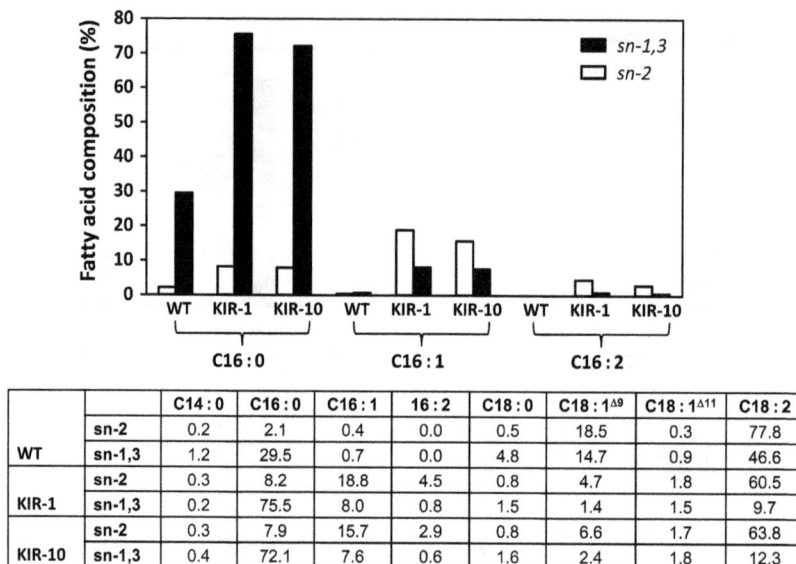

		C14:0	C16:0	C16:1	16:2	C18:0	C18:1^Δ9	C18:1^Δ11	C18:2
WT	sn-2	0.2	2.1	0.4	0.0	0.5	18.5	0.3	77.8
	sn-1,3	1.2	29.5	0.7	0.0	4.8	14.7	0.9	46.6
KIR-1	sn-2	0.3	8.2	18.8	4.5	0.8	4.7	1.8	60.5
	sn-1,3	0.2	75.5	8.0	0.8	1.5	1.4	1.5	9.7
KIR-10	sn-2	0.3	7.9	15.7	2.9	0.8	6.6	1.7	63.8
	sn-1,3	0.4	72.1	7.6	0.6	1.6	2.4	1.8	12.3

Figure 5 *sn* Positional distribution of fatty acids in TAG fraction of cotton seed oil derived from HP cotton lines, KIR-1 and KIR10, compared to the untransformed control Coker 315 (WT). *sn*-1,3 (closed box); *sn*-2 (open box). The full fatty acid composition of the specified *sn* position is tabulated below the bar diagram.

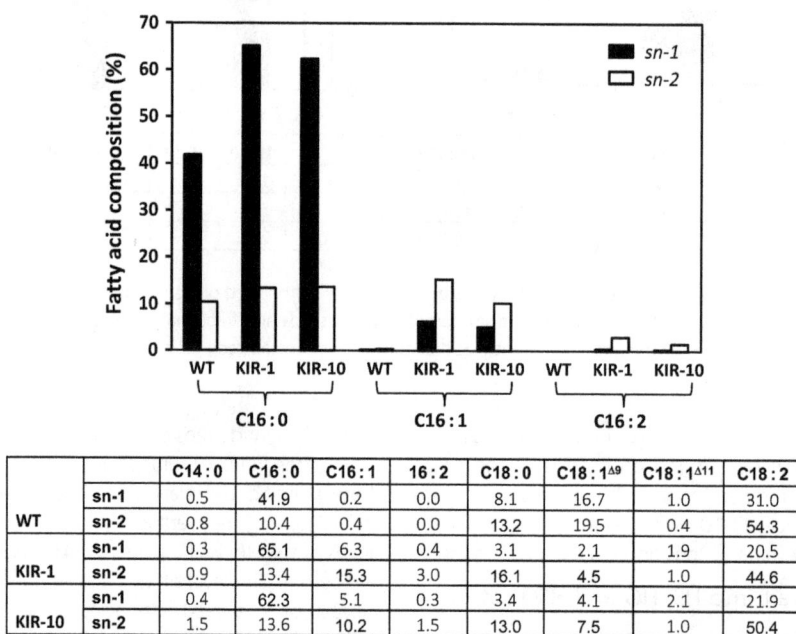

		C14:0	C16:0	C16:1	16:2	C18:0	C18:1^Δ9	C18:1^Δ11	C18:2
WT	sn-1	0.5	41.9	0.2	0.0	8.1	16.7	1.0	31.0
	sn-2	0.8	10.4	0.4	0.0	13.2	19.5	0.4	54.3
KIR-1	sn-1	0.3	65.1	6.3	0.4	3.1	2.1	1.9	20.5
	sn-2	0.9	13.4	15.3	3.0	16.1	4.5	1.0	44.6
KIR-10	sn-1	0.4	62.3	5.1	0.3	3.4	4.1	2.1	21.9
	sn-2	1.5	13.6	10.2	1.5	13.0	7.5	1.0	50.4

Figure 6 *sn* Positional distribution of fatty acids in PC fraction of cotton seed oil derived from HP cotton lines, KIR-1 and KIR10, compared to the untransformed control Coker 315 (WT). *sn*-1 (closed box); *sn*-2 (open box). The full fatty acid composition of the specified *sn* position is tabulated below the bar diagram.

through crossing HO-30 and HS-35. In its total lipids, similar to what have been observed in the HP × HO and HP × HS, the combination of HP and HO traits was evident, but the stearic acid level was only moderately increased, averaging 3.4%, marginally, yet significantly higher than WT (2.8%). Linoleic acid remained low, about 5%, a similar level to the HO and HP × HO genotypes. Such a trend was consistent in the other three lipid pools, including TAG, PL and PC fractions. This hybrid also showed the common feature with HP and HPO genotypes that in PL and PC the ratio of palmitic acid *vs* linoleic

acid was substantially lower than that in total lipids and TAG fraction.

Regiospecific distributions of fatty acids in the hybrids combining HP and other genotypes with altered fatty acids

The regiospecific distribution of fatty acids on TAG and PC molecules in the hybrids incorporating the HP trait into HS, HO and HSO lines was also analysed and shown in Tables 3 and 4, respectively. In the TAG of high saturate lines, that is HP, HS and

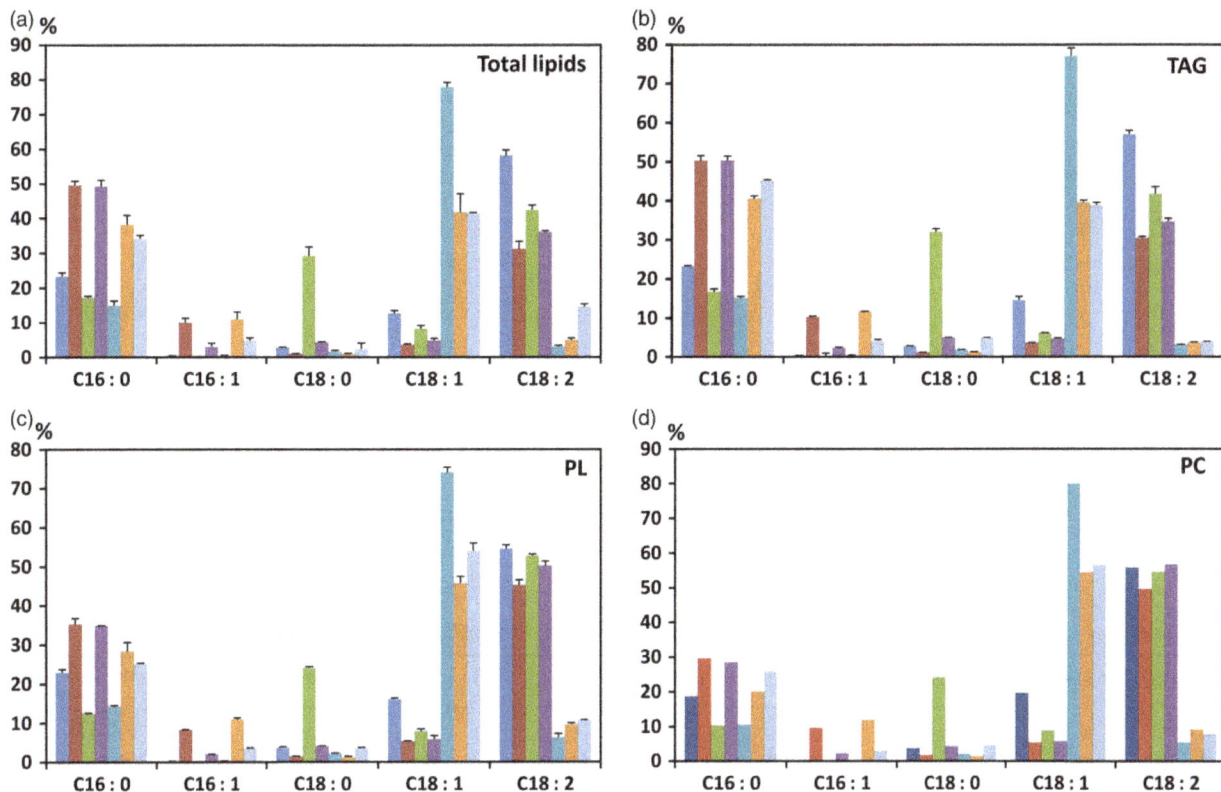

Figure 7 Fatty acid composition of total lipids (a), TAG (b), PL (c) and PC (d) in various cotton seed oil featuring Coker315 (blue), HP (red), HS (green), HP × HS (purple), HO (turquoise), HP × HO (orange), HP × (HS × HO) (grey) genotypes. Total lipids were fractionated by TLC to generate neutral and polar lipids. Data are presented as mean ± SD (n = 3).

HPS, the increases in palmitic acid and stearic acid were mainly concentrated on the *sn*-1 and *sn*-3 position (Table 3). This was consistent with PC except that stearic acid seemed to prefer *sn*-2 in the WT, HP and HO lines where it was present at low levels (Table 4). A general preference of *sn*-2 position by unsaturated fatty acids, including palmitoleic acid, C16:2, oleic acid and linoleic acid, was evident in both TAG (Table 3) and PC. However, oleic acid was found on all three *sn* positions when its accumulation reached very high level in HO-30.

Discussion

Palmitic acid is a common saturated fatty acid widely present in our diet system, and reports claiming and disputing a link between ingestion of palmitic acid and heart disease have been well documented (Chowdhury *et al.*, 2014; Martinez-Ortiz *et al.*, 2006; Narang *et al.*, 2005). Although the overall benefit or risk of HP oils has not been established, it is clear that HP vegetable oil is a better choice than *trans* fats (de Souza *et al.*, 2015). While emphasis has been placed on reducing saturate fatty acids in the main stream vegetable oils, considerable efforts have also been made in raising palmitic acid in numerous vegetable oils because of its essential functional role in structural food lipids and high oxidative stability in *trans*-free fats.

In soya bean, major alleles in at least five loci have been reported to cause an increase in palmitic acid content ranging between 4% and 40%, and the palmitic acid content has been elevated to above 40% of total fatty acids in lines combining HP mutant alleles (Stoltzfus *et al.*, 2000). Some of these HP

genotypes have been found to be associated with defective *KAS2* genes. For instance, J10 has a deletion of entire *gmKASIIA*; M22 has a small deletion in an intron of *GmKASIIB*, which results in mistranslation of approximately 80% of *GmKASIIB* transcripts (Anai *et al.*, 2012).

Arabidopsis mutant *fab1-1*, having approximately double the amount of palmitic acid compared to that of WT plants (James and Dooner, 1990), was identified as a Leu-337Phe substitution in KASII protein that causes instability due to insufficient space for accommodating the imidazole ring of the mutant Phe-337 residual (Carlsson *et al.*, 2002; Wu *et al.*, 1994). Enzyme assay on the extracts from *fab1-1* mutant line revealed a 40% reduction of KASII activity *in vitro*. Seed-specific down-regulation of *atKAS2* driven by *Phaseolin* promoter in Arabidopsis illustrated that up to 53% of palmitic acid could be accumulated in seed oil without obvious negative impact on seed development (Pid-kowich *et al.*, 2007). In this report, we have demonstrated in cotton that it is possible to raise palmitic acid level substantially and mimic an oil palm-like profile by genetically altering the expression level of *ghKAS2* gene in an otherwise HO genetic background.

KASII enzymes are highly conserved between different plant species, and there is a consensus regarding their role as the general housekeeping fatty acid synthases responsible for the production of most C18 fatty acids in plastids. The characterisa-tion of a T-DNA knockout mutant of *KAS2* in Arabidopsis suggested that a basic level of KASII activity is essential for normal seed viability because the homozygote mutant genotype was not able to be recovered (Pidkowich *et al.*, 2007).

Table 3 Positional distribution of fatty acids on TAG in different cotton genotypes containing altered fatty acid composition

Sample	Position	C14:0	C16:0	C16:1	16:2	C18:0	C18:1$^{\Delta9}$	C18:1$^{\Delta11}$	C18:2
Coker 315	sn-2	0.2	2.5	0.4	0.0	0.7	16.9	0.3	78.8
	sn-1 + 3	1.0	36.1	0.5	0.1	5.3	11.7	0.8	42.2
KIR-10	sn-2	0.3	7.9	15.7	2.9	0.8	6.6	1.7	63.8
	sn-1 + 3	0.4	72.1	7.6	0.6	1.6	2.4	1.8	12.3
HS-35	sn-2	0.5	4.9	0.3	0.0	5.5	12.9	0.0	75.3
	sn-1 + 3	0.5	26.3	0.4	0.1	41.6	6.4	0.3	21.3
HPS	sn-2	0.2	6.5	3.9	0.5	0.8	8.7	0.6	78.5
	sn-1 + 3	0.7	72.8	2.4	0.2	7.8	3.9	0.6	9.0
HO-30	sn-2	0.1	1.2	0.4	0.0	0.5	92.6	0.6	4.5
	sn-1 + 3	1.7	19.1	1.8	0.0	4.1	65.2	1.1	5.3
HPO	sn-2	0.1	2.6	17.2	0.0	0.4	69.5	2.2	7.9
	sn-1 + 3	0.3	58.4	8.9	0.0	2.2	24.8	1.8	1.8
HPSO	sn-2	0.1	3.3	6.0	0.0	0.6	80.1	1.0	8.7
	sn-1 + 3	0.5	64.6	3.5	0.0	7.5	18.9	0.5	1.4

HPS = KIR-10 × HS-35; HPO = KIR-10 × HO-30; HPSO = KIR-10 × (HS-35 × HO-30).

Table 4 Positional distribution of fatty acids on PC in different cotton genotypes containing altered fatty acid composition

Sample	Position	C14:0	C16:0	C16:1	C16:2	C18:0	C18:1$^{\Delta9}$	C18:1$^{\Delta11}$	C18:2
Coker 315	sn-1	0.7	48.1	0.1	0.0	10.7	13.1	1.1	25.6
	sn-2	2.0	18.6	0.4	0.0	21.0	14.5	0.3	41.7
KIR-10	sn-1	0.4	62.3	5.1	0.3	3.4	4.1	2.1	21.9
	sn-2	1.5	13.6	10.2	1.5	13.0	7.5	1.0	50.4
HS-35	sn-1	0.3	21.2	0.1	0.0	49.0	5.8	0.2	22.1
	sn-2	0.5	7.0	0.3	0.0	11.5	11.5	0.1	67.9
HPS	sn-1	0.3	61.2	1.0	0.0	9.4	3.7	0.7	22.8
	sn-2	0.7	11.0	2.7	0.3	12.6	7.6	0.3	63.3
HO-30	sn-1	0.5	23.4	0.4	0.0	4.8	65.2	1.4	3.9
	sn-2	0.9	9.8	0.3	0.0	15.5	68.1	0.4	4.4
HPO	sn-1	0.3	43.9	7.0	0.0	2.9	37.1	2.5	5.8
	sn-2	1.1	12.1	10.7	0.2	15.5	50.6	0.7	7.6
HPSO	sn-1	0.4	41.4	5.8	0.0	6.7	39.2	2.2	4.3
	sn-2	1.1	10.2	11.5	0.2	16.5	53.8	0.9	5.8

HPS = KIR-10 × HS-35; HPO = KIR-10 × HO-30; HPSO = KIR-10 × (HS-35 × HO-30).

During cotton transformation process, many of the somatic embryos derived from transgenic calli have been aborted at globular stage, which led to substantially fewer number of transgenic plants compared to a standard cotton transformation experiment in our laboratory. In addition to the two fertile transgenic lines we are reporting here, we have also generated a number of male sterile lines at an unusually high rate. When the male sterile lines were pollinated with WT, among a large number of aborted seeds, the recovered viable seeds contained higher level of palmitic acid than those of KIR-1 and KIR-10. But the effort has failed to establish a stable transgenic line from their progenies because of the highly variable palmitic acid content and poor seed viability that is likely attributable to the nature of high copy number of *ghKAS2-RNAi* transgene. Such a result is consistent with the observation in Arabidopsis that strong suppression *atKAS2* resulted in abortion of about one-fourth of zygotic embryos before torpedo stage, suggesting lethality at homozygosity (Pidkowich et al., 2007).

A slight yet significant reduction in oil content was observed in both KIR-1 and KIR-10 seeds. It is well established that the energy requirement for the entire seed germination process including seed imbibition, radical protrusion, elongation and initial seedling development prior to photosynthetic autotrophy is provided by the available nutritive reserves of the cotyledons (Snider et al., 2014; Turley and Chapman, 2010). As seed oil represents the most energy-dense storage compound of the quiescent cotton seed, it has been suggested that higher seed oil content would enhance seed germination rate and seedling vigour by providing more chemical energy (Bartee and Krieg, 1974; Snider et al., 2014). However, the reduction in seed oil content in both HP lines did not result in compromised seed germination at both cool and warm temperatures. It is particularly interesting that the dramatic increase of palmitic acid that has relatively high melting point compared to unsaturated fatty acids showed a neutral effect on germination rate, which is in sharp contrast to the severe impairment on germination observed in HS cotton seeds generated by RNAi down-regulation of SAD (Liu et al., 2002). There is no doubt that detailed assessments on a range of seed characteristics need to be carried out, especially in field conditions as early season soil temperatures may strongly influence seed germination and seedling establishment.

Similar to the observation in Arabidopsis, a key feature of the HP cotton lines was a significant increase in levels of the so-called ω7 fatty acids, including palmitoleic acid, C16:2 and cis-vaccenic acid when compared with WT cotton. Palmitoleate is found only in trace amounts in most plant tissues, whereas oleate is the main product of plastidial fatty acid biosynthesis. The increased production of C16 unsaturated fatty acids may suggest the presence of a broad substrate specificity of SAD that acts on palmitoyl-ACP to produce palmitoleic acid that could be either further desaturated by a FAD2 to form C16:2 or catalysed by an elongase to produce cis-vaccenic acid.

Despite of disputing evidences, palmitoleic acid is gaining attention for its potential to reduce risk factors for coronary heart disease, type 2 diabetes and other obesity-related diseases (Bernstein et al., 2014; Hiraoka-Yamamoto et al., 2004). Further, it has not only high oxidative stability, but also lower melting points than oleic acid, which makes it desirable for use as feedstock of biodegradable lubricants or as a valuable precursor for linear low density polyethylene products (Ohlrogge, 1994; Rybak et al., 2008). In sunflower, it has been reported that SAD activity towards stearoyl-ACP was about 100-fold higher than for palmitoyl-ACP (Salas et al., 2004). Therefore, it is likely that the low specificity of cotton SAD towards palmitoyl-ACP could be the bottleneck for raising palmitoleic acid level in HP cotton seed oil. This might be overcome by co-expressing a modified SAD with enhanced specificity towards palmitoyl-ACP, together with the down-regulation of ghKAS2 expression. In a KAS2-suppressed Arabidopsis, a modified plastidial desaturase and an extraplastidial palmitate Δ9 desaturase yielded a mean accumulation of ~67% palmitoleic acid and cis-vaccenic acid, with individual lines showing greater than 71% in the best engineered line (Nguyen et al., 2010). It could be envisaged that as an annual crop amendable to genetic modification, cotton seed oil rich in palmitic acid may become the launching pad for further genetic modification to produce valuable palmitoleic acid in planta.

In cotton seed, a relatively smaller proportion of C16 fatty acids are converted to their dienoic form than are the C18 fatty acids. This might be caused by the cotton FAD2 which had a strong substrate preference to oleic acid in relative to palmitoleic acid on PC. A FAD2 that has high substrate specificity for palmitoleic acid has been reported in safflower (Cao et al., 2013). However, such a palmitate preference has not yet been identified in cotton FAD2 that is comprised of only 4 gene members in contrast to a staggering 11 members as reported in safflower (Cao et al., 2013). Cis-vaccenic acid is clearly an elongation product of palmitoleic acid, but it remains to be resolved whether the residual KASII in plastids, or the cytoplasmic β-ketoacyl-CoA synthase in eukaryotic pathway is the catalysing enzyme (Nguyen et al., 2010).

In addition to chain length and degree of unsaturation of fatty acids, it is now recognised that the positional composition of TAG may be an important determinant of metabolic availability. It has been hypothesised that the saturated fatty acids on the sn-2 position are hypercholesterolaemic, but neutral on cholesterol metabolism when on the sn-1 or sn-3 positions (German and Dillard, 2004; Karupaiah and Sundram, 2007). A plausible mechanism to support this hypothesis is that pancreatic lipase and lipoprotein lipase preferentially hydrolyse the fatty acids in the sn-1 and sn-3 position and produce a 2-MAG that is preferentially transported to liver for LDL cholesterol metabolism (Berry, 2009). Therefore, despite the high level of palmitic acid, the HP cotton seed oil may not be nutritionally undesirable due to its low level of palmitic acid at the sn-2 position in TAG.

We have previously generated HO and HS cotton genotypes using RNAi down-regulation of ghFAD2-1 and ghSAD-1, respectively (Liu et al., 2002). The HS cotton seed oil, albeit having a significantly raised melting point compared to the conventional cotton seed oil, contained insufficient saturated fatty acids to provide the required texture and spread ability for direct use as a solid fat unless used in combination with another appropriate fat (unpublished data). Therefore, raising palmitic acid content in the HO and HS genotypes has a clear advantage in providing functionality required for the food industry, especially in the area of solid fat applications, such as margarine, shortening and confectionary industries. In addition, simultaneously raised levels of palmitic, stearic and oleic acids could be potentially useful for mimicking the unique TAG profile of cocoa butter that is in high demand by the ever expanding confectionary, cosmetic and pharmaceutical industries. Cocoa butter is one of the most highly valued plant lipids in commerce because of its melting curve properties conferred by symmetrical TAGs such as sn-1,3 distearoyl sn-2 oleoyl glycerol (SOS), sn-1 palmitoyl sn-2 oleoyl sn-3 stearoyl glycerol (POS) and sn-1,3 dipalmitoyl sn-2 oleoyl glycerol (POP).

We have demonstrated here that both palmitic and oleic acids could be raised simultaneously in cotton seed oil by crossing the HP and HO lines. This is consistent with the HP/HO sunflower mutant CAS-12 that produces oil with 30% palmitate and 40% oleate (Salas et al., 2004). In contrast, the combination of HP and HS trait appeared to be troublesome as the stearic acid level in the hybrid was only 4% compared to about 30% in its HS parent while palmitic acid was retained at a similar level as its HP parent. In this study, we have shown that such a biased accumulation of saturated fatty acids occurred in both PC and TAG lipids. Although the oil biosynthesis pathway is made of discrete elements, it has been established that there are interactions between such components expediting the dynamic channelling of the metabolic flux through fatty acid elongation, desaturation and TAG assembly (Roughan and Ohlrogge, 1996). Several hypotheses could be proposed to explain why the HP cotton is not able to accumulate stearic acid to a high level. These include the low enzyme specificity of downstream enzymes needed for stearic acid incorporation, and FatB's substrate preference towards palmitic acid. The competition between the two saturated fatty acids for their esterification on TAG molecule might be eased by introducing a FatB enzyme or acyltransferases with substrate preference towards stearic acid.

In summary, we report here that expression of RNAi construct targeting the down-regulation of KAS2 expression led to substantial increase in palmitic acid and its derivative C16 unsaturated fatty acids at the expense of C18 fatty acids in cotton seed oil. A slight yet significant reduction in oil content has been observed in both HP lines, but seed germination was unaffected. Simultaneous increase of both palmitic acid and oleic acid was also successful. These novel cotton genotypes may provide a temperate vegetable oil, derived from an annual crop, which could potentially replace imported palm oil in many parts of the world. Additional efforts are required for further enhancement of stearic acid in the HP/HO background that may have a clear advantage in providing functionality required for valuable cocoa butter substitute.

Experimental procedures

Plant material and growth conditions

Upland cotton (*G. hirsutum*) cv Coker315 was used for gene transformation. Homozygous transgenic cotton lines including HO-30 and HS-35 were generated as previously described (Liu *et al.*, 2002). Plants were cultivated in greenhouse conditions at 28/18 °C (day/night), with 16-h photoperiod.

Cloning cotton *ghKAS2* genes

The cotton *KASII* DNA sequences were isolated from a developing cotton seed cDNA library using a DNA fragment corresponding to the first 400 nucleotide of the coding region of *KASII* gene from Arabidopsis (AF318307) as a probe. Two different *KASII* cDNAs, *ghKASII-A* and *ghKASII-B*, were isolated from this cDNA library as previously described (Liu *et al.*, 1999).

Design of RNAi gene silencing construct targeting *ghKAS2*

We designed an RNAi cassette consisting of inverted repeat structure of partial *ghKAS2-1* DNA sequences driven by a seed-specific promoter derived from soya bean *lec1* gene (Cho *et al.*, 1995). To assemble this RNA cassette, as shown in Fig. 8, we firstly obtained the trigger DNA fragment corresponding to nucleotides from 1001 to 1620 of *ghKAS2-1* by PCR using a pair of primers incorporating the *Nco*I site (underlined) at both ends: 5'-CCATGGCTAATCGCGATGGATTTGTCATGG-3', and 5'-CCATGGCTCTTCGGCCAAATGTAAGAAA-3'. The amplified PCR fragment was then inserted at the *Nco*I site at nt-265 of *ghKAS2-1*, in an antisense orientation. A DNA fragment consisting of 400 bp of the inverted repeats in a head to head fashion, together with 750 bp adjacent DNA sequence acting as a spacer, was amplified by PCR using the following single primer that is present at the both sides of the inverted repeats and with an additional *Sma*I restriction site (underlined): 5'-CCCGGGCGTATTGCCTGT ACCGTTGC-3'. The resulting inverted repeat structure was then inserted at the *Sma*I site between the lectin promoter and terminator sequences (Cho *et al.*, 1995) in a binary vector using *NPTII* gene as the selectable marker for gene transformation.

Transformation of cotton

The above described binary construct was transformed to *Agrobacterium tumefaciens* strain AGL1 and used to transform cotton variety Coker 315 as described by Liu *et al.* (2002). Briefly,

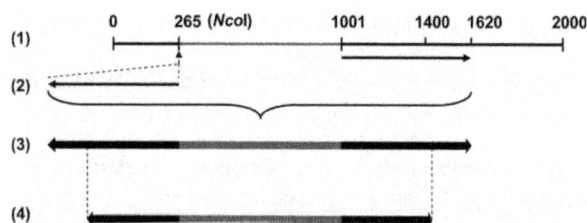

Figure 8 A diagram showing the selection process of trigger DNA sequence for the inverted repeat configuration in RNAi-*ghKAS2* construct. (1) The DNA sequence corresponding to nucleotides from 1001 to 1620 of *ghKAS2-1* was PCR amplified and (2) inserted at the *Nco*I site at nt 265 of *ghKAS2-1*, in an antisense orientation and therefore resulted in the inverted repeat structure as shown in (3). The first 400 bp of the repeated DNA sequences and its adjacent 750-bp spacer were PCR amplified to create the final RNAi trigger DNA sequence (4).

cotyledons excised from cotton seedlings were used as explants and were infected and co-cultivated with *A. tumefaciens* harbouring the RNAi-*ghKAS2-1* construct. Healthy calli derived from the cotyledon explants were selected by growing on MS medium containing kanamycin until somatic embryo formation. Plantlets developed from the somatic embryos were subsequently transferred to soil and maintained in a glasshouse once leaves and roots were developed, with 28/20 °C (day/night) growth temperature.

Real-time RT-qPCR

RNAs from developing embryos of homozygous T_4 KIR-1, KIR-10 plants were isolated using QIAEX Plant RNA Miniprep kit (Qiagen, Hilden, Germany). The gene expression patterns were studied with RT-qPCR carried out in triplicate using Platinum SYBR Green qPCR SuperMix (BioRad, Hercules, CA) and run on ABI 7900HT Sequence Detection System. Each 10-μL PCR contained 20 ng of total RNA template, 800 mM each of the forward and reverse primers, 0.25 μL of reverse transcriptase, 5 μL One-step RT-PCR master mix reagents. PCR was carried out at following conditions with an initial cycle at 48 °C for 30 min, and 95 °C for 10 min, followed by 40 cycles of 95 °C for 15 s and 60 °C for 60 s. The following primers were designed such that amplifying a fragment from the *ghKAS2-1* gene (5'-CCAACAAAGACGTGGGCTTGA-3' and 5'-GGAGGCT TCTTCTTCGTTGTAA-3') would give a 290-bp fragment; and amplifying a *ghKAS2-2* gene (5'-TCCAAACCCAAGTCCTCAA AACTC-3' and 5'-AGGAGGTTTCTGTTTTGTCATG-3') would give rise to a 393-bp fragment. Cotton unibiquitin-14 gene (*ghUBQ14*, GenBank accession number: DW505546) was used as an internal reference gene (Zhang *et al.*, 2013). The primers for *ghUBQ14* are sense: 5'-CAACGCTCCATCTTGTCCTT-3' and antisense: 5'-TGA TCGTCTTTCCCGTAAGC-3'. The calculations were made using the comparative CT method as reported (Livak and Schmittgen, 2001). The data are presented as means ± SD of three reactions performed on independent 96-well plates.

Oil content analysis, lipid separation and quantification

Seed oil content was quantified by Oxford MQC benchtop nuclear magnetic resonance analyser (NMR) (Oxford Instruments, Oxford, UK). Three-gram samples of cotton seed were measured and equilibrated to 40 °C for 2 h prior to acquisition of NMR data. The experiment was carried out in triplicate. Two-way ANOVA was conducted using Microsoft Excel, and LSD test was applied at 5% probability level to compare the differences among treatment means.

Fatty acid profiles of cotton seed oil were determined by GC-FID analysis of prepared seed oil fatty acid methyl esters (FAME). TAG and polar lipids were fractionated from total lipids on thin layer chromatography (TLC) plates Silica gel 60 (Merck Millipore, Darmstadt, Germany) using hexane/diethyl ether/acetic acid (50/50/1, by volume). Lipid bands were visualised by spraying 0.001% primuline made in 80% acetone in water, and individual lipid classes were identified by running authentic standards on the same TLC plate. Fatty acid profiles of TAG and total polar lipids were determined by incubating corresponding silica bands in 1 N methanolic HCl and by analysing corresponding FAME by GC as above.

Positional distribution of the fatty acids in TAG was analysed by treatment with TAG lipase derived from *Rhizopus arrhizus* (Fluka, Buchs, Switzerland). Purified TAG (1 mg) in chloroform was transferred to a glass vial, evaporated under nitrogen and mixed strongly using a vortex with 2 U *Rhizopus* lipase dissolved in

0.4 mL 0.1 M Tris-HCl (pH 7.7) and 5 mM $CaCl_2$ for 1 min. The reaction was stopped by adding 50 μL 6 M HCl, and lipid was extracted using 1.5 mL $CHCl_3$/MeOH (2/1, by volume). The lipid was fractionated into TAG, DAG, sn-1/3-MAG, sn-2-MAG and FFA using 15 cm 2.3% boric acid TLC (silica gel-60, Merck Millipore) and hexane/diethyl ether/acetic acid (50/50/1, by volume) solvent mixture. FAMEs of individual lipid classes were prepared directly by collecting the lipid bands in silica gel isolated from the TLC plate into glass vials and by incubation in 1 N methanolic HCl at 80 °C for 2 h prior to GC analysis.

PC was fractionated from the polar lipid pool using TLC plates Silica gel 60 (Merck Millipore) and chloroform/methanol/acetic acid/water (90/15/10/3, by volume) as the solvent system. Lipid classes were visualised and identified as above, and PC was extracted from silica using chloroform/methanol/water (2/1/1, by volume). Positional distribution of fatty acids in PC was determined by the treatment of 0.1 mg of lipid in 3 U of phospholipase A_2 derived from honey bee (Sigma-Aldrich, St. Louis, MO) as described by Williams et al. (1995). Lipid was further purified using chloroform/methanol/0.1 M KCl (2/1/1, by volume), and it was fractionated into PC, lyso-PC and free fatty acids by TLC (Silica gel 60, MERCK) in chloroform/methanol/acetic acid/water (68/22/6/4, by volume) as solvent system. FAMEs of individual lipid bands were prepared and analysed as above.

Primary assessment of seed germination rate

To evaluate seed germination response to temperature, 30 seeds from each of the two HP lines, together with the untransformed control Coker 315, were evenly spaced on moist filter paper in a 200 × 100 mm plastic tray with cover and placed in incubators with cool (18 °C) and warm (28 °C) temperatures. The filter paper was kept moist for the duration of the experiment by adding distilled water when required. Seeds were incubated for 5 days at each temperature and were counted as germinated when the radicle protruded beyond the seed coat and elongated to at least 1 cm in length. The experiment was run in quadruplicate ($n = 4$). Two-way ANOVA was conducted using Microsoft Excel, and LSD test was applied at 5% probability level to compare the differences among treatment means.

Acknowledgements

The authors wish to Geraldine Lester, Nathalie Niesner and Lijun Tian for excellent technical assistance. This study was supported by Cotton Research & Development Corporation (CRDC) Australia.

References

Anai, T., Hoshino, T., Imai, N. and Takagi, Y. (2012) Molecular characterization of two high-palmitic-acid mutant loci induced by X-ray irradiation in soybean. Breeding Sci. 61, 631.

Aslan, S., Hofvander, P., Dutta, P., Sitbon, F. and Sun, C. (2015) Transient silencing of the KASII genes is feasible in Nicotiana benthamiana for metabolic engineering of wax ester composition. Sci. Rep. 5, 11213.

Bartee, S.N. and Krieg, D.R. (1974) Cottonseed density: associated physical and chemical properties of 10 cultivars. Agron. J. 66, 433–435.

Bernstein, A.M., Roizen, M.F. and Martinez, L. (2014) Purified palmitoleic acid for the reduction of high-sensitivity C-reactive protein and serum lipids: A double-blinded, randomized, placebo controlled study. J. Clin. Lipidol. 8, 612–617.

Berry, S.E.E. (2009) Triacylglycerol structure and interesterification of palmitic and stearic acid-rich fats: an overview and implications for cardiovascular disease. Nutr. Res. Rev. 22, 3–17.

Cahoon, E.B. and Shanklin, J. (2000) Substrate-dependent mutant complementation to select fatty acid desaturase variants for metabolic engineering of plant seed oils. Proc. Natl Acad. Sci. USA, 97, 12350–12355.

Cao, S., Zhou, X.-R., Wood, C., Green, A., Singh, S., Liu, L. and Liu, Q. (2013) A large and functionally diverse family of Fad2 genes in safflower (Carthamus tinctorius L.). BMC Plant Biol. 13, 5.

Carlsson, A.S., LaBrie, S.T., Kinney, A.J., Von Wettstein-Knowles, P. and Browse, J. (2002) A KAS2 cDNA complements the phenotypes of the Arabidopsis fab1 mutant that differs in a single residue bordering the substrate binding pocket. Plant J. 29, 761–770.

Cho, M.-J., Widholm, J. and Vodkin, L. (1995) Cassettes for seed-specific expression tested in transformed embryogenic cultures of soybean. Plant Mol. Biol. Rep. 13, 255–269.

Chowdhury, R., Warnakula, S., Kunutsor, S., Crowe, F., Ward, H.A., Johnson, L., Franco, O.H. et al. (2014) Association of dietary, circulating, and supplement fatty acids with coronary risk -A systematic review and meta-analysis. Ann. Intern. Med. 160, 398–406.

Dehesh, K., Jones, A., Knutzon, D.S. and Voelker, T.A. (1996) Production of high levels of 8:0 and 10:0 fatty acids in transgenic canola by overexpression of Ch FatB2, a thioesterase cDNA from Cuphea hookeriana. Plant J. 9, 167–172.

Dormann, P., Voelker, T.A. and Ohlrogge, J.B. (2000) Accumulation of palmitate in Arabidopsis mediated by the acyl-acyl carrier protein thioesterase FATB1. Plant Physiol. 123, 637–644.

German, J.B. and Dillard, C.J. (2004) Saturated fats: what dietary intake? Am. J. Clin. Nutr. 80, 550–559.

Hayes, K.C. and Pronczuk, A. (2010) Replacing trans fat: the argument for palm oil with a cautionary note on interesterification. J. Am. Coll. Nutr. 29, 253s–284s.

Hiraoka-Yamamoto, J., Ikeda, K., Negishi, H., Mori, M., Hirose, A., Sawada, S., Onobayashi, Y. et al. (2004) Serum lipid effects of a monounsaturated (palmitoleic) fatty acid-rich diet based on macadamia nuts in healthy, young Japanese women. Clin. Exp. Pharmacol. P. 31, S37–S38.

James, D.W. Jr and Dooner, H.K. (1990) Novel seed lipid phenotypes in combinations of mutants altered in fatty acid biosynthesis in Arabidopsis. Theor. Appl. Genet. 82, 409–412.

Jones, L. and King, C. (1993) Cottonseed Oil. Memphis: National Cottonseed Products Associations, Inc. and the Cotton Foundation.

Karupaiah, T. and Sundram, K. (2007) Effects of stereospecific positioning of fatty acids in triacylglycerol structures in native and randomized fats: a review of their nutritional implications. Nutr. Metab. 4, 16–16.

L'Abbe, M.R., Stender, S., Skeaff, C.M., Ghafoorunissa, R. and Tavella, M. (2009) Approaches to removing trans fats from the food supply in industrialized and developing countries. Eur. J. Clin. Nutr. 63, S50–S67.

Liu, Q., Singh, S.P., Brubaker, C.L., Sharp, P.J., Green, A.G. and Marshall, D.R. (1999) Molecular cloning and expression of a cDNA encoding a microsomal ω-6 fatty acid desaturase from cotton (Gossypium hirsutum). Funct. Plant Biol. 26, 101–106.

Liu, Q., Singh, S.P. and Green, A.G. (2002) High-stearic and high-oleic cottonseed oils produced by hairpin RNA-mediated post-transcriptional gene silencing. Plant Physiol. 129, 1732–1743.

Livak, K.J. and Schmittgen, T.D. (2001) Analysis of relative gene expression data using real-time quantitative PCR and the ($2^{-\Delta\Delta Ct}$) method. Methods, 25, 402–408.

Martinez-Ortiz, J.A., Fung, T.T., Baylin, A., Hu, F.B. and Campos, H. (2006) Dietary patterns and risk of nonfatal acute myocardial infarction in Costa Rican adults. Eur. J. Clin. Nutr. 60, 770–777.

Martz, F., Kiviniemi, S., Palva, T.E. and Sutinen, M.L. (2006) Contribution of omega-3 fatty acid desaturase and 3-ketoacyl-ACP synthase II (KASII) genes in the modulation of glycerolipid fatty acid composition during cold acclimation in birch leaves. J. Exp. Bot. 57, 897–909.

Mozaffarian, D., Katan, M.B., Ascherio, A., Stampfer, M.J. and Willett, W.C. (2006) Trans fatty acids and cardiovascular disease. New Engl. J. Med. 354, 1601–1613.

Narang, D., Sood, S., Thomas, M., Maulik, S.K. and Dinda, A.K. (2005) Dietary palm olein oil augments cardiac antioxidant enzymes and protects against isoproterenol-induced myocardial necrosis in rats. *J. Pharm. Pharmacol.* **57**, 1445–1451.

Neff, W.E. and List, G.R. (1999) Oxidative stability of natural and randomized high-palmitic and high-stearic acid oils from genetically modified soybean varieties. *JAOCS*, **76**, 825–831.

Nguyen, H.T., Mishra, G., Whittle, E., Pidkowich, M.S., Bevan, S.A., Merlo, A.O., Walsh, T.A. *et al.* (2010) Metabolic engineering of seeds can achieve levels of ω-7 fatty acids comparable with the highest levels found in natural plant sources. *Plant Physiol.* **154**, 1897–1904.

Nzikou, J.M., Matos, L., Moussounga, J.E., Ndangui, C.B., Pambou-Tobi, N.P., Bandzouzi, E.M., Kimbonguila, A. *et al.* (2009) Study of oxidative and thermal stability of vegetable oils during frying. *Res. J. Appl. Sci.* **4**, 94–100.

O'Brien, R.D. (2002) Cottonseed oil. In *Vegetable Oils in Food Technology: Composition, Properties and Uses* (Gunstone, F.D., ed), pp. 203–230. Oxford: Blackwell Publishing.

Ohlrogge, J.B. (1994) Design of new plant products: engineering of fatty acid metabolism. *Plant Physiol.* **104**, 821–826.

Ohlrogge, J.B. and Jaworski, J.G. (1997) Regulation of fatty acid synthesis. *Annu. Rev. Plant Phys.* **48**, 109–136.

Oomen, C.M., Ocke, M.C., Feskens, E.J., van Erp-Baart, M.A., Kok, F.J. and Kromhout, D. (2001) Association between trans fatty acid intake and 10-year risk of coronary heart disease in the Zutphen Elderly Study: a prospective population-based study. *Lancet*, **357**, 746–751.

Pidkowich, M.S., Nguyen, H.T., Heilmann, I., Ischebeck, T. and Shanklin, J. (2007) Modulating seed beta-ketoacyl-acyl carrier protein synthase II level converts the composition of a temperate seed oil to that of a palm-like tropical oil. *Proc. Natl Acad. Sci. USA*, **104**, 4742–4747.

Roughan, P.G. and Ohlrogge, J.B. (1996) Evidence that isolated chloroplasts contain an integrated lipid-synthesizing assembly that channels acetate into long-chain fatty acids. *Plant Physiol.* **110**, 1239–1247.

Rybak, A., Fokou, P.A. and Meier, M.A.R. (2008) Metathesis as a versatile tool in oleochemistry. *Eur. J. Lipid Sci. Tech.* **110**, 797–804.

Salas, J.J., Martínez-Force, E. and Garcés, R. (2004) Biochemical characterization of a high-palmitoleic acid *Helianthus annuus* mutant. *Plant Physiol. Bioch.* **42**, 373–381.

Snider, J.L., Collins, G.D., Whitaker, J., Chapman, K.D., Horn, H. and Grey, T.L. (2014) Seed size and oil content are key determinants of seedling vigor in *Gossypium hirsutum*. *J. Cotton Sci.* **18**, 1–9.

de Souza, R.J., Mente, A., Maroleanu, A., Cozma, A.I., Ha, V., Kishibe, T., Uleryk, E. *et al.* (2015) Intake of saturated and trans unsaturated fatty acids and risk of all cause mortality, cardiovascular disease, and type 2 diabetes: systematic review and meta-analysis of observational studies. *BMJ*, **351**, h3978.

Stoltzfus, D.L., Fehr, W.R., Welke, G.A., Hammond, E.G. and Cianzio, S.R. (2000) A *fap5* allele for elevated palmitate in soybean. *Crop Sci.* **40**, 647–650.

Sun, J.-Y., Hammerlindl, J., Forseille, L., Zhang, H. and Smith, M.A. (2014) Simultaneous over-expressing of an acyl-ACP thioesterase (FatB) and silencing of acyl-acyl carrier protein desaturase by artificial microRNAs increases saturated fatty acid levels in *Brassica napus* seeds. *Plant Biotech. J.* **12**, 624–637.

Turley, R.B. and Chapman, K.D. (2010) Ontogeny of cotton seeds: gametogenesis, embryogenesis, germination, and seedling growth. In *Physiology of Cotton* (Stewart, J.M., Oosterhuis, D.M., Heitholt, J.J. and Mauney, J.R., eds), pp. 332–341. New York: Springer.

Williams, J.P., Khan, M.U. and Wong, D. (1995) A simple technique for the analysis of positional distribution of fatty acids on di- and triacylglycerols using lipase and phospholipase A2. *J. Lipid Res.* **36**, 1407–1412.

Wu, J., James, D.W. Jr, Dooner, H.K. and Browse, J. (1994) A mutant of Arabidopsis deficient in the elongation of palmitic acid. *Plant Physiol.* **106**, 143–150.

Zaliha, O., Chong, C.L., Cheow, C.S., Norizzah, A.R. and Kellens, M.J. (2004) Crystallization properties of palm oil by dry fractionation. *Food Chem.* **86**, 245–250.

Zhang, Y., Wang, X.F., Ding, Z.G., Ma, Q., Zhang, G.R., Zhang, S.L., Li, Z.K. *et al.* (2013) Transcriptome profiling of *Gossypium barbadense* inoculated with *Verticillium dahliae* provides a resource for cotton improvement. *BMC Genom.* **14**, 637.

15

Domestication and the storage starch biosynthesis pathway: signatures of selection from a whole sorghum genome sequencing strategy

Bradley C. Campbell[1], Edward K. Gilding[1], Emma S. Mace[2], Shuaishuai Tai[3], Yongfu Tao[4], Peter J. Prentis[5], Pauline Thomelin[6], David R. Jordan[4] and Ian D. Godwin[1]*

[1]School of Agriculture and Food Sciences, The University of Queensland, Brisbane, Qld, Australia

[2]Department of Agriculture and Fisheries (DAF), Warwick, Qld, Australia

[3]BGI-Shenzhen, Shenzhen, China

[4]Queensland Alliance for Agriculture and Food Innovation, The University of Queensland, Warwick, Qld, Australia

[5]Science and Engineering Faculty, Queensland University of Technology (QUT), Brisbane, Qld, Australia

[6]Australian Centre for Plant Functional Genomics, Glen Osmond, SA, Australia

*Correspondence
email i.godwin@uq.edu.au

Keywords: Sorghum (*Sorghum bicolor*), domestication, starch synthesis, whole-genome sequencing, selection, metabolic pathway.

Summary

Next-generation sequencing of complete genomes has given researchers unprecedented levels of information to study the multifaceted evolutionary changes that have shaped elite plant germplasm. In conjunction with population genetic analytical techniques and detailed online databases, we can more accurately capture the effects of domestication on entire biological pathways of agronomic importance. In this study, we explore the genetic diversity and signatures of selection in all predicted gene models of the storage starch synthesis pathway of *Sorghum bicolor*, utilizing a diversity panel containing lines categorized as either 'Landraces' or 'Wild and Weedy' genotypes. Amongst a total of 114 genes involved in starch synthesis, 71 had at least a single signal of purifying selection and 62 a signal of balancing selection and others a mix of both. This included key genes such as *STARCH PHOSPHORYLASE 2* (*SbPHO2*, under balancing selection), *PULLULANASE* (*SbPUL*, under balancing selection) and ADP-glucose pyrophosphorylases (*SHRUNKEN2*, *SbSH2* under purifying selection). Effectively, many genes within the primary starch synthesis pathway had a clear reduction in nucleotide diversity between the Landraces and wild and weedy lines indicating that the ancestral effects of domestication are still clearly identifiable. There was evidence of the positional rate variation within the well-characterized primary starch synthesis pathway of sorghum, particularly in the Landraces, whereby low evolutionary rates upstream and high rates downstream in the metabolic pathway were expected. This observation did not extend to the wild and weedy lines or the minor starch synthesis pathways.

Introduction

The study of the evolution of metabolic pathways is fundamental to understanding evolutionary change as well as key events in the domestication process (Fraser *et al.*, 2002; Ramsay *et al.*, 2009). By examining the evolution of genes in an integrated metabolic pathway, we can gain some understanding of the complex forces that are shaping domesticated germplasm and question whether the differential selection acting upon certain loci conforms to previous studies that sought to explain how the structure of such pathways affects evolutionary rate (Clotault *et al.*, 2012; Livingstone and Anderson, 2009; Ramsay *et al.*, 2009; Rausher *et al.*, 2008). Positional rate variation (PRV) is the theory that low evolutionary rates upstream and high rates downstream upon genes within a metabolic pathway is expected as a consequence of purifying selection. This increase in selective constraint is believed to occur because the consequence of a nonsynonymous mutation within a gene at a key branch point higher up in the synthesis process could lead to major pleiotropic effects that result in no useful end products (Rausher *et al.*, 1999). This was the case within the plant biosynthetic pathways of isoprene (Sharkey *et al.*, 2005), anthocyanin (Lu and Rausher, 2003; Rausher *et al.*, 1999, 2008), carotenoid (Clotault *et al.*, 2012; Livingstone and Anderson, 2009) and terpenoid synthesis (Ramsay *et al.*, 2009). But does this model hold for the critical process of storage starch synthesis? Unlike other pathways, the starch synthesis pathway (SSP) is not a simple unidirectional pathway but also contains several alternative branches resulting in starch as the terminal product and has to support respiration and other key processes. Similarly, the SSP contains a series of catabolic genes that encode enzymes that work in reverse. Cereal grains are the single most important source of calories in the world, predominantly in the form of starch, and comprise as much as 80% of the calorific intake for some of the poorest countries (WHO, 2003). The SSP is one of the most well-characterized pathways in plant science and has undergone strong selection during domestication (James *et al.*, 2003; Tetlow *et al.*, 2004; Whitt *et al.*, 2002; Zeeman *et al.*, 2010). With the exception of soya bean, all of the top 10 most important human food crops (cereals; roots; tubers; plantains) are eaten primarily for their starch content. Hence, the SSP is essential to human food security (http://faostat3.fao.org/browse/rankings/commodities_by_regions/E).

Starch is composed of two D-glucose homopolymers: amylose and amylopectin (Smith, 2001). Amylose consists predominantly of α-1,4 linked glucosyl monomers with low–moderate branching, while amylopectin, the more abundant polymer, contains highly branched (through α-1,6 glucosyl bonds) linear chains of various lengths (Smith, 2001).

Starch synthesis requires the function of a number of vital enzymes, primarily ADP-glucose pyrophosphorylases (AGP; SHRUNKEN2, Sh2 and BRITTLE ENDOSPERM2, Bt2), starch synthases (STARTCH SYNTHASE-I, SSI; SUGARY2, su2; STARCH SYNTHASE-IIb, SSIIb, DULL 1, du1 and WAXY, wx1) and the starch-branching (SBE; STARCH-BRANCHING ENZYME 1, sbe1 and AMYLOSE EXTENDER 1, Ae1) and debranching (DBE; SUGARY1, su1 and PULLULANASE, zpu1) enzymes (James et al., 2003; Tetlow et al., 2004). Several isoforms of these enzymes exist, leading to a highly complex synthesis process. Starch production begins with the conversion of glucose 1-phosphate to ADP-glucose utilizing the enzyme ADP-glucose pyrophosphorylase. Starch synthase enzymes then link the ADP-glucose via a α-1,4 glycosidic bond to an emerging chain of glucose residues, releasing ADP and generating amylose. The creation of branched amylopectin takes place when starch-branching enzymes introduce α-1,6 glycosidic bonds between these chains, some of which are later removed by starch debranching enzymes (Figure 1) (James et al., 2003; Smith, 2001).

The majority of studies investigating the evolution of metabolic pathways in higher plants have included only small numbers of genes that are well characterized, frequently only with partial sequence availability and/or confined to well-studied species such as Arabidopsis, maize and rice (Li et al., 2012; Slotte et al., 2011; Whitt et al., 2002).

With the advent of next-generation sequencing (NGS), all predicted gene models in the genome can be sequenced in a single run. The NGS strategy facilitates the evolutionary analysis of entire metabolic pathways by utilizing population genetic techniques (Wright and Gaut, 2005; Yu et al., 2011) and metabolic pathway data available via online databases such as www.gramene.org/pathways. Furthermore, these techniques will be integral for the discovery of key candidate genes of importance to breeding programmes, especially in the light of the terabytes of data that now require analysis and the fact remarkably little is known about the nature of biochemical pathways, mutations and genes with respect to adaptive evolution (Wright and Gaut, 2005).

Utilizing the important cereal species sorghum (Sorghum bicolor) as a model, we examine the genetic history of starch at a systems biology level through the lens of energy storage in sorghum grain. This seminal work provides a framework for understanding the changes brought about by evolution and domestication. Changes in expression and specific enzyme functionality are levels of complexity beyond the purview of this manuscript, which aims to determine how genetic changes could have adapted genes for the purpose of starch storage in the grain.

Several studies on sorghum have utilized population genetic analysis techniques to study the impacts of domestication, linkage disequilibrium and sequence polymorphism (de Alencar Figueiredo et al., 2008; Casa et al., 2005, 2006; Frere et al., 2011; Hamblin et al., 2004, 2006, 2007; Mace et al., 2013; Morris et al., 2013). Of these, only Hamblin et al. (2007), de Alencar Figueiredo et al. (2008) and Frere et al. (2011) specifically examined genes involved in the SSP. Hamblin et al. (2007) explored sequence and linkage disequilibrium (LD) variation in partial sequence in 15 genes in the SSP, while de Alencar

Figueiredo et al. (2008) investigated sequence diversity of partial sequence in six candidate sorghum homologs of known maize SSP genes for grain quality (Sh2, Bt2, Sss1, Wx, Ae1 and opaque-2). A more recent study by Frere et al. (2011) sought to detect evidence of artificial selection upon seed storage proteins and partial sequence of three starch biosynthesis loci (SSIIa, sbe1 and sbpul).

The main objective of this study was to identify whether the SSP is under strong selective pressure utilizing population genetic measures focussed upon complete gene model sequences from a diverse set of sorghum germplasm. Further, it was questioned whether selective pressure acts differentially upon genes depending on their position within the metabolic pathway.

Results

The starch biosynthesis pathway and patterns of expression

None of the 114 unique starch synthesis genes analysed showed any significant difference (posterior probability of being differentially expressed (PPDE) = ≥0.95) in expression between solar noon and solar midnight. Some gene functions prominent in the dark cycle are involved with protective activities against catabolism, whereas during the light cycle we observe the expression of gene functions involved in protein maturation, carbon metabolism and auxin signal transduction (Table S4). Of note is the increase in expression of the alpha-amylase inhibitor (Sobic.002G077500.1) and protease inhibitor (Sobic.002G078800.1) indicating that transcription might not be the level at which regulation is occurring in some genes of the grain. However, our transcriptome is primarily a tool to validate expression of genes in the pathway (Table S4).

For SSP-1, five genes were associated with PPI-1 (sucrose + UDP to UDP-D-glucose + D-fructose). SUCROSE SYNTHASE-I (SbSSI, Sobic.010G072300.1) was the most abundant SSP-1 transcript with a day average FPKM of 362.48 (Table 1 and Figure 1). The metabolism of UDP-D-glucose (PPI-2) into α-D-glucose 1-phosphate involved four genes, with the key enzyme responsible, UDP–glucose pyrophosphorylase (Sobic.002G291200.1) reaching a day average FPKM of 383.08, which was expressed ≈26× greater than the next most expressed gene (Sobic.001G424500.2).

Eight AGP-related isozyme genes showed expression at PPI-3 (α-D-glucose 1-phosphate + ATP to ADP-D-glucose + diphosphate). The genes with homology to the maize homologs Sh2 (large subunit) and Bt2 (small subunit) were Sobic.003G230500.1 and Sobic.007G101500.3, respectively. Of these eight genes, the genes with the highest levels of expression were the homologues of Sh2 (large subunit) and Bt2 (small subunit) with FPKM of 843.73 and 359.96, respectively (Table 1 and Figure 1). At PPI-4, free ADP-D-glucose is then actively transported through the ADP-glucose transporter (BRITTLE-1, SbBT1; Sobic.004G085100.1) into the amyloplast (Figure 1). This gene had a day average FPKM of 267.18.

PPI-5, the critical stage of converting ADP-D-glucose + (1,4-α-D-glucosyl)(n) to linear chains of amylose and amylopectin, demonstrated nine SS loci showing expression within immature sorghum grain. The SS with the greatest expression was Sobic.010G022600.1 (SbWX) with a day average FPKM of 1735.74, which was ≈ 9× more expressed than the next closest SS (Sobic.010G047700.1 or SbSSI at 191.09) (Tables 1, S2 and S3).

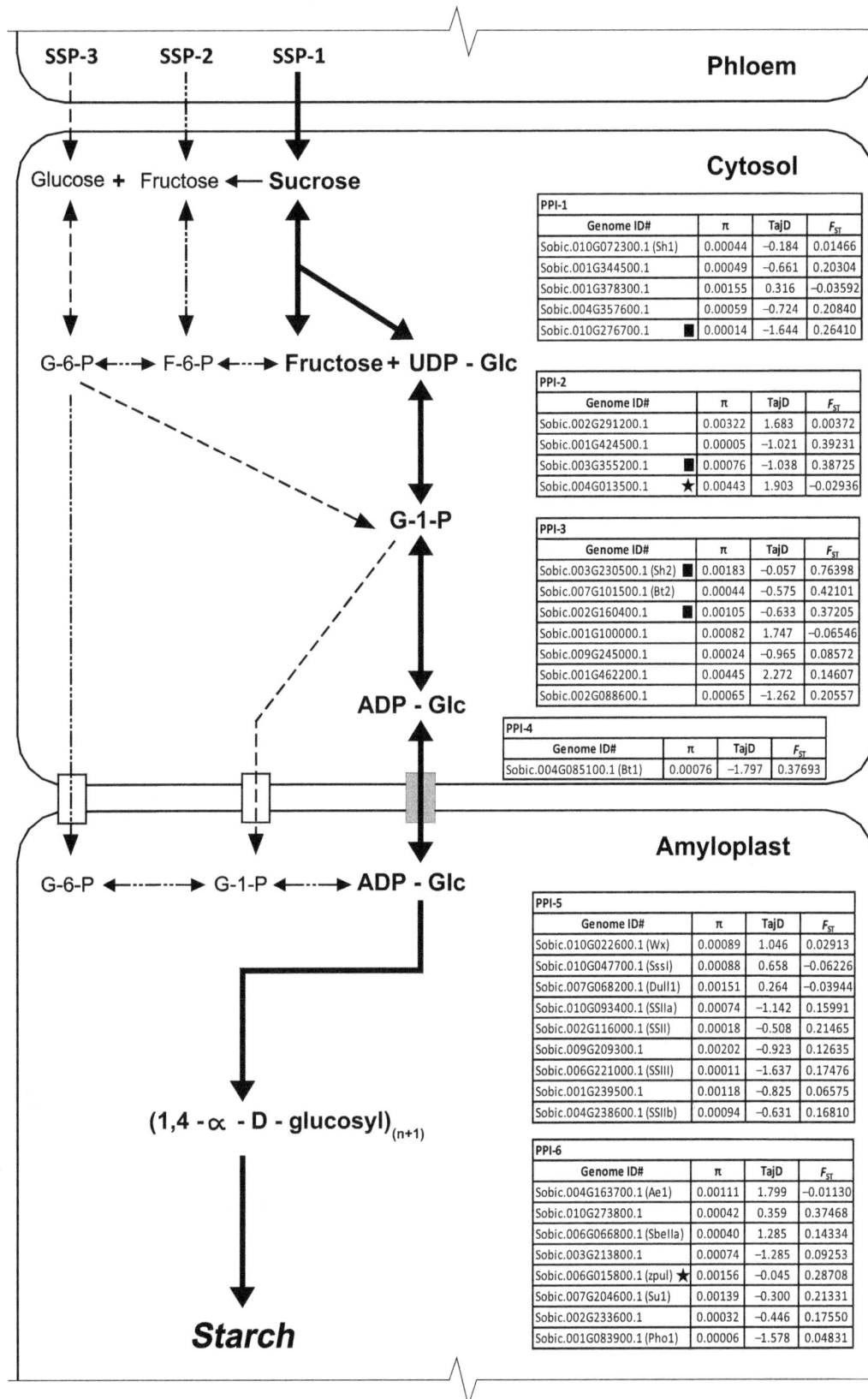

Figure 1 Plant (monocot) starch synthesis. Diagram of the reactions involved in starch synthesis. Arrows represent single or groups of enzymes that convert one metabolite to another. The primary starch synthesis pathway from sucrose to starch is represented by bold arrows (SSP-1). Other pathways examined in this study from fructose to starch (SSP-2) and glucose to starch (SSP-3) in sink tissues are represented by – – – – and –··–··–, respectively. Expressed genes for the primary pathway have been sorted according to the Pathway Pleiotropy Index (PPI; Ramsay et al., 2009), with CDS values calculated across all lines for π, Tajima's D and F_{ST} (Landraces vs wild and weedy). Genes under purifying selection are represented with a (■) and under balancing selection with a (★).

Table 1 Fragments per kilobase of exon per million fragments mapped (FPKM) of genes involved in the primary starch synthesis pathway from immature grain 16 days post anthesis, sampled at solar midday and midnight

	PPI*	Genome ID	JGI annotation	Maize homolog	FPKM* count Day average	FPKM* count Night average
CYTOSOL			**sucrose + UDP ↔ UDP-D-glucose + D-fructose**			
	1	Sobic.010G072300.1	"similar to Sucrose synthase 1"	Sh1	362.48	316.39
		Sobic.001G344500.1	"similar to Sucrose synthase 2"	-	76.05	89.69
		Sobic.001G378300.1	"similar to Sucrose synthase 3"	-	32.54	20.72
		Sobic.004G357600.1	"similar to Putative uncharacterized protein"	-	4.85	1.14
		Sobic.010G276700.1	"similar to SUS6 (Sucrose Synthase 6)"	-	0.64	0.35
			UDP-D-glucose + diphosphate ↔ α-D-glucose 1-phosphate + UTP			
	2	Sobic.002G291200.1	"similar to UTP--glucose-1-phosphate uridylyltransferase"	-	383.08	297.93
		Sobic.002G291200.2	"similar to UTP--glucose-1-phosphate uridylyltransferase"	-	8.44	0.00
		Sobic.001G424500.2	"similar to Nucleoside-diphosphate-sugar pyrophosphorylase"	-	15.00	23.60
		Sobic.001G424500.1	"similar to Nucleoside-diphosphate-sugar pyrophosphorylase"	-	7.68	5.44
		Sobic.003G355200.1	"similar to Nucleoside-diphosphate-sugar pyrophosphorylase"	-	16.63	7.33
		Sobic.004G013500.1	"similar to UDP-glucose pyrophosphorylase"	-	0.63	0.46
			α-D-glucose 1-phosphate + ATP → ADP-D-glucose + diphosphate			
	3	Sobic.003G230500.1	"similar to Glucose-1-phosphate adenylyltransferase large subunit 1"	Sh2	843.73	631.91
		Sobic.007G101500.3	"similar to Glucose-1-phosphate adenylyltransferase"	Bt2	359.96	310.98
		Sobic.007G101500.2	"similar to Glucose-1-phosphate adenylyltransferase"	Bt2	43.97	31.62
		Sobic.002G160400.1	"similar to Glucose-1-phosphate adenylyltransferase"	-	89.13	72.65
		Sobic.001G100000.1	"similar to Glucose-1-phosphate adenylyltransferase"	-	57.31	44.28
		Sobic.009G245000.1	"similar to Glucose-1-phosphate adenylyltransferase large subunit 1"	-	23.53	18.70
		Sobic.001G462200.1	"similar to ADP-glucose pyrophosphorylase family protein,"	-	18.02	13.22
		Sobic.002G088600.1	"similar to Glucose-1-phosphate adenylyltransferase"	-	2.30	3.14
			ADP-D-glucose (cytosol) → ADP-D-glucose (amyloplast)			
	4	Sobic.004G085100.1	"similar to Protein brittle-1, chloroplast precursor"	Bt1	267.18	340.81
AMYLOPLAST			**ADP-D-glucose + (1,4-α-D-glucosyl)(n) → ADP + (1,4-α-D-glucosyl)(n+1)**			
	5	Sobic.010G022600.1	"similar to Granule-bound starch synthase 1, chloroplast precursor"	Wx	1735.74	1669.27
		Sobic.010G047700.1	"similar to Plastid starch synthase I precursor"	SssI	191.09	137.98
		Sobic.010G047700.2	"similar to Plastid starch synthase I precursor"	SssI	9.61	6.61
		Sobic.007G068200.1	"similar to Starch synthase DULL1" SSIII	SS-DULL1	66.54	42.29
		Sobic.010G093400.1	"similar to Starch synthase isoform zSTSII-1"	SSIIa	49.28	39.43
		Sobic.002G116000.1	"similar to Granule bound starch synthase IIa"	SSII	13.14	10.64
		Sobic.009G209300.1	"similar to Starch synthase-like protein"	SS?	5.93	5.72
		Sobic.006G221000.1	"similar to Starch synthase IIIb-1"	SSIII	2.38	2.21
		Sobic.001G239500.1	"similar to Putative uncharacterized protein" (SSI)	-	8.50	8.59
		Sobic.004G238600.1	"similar to Starch synthase IIb-2"	SSIIb	0.88	0.74
			a 1,4-α-D-glucan → starch			
	6	Sobic.004G163700.1	"similar to 1,4-alpha-glucan-branching enzyme 2, chloroplast precursor"	Ae1	318.48	255.11
		Sobic.010G273800.1	"similar to Starch branching enzyme I precursor"	-	152.23	134.44
		Sobic.006G066800.1	"similar to Starch branching enzyme IIa"	Sbe IIa	100.91	74.01
		Sobic.006G015800.2	"similar to Pullulanase-type starch debranching enzyme"	zpul	68.12	56.58
		Sobic.006G015800.1	"similar to Pullulanase-type starch debranching enzyme"	zpul	3.16	0.00
		Sobic.007G204600.1	"similar to Su1p" (Isoamylase)	Sbsu1	41.95	37.60
		Sobic.002G233600.1	"similar to Isoamylase-type starch debranching enzyme ISO3"	-	11.73	5.37
		Sobic.002G233600.2	"similar to Isoamylase-type starch debranching enzyme ISO3"	-	0.98	3.67
		Sobic.002G233600.3	"similar to Isoamylase-type starch debranching enzyme ISO3"	-	7.57	2.88
		Sobic.003G213800.2	"similar to Putative 1,4-alpha-glucan branching enzyme"	-	0.39	0.17
		Sobic.001G083900.1	"similar to Alpha 1,4-glucan phosphorylase L isozyme"	Pho1	92.07	84.99
		Sobic.001G083900.2	"similar to Alpha 1,4-glucan phosphorylase L isozyme"	Pho1	64.87	38.70
	-	Sobic.003G358600.1	"similar to Phosphorylase"	Pho2	20.39	19.22
	* Fragments per kilobase of exon per million fragments mapped					
	* Pathway Pleiotropy Index (A "-" means no allocated position)					

Data bars (red colour) indicate the value of FPKM in relation to the highest expressed gene model.

*Fragments per kilobase of exon per million fragments mapped. Pathway Pleiotropy index (A '–' means no allocated position).

The final stage in the SSP-1 (PPI-6) involves the creation of branched amylopectin via the action of SBEs and DBEs. Of the four isozymes of SBEs identified, *SbSBEIIb* or *Ae1* (Sobic.004G163700.1) was the most highly expressed with an FPKM of 318.48. Amongst the DBEs, *SBPUL* (Sobic.006G015800.2; 68.12) was more highly expressed than the isoamylase genes. Expression of the two genes functionally related to starch phosphorylase revealed that Sobic.001G083900.1 (Alpha 1,4-glucan phosphorylase L isozyme; *SbPHO1*) was more highly expressed with a day average FPKM of 92.07 with transcriptomic data revealing an expressed alternative transcript (Table 1 and Figure 1).

SSP-2 and SSP-3 initiate the conversion of sucrose + H_2O to D-fructose + α-D-glucose (PPI-1). Six genes were expressed at this point, with the highest expression observed for beta-fructofuranosidase (neutral invertase) with day average FPKM of 10.67 (Sobic.K041200.1) (Tables S2 and S3).

At PPI-2, for both SSP-2 and SSP-3, hexokinase enzymes convert either D-fructose + ATP to fructose-6-phosphate + ADP in the case of SSP-2, or to convert α-D-glucose + ATP to α-D-glucose 6-phosphate + ADP in the case of SSP-3. Fifty-two genes were expressed at PPI-2, with several homologs of hexokinase present, in particular, the gene hexokinase-8 with a day average FPKM of 67.79 (Table S2).

The conversion of fructose-6-phosphate to β-D-glucose-6-phosphate (PPI-3) of SSP-2, involves 22 genes. This included glucose-6-phosphate isomerase (Sobic.002G230600.1; FPKM 11.29), with the remaining genes involved with sugar transport or related to integral membrane proteins (Table S2).

At PPI-5 of SSP-2 and PPI-3 of SSP-3, the conversion of α-D-glucose 6-phosphate to α-D-glucose 1-phosphate takes place, with three genes containing phosphoglucomutase or phosphomannomutase functionality identified, chief of which is Sobic.001G116500.1, which had a day average FPKM of 58.45. After this point in both SSP routes, they proceed through the pathway steps already described for SSP-1 leading to starch synthesis (Tables S2 and S3).

Ka/Ks ratio and the correlation with pathway position

The Ka/Ks ratio can be used to deduce the direction and extent of selection, with a ratio greater than one denoting positive selection, less than one indicating purifying (stabilizing) selection and a ratio equal to one inferring neutral or no selection. Up to 85.7% of genes in the SSP-1 (all lines) had a Ka/Ks ratio of less than 1, with 18 of these genes identified as being significantly different from 1 (≤0.05), and only one gene had a Ka/Ks greater than 1 (1.63; sucrose synthase-6; Sobic.010G276700.1).

This trend was clear across groups and the two other SSP routes, which had 84% of Ka/Ks ratio values <1 (39 genes which were significantly different from 1, ≤0.05) and 85% of Ka/Ks ratio values <1 (33 genes which were significantly different to 1, ≤0.05), for SSP-2 and 3, respectively. This indicated that the SSP pathway is under the influence of strong purifying selection (Figure 2). No genes had a Ka/Ks significantly greater than 1.

Kendall's τ rank correlation coefficient revealed there was a clear correlation between PPI and Ka/Ks ratio for the primary SSP of the Landraces (τ value = 0.338, P-value ≤0.023). The correlation for SSP-1 remains significant (τ value = 0.298, P-value ≤0.05) even when the highest Ka/Ks value (1.84) is removed. The correlation between PPI and Ka/Ks ratio for all the lines combined (τ value = 0.202, P-value = 0.145) and the wild and weedy lines

(τ value = 0.201, P-value = 0.203) was not significant. However, the removal of the highest Ka/Ks value (1.62) from the combined analysis of all genotypes did result in a significant correlation (τ value = 0.284, P-value ≤0.044) (Figure 2). There was no significant correlation between any derivation of the Ka/Ks ratio data and PPI for either of the other SSPs.

Sequence diversity in cultivated and wild sorghum

The primary sequence level comparisons made in this study were focussed at the coding sequence (CDS) level. The mean level of $θ_π$ for the genes in the SSP-1 was 0.00108, with a range of 5.49E-05 for a sugar pyrophosphorylase (Sobic.001G424500.1) involved in conversion of UDP-D-glucose to α-D-glucose 1-phosphate up to 0.00445 for an ADP-glucose pyrophosphorylase (Sobic.001G462200.1) (Figure 1). For SSP-2, the mean $θ_π$ was 0.00125, ranging from 0 for a sugar transporter (Sobic.008G111100.1) involved in conversion of fructose-6-phosphate to β-D-glucose-6-phosphate up to 0.00678 for a cell wall invertase (Sobic.004G166700.1) (Table S6). Mean $θ_π$ for SSP-3 was 0.00137 ranging from 5.90E-05 for *SbPHO1* up to 0.00678 for the cell wall invertase in SSP-2 (Sobic.004G166700.1) (Table S7). Clear reductions in nucleotide diversity ($θ_π$) and high levels of genetic differentiation (F_{ST}) were observed in the genes in the primary SSP between the Landraces and the wild and weedy genotypes, based both at the whole CDS and nucleotide levels (Figures 3a and 4), with a ~1.4-fold reduction in mean $θ_π$ between Landraces (0.0009933) and wild and weedy lines (0.00119). Amongst the genes which showed a substantial decline in $θ_π$ between Landraces and wild and weedy genotypes were key enzymes at pathway position branch points including phosphoglucomutase (91.8%; Sobic.001G116500.1), glucose-6-phosphate isomerase (95.4%; Sobic.002G230600.1), various hexose transporters (90.5%; e.g. Sobic.003G084000.1) and glycosyltransferases (98%; Sobic.003G094600.1), as well as several AGPases such as *SbSH2* with a 92.4% reduction; *SbBT1* with a 75.4% reduction; and *SbSSIIa* (Sobic.010G093400.1) and *SbSSIIb* (Sobic.004G238600.1) with a 73.8% and 65.5% reduction, respectively (Tables S5 and S6).

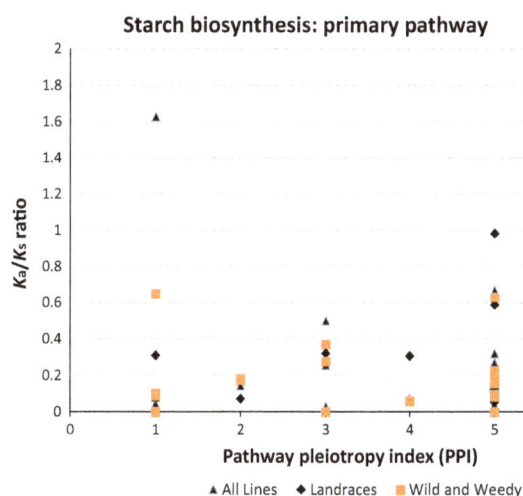

Starch biosynthesis: primary pathway

Figure 2 The relationship between K_a/K_s and Pathway Pleiotropy Index (PPI; Ramsay et al., 2009) for the Primary Starch Biosynthesis Pathway. K_a/K_s ratio values' comparisons are shown between sorghum groups 'all lines', 'Landraces' and 'wild and weedy'.

(a) **Pi (Θπ) for sorghum CDS sequences involved in the primary starch synthesis pathway between Landraces and Wild & Weedy lines**

(b) **Tajima's D for sorghum CDS sequences involved in the primary starch synthesis pathway between Landraces and Wild & Weedy lines**

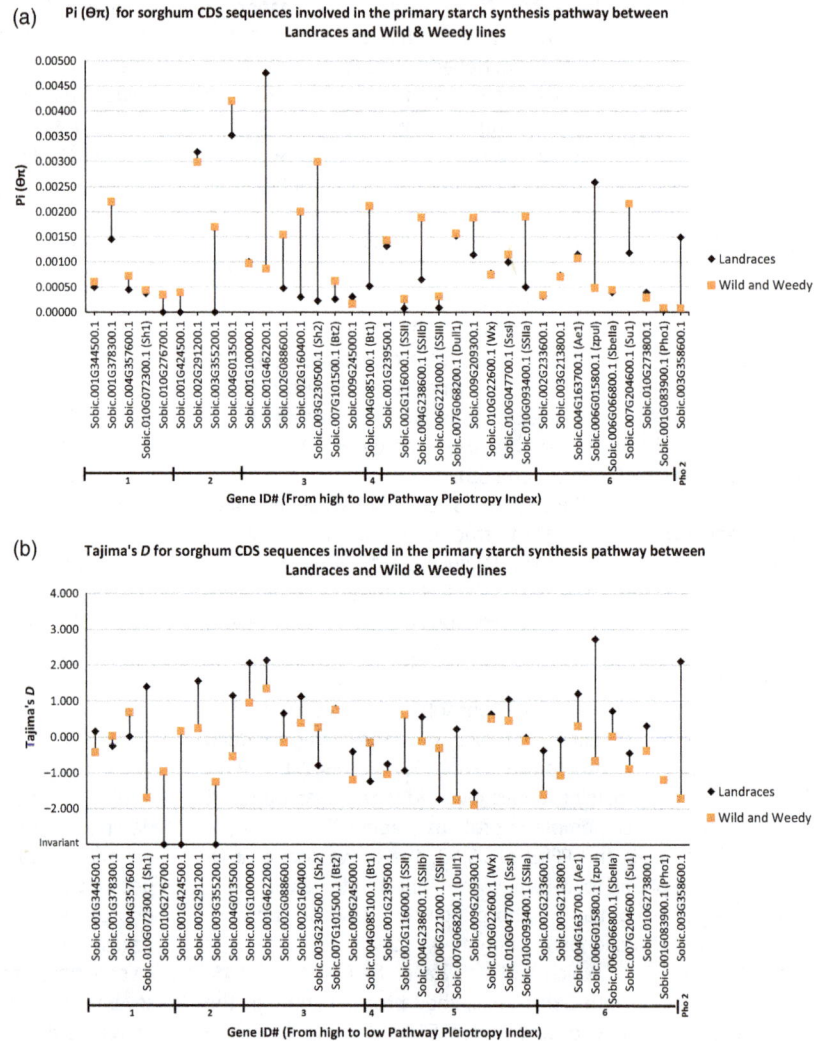

Figure 3 Comparison between Landraces and wild and weedy sorghum lines for CDS sequence of genes involved in the Primary Starch Synthesis Pathway for (a) nucleotide diversity (θ_π) and (b) Tajima's D. Tajima's D values for data bars marked with 'Invariant' are purely for graphical display only and in no way represent that actual value. Genes are sorted according to Pathway Pleiotropy Index (PPI; Ramsay *et al.*, 2009).

Conversely, several genes have shown an increase in θ_π between Landraces and wild and weedy lines, with ten genes in SSP-1, twenty-eight genes in SSP-2 and twenty-six genes in SSP-3. Chief amongst these genes for increased θ_π were *SbPHO2* (94.6%) and the DBE *SbPUL* (81.2%) (Table S5).

Two genes, encoding SEC15 (involved in vesicle trafficking) and a shikimate metabolic gene, respectively, were within the lower 5% of the distribution of nucleotide diversity, and they are not well characterized (Sobic.003G158700.1 and Sobic.010G066700.1). However, both are associated with the conversion of D-fructose to fructose-6-phosphate and had θ_π values of 8.73E-05 and 0.000173, respectively (Tables S6 and S7). Two genes have become invariant within Landraces since domestication, and these were a nucleoside-diphosphate-sugar pyrophosphorylase (Sobic.003G355200.1) and an uncharacterized gene (Sobic.010G066700.1) involved in the conversion of D-fructose to fructose-6-phosphate. Conversely, five genes were within the upper 95% of the empirical distribution of θ_π in the Landraces, including an alcohol dehydrogenase (0.0103; Sobic.010G071900.1), AGPase (0.00977; Sobic.001G462200.1), glycosyltransferase (0.00778; Sobic.010G144400.1), neutral invertase (0.00694; Sobic.004G166700.1) and a UDP-glucose pyrophosphorylase (0.00748; Sobic.004G013500.1) (Tables S5–S7).

Four genes had high levels of genetic differentiation (F_{ST}) between the Landraces and wild and weedy genotypes: *SBSH2* (0.764), hexokinase (0.63; Sobic.009G203500.1), hexose (*HEX6*) transporter (0.628; Sobic.001G297600.1) and a high-affinity potassium transporter (0.627; Sobic.006G061300.1) (Figure 4; Tables S5–S7).

Tests of selection at the whole gene and base pair level

In total, 71 genes were identified with signatures of purifying selection versus 62 genes with a signal of balancing selection, including *SbSH2* (purifying selection) and *SbPUL* (balancing selection) (Figures 5 and 6). These genes included several important classes such as AGPases, SS, SBE, DBE as well as genes that code for proteins at key branch points, for example sucrose synthase (Sobic.001G344500.1). Of these, 43 genes also had coding regions under both purifying and balancing selection, including hexokinases, sugar transporters, AGPases (e.g. *SbSH2*), SBE (e.g. *SbAE1*) and DBE (e.g. *SbPUL*) enzymes (Tables S9 and S10).

Fifteen sites in phosphoglucomutase (Sobic.001G116500.1) showed signatures of purifying selection, all synonymous, whereas only a single site was identified within 28 genes such as *SbAE1* (Table S9). In contrast, the number of codons identified as under balancing selection per gene ranged from 18 in both an alcohol dehydrogenase (seven nonsynonymous sites; Sobic.010G071900.1) and glycosyltransferase (nine nonsynonymous sites; Sobic.010G144400.1) through to only single sites identified within 23 genes such as SS *SbDU1* (Sobic.007G068200.1) (Table S10).

Within well-characterized genes, particularly those involved in the SSP after the conversion of glucose 1-phosphate to ADP-glucose, nonsynonymous SNPs were largely occurring in coding regions outside of known protein domains and/or substrate binding sites. For example, *SbSH2* has four nonconservative mutations under purifying selection, which results in changes to amino acid polarity and possibly in protein folding.

Six genes in the SSP were identified with signatures of purifying selection at the gene level. Four of these genes were detected within the primary SSP and include a nucleoside-diphosphate-sugar pyrophosphorylase (Sobic.003G355200.1), two AGPases including *SbSH2* (Sobic.003G230500.1 and Sobic.002G160400.1) and a sucrose synthase (Sobic.010G276700.1). The remaining genes included a hexokinase (Sobic.009G203500.1) and uncharacterized gene (Sobic.010G066700.1), which are located at PPI-2 of both the SSP-2 and SSP-3 pathways and play a part in the conversion of D-fructose into fructose-6-phosphate or the conversion of α-D-glucose into α-D-glucose 6-phosphate (Figure 4).

Nine genes were identified with signatures of balancing selection. The genes *SbPHO2* and *SbPUL* were identified at the lower PPI of the SSP. The other genes with signatures of balancing selection were detected higher up in the SSP, the majority at PPI-2 of both the SSP-2 and SSP-3 pathways, and included glycosyltransferase and alcohol dehydrogenase functions (Sobic.010G144400.1; Sobic.002G195700.1; Sobic.005G225500.1; Sobic.010G216 700.1; Sobic.010G071900.1; Sobic.006G211900.1).

Utilizing the mlHKA test to validate gene level domestication candidates for patterns of genetic variation (positive or balancing selection) showed that a model of directional selection best explained the patterns of polymorphism to divergence of the five candidate genes under purifying selection relative to 34 neutral loci (mean log likelihood ratio test statistic = 283.8; $P < 0.0001$ for all comparisons; Table S11). Likewise, a model of balancing selection best explained the patterns of genetic variation in the nine candidate genes under balancing selection (mean log likelihood ratio test statistic = 310.13; $P < 0.0001$ for all comparisons; Table S11).

Matches between orthologs from maize identified as domestication candidates (Hufford *et al.,* 2012; Jiao *et al.,* 2012) and sorghum (gene and base pair level) revealed eight orthologous genes involved in starch biosynthesis with evidence of selection signatures in both maize and sorghum (Tables S9 and S10), including genes such as SS (*SbSSIIa*), isoamylase (Sobic.007G204600.1), *SbSBEI* (Sobic.010G273800.1) and hexokinase-8 (Sobic.003G035500.1). For rice (He *et al.,* 2011; Huang *et al.,* 2012; Xu *et al.,* 2011), a hexose sugar transporter (*HEX6*) (Sobic.001G297600.1) and alcohol dehydrogenase (Sobic.006G211900.1) were identified with signatures of selection in both sorghum and rice.

The potential for genetic hitchhiking or selective sweeps, whereby strong positive selection on a favourable new mutation can lead to neighbouring chromosomal regions becoming fixed over time, was also examined. Fifty-two genes were analysed for selective sweeps based on their signatures of selection, association with grain quality traits or presence within SSP-1 (Table S8). Of these, 48 loci had evidence of selective sweeps with genes identified as being under selection or fixed (invariant) within sorghum groupings. The functional classes of genes under selection were broad and variable, with some of known importance to agronomic traits or domestication. These included *SbGhd7* (Sobic.001G298400.1), in the vicinity of a hexose transporter (Sobic.001G297600.1), which may confer differences in photoperiod sensitivity and flowering times. *SbRd* (Sobic.003G230900.1), located in the vicinity of the AGP (*SbSH2*), is thought to be involved in proanthocyanidin synthesis or pericarp colour. *Psy1* (Sobic.003G230900.1), which belongs to the phytoene synthase family of genes and is linked to carotenoid content and grain colour, was located near a sucrose synthase-like gene (Sobic.010G276700.1). And finally, *qSH1* (Sobic.003G356200.1), situated near a GDP-mannose pyrophosphorylase (Sobic.003G355200.1), involved in conversion of UDP-D-glucose to α-D-glucose 1-phosphate, which codes for a BEL1-type homeobox-containing protein involved in seed shattering in rice. Of all the SSP genes under selection, a UDP-glucose pyrophosphorylase (Sobic.004G013500.1; balancing selection) had the highest number of genes (6) clustered (four balancing and two purifying) around its genomic location, with flanking genes 0.073 and 0.063 Mb either side. The protein function of the genes surrounding the UDP-glucose pyrophosphorylase

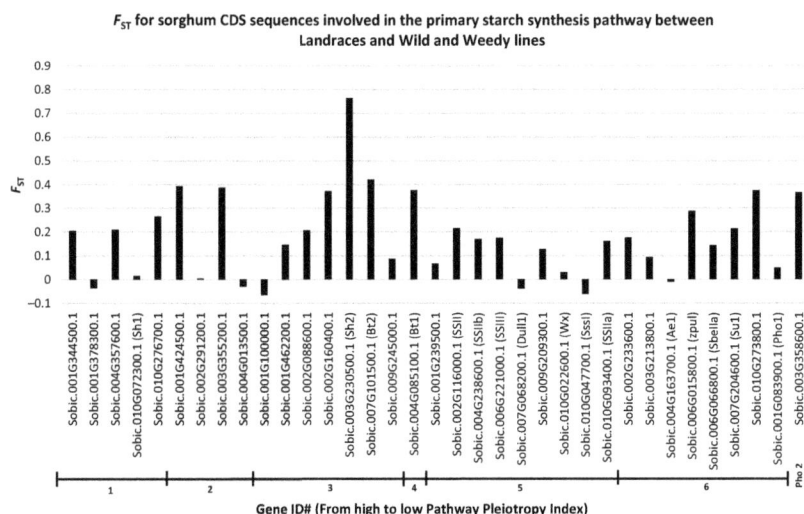

Figure 4 The F_{ST} of CDS sequence and genes of the primary Starch Synthesis Pathway of sorghum. F_{ST} value comparisons are shown between sorghum groups 'Landraces' and 'wild and weedy'. Genes are sorted according to Pathway Pleiotropy Index (PPI; Ramsay *et al.,* 2009).

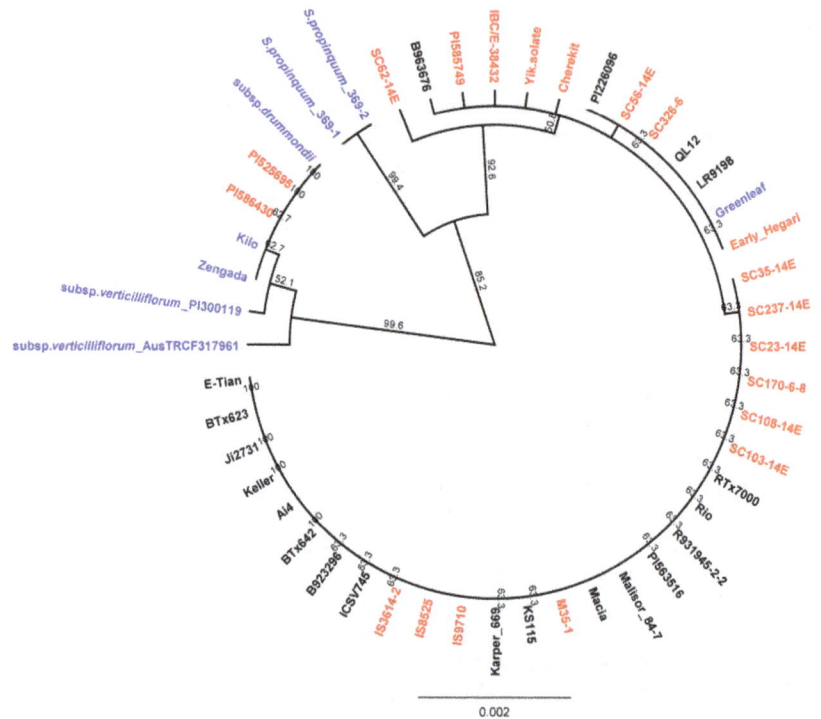

Figure 5 Phylogenetic tree for the whole gene sequence of the large subunit ADP-glucose pyrophosphorylase (*Sh2*; Sobic.003G230500.1) constructed utilizing the Unweighted Pair Group Method with Arithmetic mean (UPGMA) algorithm. Improved Inbred lines are labelled in 'Black'; Landraces are labelled in 'Red'; and Wild and Weedy lines are labelled in 'Blue'.

included a peptidyl-prolyl cis-trans isomerase (Sobic.004G013400.1) and putative glutamate receptor (Sobic.004G013300.1).

To measure any deviation from the standard coalescent model, which assumes at its most basic level, that no recombination, natural selection or gene flow is occurring within a population, the statistical test Tajima's D was employed. The results revealed a mean Tajima's D value for genes in the primary SSP of −0.144, ranging from −1.80 to 2.27 for *SbBT1* and AGPase (Sobic.001G462200.1), respectively (Figure 3b; Table S5). Likewise, the mean level of Tajima's D for SSP-2 and SSP-3 was −0.332 and −0.254, respectively (Tables S6 and S7). The lowest Tajima's D value for these pathways was for a sugar transporter (Sobic.008G111100.1), which was invariant, and the highest an alcohol dehydrogenase (Sobic.006G211900.1) with a value of 3.12. The mean level of Tajima's D for all the pathways deviates from the standard neutral model (< or >1), with the magnitude of the statistic trending towards purifying selection.

Several genes had a clear reduction in Tajima's D between the Landraces and wild and weedy genotypes, including granule bound starch synthase II (−0.93 vs 0.63; *SbGBSSII*, Sobic.002G116000.1), *SbSSIII* (−1.74 vs −0.30; Sobic.006G221000.1), *SbBT1* (−1.24 vs −0.15), *SbSH2* (−0.79 vs 0.27) and sucrose synthase-3 (−0.25 vs 0.04; Sobic.001G378300.1) (Figure 3b; Table S5). Genes specific to SSP-2 and SSP-3 with a clear reduction in Tajima's D between the groups included key hexokinases such as hexokinase-8 (−0.50 vs 0.12; Sobic.003G035500.1) and hexokinase-3 (−0.87 vs 0.54; Sobic.003G421200.1) as well as sugar transporters such as this mannitol transporter (invariant vs 0.17; Sobic.001G469600.1) and hexose transporter (−1.19 vs −0.30; Sobic.003G084000.1) (Tables S6 and S7). Twenty-six genes showed a clear increase in Tajima's D in the Landrace genotypes in comparison with the wild and weedy genotypes in SSP-1. This pattern was also observed for 59 genes in SSP-2 and 52 genes in SSP-3 and included such genes as *SbPHO2*, *SbPUL*, sucrose synthase-1 (Sobic.010G072300.1),

cathepsin B-like cysteine protease (Sobic.002G315800.1) and glycosyltransferase (Sobic.010G144400.1) (Tables S5–S7).

Discussion

Selective constraints were found to be unevenly distributed within the primary SSP of sorghum. The level of purifying selection generally correlates with the hierarchical position of the genes, with upstream genes invariably being the most constrained (Waxman and Peck, 1998). This observation has also been made within unidirectional plant metabolic pathways such as isoprene (Sharkey *et al.*, 2005), anthocyanin (Lu and Rausher, 2003; Rausher *et al.*, 1999, 2008), carotenoid (Clotault *et al.*, 2012; Livingstone and Anderson, 2009) and terpenoid biosynthesis (Ramsay *et al.*, 2009). Several of the gene families involved in starch synthesis at the lower pathway position levels (from the conversion of glucose 1-phosphate to ADP-glucose) encode key enzymes responsible for the diversity of different starch qualities and quantities for a multitude of end uses. Hence, human selection is certain to play a role in their fixation or diversification. Many of these genes are the products of ancient whole-genome duplications, for example AGPase, SS (Paterson *et al.*, 2004), which have led to subfunctionalization as well as the formation of protein complexes (Tetlow *et al.*, 2008). Particular homologs of these genes that undergo higher evolutionary rates (Li *et al.*, 2012), in some cases, asymmetrically (Corbi *et al.*, 2011; Georgelis *et al.*, 2007) were observed. In contrast, purifying selection largely affected genes higher up in the SSP. This may explain why PRV can occur within the primary SSP regardless of the bidirectional and/or cyclic action of many of the downstream enzymes.

However, this result does not mirror the observations made in rice and its wild ancestor (*O. rufipogon*). Yu *et al.* (2011) did not detect any correlation between the levels or patterns of diversity and pathway position within the SSP. Our study extended analysis

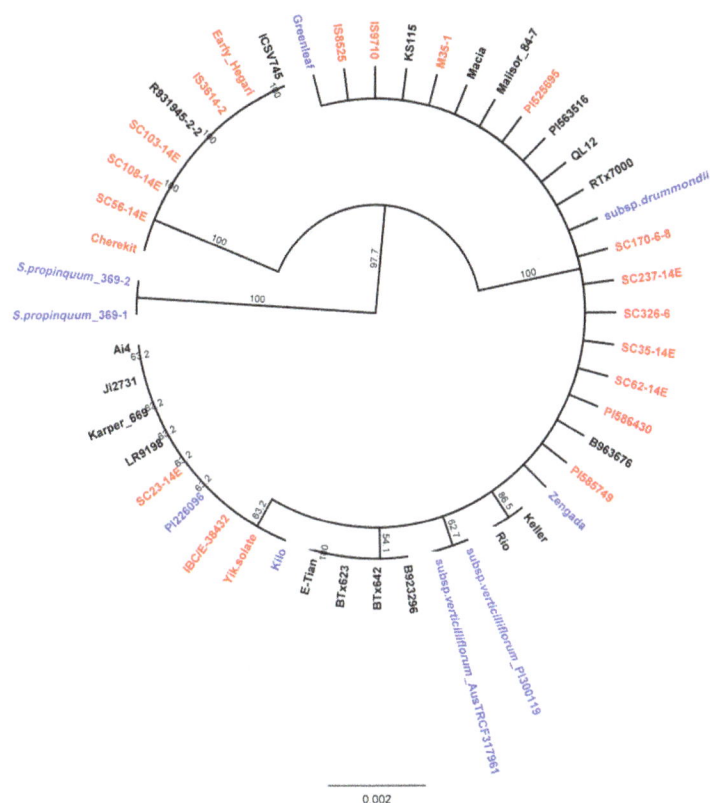

Figure 6 Phylogenetic tree for the whole gene sequence of starch debranching enzyme pullulanase (Sobic.006G015800.1) constructed utilizing the Unweighted Pair Group Method with Arithmetic mean (UPGMA) algorithm. Improved Inbred lines are labelled in 'Black'; Landraces are labelled in 'Red'; and Wild and Weedy lines are labelled in 'Blue'.

of pathway position prior to the conversion of glucose 1-phosphate to ADP-glucose, which may explain these differences.

There were 47 genes present for the conversion of D-fructose to fructose-6-phosphate, which may explain our observation that PRV was not occurring for the minor routes leading to starch synthesis (SSP-2 & SSP-3). Without more experimental evidence, it is impossible to rule candidate genes in or out for many of the stages higher up in SSP and this result should be re-examined in future when such data become available.

Distinct signatures of selection amongst genes involved in the SSP during domestication and evident under both natural and human selection conditions reiterate the findings that the genes involved in the SSP are remarkably conserved amongst grasses (Li et al., 2012). This changing flux in purifying selection revealed several genes to have notable reductions in nucleotide diversity including isozymes of ADP-glucose pyrophosphorylase (e.g. SbSH2), SbBT1, SbSSIIa, as well as phosphoglucomutase, glucose-6-phosphate isomerise, hexose transporters and glycosyltransferases.

In concordance with Hamblin et al. (2007), there were clear divergences in the level of selection between certain sorghum genes and their ortholog in maize (da Fonseca et al., 2015; Whitt et al., 2002) and rice genes (Yu et al., 2011). The most prominent examples were the increased nucleotide diversity of SBEIIb (Ae1) and isoamylase (su1). Correspondingly, the reduction in nucleotide divergence for Wx, Bt2 and Sh2 compared with maize was also detected (de Alencar Figueiredo et al., 2008; Hamblin et al., 2007). This underlined that the selection acting upon particular genes within the SSP through domestication, essentially leading to larger grain size, higher starch content and different starch structural properties, can take alternative paths.

The central importance of starch production to higher plants as a means to store high-density energy in the form of carbohydrate, for use with essential growth and metabolic requirements, is already well documented (Tetlow, 2011). Likewise, the fundamental importance of cereal starch for human needs and industrial uses (Burrell, 2003) explains why signatures of purifying selection have been found not only in genes present both upstream and downstream within the starch synthesis pathway of the Landraces but also within the wild and weedy lines. The interfertility within Sorghum also means that many wild relatives can readily cross-pollinate with improved germplasm (e.g. Sagnard et al., 2011).

This study identified as many as 90 genes under some form of selection. Moreover, 43 upstream genes have showed evidence of differential selection, whereby specific genes have several base pairs showing signatures of purifying selection and other regions showing balancing selection.

The phosphoglucomutase and the AGPase subunit Sh2 genes encode enzymes which work sequentially in the SSP via generation of Glc-1-P from Glc-6-P and then by Glc-1-P and ATP to generate ADPGlc and inorganic pyrophosphate (Tetlow, 2011). Both genes were under purifying selection. While all enzymes of a pathway contribute to the control of metabolic flux into a product, strong evidence exists, particularly in potato tubers, that both of these enzymes make a significant contribution to the flux control conversion of sucrose to starch (Geigenberger et al., 2004). While the contributions of these enzymes to the control of flux into starch have not been quantified in cereal endosperm, mutants of AGPase do lead to a low-starch phenotype (Stark et al., 1992) while overexpression provides the essential substrate needed for greater starch content and seed size in domesticated cereals (Tuncel and Okita, 2013). Previously identified as being a target for domestication in sorghum (de Alencar Figueiredo et al., 2008; Hamblin et al., 2007), certain allelic variants of Sh2 and Bt2, which work synergistically as a protein complex, have been shown to be associated with yield in sorghum (de Alencar Figueiredo et al., 2010). Interestingly, the enzyme believed to make the greatest contribution to flux control during starch accumulation (Geigenberger et al., 2004), the ADP-glucose transporter (Bt1), which is the primary route for transport of ADP-glucose into the amyloplast, did not generate the strongest signals of purifying selection. However, clear reductions in nucleotide diversity and Tajima's D between Landraces and wild and weedy genotypes were observed, combined with purifying selection at the codon level of three sites (two of which were nonsynonymous) and balancing selection at two sites, suggesting that domestication has made identifiable alterations to this key protein.

Nine genes were under balancing selection at the whole gene level within the SSP, including such genes as SbPHO2, SbPUL, UDP-glucose pyrophosphorylase (Sobic.004G013500.1) and various uncharacterized glycosyltransferase and alcohol dehydrogenase genes. The functional role of pullulanase and its influence upon starch structural properties is not well understood. Knockout mutations for this gene in maize and rice has not revealed clear grain phenotypes (Dinges et al., 2003; Fujita et al., 2009), but a distinct relationship between a particular pullulanase allelic variant and increased sorghum grain digestibility has been demonstrated (Gilding et al., 2013). While no clear connection between the high digestibility allele type (SbPUL-RA) and end use has been established, the well-defined partition in the gene tree between genotypes carrying this allele and the others carrying its less digestible counterpart (SbPUL-GD) could be a reason for the balancing selection acting upon this gene (Gilding et al., 2013).

UDP-glucose pyrophosphorylase is a key enzyme for carbohydrate metabolism, catalysing the reversible production of UDP-glucose and pyrophosphate from Glc 1-P and UTP. The UDP-glucose resulting from this reaction also serves as a key substrate in the biosynthesis of cell wall polysaccharides (Gibeaut, 2000). While it has been shown that reducing UGPase activity via mutation has little effect upon carbohydrate content under controlled conditions (Meng et al., 2009; Zrenner et al., 1995), Arabidopsis mutants had significantly decreased fitness under field conditions evidenced by a 50% reduction in seed set than wild-type plants (Meng et al., 2009). A proposed mechanism for the reduced fitness was the observation in mutant studies of rice, that the UGPase enzyme was essential for pollen viability (Chen et al., 2007; Woo et al., 2008) as well as the need for starch during pollen development (Mu et al., 2009). While UGPase is believed not to be a rate-limiting enzyme in SSPs, it is possible that balancing selection may be responding to the interacting selective pressures of larger seed size as well as overall yield, which is part of the domestication syndrome, or its essential role in providing UDP-glucose for cell wall biogenesis means that in actively dividing cells, such as developing endosperm, its activity would likely be an important yield determining factor at early stages of development during active cell division prior to grain filling. Conversely, other isozymes of this gene family are shown to be under purifying selection in sorghum (Sobic. 003G355200.1), illustrating the complex dynamics of selection on this pathway.

Understanding why balancing selection has maintained polymorphism in starch phosphorylase (SbPHO2) is less clear. Starch phosphorylase catalyses the reversible transfer of glucosyl units from glucose-1-phosphate to the nonreducing end of α-1-4

linked glucan chains and can serve in either a degradative or synthetic role depending on the relative concentration of its substrates (Tetlow, 2011). Two major isoforms of starch phosphorylase enzyme are known to exist in higher plants and have been termed plastidic (chloroplast/amyloplast, Pho1) and cytosolic (Pho2), respectively (Nakano and Fukui, 1986). In terms of starch synthesis, the exact action of Pho1 is unclear; however, several groups have shown that its gene expression correlates with biosynthesis in developing endosperm of maize (Yu et al., 2001) and rice (Satoh et al., 2008) as well as in this study. Further, Satoh et al. (2008) demonstrated via Pho1 rice mutants that this gene plays a central role in starch synthesis, content and structure during low temperatures. There is no known role for Pho2 in starch metabolism as starch granules are bound within the chloroplast/amyloplasts, and its relative expression compared with Pho1 was lower in this study and the Morokoshi database (Makita et al., 2015). Hypothetical roles for Pho2 extend to the possible degradation of reserve starch in plant organs in which the starch-containing cells have lost their compartmental integrity (Buchner et al., 1996; Schupp and Ziegler, 2004) and/or metabolizing products of starch degradation as well as regulating cytosolic Glc-1-P levels (Rathore et al., 2009).

Sugar transporters are expressed during sorghum grain fill, yet, due to a lack of experimental information, we were unable to integrate those genes into the primary SSP and had to rely on the information detailed in the SorghumCyc online database. In fact, the full list of genes analysed in this study may not be exhaustive or conversely and contain members which do not significantly participate in starch synthesis; for example, several genes had FPKM expression close to zero, particularly within PPI-2 of SSP-2 and SSP-3. Bioinformatic analysis of differential expression within the RNA-Seq data also showed a lack of statistically significant differences in diurnal expression for any of the SSP genes. SBEs have been reported to be diurnally expressed (Mutisya et al., 2009), and our data show that the SBEs are ≈30% downregulated during the night; however, the two approaches are not directly comparable since our analysis is quantitative versus the semiquantitative analysis conducted by Mutisya et al. (2009). Ultimately, relative differences in transcript abundance do not provide any information regarding enzyme activity which is more reliant on the amount of enzyme protein synthesized and the extent to which this does or does not turn over within the cell.

The allelic diversity present within a number of starch synthesis loci was substantial amongst the diverse sorghum set analysed in this study and included nonsynonymous SNPs that may lead to altered protein function, providing an opportunity to conduct further starch structural studies and enhance our understanding of this complex synthesis process. This would be particularly useful for the study of the SBEs, which may possess allelic variants that could generate high amylose starches which are not currently available in sorghum and are valuable for their human health attributes and utility in the manufacture of biomaterials. Given the levels of allelic diversity observed within our restricted sampling of wild relatives (both verticilliflorum and propinquum), and the guinea-margaritiferums, germplasm from these sources represent diverse and valuable sources of new genetic variation.

Materials and methods

Plant materials

Resequencing data from accessions of Sorghum bicolor, representing all races of cultivated sorghum, in addition to its progenitors and *S. propinquum* were studied (Mace *et al.*, 2013) (Table S1). The NGS data achieved an average final effective mapping depth of ≈22× per line (16× to 45×) with a SNP calling accuracy of 99.85% and 99.72% as validated via targeted sequencing and whole-genome *de novo* assembly of representative lines, respectively.

Sample preparation for RNA-Seq and subsequent analysis

The RNA-Seq experiment generated transcriptomic data from grain heads sampled 16 days after anthesis, within the critical window for peak grain fill. In brief, samples were run through a pipeline consisting of the RNA extraction protocol of Li and Trick (2005) followed by standard sample preparation for RNA-seq on the Ion Torrent platform (Thermo Fisher Scientific, Waltham, MA). RSEM, a software package for determining expression counts for transcripts, was employed to derive the FPKM values used and fed into a differential expression analysis using EBSeq (Detailed summary; Method S1).

Identification of starch biosynthesis genes

A multistage process was employed to identify a list of high-confidence starch biosynthesis genes. The first stage involved utilizing the SorghumCyc online database (http://pathway. gramene.org/gramene/sorghumcyc.shtml), to identify the complete set of genes believed to be components of the SSP based on known or predicted gene identification, using published resources or homology and/or hidden Markov models (HMM) calculations. There are several SSP routes, and this study mainly focused on the primary route in monocots (SSP-1), which initiates with the breakdown of sucrose into UDP-D-glucose + D-fructose and continues with the conversion of UDP-D-glucose into ADP-D-glucose, which is primarily transported by an ADP-glucose transporter (*Bt1*) into the amyloplast for final starch synthesis (Figure 1). SSP schematics via metabolism of D-fructose (SSP-2) and α-D-glucose (SSP-3) were also analysed, but due to the dynamic process of starch synthesis these pathways are not the only possible routes (Figure 1). In the second stage, this list of genes was then compared to genes shown to be expressed in developing endosperm via the RNA-Seq experiment, identifying 35 genes in SSP-1, 106 genes in SSP-2 and 86 genes in SSP-3.

In addition to the genes confirmed to be expressed via the RNA-Seq data, a further 3 genes were added to the analysis list, based on published data that suggested they played a prominent role in starch synthesis. These genes were *PULLULANASE* (*SbPUL*, Sobic.006G015800.1), a starch DBE with certain allelic variants shown to contribute to sorghum grain *in vitro* digestibility (Gilding *et al.*, 2013), and two genes consisting of the plastidial *STARCH PHOSPHORYLASE 1* (*SbPHO1*, Sobic.001G083900.1) and cytosolic *STARCH PHOSPHORYLASE 2* (*SbPHO2*, Sobic.003G358600.1), which are functionally related to starch phosphorylase (alpha 1,4-glucan phosphorylase) and have been shown to play a prominent role in starch metabolism (Blennow *et al.*, 2002) and is functionally expressed (Ohdan *et al.*, 2005). All genes were expressed in immature sorghum grain (Tables 1, S2 and S3). Genomic sequence of the 114 identified starch metabolism genes in sorghum was extracted from the NGS data (Mace *et al.*, 2013) with a focus on Landraces and wild and weedy groups (Table S1). These sorghum genes were also contrasted with orthologs from maize (Hufford *et al.*, 2012; Jiao *et al.*, 2012) and rice (He *et al.*,

2011; Huang *et al.*, 2012; Xu *et al.*, 2011) that have been characterized as under selection.

Assignation of pathway position and assessments of selection

Pathway position for the 114 genes found within the 3 SSP routes was assigned to their location based on the Pathway Pleiotropy Index (PPI) (Ramsay *et al.*, 2009), which ascribes 'single' or 'groups' of enzymes that act between pathway branch points to their set position in the metabolic pathway and numbers them from upstream to downstream positions. Selection of PPI to assess PRV was made due to its superiority to track changes in pleiotropy along a pathway and avoid the influence of factors that can mitigate the role of pathway structure upon evolutionary rate variation such as gene duplication.

The degree of selective pressure acting on protein coding regions was measured utilizing the Ka/Ks ratio. Calculating Ka and Ks substitution rates was conducted using the software package KaKs_Calculator1.2 (MYN method) (Zhang *et al.*, 2006). Correlation between Ka/Ks ratio and PPI was calculated utilizing Kendall's τ rank correlation coefficient (Ramsay *et al.*, 2009).

Population genetic analyses

Nucleotide diversity per site (θ_π) (Nei, 1987), Watterson's estimator (θ_W) (Watterson, 1975) and the neutrality test Tajima's D (Tajima, 1989) were estimated for (CDS and gene) each of the 114 starch synthesis genes (Mace *et al.*, 2013). These parameters were directly calculated for each genetic component using a BioPerl module and an in-house perl script. F_{ST} was calculated on the same genetic components to measure population differentiation using an alternative BioPerl module (Mace *et al.*, 2013).

Genes with signatures of purifying selection were identified by the criteria utilized in Mace *et al.* (2014): θ_π and θ_W in the lower 5% of the empirical distribution in the descendent population, F_{ST} values >95% of the empirical distribution and negative Tajima's D. Likewise, genes with signatures of balancing selection were identified as follows: θ_π and θ_W in the upper 25% of the empirical distribution in the descendent population, Tajima's D in the upper 5% of the empirical distribution and F_{ST} values <90% of the population pairwise distribution.

To validate genes identified as under selection at the whole gene level, five candidate genes under purifying selection and nine candidate genes under balancing selection were used as input, together with 34 neutral genes, for the mlHKA test (Wright and Charlesworth, 2004). The mlHKA program was run under a neutral model (numselectedloci = 0) and then under a selection model (numselectedloci > 0). Significance was assessed by the mean log likelihood ratio test statistic, where twice the difference in log likelihood between the models is approximately chi-squared distributed with df equal to the difference in the number of parameters.

To calculate selection at the base pair level, θ_π, Tajima's D and F_{ST} were calculated from CDS sequences of all genes using PopGenome (Pfeifer *et al.*, 2014). Criteria to identify sites of purifying selection were as follows: (i) reduction in diversity in the pairwise ancestor/descendant population comparison, greater than the mean gene diversity in the whole genome; (ii) $F_{ST} > 0$; and (iii) negative Tajima's D. Criteria to identify sites of balancing selection were as follows: (i) increase in diversity in the pairwise ancestor/descendant population comparison, greater than the mean gene diversity in the whole genome; (ii) $F_{ST} > 0$; and (iii) positive Tajima's D.

Acknowledgements

This work was partially supported by the Grains Research and Development Corporation (GRDC) project UQ76 and the Australian Research Council (ARC) projects DP0986043, DP140102505 and LP0990626. We also acknowledge funding support from the University of Queensland, the Department of Agriculture, Fisheries and Forestry—Queensland.

References

de Alencar Figueiredo, L.F., Calatayud, C., Dupuits, C., Billot, C., Rami, J.F., Brunel, D., Perrier, X. et al. (2008) Phylogeographic evidence of crop neodiversity in sorghum. Genetics, 179, 997–1008.

de Alencar Figueiredo, L., Sine, B., Chantereau, J., Mestres, C., Fliedel, G., Rami, J.F., Glaszmann, J.C. et al. (2010) Variability of grain quality in sorghum: association with polymorphism in Sh2, Bt2, SssI, Ae1, Wx and O2. Theor. Appl. Genet. 121, 1171–1185.

Blennow, A., Nielsen, T.H., Baunsgaard, L., Mikkelsen, R. and Engelsen, S.B. (2002) Starch phosphorylation: a new front line in starch research. Trends Plant Sci. 7, 445–450.

Buchner, P., Borisjuk, L. and Wobus, U. (1996) Glucan phosphorylases in Vicia faba L.: cloning, structural analysis and expression patterns of cytosolic and plastidic forms in relation to starch. Planta, 199, 64–73.

Burrell, M.M. (2003) Starch: the need for improved quality or quantity – an overview. J. Exp. Bot. 54, 451–456.

Casa, A.M., Mitchell, S.E., Hamblin, M.T., Sun, H., Bowers, J.E., Paterson, A.H., Aquadro, C.F. et al. (2005) Diversity and selection in sorghum: simultaneous analyses using simple sequence repeats (SSRs). Theor. Appl. Genet. 111, 23–30.

Casa, A.M., Mitchell, S.E., Jensen, J.D., Hamblin, M.T., Paterson, A.H., Aquadro, C.F. and Kresovich, S. (2006) Evidence for a selective sweep on chromosome 1 of cultivated sorghum. Plant Genome, 46, S27–S40.

Chen, R.Z., Zhao, X., Shao, Z., Wei, Z., Wang, Y.Y., Zhu, L.L., Zhao, J. et al. (2007) Rice UDP-glucose pyrophosphorylase1 is essential for pollen callose deposition and its cosuppression results in a new type of thermosensitive genic male sterility. Plant Cell, 19, 847–861.

Clotault, J., Peltier, D., Soufflet-Freslon, V., Briard, M. and Geoffriau, E. (2012) Differential selection on carotenoid biosynthesis genes as a function of gene position in the metabolic pathway: a study on the carrot and dicots. PLoS One, 7, e38724.

Corbi, J., Debieu, M., Rousselet, A., Montalent, P., Le Guilloux, M., Manicacci, D. and Tenaillon, M.I. (2011) Contrasted patterns of selection since maize domestication on duplicated genes encoding a starch pathway enzyme. Theor. Appl. Genet. 122, 705–722.

Dinges, J.R., Colleoni, C., James, M.G. and Myers, A.M. (2003) Mutational analysis of the pullulanase-type debranching enzyme of maize indicates multiple functions in starch metabolism. Plant Cell, 15, 666–680.

da Fonseca, R.R., Smith, B.D., Wales, N., Cappellini, E., Skoglund, P., Fumagalli, M., Samaniego, J.A. et al. (2015) The origin and evolution of maize in the Southwestern United States. Nat. Plants, 1, 14003. doi:10.1038/NPLANTS.2014.3.

Fraser, H.B., Hirsh, A.E., Steinmetz, L.M., Scharfe, C. and Feldman, M.W. (2002) Evolutionary rate in the protein interaction network. Science, 296, 750.

Frere, C.H., Prentis, P.J., Gilding, E.K., Mudge, A.M., Cruickshank, A. and Godwin, I.D. (2011) Lack of low frequency variants masks patterns of non-neutral evolution following domestication. PLoS One, 6, e23041.

Fujita, N., Toyosawa, Y., Utsumi, Y., Higuchi, T., Hanashiro, I., Ikegami, A., Akuzawa, S. et al. (2009) Characterization of pullulanase (PUL)-deficient mutants of rice (Oryza sativa L.) and the function of PUL on starch biosynthesis in the developing rice endosperm. J. Exp. Bot. 60, 1009–1023.

Geigenberger, P., Stitt, M. and Fernie, A.R. (2004) Metabolic control analysis and regulation of the conversion of sucrose to starch in growing potato tubers. Plant Cell Environ. 27, 655–673.

Georgelis, N., Braun, E.L., Shaw, J.R. and Hannah, L.C. (2007) The two AGPase subunits evolve at different rates in angiosperms, yet they are equally sensitive to activity-altering amino acid changes when expressed in bacteria. Plant Cell, 19, 1458–1472.

Gibeaut, D.M. (2000) Nucleotide sugars and glucosyltransferases for synthesis of cell wall matrix polysaccharides. Plant Physiol. Biochem. 38, 69–80.

Gilding, E.K., Frère, C.H., Cruickshank, A., Rada, A.K., Prentis, P.J., Mudge, A.M., Mace, E.S. et al. (2013) Allelic variation at a single gene increases food value in a drought-tolerant staple cereal. Nat. Commun. 4, 1483.

Hamblin, M.T., Mitchell, S.E., White, G.M., Gallego, J., Kukatla, R., Wing, R.A., Paterson, A.H. et al. (2004) Comparative population genetics of the panicoid grasses: sequence polymorphism, linkage disequilibrium and selection in a diverse sample of Sorghum bicolor. Genetics, 167, 471–483.

Hamblin, M.T., Casa, A.M., Sun, H., Murray, S.C., Paterson, A.H., Aquadro, C.F. and Kresovich, S. (2006) Challenges of detecting directional selection after a bottleneck: lessons from Sorghum bicolor. Genetics, 173, 953–964.

Hamblin, M.T., Salas Fernandez, M.G., Tuinstra, M.R., Rooney, W.L. and Kresovich, S. (2007) Sequence variation at candidate loci in the starch metabolism pathway in sorghum: prospects for linkage disequilibrium mapping. Plant Genome, 2, 125–134.

He, Z., Zhai, W., Wen, H., Tang, T., Wang, Y., Lu, X., Greenberg, A.J. et al. (2011) Two evolutionary histories in the genome of rice: the roles of domestication genes. PLoS Genet. 7, e1002100.

Huang, X., Kurata, N., Wei, X., Wang, Z.X., Wang, A., Zhao, Q., Zhao, Y. et al. (2012) A map of rice genome variation reveals the origin of cultivated rice. Nature, 490, 497–501.

Hufford, M.B., Xu, X., van Heerwaarden, J., Pyhäjärvi, T., Chia, J.M., Cartwright, R.A., Elshire, R.J., et al. (2012) Comparative population genomics of maize domestication and improvement. Nat. Genet. 44, 808–811.

James, M.G., Denyer, K. and Myers, A.M. (2003) Starch synthesis in the cereal endosperm. Curr. Opin. Plant Biol. 6, 215–222.

Jiao, Y., Zhao, H., Ren, L., Song, W., Zeng, B., Guo, J., Wang, B. et al. (2012) Genome-wide genetic changes during modern breeding of maize. Nat. Genet. 44, 812–815.

Li, Z.W. and Trick, H.N. (2005) Rapid method for high-quality RNA isolation from seed endosperm containing high levels of starch. Biotechniques, 38, 872–876.

Li, C., Li, Q., Dunwell, J.M. and Zhang, Y. (2012) Divergent evolutionary pattern of starch biosynthetic pathway genes in grasses and dicots. Mol. Biol. Evol. 29, 3227–3236.

Livingstone, K. and Anderson, S. (2009) Patterns of variation in the evolution of carotenoid biosynthetic pathway enzymes of higher plants. J. Hered. 100, 754–761.

Lu, Y. and Rausher, M.D. (2003) Evolutionary rate variation in anthocyanin pathway genes. Mol. Biol. Evol. 20, 1844–1853.

Mace, E.S., Tai, S., Gilding, E.K., Li, Y., Prentis, P.J., Bian, L., Campbell, B.C. et al. (2013) Whole-genome sequencing reveals untapped genetic potential in Africa's indigenous cereal crop sorghum. Nat. Commun. 4, 2320. doi:10.1038/ncomms3320.

Mace, E., Tai, S., Innes, D., Godwin, I., Hu, W., Campbell, B., Gilding, E. et al. (2014) The plasticity of NBS resistance genes in sorghum is driven by multiple evolutionary processes. BMC Plant Biol. 14, 253.

Makita, Y., Shimada, S., Kawashima, M., Kondou-Kuriyama, T., Toyoda, T. and Matsui, M. (2015) MOROKOSHI: transcriptome database in Sorghum bicolor. Plant Cell Physiol. 56, e6.

Meng, M., Geisler, M., Johansson, H., Harholt, J., Scheller, H.V., Mellerowicz, E.J. and Kleczkowski, L.A. (2009) UDP-glucose pyrophosphorylase is not rate-limiting, but is essential in Arabidopsis. Plant Cell Physiol. 50, 998–1011.

Morris, G.P., Ramu, P., Deshpande, S.P., Hash, C.T., Shah, T., Upadhyaya, H.D., Riera-Lizarazu, O. et al. (2013) Population genomic and genome-wide association studies of agroclimatic traits in sorghum. Proc. Natl Acad. Sci. USA, 110, 453–458.

Mu, H., Ke, J.H., Liu, W., Zhuang, C.X. and Yip, W.K. (2009) UDP-glucose pyrophosphorylase2 (OsUgp2), a pollen-preferential gene in rice, plays a critical role in starch accumulation during pollen maturation. Chinese Sci. Bull. 54, 234–243.

Mutisya, J., Sun, C., Rosenquist, S., Baguma, Y. and Jansson, C. (2009) Diurnal oscillation of SBE expression in sorghum endosperm. J. Plant Physiol. 166, 428–434.

Nakano, K. and Fukui, T. (1986) The complete amino acid sequence of potato alpha-glucan phosphorylase. *J. Biol. Chem.* **261**, 8230–8236.

Nei, M. (1987) *Molecular Evolutionary Genetics*, p. 512. New York: Columbia University Press.

Ohdan, T., Francisco, P.B. Jr, Sawada, T., Hirose, T., Terao, T., Satoh, H. and Nakamura, Y. (2005) Expression profiling of genes involved in starch synthesis in sink and source organs of rice. *J. Exp. Bot.* **56**, 3229–3244.

Paterson, A.H., Bowers, J.E. and Chapman, B.A. (2004) Ancient polyploidization predating divergence of the cereals, and its consequences for comparative genomics. *Proc. Natl Acad. Sci. USA*, **101**, 9903–9908.

Pfeifer, B., Wittelsbürger, U., Ramos-Onsins, S.E. and Lercher, M.J. (2014) PopGenome: an efficient Swiss army knife for population genomic analyses in R. *Mol. Biol. Evol.* **31**, 1929–1936.

Ramsay, H., Rieseberg, L.H. and Ritland, K. (2009) The correlation of evolutionary rate with pathway position in plant terpenoid biosynthesis. *Mol. Biol. Evol.* **26**, 1045–1053.

Rathore, R., Garg, N., Garg, S. and Kumar, A. (2009) Starch phosphorylase: role in starch metabolism and biotechnological applications. *Crit. Rev. Biotechnol.* **29**, 214–224.

Rausher, M.D., Miller, R.E. and Tiffin, P. (1999) Patterns of evolutionary rate variation among genes of the anthocyanin biosynthetic pathway. *Mol. Biol. Evol.* **16**, 266–274.

Rausher, M.D., Lu, Y. and Meyer, K. (2008) Variation in constraint versus positive selection as an explanation for evolutionary rate variation among anthocyanin genes. *J. Mol. Evol.* **67**, 137–144.

Sagnard, F., Deu, M., Dembélé, D., Leblois, R., Touré, L., Diakité, M., Calatayud, C. *et al.* (2011) Genetic diversity, structure, gene flow and evolutionary relationships within the *Sorghum bicolor* wild-weedy-crop complex in a western African region. *Theor. Appl. Genet.* **123**, 1231–1246.

Satoh, H., Shibahara, K., Tokunaga, T., Nishi, A., Tasaki, M., Hwang, S.-K., Okita, T.W. *et al.* (2008) Mutation of the plastidial α-glucan phosphorylase gene in rice affects the synthesis and structure of starch in the endosperm. *Plant Cell*, **20**, 1833–1849.

Schupp, N. and Ziegler, P. (2004) The relation of starch phosphorylases to starch metabolism in wheat. *Plant Cell Physiol.* **45**, 1471–1484.

Sharkey, T.D., Yeh, S., Wiberley, A.E., Falbel, T.G., Gong, D. and Fernandez, D.E. (2005) Evolution of the isoprene biosynthetic pathway in kudzu. *Plant Physiol.* **137**, 700–712.

Slotte, T., Bataillon, T., Hansen, T.T., St Onge, K.R., Wright, S.I. and Schierup, M.H. (2011) Genomic determinants of protein evolution and polymorphism in Arabidopsis. *Genome Biol. Evol.* **3**, 1210–1219.

Smith, A.M. (2001) The biosynthesis of starch granules. *Biomacromolecules*, **2**, 335–341.

Stark, D.M., Timmerman, K.P., Barry, G.F., Preiss, J. and Kishore, G.M. (1992) Regulation of the amount of starch in plant tissues by ADP-glucose pyrophosphorylase. *Science*, **258**, 287–292.

Tajima, F. (1989) Statistical method for testing the neutral mutation hypothesis by DNA polymorphism. *Genetics*, **123**, 585–595.

Tetlow, I.J. (2011) Starch biosynthesis in developing seeds. *Seed Sci. Res.* **21**, 5–32.

Tetlow, I.J., Morell, M.K. and Emes, M.J. (2004) Recent developments in understanding the regulation of starch metabolism in higher plants. *J. Exp. Bot.* **55**, 2131–2145.

Tetlow, I.J., Beisel, K.G., Cameron, S., Makhmoudova, A., Liu, F., Bresolin, N.S., Wait, R. *et al.* (2008) Analysis of protein complexes in wheat amyloplasts reveals functional interactions among starch biosynthetic enzymes. *Plant Physiol.* **146**, 1878–1891.

Tuncel, A. and Okita, T.W. (2013) Improving starch yield in cereals by over-expression of ADPglucose pyrophosphorylase: expectations and unanticipated outcomes. *Plant Sci.* **211**, 52–60.

Watterson, G.A. (1975) On the number of segregating sites in genetical models without recombination. *Theor. Popul. Biol.* **7**, 256–276.

Waxman, D. and Peck, J.R. (1998) Pleiotropy and the preservation of perfection. *Science*, **279**, 1210–1213.

Whitt, S.R., Wilson, L.M., Tenaillon, M.I., Gaut, B.S. and Buckler, E.S. (2002) Genetic diversity and selection in the maize starch pathway. *Proc. Natl Acad. Sci. USA*, **99**, 12959–12962.

WHO. (2003) Diet, nutrition and the prevention of chronic diseases. *WHO Technical Report Series, Geneva.* **916**, 1–150.

Woo, M.O., Ham, T.H., Ji, H.S., Choi, M.S., Jiang, W., Chu, S.H., Piao, R. *et al.* (2008) Inactivation of the *UGPase1* gene causes genic male sterility and endosperm chalkiness in rice. *Plant J.* **54**, 190–204.

Wright, S.I. and Charlesworth, B. (2004) The HKA test revisited: a maximum-likelihood ratio test of the standard neutral model. *Genetics*, **168**, 1071–1076.

Wright, S.I. and Gaut, B.S. (2005) Molecular population genetics and the search for adaptive evolution in plants. *Mol. Biol. Evol.* **22**, 506–519.

Xu, X., Liu, X., Ge, S., Jensen, J.D., Hu, F., Li, X., Dong, Y. *et al.* (2011) Resequencing 50 accessions of cultivated and wild rice yields markers for identifying agronomically important genes. *Nat. Biotechnol.* **30**, 105–111.

Yu, Y., Mu, H.H., Wasserman, B.P. and Carman, G.M. (2001) Identification of the maize amyloplast stromal 112-kD protein as a plastidic starch phosphorylase. *Plant Physiol.* **125**, 351–359.

Yu, G., Olsen, K.M. and Schaal, B.A. (2011) Molecular evolution of the endosperm starch synthesis pathway genes in rice (*Oryza sativa* L.) and its wild ancestor, *O-rufipogon* L. *Mol. Biol. Evol.* **28**, 659–671.

Zeeman, S.C., Kossmann, J. and Smith, A.M. (2010) Starch; its metabolism, evolution and biotechnological modification in plants. *Annu. Rev. Plant Biol.* **61**, 209–234.

Zhang, Z., Li, J., Zhao, X.Q., Wang, J., Wong, G.K. and Yu, J. (2006) KaKs calculator: calculating Ka and Ks through model selection and model averaging. *Genomics Proteomics Bioinformatics*, **4**, 259–263.

Zrenner, R., Salanoubat, M., Willmitzer, L. and Sonnewald, U. (1995) Evidence for the crucial role of sucrose synthase for sink strength using transgenic potato plants (*Solanum tuberosum* L.). *Plant J.* **7**, 97–107.

16

Enrichment of genomic DNA for polymorphism detection in a non-model highly polyploid crop plant

Peter C. Bundock[1,*], Rosanne E. Casu[2] and Robert J. Henry[1,†]

[1]Co-operative Research Centre for Sugar Industry Innovation through Biotechnology, Southern Cross Plant Science, Southern Cross University, Lismore, NSW, Australia

[2]Co-operative Research Centre for Sugar Industry Innovation through Biotechnology, CSIRO Plant Industry, Queensland Bioscience Precinct, St Lucia, Qld, Australia

*Correspondence
email peter.bundock@scu.edu.au
†Present address: Centre for Plant Science, Queensland Alliance for Agriculture and Food Innovation, University of Queensland, Australia.

Keywords: sugarcane, next-generation sequencing, single nucleotide polymorphisms, second-generation sequencing.

Summary

Large polyploid genomes of non-model species remain challenging targets for DNA polymorphism discovery despite the increasing throughput and continued reductions in cost of sequencing with new technologies. For these species especially, there remains a requirement to enrich genomic DNA to discover polymorphisms in regions of interest because of large genome size and to provide the sequence depth to enable estimation of copy number. Various methods of enriching DNA have been utilised, but some recent methods enable the efficient sampling of large regions (e.g. the exome). We have utilised one of these methods, solution-based hybridization (Agilent SureSelect), to capture regions of the genome of two sugarcane genotypes (one *Saccharum officinarum* and one *Saccharum* hybrid) based mainly on gene sequences from the close relative *Sorghum bicolor*. The capture probes span approximately 5.8 megabases (Mb). The enrichment over whole-genome shotgun sequencing was 10–11-fold for the two genotypes tested. This level of enrichment has important consequences for detecting single nucleotide polymorphisms (SNPs) from a single lane of Illumina (Genome Analyzer) sequence reads. The detection of polymorphisms was enabled by the depth of sequence at or near probe sites and enabled the detection of 270 000–280 000 SNPs within each genotype from a single lane of sequence using stringent detection parameters. The SNPs were present in 13 000–16 000 targeted genes, which would enable mapping of a large number of these chosen genes. SNP validation from 454 sequencing and between-genotype confirmations gave an 87%–91% validation rate.

Introduction

The advent of second-generation DNA sequencing methods has increased sequencing throughput to such an extent that re-sequencing of some genomes via the whole-genome shotgun (WGS) approach has become relatively trivial (e.g. Arabidopsis where currently 240 strains have been sequenced as part of the 1001Genomes Project from a total of 1001 planned—http://1001genomes.org/). However, for the discovery of DNA polymorphisms across populations and/or in regions of interest in large genomes, the WGS approach is currently too inefficient and costly. For species without an assembled genome sequence where the exact target sequence is unknown, there is an additional level of complexity to assemble reads from the region(s) of interest.

Early methods developed for selecting DNA of interest involved screening libraries of cloned DNA using hybridisation-based assays (Grunstein and Hogness, 1975). This can be a laborious task, especially for libraries from large target genomes such as the human genome (Bell et al., 1980), and especially because screening of libraries is usually carried out using one or a few probes at a time. Another method in use well before the advent of next-generation sequencing (NGS) is PCR, which has been an extremely successful approach for enrichment or capture of genomic targets (Mullis and Faloona, 1987). PCR-based approaches can be effective where a population is being screened to detect polymorphisms and generally where the size of target regions is relatively short and the number of regions relatively small in number (e.g. Bundock et al., 2009; Malory et al., 2011). However, a highly parallel PCR method, based on microdroplets and using specialised technology, has been used successfully to target a large number of regions (Tewhey et al., 2009b).

Circularising probe systems that capture target genomic DNA combine specific probe sequences with universal PCR sequences to enable the multiplex amplification of large numbers of captured fragments for next-generation sequencing (NGS) (Dahl et al., 2005, 2007). These have evolved from the padlock probe system used for single nucleotide polymorphisms (SNP) genotyping (Hardenbol et al., 2003, 2005). A recent iteration called molecular inversion probes (MIPs) (reviewed in Mamanova et al., 2010) has been used to capture target genomic DNA, which can then be sequenced directly (without additional library preparation) using a NGS instrument and has been demonstrated to capture 1 Mb of sequence using 13 000 MIPs (Turner et al., 2009).

To overcome PCR limitations, hybridisation capture-based techniques have been successfully applied to enrich for target sequences. The technique of 'direct selection' was used initially for enriching cDNA fractions using large genomic clones (Lovett et al., 1991; Morgan et al., 1992) but was later developed for selecting genomic DNA using biotinylated BAC DNA (Bashiardes et al., 2005). A recent approach has been the use of microarrays to capture targets through preferential hybridisation. The captured DNA is eluted from the microarray and sequenced (Albert et al., 2007; Okou et al., 2007). A modification of this approach has been to carry out the hybridisation in solution using the microarray as a platform for oligonucleotide synthesis (Gnirke et al., 2009; Tewhey et al., 2009a).

The human exome has been the target for enrichment strategies by some groups, and this large and dispersed subset of the human genome has been successfully targeted using microarray-based capture (Hodges et al., 2007). One goal for the capture and sequencing of human exomes has been to identify the causes of genetic disorders, which has been demonstrated in principle by Ng et al. (2009) and then demonstrated for two Mendelian-inherited disorders—Millers syndrome (Ng et al., 2010b) and Kabuki syndrome (Ng et al., 2010a). An example of array-based capture used successfully on a plant species is described by Fu et al. (2010) where two sequential array hybridizations were carried out on maize. The first hybridisation was carried out with a repeat-based array, and the second with the target array. This was designed to create a blocker-free capture protocol because a large proportion of the maize genome is composed of repeat sequences.

For many plant species of economic importance, a complete genome sequence is not yet available, so there is no reference genome for the design of capture probes. For those wishing to discover DNA sequence polymorphisms en masse, obtaining sufficient coverage of selected regions using WGS sequencing could be prohibitively expensive. An alternative solution is explored in this paper, using a genome sequence from a close relative both for the design of probe sequences for targeted enrichment and for the alignment of the resulting enriched reads. This approach greatly extends the number of species to which targeted enrichment strategies will be possible.

Results and discussion

Creation of sequencing libraries and mapping of sequencing reads

Two sugarcane genotypes were chosen for the SureSelect targeted enrichment procedure: IJ76-514, a pure Saccharum officinarum clone originally sourced from Irian Jaya, and Q165, a commercial cultivar that is a backcrossed hybrid of S. officinarum with Saccharum spontaneum. These genotypes are parents of the PJ2 mapping population (Aitken et al., 2005). A WGS

library for Illumina Genome Analyser (GA) analysis was constructed for each genotype with each library run on a single GA lane producing 76-bp paired-end sequence reads (Table 1). The proven WGS libraries were then subjected to the Agilent SureSelect enrichment procedure and again each run on a single GA lane with 76-bp paired-end reads on an Illumina GAIIx slide. After trimming and discarding low-quality sequence and depleting those reads that matched to chloroplast, mitochondrial or known repeat sequences, there were between 26 and 41 million reads acquired for each genotype/library (1.65 to almost 3 gigabases of sequence) (Table 1). A library derived from a Sorghum bicolor genotype (R931945-2-2) was enriched using the same probes for comparative purposes. All sugarcane sequences have been lodged at the Short Read Archive at the National Center for Biotechnology Information (NCBI, SRA http://www.ncbi.nlm.nih.gov/) with the SRA accession number SRA051387.1.

The WGS sequences from the two sugarcane genotypes were aligned to the sorghum genome with around 60% of WGS reads from both genotypes mapping to this genome and more than 30% mapping to genes using low stringency thresholds (Table 2). This indicates the high degree of similarity between the sugarcane and sorghum genomes and adds weight to the plausibility of enriching the sugarcane genome based on probes designed to sorghum DNA sequence. The subsequent enrichment of these libraries using probes designed mainly to sorghum coding regions increased the proportion of reads that mapped to the sorghum genome, particularly the genes, where more than 50% of reads aligned at low stringency (Table 2). Under stringent alignment conditions, the percentage of reads mapping to the sorghum genome was twice as high for the enriched libraries compared with the WGS libraries, and for mapping to the genic regions, there was at least a threefold increase when the enriched library was compared with the corresponding WGS library for both genotypes (Table 2). This was an initial indication that the enrichment process was successful in capturing the targeted genic regions.

The extent to which the method of enrichment has succeeded in capturing sequences with homology to the probe sequences can be partly judged by the enrichment factor. This was calculated empirically for the sugarcane genotypes as the ratio of the proportion of the reads from the SureSelect enriched library that mapped to a probe sequence to the proportion of reads from the corresponding WGS library that mapped to a probe sequence. Using reads after quality

Table 1 Summary of DNA sequencing of two sugarcane genotypes (Q165 and IJ76-514) and a sorghum genotype

Genotype/Library type	No. of reads obtained (millions)	Total length of sequences (Gb)	No. of reads after quality trimming (millions)	No. of reads after depletion of repeats	Total no. of bases after depletion (Gb)
IJ76-514 WGS	30.9	2.06	27.2	26.6	1.65
Q165 WGS	38.7	2.48	32.7	31.9	1.86
IJ76-514 enriched	57.9	4.29	56.2	41.0	2.98
Q165 enriched	51.7	3.81	50.3	38.0	2.77
Sorghum enriched	66.7	3.79	62.2	46.9	2.59

WGS, whole-genome shotgun.

Enriched—library created using the Agilent SureSelect targeted enrichment process.

Table 2 Mapping of reads to the sorghum genome sequence

Genotype/Library type	Percentage of reads mapping to sorghum genome*	Percentage of reads mapping to sorghum genes
IJ76-514 WGS	57 (16)	31 (9)
Q165 WGS	60 (17)	34 (11)
IJ76-514 enriched	75 (35)	53 (29)
Q165 enriched	76 (39)	58 (33)
Sorghum enriched	98 (67)	83 (54)

WGS, whole-genome shotgun.

Enriched—library created using the Agilent SureSelect targeted enrichment process.

*Percentages in parentheses are for reads that mapped uniquely under stringent mapping conditions of length = 0.9 and similarity = 0.9; otherwise, default parameters of mapping non-unique matches, length = 0.5 and similarity = 0.8 were used.

trimming, the enrichment factor for IJ76-514 was 10.0-fold and for Q165 was 11.3-fold based on read mapping to all probes (Table 3). To determine the success of the strategy to use probes based on sorghum gene sequences to capture sugarcane sequences, the enrichment factor was calculated for a set of these probes alone. For IJ76-514, the enrichment was nine-fold, and for Q165 it was 12-fold (Table 3). Thirty-four per cent of sorghum R931945 reads matched to the sorghum baits compared with around 25% for Q165 and 19% for IJ76-514 enriched libraries (Table 3). The enrichment of the sorghum sample could not be calculated empirically as there was no WGS sequence for comparison; however, the fraction of the sorghum genome covered by these sorghum probe sequences is 0.6%. As the length fraction parameter for read mapping used here was 0.5, 76-bp reads will map if they overlap the 120mer probe sequences by 38 bp of sequence. This effectively extends the probe coverage by 76 bp (38 bp at each end) to 196 bp. This extends the coverage of the probes to 1% of the genome and the fold enrichment would be 34-fold, with a theoretical maximum of 100-fold enrichment. None of these calculations take into account the length of the inserts that have

been captured, which had a mean length of 230 bp for IJ76-514, 180 bp for Q165 and 280 bp for sorghum. In the case of IJ76-514 and sorghum, the mean inserts would further extend the target region on either side of each probe and so the estimates for enrichment are conservative. For the probes designed to tile across sucrose metabolism genes and 242 gene fragments previously targeted using 454 sequencing (Bundock et al., 2009), the enrichment for both IJ76-514 and Q165 was 35-fold (Table 3). Clearly relative to the enriched sorghum library, the level of enrichment was greater in the sugarcane genotypes for these genes, likely due to higher homology. The level of enrichment was also greater than that found for the probes designed to sorghum sequences but also possibly assisted by the tiling strategy (see Experimental procedures) and the fact that there were extra copies of these probes for hybridisation in the SureSelect Oligo capture library (see Experimental procedures).

Discovery of single nucleotide polymorphisms

The main aim of this project was to discover SNPs in a large number of sugarcane genes and in addition to reads covering probes, sequence from either side of the target probe is also expected to be captured during enrichment. As the majority of the SureSelect probes were based on the sorghum genome, this genome was chosen as the main reference for aligning the Illumina reads from all libraries for SNP detection. SNP discovery was carried out on the reads aligned to the ten chromosomes of the sorghum genome sequence after depleting reads that matched to chloroplast, mitochondrial or known repeat sequences. The main focus was to find SNPs between the homologous loci within each sugarcane genotype as these provide the potential for mapping (Bundock et al., 2009). Large numbers of candidate SNPs were identified from the sequence alignments and although fixed SNPs between sugarcane and sorghum were readily identified, interestingly these were in the minority (Table 4). Even with stringent parameters for SNP detection, an enormous number of putative SNPs within each sugarcane genotype were detected, particularly from the enriched libraries with more than a quarter of a million SNPs detected as polymorphisms within each sugarcane genotype (Table 4). For the enriched libraries, the SNPs were found in reads mapping to a large number of sorghum genes (almost

Table 3 Reads mapping to 120-bp target probe sequences used for SureSelect enrichment

Genotype/Library type	Percentage of reads mapping to all probe sequences*	Percentage of reads mapping to 34 105 probes designed to sorghum genes only	Percentage of reads mapping to 961 probes 2× tiled across sugarcane genes	Percentage of reads mapping to 7263 SureSelect probes designed to sugarcane ESTs
IJ76-514 WGS	5.1	2.2	0.7	2.3
Q165 WGS	5.1	2.1	0.6	2.4
IJ76-514 enriched	50 (×10)	19 (×8.9)	23 (×33)	12 (×5.0)
Q165 enriched	57 (×11.2)	25 (×11.9)	21 (×35)	14 (×5.8)
Sorghum enriched	42	34 (×34)	3.4	2.7

WGS, whole-genome shotgun.

Enriched—library created using the Agilent SureSelect targeted enrichment process.

*Figures in parentheses are the enrichment factor obtained by dividing the enriched percentage by the corresponding WGS percentage, or for sorghum calculated from expected coverage.

Table 4 Putative SNPs detected for each of the two parental sugarcane genotypes for the two different libraries (WGS versus enriched) after reads were aligned to the *Sorghum bicolor* genome sequence

Genotype/ Library type	Library size (millions of reads)	Total no. of putative SNPs*	No. of SNPs within sugarcane reads only	Sugarcane SNPs within annotated genes	Genes with one or more SNP sites	No. of predicted amino acid changes
IJ76-514 WGS	30.9	59 417	50 947	3169	384	341
Q165 WGS	38.7	65 808	56 839	3276	392	334
IJ76-514 enriched	57.9	434 633	268 797	139 725	12 963	51 843
Q165 enriched	51.7	430 549	285 836	168 090	15 799	64 098

WGS, whole-genome shotgun; SNPs, single nucleotide polymorphisms.

Enriched—library created using the Agilent SureSelect targeted enrichment process.

*SNP numbers are from SNP discovery analyses using minimum read coverage at a site of 20 and minimum 5% frequency of alternative allele.

13 000 for IJ76-514 and nearly 16 000 for Q165. This compares to a much smaller number of genes with SNPs for the WGS libraries (around 400), indicating that the enrichment procedure has worked to enrich for reads from the targeted genes (Table 4). When the reads from the Q165 and IJ76-514 enriched libraries were combined and mapped to the sorghum genome, the number of SNPs detected increased to 446 329 for SNPs within sugarcane, with 230 692 of these in predicted genes. There were in addition 232 270 fixed SNPs between sugarcane reads and the sorghum reference.

To ensure that the large differences in the number of SNPs detected between WGS and enriched libraries was not due mainly to differences in the number of reads obtained from each library and also to equitably compare SNP numbers between genotypes, 25 million reads were randomly selected from each library and used for SNP detection (Table 5). This clearly showed that there was a large increase in the number of SNPs (4–5.5-fold) detectable in the reads from the enriched libraries compared to WGS libraries, particularly in genic regions (36–60-fold). As this could not be attributed to differences in the number of reads in this case, it is likely to be due to the targeting of these reads as a result of the enrichment process. It could also be seen that although there were a larger number of SNPs detected in IJ76-514 from WGS compared with Q165

(15% more), the Q165 enriched library provided a larger number of SNPs than the IJ76-514 enriched library (18% more). About 61% of SNPs detected from the Q165 enriched library occurred within annotated genes compared with 54.6% for the IJ76-514 enriched library. For the WGS libraries, the corresponding proportions are 5.5% and 6%, respectively. The larger number of SNPs detected in IJ76-514 compared with Q165 from the same number of WGS reads (Table 5) may not necessarily be due to a higher frequency of SNPs within this genotype. A difference in genome size may contribute so that fewer SNPs cross the sequence depth threshold for Q165. The larger number of SNPs in Q165 from enrichment may also not be due to greater SNP frequency but may be largely due to higher proportion of capture of targets owing to greater similarity between designed probes and the Q165 genome, possibly due to the *S. spontaneum* DNA represented in this genome being more similar to sorghum sequence. Evidence for this interpretation comes from the fact that a higher proportion of reads from Q165 (WGS) map to the sorghum genome (Table 2).

A random selection of reads, for subsets of different sizes, was sampled from both the WGS and enriched Q165 libraries and used for SNP analysis. A plot of the number of SNPs detected versus the number of reads in the subset indicated the different trends for the two libraries (Figure 1). The trend line for the enriched library indicates that the number of SNPs detected is approaching saturation from the single lane of sequence, whilst the trend line for the WGS library is still climbing steeply. This indicates that for the SNPs detected from the enriched library, which constituted SNPs mainly within the targeted genic region, most had been detected, whilst for the WGS library, SNPs from across the whole genome are sampled randomly and constitute a much larger total pool of SNPs that would require many lanes of sequence before saturation is observed.

An analysis was carried out to determine the extent to which earlier 454 sequencing of PCR products and cross-genotype SNP discovery could contribute to SNP validation. As part of the SureSelect probe design, a set of probes were tiled across short products (approximately 200–400 bp) that had been sequenced from the same two genotypes using 454 technology (Bundock *et al.*, 2009). Illumina SureSelect reads, and separately 454 reads, from both sugarcane genotypes were aligned to 205 of these sequences (derived from 242 which had been formed into contig consensus sequences). For Q165, 1906 SNPs were detected within these regions using the same stringent SNP

Table 5 Putative SNPs detected from 25 million randomly selected reads for each of the two parental sugarcane genotypes for the two different libraries (WGS versus enriched)

Genotype/ Library type	Putative SNPs within sugarcane genotype only*	Putative SNPs in annotated genes
IJ76-514 WGS	46 616	2825
Q165 WGS	40 582	2246
IJ76-514 enriched	188 340	102 892
Q165 enriched	222 666	136 042

WGS, whole-genome shotgun; SNPs, single nucleotide polymorphisms.

Enriched—library created using the Agilent SureSelect targeted enrichment process.

*SNP numbers are from stringent SNP discovery analyses using minimum read coverage (20) with minimum of 5% frequency of alternative allele.

(a)

Q165 SS SNPs

(b)

Q165 WGS SNPs

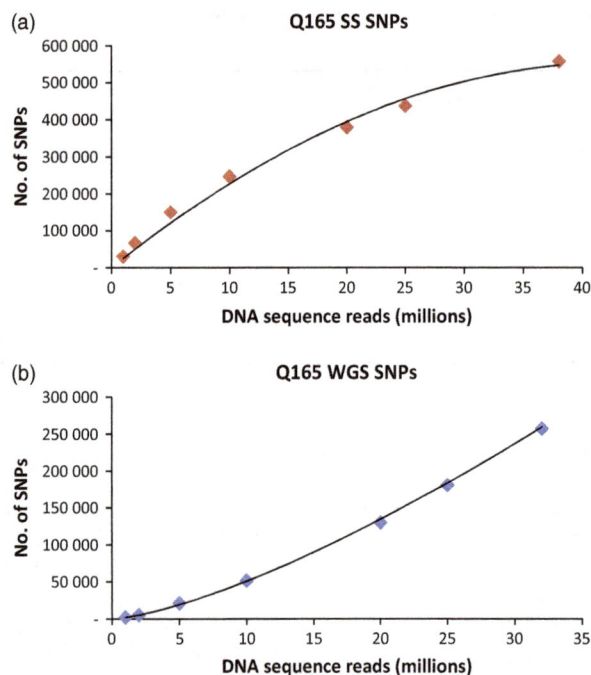

Figure 1 Plot of the number of SNPs discovered within the DNA sequence reads from the sugarcane genotype Q165. Reads were selected randomly from the total set in subsets of various sizes and aligned to the sorghum genome for SNP discovery. SNP numbers are from SNP discovery analyses using low minimum read coverage (4) with minimum of 1% frequency of alternative allele. (a) Number of SNPs discovered in reads from the SureSelect enriched library for Q165. The trend line indicates that the curve is flattening out and reaching saturation. (b) SNPs discovered from the whole-genome shotgun (WGS) reads from Q165. The trend line is concave up, indicating that as expected the genome is far from saturated with reads. Most WGS SNPs are not located in target regions.

detection parameters used for the discovery of SNPs from reads aligned to the sorghum genome. One thousand three hundred and 23 of these SNPs were discovered as sequence variants within Q165 from the 454 reads (69.4%). A further 332 of the 1906 Q165 SNPs were discovered within IJ76-514 Illumina reads, and a further seven SNPs could be found in IJ76-514 454 reads. So of the 1906 SNPs found in Q165 SureSelect sequence, there was independent sequence validation for 87% (1653). For IJ76-514, 1428 SNPs were discovered from aligning the Illumina SureSelect reads to the 205 consensus sequences, with 1298 of these SNPs confirmed as present in other sequence—mostly 454 sequence from IJ76-514—providing validation for 91%. These high validation percentages for these subsets, 87% for Q165 and 91% for IJ76-514, should be applicable to, and give a high degree of confidence in, the SNPs discovered from alignment of SureSelect reads to sorghum genes generally.

A cross-genotype validation, for all the SNPs discovered in genes from the Illumina SureSelect reads, was carried out for both genotypes. From the Q165 enriched library, of the 168 090 SNPs found to be within sorghum-defined genes (Table 4), 81 368 (48.4%) were also identified as SNPs in the IJ76-514 enriched library. These shared SNPs constitute almost 50% of the Q165 SNPs in genes and almost 60% of IJ76-514

SNPs in genes. There are a total of 226 447 different SNP sites found in genes when results from separate genotype analyses are combined. Again this cross-genotype validation provides confidence in the SNPs discovered.

Some proportion of these SNPs may represent sequence differences between paralogous loci and not be allelic. One way to remove most of the SNPs that are due to sampling two paralogs is to set a threshold of more than 55% sequence representation for the major allele (for bi-allelic SNPs). This will remove those SNPs where there is almost equal representation from the two alleles which would occur where two paralogs have been sampled (assuming equal number of chromosome copies for each paralog). For Q165, of the 163 248 bi-allelic SNPs in genes, 154 733 (approximately 95%) have a major allele with >55.0% frequency, and thus only a small proportion would be excluded using this suggested threshold.

For all the bi-allelic SNPs detected within predicted genes, the frequency of the minor allele was calculated, and a frequency histogram created for each genotype (Figure 2). SNPs with a coverage of 50–400 were selected for this because at low coverage artifactual peaks arise owing to the large number of SNPs in this category and the small number of possible frequencies. At the high end of coverage (400 used arbitrarily here), SNPs are more likely to result from sequences that are repeated in the *Saccharum* genome but are unique in the *S. bicolor* genome and will tend to have low minor allele frequency. There is a higher proportion of SNPs for Q165 than for

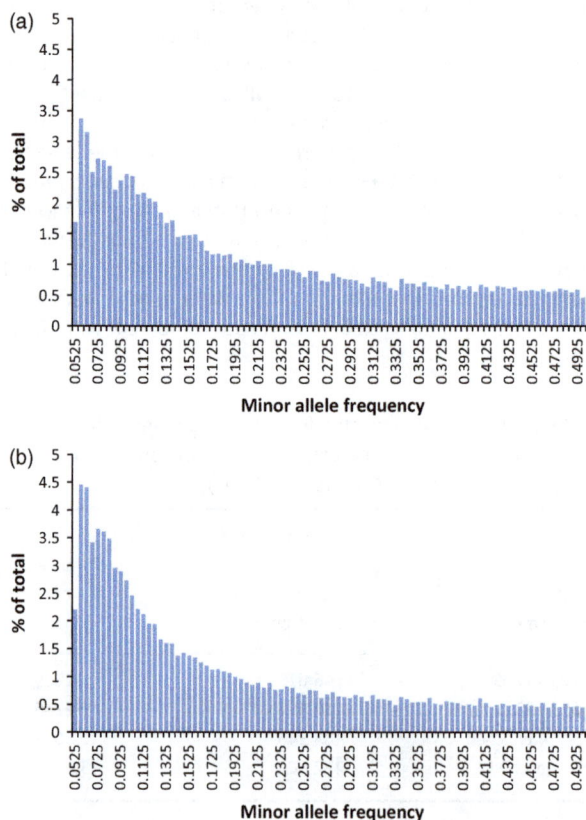

(a)

(b)

Figure 2 Histogram showing the percentage of SNPs with the corresponding minor allele frequency. Pertains to bi-allelic SNPs found at a coverage threshold of 50–400 within genes from the (a) IJ76-514 enriched library (b) Q165 enriched library.

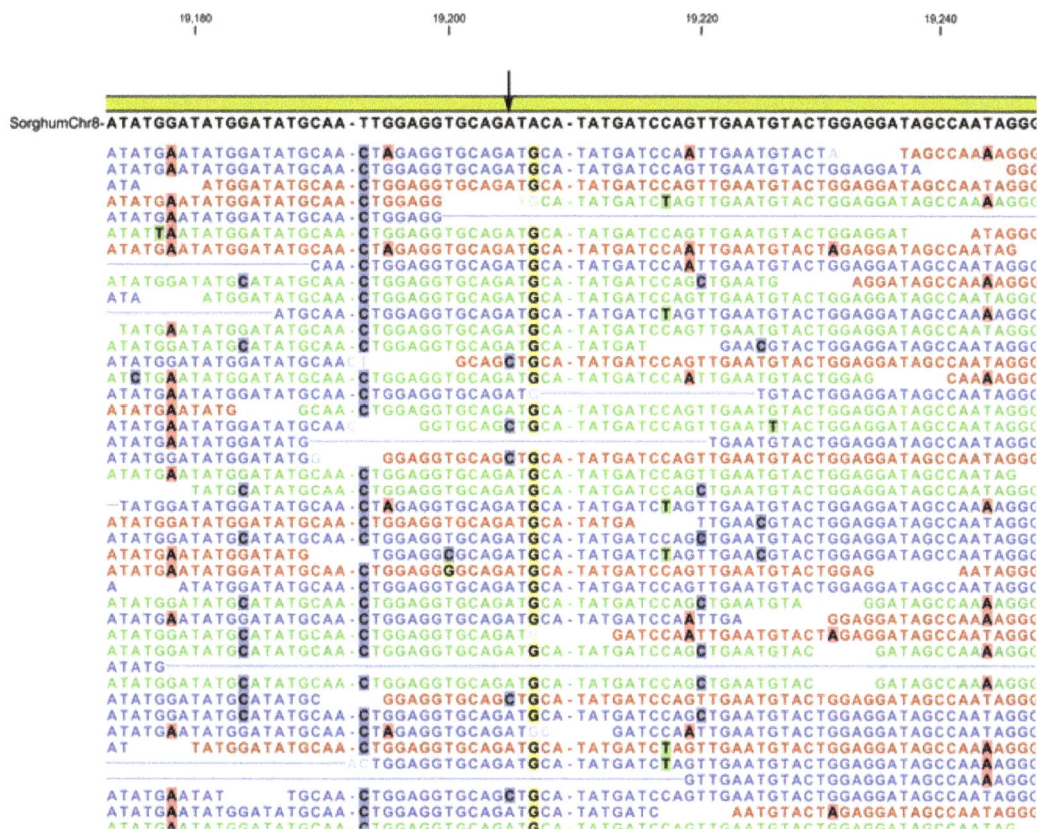

Figure 3 An alignment of reads from the SureSelect enriched library of Q165 to a region of sorghum chromosome 8 with SNPs highlighted. There is a C/A SNP at position 19 205 (see arrow) where the cytosine minor allele (C in blue) is at 8.6% frequency (34/360). This is a good candidate SNP for single-dose marker development as the likely ratio of minor allele to major is 1:11, and thus, the SNP is likely to be present on one chromosome copy with 11 copies having the alternative allele. Two bases upstream at 19 207, there appears to be a difference between sugarcane (G in yellow) and the sorghum reference (A).

IJ76-514 where the minor allele is at the rare end of the spectrum (Figure 2)—with more than 43% of SNPs having a minor allele frequency of 0.05–0.127 in Q165 compared with 36% in IJ76-514. This higher proportion of rare SNPs was observed in an earlier study of SNPs in gene fragments in these same sugarcane genotypes (Bundock *et al.*, 2009). A possible explanation for this observation is that the *S. spontaneum* chromosomes of the Q165 genome would be likely to have many sequence differences from the *S. officinarum* chromosomes providing numerous SNPs on a minority of chromosomes. In addition, the number of chromosome copies is likely to be larger for Q165 than for IJ76-514. This would lead to a larger proportion of SNPs with the minor allele being rare compared with IJ76-514.

The SureSelect approach has been highly successful with regard to the discovery of SNPs from the enriched reads aligning to genes. With an enrichment of 10–11-fold, this method is economical, as the cost of library enrichment per library is much less than the cost required to produce the equivalent amount of sequence to cover the genic regions to an equivalent read depth. The method has allowed discovery of very large numbers of SNPs and should allow the mapping of a significant proportion of the genes in this complex genome. This strategy will enable high-density SNP maps of the sugarcane genome to be constructed, which may be the key to successful assembly of a reference genome sequence for sugarcane (Souza *et al.*, 2011).

As candidates for mapping, the most useful SNPs are single dose in one parent. This occurs where one allele is represented on one chromosomal copy only and the alternative allele occurs on the other copies in the parent genome, with the second parent having only the major allele present on all chromosome copies. Ideally, the rare allele is absent from the other parent, and a single-dose marker inherited from one parent can be developed, which can segregate 1 : 1 in the progeny and be fully informative for mapping (i.e. all progeny are informative). Selecting SNPs for assay development that have one allele at a low frequency is a prudent strategy for developing these single-dose markers (Figure 3). Ideally, the frequency would correspond to the reciprocal of the number of chromosomal copies—which for commercial sugarcane varieties is usually unknown. As the DNA sequences analysed here are publicly available, this now represents hundreds of thousands of sugarcane SNPs that are available for sugarcane researchers to use in mapping and association studies.

Design of probe sequences for hybridisation during capture

There were five different sources of probe sequence design for the baits used for capture during enrichment (Table 6). The most numerous were single probes designed to the coding region of predicted sorghum genes, with the next most numerous being probes designed to sugarcane expressed sequence tags (ESTs) that appeared not to be represented in the predicted sorghum genes. A third group of probes were designed to genomic sequence corresponding to selected fragments

Table 6 Design of probes used for enrichment of sugarcane and sorghum genomic libraries

Target sequence	Number of probes	Number of times represented on manufacturing array	Design strategy
Predicted sorghum coding sequences— 28 008 in total	44 429	1	Either a probe at both the 5′ and 3′ ends, or a probe at one end or mid sequence
Sugarcane ESTs	7263	1	1 centrally located probe
Sucrose metabolism genes—genomic sequence—sugarcane	757	4	2× tiling
Sucrose metabolism genes—genomic sequence—sorghum	204	4	2× tiling
240 sugarcane gene fragments from 454 genomic DNA sequence	595	3	1× tiling
Total	53 248		

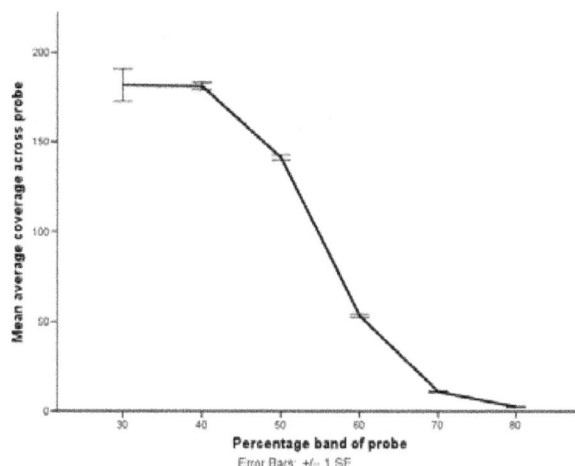

Figure 4 Graph of the GC content versus the average coverage of probes (baits) with reads mapped from the Q165 enriched library. The mean of the average coverage across the 120mer probes for each 10% band (25%–35%, 35%–45% etc.) has been calculated. Error bars are ± 1 SE of the mean. It can be observed that high GC content probes tend to have reduced coverage compared to probes with 25%–45% GC content which are optimal.

(202–445 bp in length) of 240 sugarcane genes included in a previous study. To compare the effectiveness of probe design, each probe sequence was used as a reference for read mapping. The GC content of probe designs was found to be highly significantly associated with the average read coverage for a probe. Probes with a GC content of around 30%–40% had the highest average read coverage (Figure 4), a property that is probably directly related to DNA capture performance.

For the probes designed to the sorghum coding regions, there were three locations used for design: extreme 5′ end, middle and extreme 3′ end. An analysis was undertaken to determine whether there was any effect of position on the ability of probes to capture sequences from hybridisation. However, it was found that probes designed at the start of a predicted coding region (5′ end) had a significantly higher GC content ($P < 0.0001$, ANOVA) than those designed to the middle or the 3′ end of the coding region (Figure 5). So a model that included probe position, percentage GC content of probe and the interaction was fitted for the sorghum probes to explain the average coverage for each probe and used to analyse all three enriched libraries. For the enriched sorghum library, only GC content was found to be a statistically significant influence on the capture efficiency, and the position of the probe within the coding DNA sequence (CDS) and the interaction were not significant. For both of the sugarcane genotypes, all three terms were significant and probes designed at the extreme 5′ end had higher average coverage when GC content was taken into account than probes at the extreme 3′ end. Based on a smaller number of probes designed to the middle of sorghum coding regions, there was no significant difference in performance to probes at the 3′ end for either sugarcane genotype.

The likely explanation for these observations is that 5′ end probes tend to represent more conserved sequences than probes at the 3′ end, so that when there are sequence differences between probe and target, the capture of targets and/or read alignment during mapping is reduced. This is supported by the fact that an analysis of SNP distributions indicated that there was a very highly statistically significant difference between SNP abundance between the three locations with the 3′ end having a much higher number of SNPs than expected and the 5′ end and middle having a smaller number of SNPs for all three enriched libraries (chi-square test, $P \ll 0.0001$). Even though probes at the 5′ end had a higher coverage when GC content was taken into account, the GC content of 5′ end probes tends to be too high for the design of optimal probes (and they also miss out on capturing as much variation as 3′ end probes). The best recommendation therefore is to design a 3′ end probe as a matter of course because it is likely to have a GC content closer to optimum and also likely to encompass regions with variability. However, as insurance, if possible, a 5′ end probe could be designed to allow for those situations where there is too much variability at the 3′ end and there may be sufficient capture to enable alignment and detection of sequence variation. The caveat would be to design both probes, if possible, to a location where the GC content is below 55%, to reduce the number of poorly performing probes. If WGS sequence of the target species has been obtained, it could be used profitably during the probe design stage to determine whether any potential probe sequences have captured repetitive elements by mapping WGS reads to prospective probe sequences and culling those with high or very high coverage as belonging to repetitive elements.

As expected, the capture of sorghum sequences from the sorghum library with probes designed to the coding region of sorghum genes was more effective than the capture of sequences with probes designed to sugarcane ESTs and vice versa (Figure 6). The same effect was observed for the probes tiled across sucrose metabolism genes—those designed to

(a)

(b)

(a)

(b)

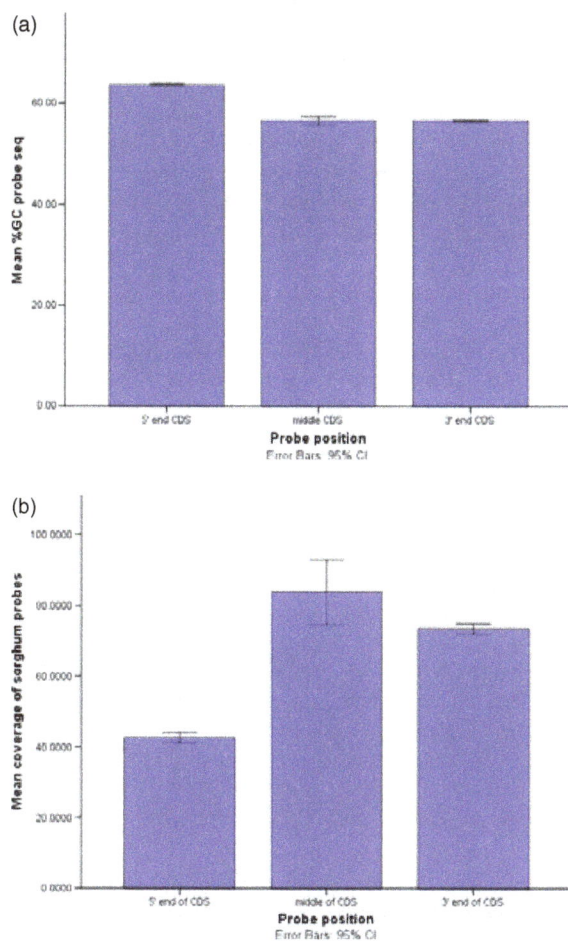

Figure 5 For probes designed to the coding region of predicted sorghum genes (the majority of probes), the region to which the probe was designed was found to significantly influence the GC content of probes on average, with 5′ end sequences leading to probes with higher GC content on average (a). These probes were also less effective on average at capturing sequence as judged by the number of reads that mapped to the probe sequences after sequencing enriched libraries. The case for sorghum is presented (b) with a similar result found for both the sugarcane enriched libraries. Error bars are 95% confidence intervals. This analysis was carried out after removing the 1% of probes with the highest coverage to reduce undue influence of very highly repeated regions.

Figure 6 The read coverage, after mapping reads to the probe sequences used for sequence capture for enrichment, is strongly influenced by the GC content of the probe sequences. After removing the effect of GC content of probes, there is a strong effect in the residuals, of probe sequence origin—whether sorghum or sugarcane. For the enriched sorghum library, the read coverage is significantly higher for probes derived from sorghum coding DNA sequence (CDS) compared to sugarcane ESTs (a). For both sugarcane libraries, the opposite is the case—probes designed to sugarcane ESTs have higher coverage on average than those designed to sorghum CDS (b—the Q165 genotype). An almost identical graph to b was obtained for IJ76-514. The error bars represent 95% confidence intervals for the means.

sorghum genes had a higher coverage in the sorghum enriched library than those designed to sugarcane and vice versa for the sugarcane libraries (data not shown). The probes designed to tile across genes or fragments had a higher average probe coverage than the non-tiled probes although this effect is confounded with the additional number of times these probes were represented on the array for the manufacture of the baits.

The sequencing of gene-enriched genomic DNA complements other strategies that employ sequencing to discover genome variations. Amplicon sequencing has been shown to allow the analysis of variation in many loci in genotypes of interest (Bundock et al., 2009; Kharabian-Masouleh et al., 2011), providing efficient SNP discovery. This more targeted SNP discovery approach is limited to much smaller numbers of loci than possible with the method reported here. Transcriptome sequencing (Winfield et al., 2010) is useful but will require analysis of many different tissues and developmental stages to achieve the level

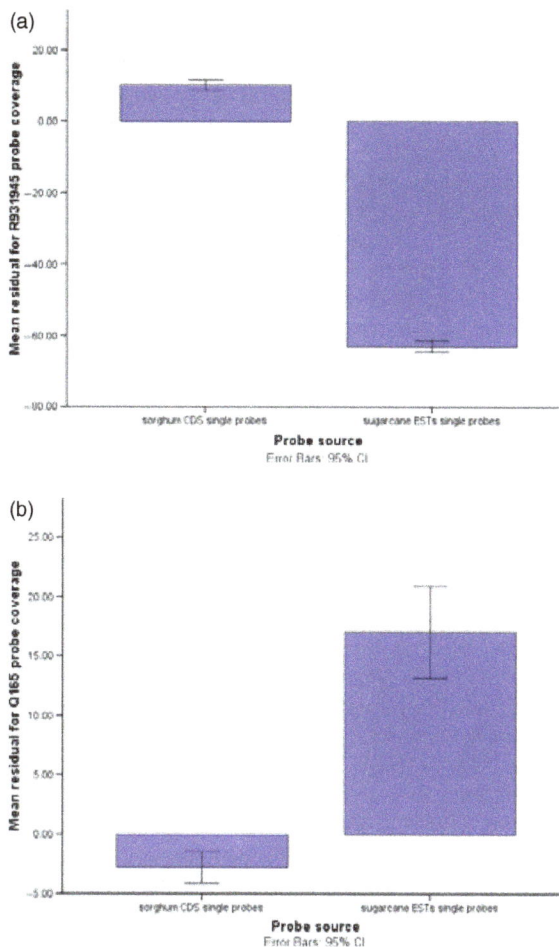

of genome coverage possible with enriched genome sequencing.

In conclusion, we have demonstrated the application of a solution hybridisation capture-based method targeted to enrich genes in the sugarcane genome based mainly on gene sequences found in the recently sequenced sorghum genome. The use of related genomes for design of probes might allow this approach to be widely applied in polymorphism discovery in poorly described species for which a closely related reference genome is available.

Experimental procedures

Design of probes for enrichment

Predicted coding regions from *S. bicolor* gene models were sourced from Phytozome v4.0 (http://www.phytozome.org/).

Sugarcane tentative consensus sequences (TCs), representing sugarcane transcripts, were downloaded from the Sugarcane Gene Index 2.2 (http://compbio.dfci.harvard.edu/tgi/cgi-bin/tgi/gimain.pl?gudb=s_officinarum). Additional sugarcane sequences were retrieved from the Representative Public identifiers corresponding to the probe sets on the Sugar Cane Whole Genome Array (http://www.affymetrix.com). The 120 base sequences were extracted from each sequence using text functions in Microsoft Excel. Simple repeats and low complexity sequences with each 120-base pair sequence were identified using Repeatmasker (http://www.repeatmasker.org). Homology searching using the BLAST algorithm (Altschul *et al.*, 1990) was performed on local customised versions of the Sugarcane Gene Index 2.1, the *Oryza sativa* (rice) coding regions and the *S. bicolor* gene models, hosted at the CSIRO Bioinformatic Facility, Canberra, ACT, Australia.

All 36 338 *S. bicolor* gene model coding regions (Sb_cds) were retrieved from Phytozome v4.0. The first 120 bases and last 120 bases (left_120mer and right_120mer, respectively) were extracted from each. Each 120mer was assessed for simple repeats and low complexity as well as for the presence of Ns in the sequence. 33 152 of the left_120mers and 35 165 of the right_120mers did not contain either Ns or low complexity sequences. Identical oligos from highly related Sb_cds were removed from both sets. This resulted in the retention of 31 770 left_120mers and 33 829 right_120mers, which represented a total of 34 965 Sb_cds. After filtration of the results of BLASTn homology against the Sugarcane Gene Index 2.2 and the *Oryza sativa* (rice) coding regions, it was determined that 7936 of the *S. bicolor* gene models were not represented by a suitable 120mer.

These gene models were targeted by homology searching their sequences against the Sugarcane Gene Index. 1080 of the gene models had alignments in excess of 120 bases long. Matching 120mer oligonucleotides were designed, commencing at the start of the matching sequence, reducing to 981 after filtration to remove 120mers containing Ns (Table S1).

Two additional approaches were used to maximise the number of individual useful probes designed. The first involved using the Representative Public identifiers corresponding to the probe sets on the Affymetrix Sugar Cane Whole Genome Array (depleted for control and chloroplast-related probe sets) to identify 7328 unique sugarcane sequences for further examination. These were homology searched against the *S. bicolor* gene models that had not yet been represented by a 120mer. Of the 770 matches returned, once alignment length and duplicate matches had been taken into account, 284 additional 120mers were designed, commencing at the start of the match to *S. bicolor* sequence. The second approach targeted the 27 728 singletons and 7263 TCs that were not homologous to the 120mers designed so far. An attempt was made to design 120mers to each of the sequences, commencing at base 101 to avoid possible poorer-quality sequence at the extreme 5′ end of each sequence. Once the resulting oligonucleotides were filtered to remove sequences that were <120 bases long, 7134 new 120mers remained (Table S2). The number of 120-base *S. bicolor* and sugarcane oligonucleotides designed and validated from all approaches described above was 51 847, reducing to 51 692 after a final check to detect oligonucleotides with inadvertent ambiguities.

A set of probes were also designed to tile across (two-times tiling) eight genes for sucrose metabolism—the five genes for sucrose phosphate synthase I–V (*SPS I–V*), soluble acid invertase (*SAI*) and sucrose synthase 1 and 2 (*SuSy*). Two-times tiling involves designing a set of probes that lie end to end and span the target sequence with a second set of probes offset (overlapping the first set) by 50% of the length of a probe (60 bases in this case) also lying across the length of the target. The genomic DNA sequence of four sugarcane *SPS* genes (*SPS II–V*) was used from sequences obtained at Southern Cross University. The sequence used for the design of probes for *SPS I* was obtained from sorghum along with *SuSy1*. A set of 240 gene fragments amplified in an earlier experiment and sequenced using 454 technology (Bundock *et al.*, 2009) were also included for probe design using a two-times tiling approach (Table S3). Table 6 shows the number of probes designed for each category, and the number of times each probe was represented on the array for manufacture, to create the bait sequences for hybridisation.

Preparation and sequencing of sugarcane genomic DNA libraries

Whole-genome shotgun libraries from the genomic DNA of Q165 and IJ76-514 were prepared using the Illumina DNA paired-end library prep kit. The Covaris S2 was used to shear the genomic DNA before adapter ligation. Size selection was carried out on an Invitrogen E-gel 2% with a target size of around 320 bp. Analysis on an Agilent Bioanalyser indicated that the size range of the vast majority of fragments was from 290 to 410 bp for IJ76-514, whilst for Q165 the size range was from 220 to 380 bp post-PCR (i.e. including 120 bases of adapter sequence).

Aliquots of WGS libraries for Q165, IJ76-514 and Sorghum R931945-2-2 were used for the SureSelect procedure that was carried out according to the protocol from Agilent (SureSelect Target Enrichment System, Illumina Paired-End Sequencing Platform Library Prep, Protocol Version 1.0, September 2009) and using the SureSelect reagents supplied by Agilent (Agilent Technologies Inc., Santa Clara, CA).

Paired-end sequencing was carried out using the Illumina GA IIx with 76-base read length.

Analysis of illumina sequence data

Illumina sequence data were analysed with CLC Genomics Workbench software, version 4 (CLC bio, Aarhus N, Denmark). Paired-end read data with quality scores was quality trimmed to remove bases with lower than 0.01 error probability and remove sequences shorter than 30 bases and sequence with unknown bases. For low stringency mapping of reads to reference sequences, the default parameters were used, which included length fraction = 0.5, similarity = 0.8 with mismatch cost = 2, insertion cost = 3, deletion cost = 3 and non-specific matches handled by random assignment. High stringency read mapping was used for SNP detection and for comparative purposes with the length fraction = 0.9, similarity = 0.9 and non-specific matches ignored (i.e. only unique matches mapped), otherwise the same as default parameters.

For SNP detection, the trimmed sequences were depleted of reads matching repeated regions by mapping with default parameters to known repetitive regions. These were sorghum repeats (TIGR_Sorghum_Repeats.v3.0_0_0.fsa.txt; http://plantrepeats.plantbiology.msu.edu/downloads.html), the sugarcane (SP80-3280) chloroplast genome (NCBI Reference Sequence:

NC_005878.2, http://www.ncbi.nlm.nih.gov/), the *S. bicolor* mitochondrial genome (GenBank: DQ984518.1; http://www.ncbi.nlm.nih.gov/) and an intron from our sugarcane sucrose phosphate synthase 3 (*SPS3*) sequence that was found to be highly repetitive based. Reads not matching to these repeat sequences were saved and mapped, using high stringency mapping parameters, to the ten chromosomes of the annotated *S. bicolor* genome sequence (Paterson *et al.*, 2009) that had previously been downloaded from NCBI (http://www.ncbi.nlm.nih.-gov/genome?term=sorghum%20bicolor) and imported into CLC Workbench. SNP detection using CLC Bio Workbench was carried out on these mapped reads using the following stringent parameters: window length = 21, maximum gaps and mismatches = 2, minimum central base quality score = 30, minimum average quality score = 20, minimum coverage at SNP site = 20, Minimum variant frequency = 5.0%, ploidy = 4, maximum coverage = 1000, sufficient variant count = 50 and required variant count = 2.

Acknowledgements

Genomic DNA samples of IJ76-514 and Q165 were obtained from Dr Karen Aitken, CSIRO Plant Industry, Queensland Bioscience Precinct, St Lucia, Qld, Australia. Sorghum R931945-2-2 genomic DNA was supplied by Dr Emma Mace, Crop and Food Science, DEEDI Hermitage Research Facility, Warwick Qld.

References

Aitken, K.S., Jackson, P.A. and McIntyre, C.L. (2005) A combination of AFLP and SSR markers provides extensive map coverage and identification of homo(eo)logous linkage groups in a sugarcane cultivar. *Theor. Appl. Genet.* **110**, 789–801.

Albert, T.J., Molla, M.N., Muzny, D.M., Nazareth, L., Wheeler, D., Song, X.Z., Richmond, T.A., Middle, C.M., Rodesch, M.J., Packard, C.J., Weinstock, G.M. and Gibbs, R.A. (2007) Direct selection of human genomic loci by microarray hybridization. *Nat. Methods,* **4**, 903–905.

Altschul, S.F., Gish, W., Miller, W., Myers, E.W. and Lipman, D.J. (1990) Basic local alignment search tool. *J. Mol. Biol.* **215**, 403–410.

Bashiardes, S., Veile, R., Helms, C., Mardis, E.R., Bowcock, A.M. and Lovett, M. (2005) Direct genomic selection. *Nat. Methods,* **2**, 63–69.

Bell, G.I., Pictet, R.L., Rutter, W.J., Cordell, B., Tischer, E. and Goodman, H.M. (1980) Sequence of the human insulin gene. *Nature,* **284**, 26–32.

Bundock, P.C., Eliott, F.G., Ablett, G., Benson, A.D., Casu, R.E., Aitken, K.S. and Henry, R.J. (2009) Targeted single nucleotide polymorphism (SNP) discovery in a highly polyploid plant species using 454 sequencing. *Plant Biotechnol. J.* **7**, 347–354.

Dahl, F., Gullberg, M., Stenberg, J., Landegren, U. and Nilsson, M. (2005) Multiplex amplification enabled by selective circularization of large sets of genomic DNA fragments. *Nucleic Acids Res.* **33** (8), e71.

Dahl, F., Stenberg, J., Fredriksson, S., Welch, K., Zhang, M., Nilsson, M., Bicknell, D., Bodmer, W.F., Davis, R.W. and Ji, H. (2007) Multigene amplification and massively parallel sequencing for cancer mutation discovery. *Proc. Natl Acad. Sci. USA,* **104**, 9387–9392.

Fu, Y., Springer, N.M., Gerhardt, D.J., Ying, K., Yeh, C.T., Wu, W., Swanson-Wagner, R., D'Ascenzo, M., Millard, T., Freeberg, L., Aoyama, N., Kitzman, J., Burgess, D., Richmond, T., Albert, T.J., Barbazuk, W.B., Jeddeloh, J.A. and Schnable, P.S. (2010) Repeat subtraction-mediated sequence capture from a complex genome. *Plant J.* **62**, 898–909.

Gnirke, A., Melnikov, A., Maguire, J., Rogov, P., LeProust, E.M., Brockman, W., Fennell, T., Giannoukos, G., Fisher, S., Russ, C., Gabriel, S., Jaffe, D.B., Lander, E.S. and Nusbaum, C. (2009) Solution hybrid selection with ultra-long oligonucleotides for massively parallel targeted sequencing. *Nat. Biotechnol.* **27**, 182–189.

Grunstein, M. and Hogness, D.S. (1975) Colony hybridization: a method for the isolation of cloned DNAs that contain a specific gene. *Proc. Natl Acad. Sci. USA,* **72**, 3961–3965.

Hardenbol, P., Baner, J., Jain, M., Nilsson, M., Namsaraev, E.A., Karlin-Neumann, G.A., Fakhrai-Rad, H., Ronaghi, M., Willis, T.D., Landegren, U. and Davis, R.W. (2003) Multiplexed genotyping with sequence-tagged molecular inversion probes. *Nat. Biotechnol.* **21**, 673–678.

Hardenbol, P., Yu, F.L., Belmont, J., MacKenzie, J., Bruckner, C., Brundage, T., Boudreau, A., Chow, S., Eberle, J., Erbilgin, A., Falkowski, M., Fitzgerald, R., Ghose, S., Iartchouk, O., Jain, M., Karlin-Neumann, G., Lu, X.H., Miao, X., Moore, B., Moorhead, M., Namsaraev, E., Pasternak, S., Prakash, E., Tran, K., Wang, Z.Y., Jones, H.B., Davis, R.W., Willis, T.D. and Gibbs, R.A. (2005) Highly multiplexed molecular inversion probe genotyping: over 10,000 targeted SNPs genotyped in a single tube assay. *Genome Res.* **15**, 269–275.

Hodges, E., Xuan, Z., Balija, V., Kramer, M., Molla, M.N., Smith, S.W., Middle, C.M., Rodesch, M.J., Albert, T.J., Hannon, G.J. and McCombie, W.R. (2007) Genome-wide in situ exon capture for selective resequencing. *Nat. Genet.* **39**, 1522–1527.

Kharabian-Masouleh, A., Waters, D.L.E., Reinke, R.F. and Henry, R.J. (2011) Discovery of polymorphisms in starch related genes in rice germplasm by amplification of pooled DNA and deeply parallel sequencing. *Plant Biotechnol. J.* **9**, 1074–1085.

Lovett, M., Kere, J. and Hinton, L.M. (1991) Direct selection—a method for the isolation of cDNAs encoded by large genomic regions. *Proc. Natl Acad. Sci. USA,* **88**, 9628–9632.

Malory, S., Shapter, F.M., Elphinstone, M.S., Chivers, I.H. and Henry, R.J. (2011) Characterizing homologues of crop domestication genes in poorly described wild relatives by high-throughput sequencing of whole genomes. *Plant Biotechnol. J.* **9**, 1131–1140.

Mamanova, L., Coffey, A.J., Scott, C.E., Kozarewa, I., Turner, E.H., Kumar, A., Howard, E., Shendure, J. and Turner, D.J. (2010) Target-enrichment strategies for next generation sequencing. *Nat. Methods,* **7**, 111–118.

Morgan, J.G., Dolganov, G.M., Robbins, S.E., Hinton, L.M. and Lovett, M. (1992) The selective isolation of novel cDNAs encoded by the regions surrounding the human interleukin-4 and interleukin-5 genes. *Nucleic Acids Res.* **20**, 5173–5179.

Mullis, K.B. and Faloona, F.A. (1987) Specific synthesis of DNA in vitro via a polymerase-catalyzed chain-reaction. *Methods Enzymol.* **155**, 335–350.

Ng, S.B., Turner, E.H., Robertson, P.D., Flygare, S.D., Bigham, A.W., Lee, C., Shaffer, T., Wong, M., Bhattacharjee, A., Eichler, E.E., Bamshad, M., Nickerson, D.A. and Shendure, J. (2009) Targeted capture and massively parallel sequencing of 12 human exomes. *Nature,* **461**, 272–276.

Ng, S.B., Bigham, A.W., Buckingham, K.J., Hannibal, M.C., McMillin, M.J., Gildersleeve, H.I., Beck, A.E., Tabor, H.K., Cooper, G.M., Mefford, H.C., Lee, C., Turner, E.H., Smith, J.D., Rieder, M.J., Yoshiura, K., Matsumoto, N., Ohta, T., Niikawa, N., Nickerson, D.A., Bamshad, M.J. and Shendure, J. (2010a) Exome sequencing identifies MLL2 mutations as a cause of Kabuki syndrome. *Nat. Genet.* **42**, 790–793.

Ng, S.B., Buckingham, K.J., Lee, C., Bigham, A.W., Tabor, H.K., Dent, K.M., Huff, C.D., Shannon, P.T., Jabs, E.W., Nickerson, D.A., Shendure, J. and Bamshad, M.J. (2010b) Exome sequencing identifies the cause of a mendelian disorder. *Nat. Genet.* **42**, 30–35.

Okou, D.T., Steinberg, K.M., Middle, C., Cutler, D.J., Albert, T.J. and Zwick, M.E. (2007) Microarray-based genomic selection for high-throughput resequencing. *Nat. Methods,* **4**, 907–909.

Paterson, A.H., Bowers, J.E., Bruggmann, R., Dubchak, I., Grimwood, J., Gundlach, H., Haberer, G., Hellsten, U., Mitros, T., Poliakov, A., Schmutz, J., Spannagl, M., Tang, H., Wang, X., Wicker, T., Bharti, A.K., Chapman, J., Feltus, F.A., Gowik, U., Grigoriev, I.V., Lyons, E., Maher, C.A., Martis, M., Narechania, A., Otillar, R.P., Penning, B.W., Salamov, A.A., Wang, Y., Zhang, L., Carpita, N.C., Freeling, M., Gingle, A.R., Hash, C.T., Keller, B., Klein, P., Kresovich, S., McCann, M.C., Ming, R., Peterson, D.G., Mehboob-Ur-Rahman, Ware, D., Westhoff, P., Mayer, K.F., Messing, J. and Rokhsar, D.S. (2009) The *Sorghum bicolor* genome and the diversification of grasses. *Nature,* **457**, 551–556.

Souza, G.M., Berges, H., Bocs, S., Casu, R., D'Hont, A., Ferreira, J.E., Henry, R., Ming Potier, R.B., Van Sluys, M.A., Vincentz, M. and Paterson, A.H. (2011) The sugarcane genome challenges: strategies for sequencing a highly complex genome. *Trop. Plant Biol.* **4**, 145–156.

Tewhey, R., Nakano, M., Wang, X.Y., Pabon-Pena, C., Novak, B., Giuffre, A., Lin, E., Happe, S., Roberts, D.N., LeProust, E.M., Topol, E.J., Harismendy, O. and Frazer, K.A. (2009a) Enrichment of sequencing targets from the human genome by solution hybridization. *Genome Biol.* **10**, R116.

Tewhey, R., Warner, J.B., Nakano, M., Libby, B., Medkova, M., David, P.H., Kotsopoulos, S.K., Samuels, M.L., Hutchison, J.B., Larson, J.W., Topol, E.J., Weiner, M.P., Harismendy, O., Olson, J., Link, D.R. and Frazer, K.A. (2009b) Microdroplet-based PCR enrichment for large-scale targeted sequencing. *Nat. Biotechnol.* **27**, 1025–1031.

Turner, E.H., Lee, C.L., Ng, S.B., Nickerson, D.A. and Shendure, J. (2009) Massively parallel exon capture and library-free resequencing across 16 genomes. *Nat. Methods*, **6**, 315–316.

Winfield, M.O., Lu, C.G., Wilson, I.D., Coghill, J.A. and Edwards, K.J. (2010) Plant responses to cold: transcriptome analysis of wheat. *Plant Biotechnol. J.* **8**, 749–771.

17

Gene targeting in maize by somatic ectopic recombination

Ayhan Ayar[1,2], Sophie Wehrkamp-Richter[1,2], Jean-Baptiste Laffaire[1], Samuel Le Goff[1], Julien Levy[1], Sandrine Chaignon[1], Hajer Salmi[1], Alexandra Lepicard[1], Christophe Sallaud[1], Maria E. Gallego[2], Charles I. White[2] and Wyatt Paul[1,*]

[1]Biogemma, Centre de Recherche de Chappes, Chappes, France
[2]CNRS UMR 6293,Clermont Université INSERM U1103, Aubière, France

*Correspondence
email wyatt.
paul@biogemma.com

Summary

Low transformation efficiency and high background of non-targeted events are major constraints to gene targeting in plants. We demonstrate here applicability in maize of a system that reduces the constraint from transformation efficiency. The system requires regenerable transformants in which all of the following elements are stably integrated in the genome: (i) donor DNA with the gene of interest adjacent to sequence for repair of a defective selectable marker, (ii) sequence encoding a rare-cutting endonuclease such as I-SceI, (iii) a target locus (TL) comprising the defective selectable marker and I-SceI cleavage site. Typically, this requires additional markers for the integration of the donor and target sequences, which may be assembled through cross-pollination of separate transformants. Inducible expression of I-SceI then cleaves the TL and facilitates homologous recombination, which is assayed by selection for the repaired marker. We used bar and gfp markers to identify assembled transformants, a dexamethasone-inducible I-SceI::GR protein, and selection for recombination events that restored an intact nptII. Applying this strategy to callus permitted the selection of recombination into the TL at a frequency of 0.085% per extracted immature embryo (29% of recombinants). Our results also indicate that excision of the donor locus (DL) through the use of flanking I-SceI cleavage sites may be unnecessary, and a source of unwanted repair events at the DL. The system allows production, from each assembled transformant, of many cells that subsequently can be treated to induce gene targeting. This may facilitate gene targeting in plant species for which transformation efficiencies are otherwise limiting.

Keywords: double-strand break, meganuclease, ectopic recombination, gene targeting, nptII, Zea mays.

Introduction

Transgenesis offers the possibility to insert a known DNA sequence into the genome of an organism to introduce new heritable characters. It is commonly used in research to investigate gene function and in biotechnology to improve agronomic traits. However, random insertion of the transgene into the genome can result in mutations caused by the insertion into an endogenous gene (Krysan et al., 1999), potential production of unintended peptides or variable expression due to the genomic environment of the transgene (Matzke and Matzke, 1998). Thus, there are currently considerable efforts worldwide to develop efficient technologies for gene targeting (GT) to produce genetically modified (GM) crops with transgenes located at predetermined positions in the plant genome. Exploiting the cellular homologous recombination (HR) machinery, GT allows the exchange of genetic information between homologous DNA sequences and can be used to precisely modify the genome. The integration of transgenes flanked by sequences homologous to the desired genomic insertion site permits efficient and routine gene targeting in prokaryotes and fungi, but GT is very inefficient in higher plants with frequencies of the order of 10^{-4} per transformant (Cotsaftis and Guiderdoni, 2005; Hanin et al., 2001; Paszkowski et al., 1988).

HR and non-homologous end-joining (NHEJ) are triggered to repair double-strand breaks (DSBs) of DNA. These lesions are formed accidentally by genotoxic stresses (Hanin and Paszkowski,

2003; Khanna et al., 2001; Tuteja et al., 2009) or in a programmed manner, for example during meiosis by the Spo11 complex (Grelon et al., 2001). Repair through NHEJ links the two ends of the DSB and is frequently accompanied by the creation of mutations at the site of the repair. HR copies an endogenous (different allele or stably inserted transgene) or exogenous (non inserted transgene) sequence template with homology on either side of the break and allows a precise modification of the genome (Puchta et al., 1996). Transgene integration is believed to generally involve insertion via NHEJ into a pre-existing DSB (Tzfira et al., 2004) occurring randomly in the plant genome. A DSB at a precise genomic location presenting homologous sequence to the transgene significantly increases the recombination rate at this site (Puchta et al., 1993; Szostak et al., 1983). This has led to the development of tools such as meganucleases, zinc-finger nucleases and transcription activator-like effector nucleases for gene targeting (Christian et al., 2010; Shukla et al., 2009; Tzfira et al., 2012). These endonucleases create a DSB at the target locus (TL) and have been used to modify the TL by mutation using NHEJ (De Pater et al., 2009; Yang et al., 2009) or by precise sequence modification using HR (Tzfira and White, 2005). For example, the mitochondrial I-SceI meganuclease from Saccharomyces cerevisiae (Jacquier and Dujon, 1985) has been successfully used in plants to perform GT (D'Halluin et al., 2008; Puchta et al., 1996). In tobacco, cleavage of the TL containing an I-SceI restriction site by I-SceI increases recombination between the TL and the transforming T-DNA around 100-fold (Puchta et al., 1996). The enzyme required to produce the DSB can be introduced into the organism or cell via stable or transient transformation. For example, I-SceI has been introduced into plants via *Agrobacterium*-mediated retransformation of a *TL* line or by crossing lines stably expressing I-SceI to a *TL* line. In the latter case, the use of an inducible I-SceI can allow the creation of the DSB at a predetermined moment. For example, application of the glucocorticoid, dexamethasone, induced the activity of an I-SceI protein fused to the rat glucocorticoid receptor (GR) domain in *Arabidopsis* (Wehrkamp-Richter *et al.*, 2009). The GR domain sequesters the I-SceI::GR complex in the cytosol. The addition of dexamethasone allows the dissolution of the complex (Aoyama and Chua, 1997), liberating the I-SceI::GR protein which can move to the nucleus and produce a DSB at the *TL*. In plants, an inducible I-SceI was used to enhance intrachromosomal recombination (Wehrkamp-Richter *et al.*, 2009) and to perform targeted mutagenesis (Yang *et al.*, 2009).

Plant GT strategies are generally based on the positive selection for GT events which repair a defective selectable marker. A DSB is induced at the *TL* inducing HR between a defective *TL* selectable marker gene and the repair DNA. For example in *Zea mays* (maize), D'Halluin *et al.* (2008) re-transformed *TL* lines with a repair DNA and a construct encoding I-SceI, either delivering the DNA via particle bombardment or *Agrobacterium*. The frequency of GT versus random insertion, measured by the acquisition of resistance to the herbicide BASTA, was up to 30% via particle bombardment and 3.7% using *Agrobacterium*. Shukla *et al.* (2009) have also reported efficient GT in maize using zinc-finger nucleases. Although these studies show that GT is now possible in a major crop plant, there is still the need to optimize GT to minimize the effort required to produce and identify GT events before GT becomes a routine tool for GM production. A major limiting factor is the need to deliver the repair DNA and nuclease-encoding sequence efficiently into a large number of cells, which in the case of maize transformation can involve the transformation of many thousands of immature embryos or calli to obtain a few GT events. An attractive alternative is to create a few transformation events where the repair DNA and I-SceI-encoding sequence are stably integrated into the genome. The repair DNA is then controllably excised from the genome and acts as a template for GT at the *TL*. This system has the advantage that every cell contains the repair template, and thus, a single transformed individual can yield a potentially unlimited population of cells for GT. Such a GT strategy has been successfully implemented in *Drosophila*, with the repair DNA being excised from the genome using the FLP recombinase and then linearized using I-SceI (Huang *et al.*, 2008; Rong, 2002) and has recently been reported also in *Arabidopsis* (Fauser *et al.*, 2012). The goal of the work presented here was to test a similar GT system in maize, using a dexamethasone-inducible I-SceI both to excise the repair DNA from the genome and to induce a DSB at the *TL*.

Results

The GT test system

Two plant transformation constructs, the *TL* construct and the donor locus (*DL*) construct, were developed to test the GT strategy. The T-DNA of the *TL* construct contains the plant transformation selectable marker phosphinothricin acyl transferase (*bar*) gene followed by an I-SceI restriction site and the 3' part of the neomycin phosphotransferase II (*nptII*) gene (Figure 1b). The T-DNA of the *DL* construct contains a dexamethasone-

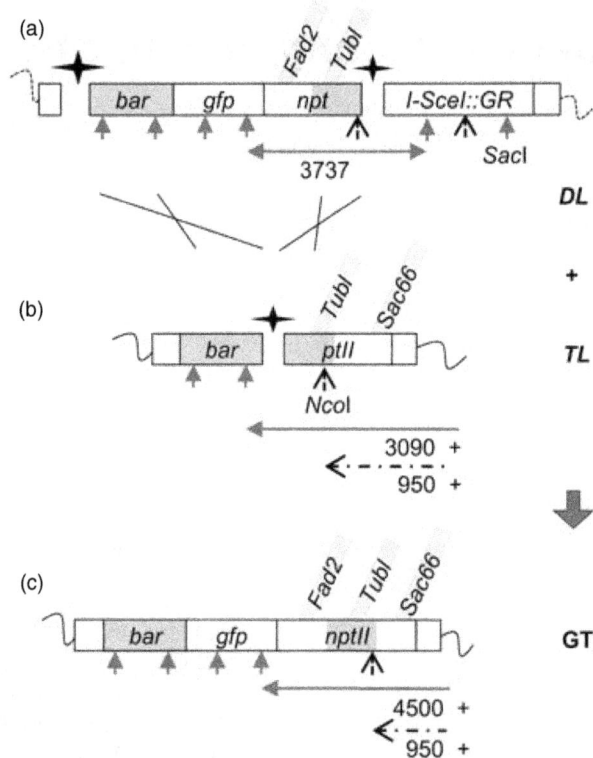

Figure 1 Maps of the transgenic loci (not to scale). Structures of the donor locus (a), the target locus (b) and the expected gene targeting event (c) with the position of the I-SceI cutting sites (black stars). For Southern blot analyses, the sizes of DNA fragments hybridized with *intTubI* (*TubI*), *intFad2* (*Fad2*) and *terSac66* (*Sac66*) are shown for digestion with *Sac*I (grey arrows) or *Nco*I (black dashed arrows). '+' indicates that the size of the fragment will be greater than indicated as it will extend to the first relevant restriction site in the flanking genomic DNA. The left border of the T-DNA is adjacent to the *bar* gene. Fragment lengths are given as bp.

activatable, maize codon-optimized I-SceI (*I-SceI::GR*) gene and an *nptII* repair region bordered by two I-SceI restriction sites (Figure 1a). The *nptII* repair region contains the *bar* gene, the green fluorescent protein (*gfp*) gene and a 5' part of the *nptII* gene. The *gfp* gene here serves as a mock gene of interest to be inserted at the *TL* and additionally allows easy identification of *DL*-containing plants. The *nptII* repair region and the *TL* share common sequences of 2992 bp in the *bar* region and 1200 bp in the *nptII* region, provided largely by the insertion of a rice tubulin gene intron (*intTubI*) into the defective *nptII* genes. This homology should allow homologous recombination between these two sequences and the consequent repair of the *nptII* gene, resulting in kanamycin resistance (Figure 1c). The *TL* and the *DL* constructs were independently transformed into maize to generate *TL* and *DL* lines, respectively. Two intact *TL* (*TL1* and *TL2*) lines and one *DL* line, each containing a single copy of the transgene, were selected by Southern blotting analysis (not shown) and their genomic flanking sequences isolated (Figure S1). The *DL* line expressed both *gfp* and the *I-SceI::GR* transcript. The two *TL* lines were then selfed in order to isolate homozygous descendants which were then crossed with the *DL* line (Figure 2a). The F1 progenies and their descendents were selfed (Figure 2b). The segregation of the *TL* and the *DL* indicates that the two constructions were not genetically linked.

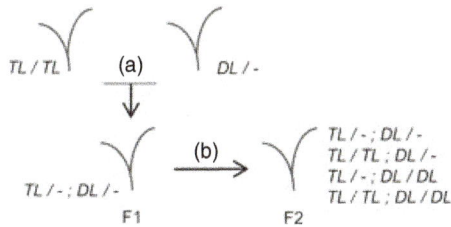

Figure 2 Genealogy of maize Lines. The two homozygous target locus (*TL*) lines were crossed with a heterozygous donor locus (*DL*) line to obtain the F1 (a). The F1 progeny containing the *DL* and the *TL* were self-fertilized to obtain F2 descendants (b). '−' represents the original wild-type loci. Only the genotypes of interest (containing *TL* and *DL*) are shown.

Detection of somatic repair of *nptII* in *TL/DL* leaves

For each *TL* line crossed with the *DL* line, the F1 progeny containing the *TL* and the *DL* were identified by PCR analysis and separated in two groups, seed of one group was treated with dexamethasone to induce I-*Sce*I activity and the other not (see Experimental procedures). To detect excision of the repair DNA from the *DL*, PCR was performed using primers positioned on either side of the *DL* I-*Sce*I restriction sites. A total of 12 untreated and seven dexamethasone-treated plants were analysed and three excision events were detected for each condition, indicating a basal activity of I-*Sce*I::GR and that dexamethasone treatment does not significantly induce I-*Sce*I::GR in these conditions.

The analysed F1 plants were then selfed to identify kanamycin-resistant plants among the F2 descendants. To detect the presence of the *TL* and the *DL*, 176 F2 (42 for the *TL1/DL* line and 134 for the *TL2/DL* line) plants were analysed by PCR; 21 *TL1/DL* and 55 *TL2/DL* descendants contained both *TL* and *DL*. Kanamycin was applied to the apical meristematic region of wild-type (WT) and the F2 plants. On WT plants and F2 descendants containing only either the *TL* or the *DL*, this resulted in bleaching of the developing leaf (Figure 3b). However, leaves with green sectors within the kanamycin-bleached zones (Figure 3c) were observed in 60% of plants carrying both the *TL* and *DL*, corresponding to 38% of *TL1/DL* and 70% of *TL2/DL* plants (Figure 3d). PCR analysis performed on the DNA extracted from the green sectors permitted the amplification and sequencing of the repaired *nptII* gene, and this was not so for DNA extracted from bleached or untreated (Figure 3a) leaf sectors. Other batches of F2 seeds were sown, and none of the additional 504 descendants were fully kanamycin-resistant; however, green kanamycin-resistant sectors were again observed in *TL/DL* plants.

Recovery of fully kanamycin-resistant plants via *in vitro* tissue culture

Notwithstanding the presence of kanamycin-resistant leaf sectors (and thus GT), no fully kanamycin-resistant progeny were identified in 680 F2 plantlets. We thus used a tissue culture approach. Plant regeneration from maize leaves has not been reported, but calli derived from immature maize embryos are routinely used to regenerate plants (Lu *et al.*, 1983). Embryos isolated from immature kernels of selfed F2 plants containing the *TL* and the *DL* were placed on callus induction medium with dexamethasone at 0 μM (control), 30 or 50 μM (Figure 4). From 2356 extracted embryos (619 from the *TL1/DL* and 1737 from the *TL2/DL* plants), seven independent kanamycin-resistant GT events

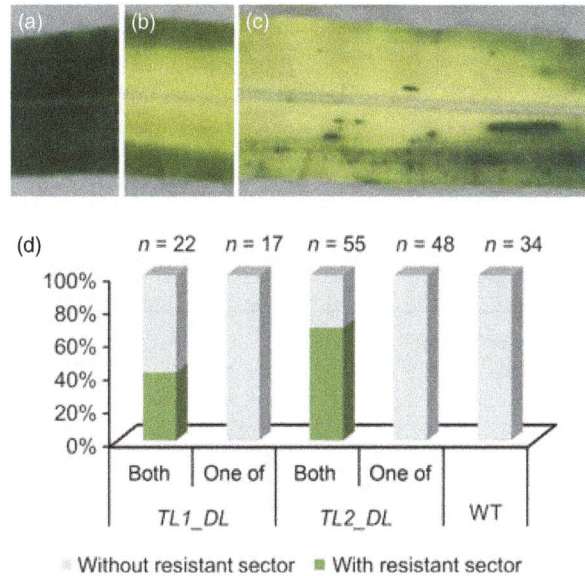

Figure 3 Frequency of green sectors in kanamycin-treated plants. Leaves of F2 maize plants untreated with kanamycin (a), sensitive to kanamycin (b) and sensitive to kanamycin with kanamycin-resistant green sectors (c). The histogram (d) shows the percentage of sensitive plants with green kanamycin-resistant sectors for the different genotypes (n is the number of plants analysed). 'Both' indicates plants possessing a target locus (*TL*) and a donor locus (*DL*); 'one of' indicates plants possessing either only a *TL* or a *DL*.

(Table 1) were recovered and shown to carry a repaired *nptII* gene which was amplified by PCR and sequenced. Two were obtained from the *TL1/DL* embryos (GT1 and GT2) and five from the *TL2/DL* embryos (GT3, GT4, GT5, GT6 and GT7). GT efficiencies calculated as the number of GT events per immature embryo range from 0.13% to 0.55% (Table 1).

Only one of the seven GT events was obtained from dexamethasone non-treated control embryos; thus, dexamethasone treatment appears to increase the number of GT events. The number of events are, however, low and as we observed somatic recombination during the development of F2 plants in the absence of dexamethasone treatment, some GT events probably come from a basal, leaky I-*Sce*I::GR activity.

To clarify the question of the inducibility of I-*Sce*I::GR activity, we tested the effect of dexamethasone treatment on *TL* DSB

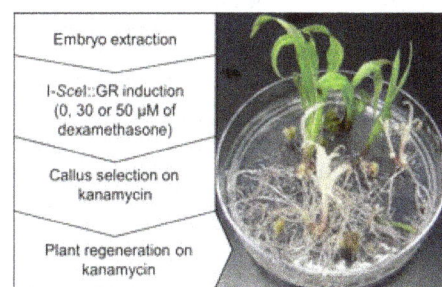

Figure 4 Strategy for *in vitro* culture and regeneration. Immature embryos were collected and separated into three groups, non-treated, treated with 30 μM and treated with 50 μM of dexamethasone. Kanamycin selection allowed the regeneration of kanamycin-resistant plants containing a repaired *nptII* gene (green plantlets).

Table 1 Summary of gene targeting experiments

Lines	Extracted embryos	Dexamethasone concentration (μM)	Number of kanamycin-resistant events	Name of kanamycin-resistant events	GT frequency (kanamycin-resistant/ total embryos)
TL1 × *DL*	183	0	0	/	0%
	255	30	1	GT1	0,39%
	181	50	1	GT2	0,55%
TL2 × *DL*	800	0	1	GT4	0,13%
	563	30	3	GT3, GT5, GT6	0,53%
	374	50	1	G7 (+)	0,27%

+ indicates a sterile plant.
GT, gene targeting; *TL*, target locus; *DL*, donor locus.

induction through the measurement of mutations in the *TL* I-*Sce*I site. Embryos of F2 plants were extracted (14 from *TL1/DL* and 69 from *TL2/DL* lines) for somatic embryogenesis. A sample of the callus formed from each embryo was analysed by PCR to determine the presence of the *DL* and the *TL*. Each callus was divided into three parts, one part placed for 1 week on medium without dexamethasone, one on medium with 30 μM and one with 50 μM dexamethasone. Samples of *TL/DL* calli from each treatment and of the *TL* calli were pooled separately for DNA extraction (Table 2). A 400-bp region around the I-*Sce*I *TL* restriction site was amplified from each pool and sequenced by 454 sequencing (Genome Sequencer FLX by Roche). Approximately 15 000 sequences were obtained and analysed to estimate the mutation rate at the I-*Sce*I restriction site due to NHEJ repair. No mutations were detected in the absence of the *DL* (*TL* controls that do not carry the *I-Sce*I gene). In the *TL/DL* lines, mutations of the *TL* I-*Sce*I site were detected, with the number of independent mutations increasing 3.5 to 5-fold with 30 μM and 5 to 6-fold with 50 μM of dexamethasone. These data thus confirm a basal activity of I-*Sce*I::GR in inducing mutations in the I-*Sce*I target site and that dexamethasone treatment increases I-*Sce*I::GR activity in calli (Table 2). The presence of mutations in the target in the absence of dexamethasone, however, confirms the leakiness in this system.

Analysis of GT obtained from *TL1/DL* plants

Two GT events were identified from calli from *TL1/DL* plants. These were regenerated to give plants GT1 and GT2. Southern blot analysis was carried out on *Sac*I-digested genomic DNA of these plants, the parent line (*TL1/DL*) and control lines carrying only the target locus (*TL1*) or the *DL*. Three different probes were used: *Arabidopsis AtFad2* gene intron (*intFad2*, present in the *DL* and predicted to be present in a GT locus), *intTubI* (common to

the *DL*, *TL* and predicted to be in the GT locus) and *Arabidopsis AtSac66* terminator (*terSac66*, present in the *TL* and predicted to be in the GT locus). The Southern blot results with the *intTubI* probe are shown in Figure 5b. A band of 3.7 kbp was detected in the *DL* lane and a 4.1-kbp band in the *TL1* lane; both bands were observed in the *DL/TL1* control lane. As expected, in GT1 and GT2 lanes, the *TL1* band disappeared and a new 5.5-kbp band, also observed with *intFad2* and *terSac66* probes (Figure S2), was detected, confirming *nptII* repair at the *TL*. The non-excised *DL* band was observed at 3.7 kbp with an intensity consistent with a homozygous state.

The GT1 and GT2 plants were backcrossed twice to wild-type plants, and kanamycin resistance was inherited as a single Mendelian locus. In the first backcross, 53% and 55% of GT1 and GT2 descendants were kanamycin resistant and all presented the non-excised *DL*, confirming that GT1 and GT2 are heterozygous for the reconstructed (by GT) *nptII* gene at the *TL*. For the second backcross, 35% and 36% of GT1 and GT2 descendants were kanamycin resistant. All resistant plants expressed *gfp*, and PCR amplification confirmed both the presence of the *intFad2* and *terSac66* regions and the absence of the *TL1*-specific fragment containing the I-*Sce*I *TL1* site (Figure S3), which together confirm the expected reconstitution of *nptII*. Among the kanamycin-resistant descendants, 61% of GT1 and 55% of GT2 also contained all sequences specific to the *DL* [left border (LB) I-*Sce*I site, right border (RB) I-*Sce*I site and *I-Sce*I::GR].

Finally, to confirm that the modified *TL* in plants GT1 and GT2 are the result of homologous recombination on both sides of the break in the *TL* with the donor, we also sequenced the junction fragments amplified by PCR with primers to the *terSac66* and the genomic flanking sequence of the *TL1* LB. PCR fragments were amplified from kanamycin-resistant GT1 and GT2 plants containing only the GT locus and sequenced. The sequence of the

Table 2 Quantification of mutations at the I-*Sce*I site of target locus (*TL*)

Lines	Number of Embryos	Genotype	Number of reads	Dexamethasone concentration (μM)	Mutations in *TL* I-*Sce*I site
TL1 × *DL*	14	*TL1*	15370	/	0
		TL1/DL (*I-Sce*I::GR)	11575	0	15
			17685	30	76
			23279	50	76
TL2 × *DL*	69	*TL2*	15597	/	0
		TL2/DL (*I-Sce*I::GR)	16293	0	6
			19583	30	15
			17931	50	36

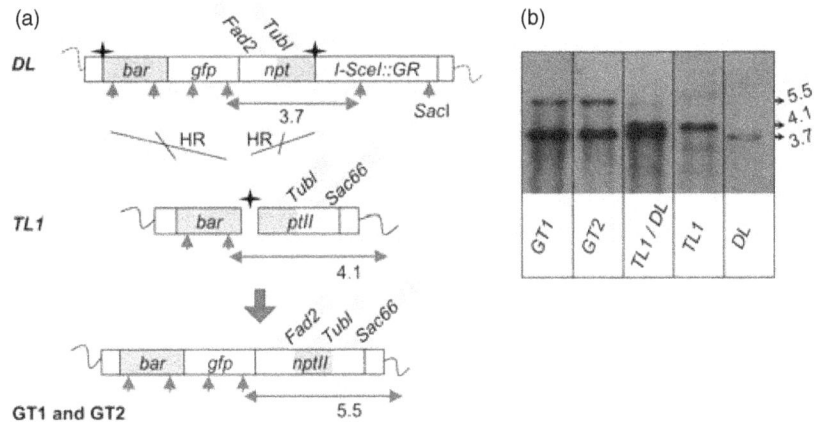

Figure 5 Southern blot analysis of gene targeting (GT)1 and GT2 events. (a) Schema of the GT event occurring in the GT1 and GT2 events. The black stars show I-SceI restriction sites, and cleavage of the site in target locus (TL)1 is indicated by the separation of the two halves of the locus. Grey arrows indicate SacI restriction sites and the size of DNA fragments detected by the intTubI probe. (b) Southern blot analysis of control parental plants (TL1, donor locus (DL), TL1/DL plants) and of recombinant GT plants from TL1/DL tissue culture (GT1 and GT2 lines) with SacI-digested DNA and the intTubI probe. A fragment around 3.7 kbp, indicative of the original DL, was detected in the GT1 and GT2 events. A 4.1-kbp fragment, indicative of TL1, was not observed in the GT1 and GT2 events. The expected fragment of 5.5 kbp, indicative of npII reconstitution due to HR between the DL and TL1, was detected in the GT1 and GT2 events. The DL band is more intense than the TL1 band for the lanes GT1 and GT2, indicating the homozygous state of the DL and the non-excision of the DL. Fragment lengths are given as kbp.

amplified fragment obtained (Figure S4) was identical to that predicted for HR between the DL and TL1 resulting in the repair of nptII at the TL1.

Analyses of the GT1 and GT2 events thus showed that they are true GT events at the TL and that the GT was not associated with excision of the donor sequence from the DL in either case, suggesting that they arose through ectopic recombination (Puchta, 1999) between the TL and DL (Figure 5a).

Analysis of GT from TL2/DL plants

Five GT events were identified from calli from TL2/DL plants. These were regenerated to give plants GT3, GT4, GT5, GT6 and GT7. Analysis of these plants revealed a second class of GT events involving reconstitution of nptII at the DL, rather than at the TL (Figure 6a).

Southern blot analysis was carried out on SacI-digested genomic DNA of these plants, the parent line (TL2/DL) and control lines carrying only the target locus (TL2) or the DL, hybridized with intFad2, intTubI and terSac66 probes. The results with the intTubI probe are presented in Figure 6b. The original DL and TL2 bands of 3.7 and 5.7 kbp respectively were detected in the GT samples, except for GT3 that lacked the original DL. However, the predicted GT-specific band of 7.1 kbp for GT at the TL (4.5-kbp plus 2.6-kbp TL2 flanking sequence) was not detected in the GT lanes. Instead, a band was observed at 5.5 kbp for GT3 and around 4.6 kbp for GT4, GT5, GT6 and GT7, indicating a different mechanism of nptII repair. This band was also observed with the intFad2 and terSac66 probes (Figure S2). The SacI digestion results were confirmed by Southern blotting of NcoI-digested DNA hybridized with the terSac66 probe (Figure 6c). The TL band of 2.0 kbp was found unchanged in all GT lanes, and an additional band was observed in GT3, GT5 and GT6 lanes. The presence of the terSac66 on two different DNA fragments indicates that either one copy of a potentially homozygous TL2 was modified, but not via the expected double crossover, or that the TL2 was used as template by HR to repair an I-SceI::GR-induced DSB in the DL (Figure 6a). The fact that PCR of the GT3

line could not detect a DL lacking the nptII repair fragment (data not shown) and that GT3 lacks a band specific to the original DL supports the idea that in GT3 at least, the DL has been modified.

To resolve this question, the GT events were backcrossed twice to the wild type and analysed by PCR (Figure S3). Kanamycin resistance was inherited as a single locus and was not correlated with the presence of the LB TL amplicon, which is specific to the TL and predicted to be present in a true GT event at the TL. Amplification of the LB TL and TL I-SceI amplicons in 46% of the GT4, 44% of the GT5 plants and about 17% of the GT6 events can thus be attributed to the presence of a segregating unmodified TL in these plants. Kanamycin resistance was strictly correlated with the presence of DL sequences on either side of the I-SceI restriction site next to the defective nptII in the DL. However, a PCR fragment of the expected size across this I-SceI site could not be amplified. This suggests either deletion around this I-SceI site or the insertion of a sequence including the TL terSac66 into this I-SceI site. This latter hypothesis was confirmed by amplification and sequencing of the GT loci using primers located in the intFad2 and in the I-SceI::GR gene. Analysis of the amplified sequence (Figure S4) showed an HR event restoring the nptII gene on the one side and a NHEJ or microhomology-mediated end-joining (MMEJ) event copying and linking a part of the TL2 flanking sequence to the I-SceI::GR promoter on the other side. For the GT3 event, after the region of homology in the nptII gene, 909 bp of the TL2 corresponding to the missing part of the defective nptII (including the terSac66) and 502 bp of the genomic flanking sequence of the TL2 RB were linked by non-homologous recombination to the other side of the break, which had lost 52 bp of DL sequence. For GT5, the event is similar to the GT3 event, but only 82 bp of the TL2 flanking sequence was copied and 9 bp of the break was deleted to repair this side by NHEJ including 115 bp of mitochondrial DNA in the junction. In the GT4 event, a microhomology of 4 bp is present at the junction, and 882 bp of the TL2 comprising the missing part of nptII (including the terSac66) was copied into the repair sequence with 55 bp deleted from the DL. The lengths of these sequences

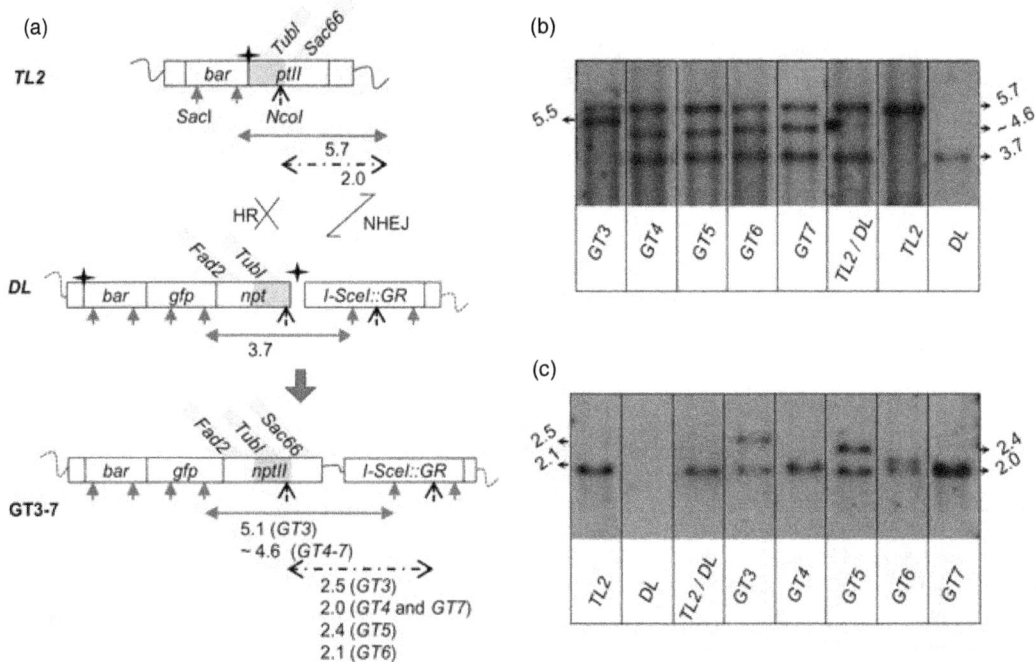

Figure 6 Southern blot results of gene targeting (GT)3-7 events. (a) Schema of the GT events occurring in GT3-7. The black stars show I-SceI restriction sites, and cleavage of the site in donor locus (*DL*) is indicated by the separation of the two halves of the locus. The grey arrows and black dashed arrows indicate respectively the SacI and NcoI restriction site positions and the size of DNA fragments detected by the probes used. (b) Southern blot analyses of control parental plants [target locus (*TL*)2, *DL*, *TL2/DL*] and of kanamycin-resistant events derived from the *TL2/DL* (GT3-7) with SacI-digested genomic DNA and the *intTubI* probe. A 3.7-kbp band, indicative of the non-excised *DL*, was detected in the GT4-7 events, but not in the GT3 event. The 5.7-kbp fragment indicative of native *TL2* was observed in the GT3–7 events. All putative GT events show an additional band around 5.5 kbp for GT3 and around 4.6 kbp for GT4–7 events (but not at the expected size of 7.1 kbp). (c) Southern blot analysis of the same events with NcoI-digested DNA blotted with a *terSac66*-specific probe. The signal at 2.0 kbp was observed for the *TL2* and *DL/TL2* controls. For the GT3, GT5 and GT6 events, unexpected bands appeared at respectively 2.5, 2.4 and 2.1 kbp. Fragment lengths are given as kbp.

of these GT events correspond to the observed sizes of bands seen on the Southern blots.

Analyses of the GT3-7 events thus showed that they are GT events, but involve the modification of the *DL* using the *TL* as template. Cleavage of the I-SceI site of the *DL* adjacent to the *nptII* sequence, followed by recombination of the *nptII* side of the break with the homologous *TL2* as donor, creates a functional *nptII* at the *DL*. The other side of the break in the *DL* does not carry homology to the *TL2*, and thus must be repaired by NHEJ or MMEJ, resulting in variable lengths of *TL2* sequence integrated into the *DL* (Figure 6a).

Discussion

The goal of this study was to develop a tool for precise remobilization of a transgene randomly inserted into the maize genome by its excision and insertion into a defined genomic site using homologous recombination. This strategy was tested by crossing of stably transformed *TL* and *DL* maize lines containing 3′ and 5′ overlapping regions of an *nptII* gene, respectively. Induction of I-SceI activity in these lines with dexamethasone was used both to create a DSB at the *TL* and also to release the *nptII* repair DNA from the *DL*. HR of the liberated *nptII* repair DNA with the *TL* would then reconstitute the *nptII* gene and also mobilize a *gfp* gene into the *TL*. Kanamycin selection allows the selection of putative GT events.

Testing of 680 F2 progeny carrying the *TL* and *DL* did not permit the identification of any kanamycin-resistant plants,

suggesting that germinal or early meristematic GT events are very rare under the conditions tested. However, in the course of testing these plants for kanamycin resistance, we noted the presence of green kanamycin-resistant sectors on the kanamycin-bleached leaves, suggesting the presence of somatic HR events between the *TL* and *DL*. DNA extracted from these green sectors, but not bleached leaf regions, could be used to amplify a restored functional *nptII* gene. Such green kanamycin-resistant sectors on bleached plants have previously been described in tobacco plants carrying an intrachromosomal HR reporter based on *nptII* reconstitution (Peterhans et al., 1990) and also in *Arabidopsis* (Assaad and Signer, 1992). Other studies of GT based on *nptII* restoration and selection of resistant plants through the addition of kanamycin to the culture medium in tobacco (Puchta, 1999) and *Arabidopsis* (Vergunst et al., 1998) did not, however, report green kanamycin-resistant sectors. In maize, we show here that application of kanamycin to the apex permits the detection and quantification of somatic GT events in leaves without affecting the survival of the sensitive plants. Multiple kanamycin treatments are possible and progeny can be obtained from treated plants. This assay, which should be applicable to other plant species, is currently being used to test and optimize GT frequencies.

In tobacco lines containing the equivalent of our *TL*, retransformed with a repair sequence and constitutive I-SceI, the observed GT frequency increased proportionally with the expression level of I-SceI (Puchta et al., 1996). Similarly, in our maize plants, the frequency of green kanamycin-resistant sectors gives direct information about I-SceI::GR activity. Given that we observed *nptII* repair sequence excision from *DL* in equivalent proportions from maize plants grown in the absence or the presence of dexamethasone treatment, there is clearly basal activity of I-*Sce*I::GR in the maize leaves and dexamethasone does not further induce I-*Sce*I::GR in the tested conditions. In our previous study with I-*Sce*I::GR in *Arabidopsis*, basal activity was found, but the expression could be induced around 25- to 200-fold when dexamethasone was supplied in the growth medium (Wehrkamp-Richter *et al.*, 2009). We speculate that the dexamethasone applied to maize germinating seed does not penetrate into the seed in sufficient quantities to further induce I-*Sce*I::GR activity. Dexamethasone treatment does, however, induce I-*Sce*I::GR activity when added to the callus growth medium, where a 3.5 to 6.0-fold increase in the number of mutations at the *TL* was observed with dexamethasone (Table 2).

Notwithstanding the GT observed in somatic tissues, no kanamycin-resistant plants were found in the 680 tested F2 progeny of the *TL/DL* lines. We thus tested a strategy based on tissue culture selection and regeneration of kanamycin-resistant plants from *TL/DL* calli. This approach permitted the selection of seven independent GT events in two separate experiments involving a total of 2356 embryos (Table 1). Two of these, GT1 and GT2, were generated from embryos from *TL1/DL* plants; molecular and genetic analyses confirmed that they are true GT in which the *TL* has been modified by ectopic recombination using the *DL* as template on both sides. The overall frequency of obtaining true GT events at the *TL* from the two experiments was therefore 0.085% (29% of recombinants). The remaining five events (GT3-7) were generated from *TL2/DL* line embryos, and Southern blot and sequence analyses showed that they result from the modification of the *DL*, using the *TL* as template. The mechanism appears to be the creation of a DSB by I-*Sce*I::GR in the *DL* I-*Sce*I site next to the 5′ *nptII* region. Recombination of the *nptII* side of the break with the homologous *TL2* region as donor creates a functional *nptII* at the *DL*. However, the other side of the break does not carry homology to the *TL2*, and thus must be repaired by NHEJ or MMEJ (GT4), resulting in variable lengths of *TL2* sequence integrated into the *DL*. Such HR + NHEJ gene

conversion events have been previously reported in plants (Puchta, 1999).

This surprising difference in the nature of the GT events identified in the calli from the two parent lines led us to resequence the *TL* and *DL* of these lines. This analysis identified a mutation which eliminates the right side I-*Sce*I cut site of the *DL* in the F1 *TL1/DL* plant (between *nptII* and I-*Sce*I::GR – see Figure 7a). In the *TL1/DL* calli therefore, and in contrast to the *TL2/DL* calli, I-*Sce*I::GR can only cleave the *DL* once (to the left of the *bar* marker). Although the numbers of GT events analysed are low, this difference very probably explains the different types of GT events identified in calli from the two lines. In the *TL1/DL* calli, recombination initiated by I-*Sce*I cleavage of the *DL* would not generate a functional *nptII* gene and so only events initiated by cleavage in the *TL* would be selected. In the *TL2/DL* calli however, recombination initiation through cleavage adjacent to the *nptII* sequences in either the *TL* or the *DL* would result in the reconstruction of *nptII* (Figure 7b). In the *TL2/DL* calli, identification of recombination events in which only the *DL* was recipient clearly shows that single, incomplete I-*Sce*I cleavage of the *DL* is frequent in these cells.

These data thus show that only cleavage of the *TL* is needed for successful GT in these plants, as well as providing a clear illustration of the risk of including multiple I-*Sce*I restriction sites in plants in which I-*Sce*I expression or activity is limiting. The basal level of I-*Sce*I cleavage in the absence of dexamethasone induction further compounds this risk, through increasing levels of mutation in the I-*Sce*I sites of the *DL*. The dependence of this problem on limited I-*Sce*I activity would thus explain the difference with the recent study in *Arabidopsis* using a comparable strategy, in which only clean GT events were found (Fauser *et al.*, 2012). They observed efficient repair DNA excision, probably due to efficient activity of the I-*Sce*I, and GT was observed in up to 1% of the progeny. Limiting endonuclease activity is, however, a common problem in experiments of this type (Puchta *et al.*, 1996). In *Drosophila*, Gong *et al.*, (Gong and Golic, 2003) also reported low I-*Sce*I-mediated repair fragment excision and estimated that excision occurred in 7% of cells. They

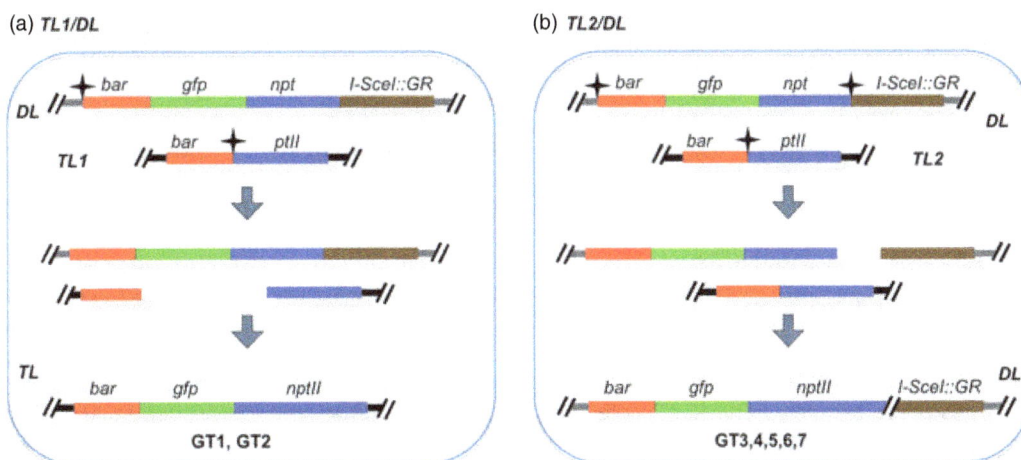

Figure 7 Model. (a) For the target locus (*TL*)1/donor locus (*DL*)-derived embryos, I-*Sce*I cleavage in the *TL* initiated recombination with the *DL*, using homology in both *bar* and *nptII* regions. This resulted in gene targeting (GT) at the *TL*, using the *DL* as template. (b) For the *TL2/DL*-derived embryos, I-*Sce*I cleavage in the *DL* (right I-*Sce*I site in the *DL* between the *nptII* and I-*Sce*I::GR cassette) initiated recombination with the *TL*, using homology in only the *nptII* region. The other side of the break was repaired by non-homologous recombination. This resulted in GT at the *DL* using the *TL* as template. The I-*Sce*I sites are shown as black stars. The rightmost I-*Sce*I in the *DL* is mutated in the *TL1/DL* line, as indicated by the absence of the corresponding star.

thus used the FLP recombinase to excise a circular repair DNA from the genome that was subsequently linearized by I-SceI. This system gave good GT rates in *Drosophila* but has not thus far been shown to work in maize (Yang *et al.*, 2009).

The goal of this work was to test in maize a GT system based on I-SceI-mediated cleavage of the target, and excision of the *nptII* repair region from the genome. In previously described GT systems in maize, GT can occur only within a limited period after transformation of donor sequences and site-specific nucleases (D'Halluin *et al.*, 2008; Shukla *et al.*, 2009). In contrast, in our GT system, transformation of components required for GT is uncoupled from recombination and GT. This allows multiplication of the cells carrying the *TL*, the *DL* and I-SceI-encoding sequences and permits the selection of rare ectopic recombination events. With this approach, a few stable transformation events can be used to generate a large population of cells from which to select GT events, of particular interest in cases where plant transformation frequencies are a limiting factor. In the work reported here, GT efficiencies range from 0.13% to 0.55% GT events per immature embryo. The actual number of cells screened is of course much higher than the number of calli, but the calculation with respect to calli expresses best the employed human effort. Our results furthermore show that cleavage of the *DL* is both unnecessary for targeted recombination, and a source of unwanted events when endonuclease cleavage is limiting.

Experimental procedures

Production of GT constructs and maize transgenic lines

The binary vectors for the creation of the target locus, pBIOS-*TL*, and donor locus, pBIOS-*DL*, were constructed in the following manner. First in order to extend the region of homology between the truncated *nptII* genes in the *TL* and *DL* lines, an 886-bp rice tubulin intron (GenBank, *AJ488063*) was introduced into the coding sequence of the *nptII* gene at position 204 bp downstream of the ATG. A 5′ truncated *nptII-intTubI* fragment lacking the first 150 bp of the *nptII* coding region was cloned between an I-SceI site and in front of the *Arabidopsis Sac66* polyadenylation sequence (GenBank, *AJ002532*). The I-SceI-3′nptII-intTubI-terSac66 fragment was then cloned into an SB11-based plant binary vector (Komari *et al.*, 1996) containing a rice actin promoter (*pAct*) (McElroy *et al.*, 1991) linked to the *bar* selectable marker gene (White *et al.*, 1990) and a nopaline synthase terminator (*terNos*), forming pBIOS-*TL*. To produce pBIOS-*DL*, a 3′ truncated *nptII-intTubI* fragment lacking the last 227 bp of the *nptII* coding region was cloned behind the constitutive *SC4* promoter (*pSC4*) (Schünmann *et al.*, 2003). A pSB11-based binary vector was created that contained the *pAct-bar-terNos* gene cassette and a cassava vein mosaic virus (*CsVMV*) promoter (Verdaguer *et al.*, 1998) linked to *gfp*, with both gene cassettes flanked by I-SceI restriction sites. The *pSC4-5′nptII-intTubI* fragment was then cloned between the terminator of the *gfp* gene and the 3′ I-SceI site to complete the *nptII* repair region. Next, the *NLS::I-SceI::GR* gene (Wehrkamp-Richter *et al.*, 2009), codon optimized for maize expression, was cloned between a *CsVMV* promoter and 35S cauliflower mosaic virus terminator. This cassette was then cloned between the *nptII* repair region 3′ I-SceI site and the *RB* to form pBIOS-*DL*. *Agrobacterium tumefaciens* strain LBA 4404 (pSB1) (Hoekema *et al.*, 1983) was transformed with pBIOS-*TL* and pBIOS-*DL*. For each construction, a clone containing the recombinant plasmid was selected. Embryos of maize inbred A188 were transformed with each construction and transformed

plants were regenerated according to Ishida *et al.* (Ishida *et al.*, 1996) using glufosinate selection.

Plant analysis

Genomic DNA was extracted from the leaves by using the DNeasy 96 plant kit (Qiagen, Valencia, CA). Genomic DNA (10 μg) was digested, separated on 1% agarose gel by electrophoresis, transferred to nylon membrane and hybridized to ^{32}P-marked probes following standard procedures (Sambrook and Russell, 2006). The genomic sequences flanking the transgenes were amplified using an adapter-anchor PCR method according to the method of Balzergue *et al.* (Balzergue *et al.*, 2001), with previously described modifications (Sallaud *et al.*, 2003), using DNA digested with *SspI* or *PvuII*. Plant genotyping was performed by PCR. To amplify the fragments longer than 2.0 kbp, a Takara La Taq kit (Takara Bio, Shiga, Japan) was used. GFP fluorescence of sampled plant leaves was visualized under a fluorescence stereomicroscope (Leica MZ16F) using a GFP2 (Leica, Bannockburn, IL) filter.

Crossing, culture and treatment of transformed plants

Plants were grown in the glasshouse with a 16-h day at 26 °C, 400 μE/m^2/s and an 8-h night at 18 °C. Dexamethasone treatments on seed were performed by immersion of the seed during 2 days in an aqueous solution of 30 μM dexamethasone. Kanamycin treatments were performed by the application of 50 μL of a solution at 200 mg/L kanamycin and 1% (v/v) Tween-20 on the apical region of 2-week-old plants.

Somatic embryogenesis

Embryos isolated from selfed plants containing the *TL* and *DL* were placed onto medium according to Ishida *et al.* (Ishida *et al.*, 1996), lacking the kanamycin selective agent. For the first experiment, LS-AS medium was complemented by 0, 30 or 50 μM dexamethasone, and after 1 week, plantlets were transferred sequentially to LSD1.5, LSZ and 1/2LSF media lacking dexamethasone and containing 50 mg/L of kanamycin. For the second experiment, callus was developed for 3 days on LS-AS, 1 week on LSD1.5 and 3 weeks on LSZ medium. Then, callus was cultivated 1 week on LSZ medium containing 0, 30 or 50 μM of dexamethasone. GFP fluorescence of calli was visualized under the fluorescence stereomicroscope (Leica MZ16F) with a GFP2 filter.

Callus analysis

PCR was performed directly on callus tissues using Terra direct PCR polymerase (Clontech Inc., Palo Alto, CA) in 20 μL with specific *TL* (forward: GTGGCGGACCGCTATCAG and reverse: ACATGTATT-AAGAAGCAATGCATGTAGTAC) and *DL* (forward: TGGCAATC-CCTTTCACAACC and reverse: CCCAGTCATAGCCGAATAGCC) primers. Genomic DNA was extracted from pooled calli, of the same genotype and dexamethasone treatment, with the DNeasy 96 plant kit (Qiagen). Primers were designed according to GS FLX Titanium emPCR LIBL kit (Roche Applied Science, Mannheim, Germany) with a specific TAG for each condition (forward: CCATCTCATCCCTGCGTGTCTCCGACTCAG-X-TCATCCCTACCC-GTTCGTT and reverse: CCTATCCCCTGTGTGCCTTGGCAGTC-TCAG-X-ATCACCCAGATCCACCCA, X represents the specific TAG of 10 bp for each condition). PCR was performed using Platinum Taq DNA Polymerase High Fidelity (Invitrogen, Carlsbad, CA). Emulsion PCR was realized on the obtained PCR products with the emPCR Emulsion kit (Roche). PCR products were sequenced with the emPCR sequencing kit (Roche) with Genome Sequencer FLX (Roche). Sequences from each condition were independently assembled and aligned to the reference sequence containing the non-mutated I-SceI site. Sequence differences of 1 or 2 bp with the reference sequence found in the *TL* genotype, lacking *I-SceI::GR*, were discarded as these are likely to be sequencing errors. Sequence differences of three or more base pairs encompassing the I-SceI site were identified and manually regrouped per condition to identify the number of independent mutations per condition.

Acknowledgements

This work was supported by the EU FP6 project TAGIP; by the EU FP7 project RECBREED; and by an ANRT CIFRE grant to AA. We thank Friedrich Fauser and Holger Puchta for sending us their manuscript prior to publication. The authors are grateful for the support of the Biogemma Cloning, Transformation, Analysis, Greenhouse and Upstream Genomic teams and to Dr. O. Da Ines and Dr. S. Amiard for reading of the manuscript.

References

Aoyama, T. and Chua, N.-H. (1997) A glucocorticoid-mediated transcriptional induction system in transgenic plants. *Plant J.*, **11**, 605–612.

Assaad, F.F. and Signer, E.R. (1992) Somatic and germinal recombination of a direct repeat in *Arabidopsis*. *Genetics*, **132**, 553–566.

Balzergue, S., Dubreucq, B., Chauvin, S., Le-Clainche, I., Le Boulaire, F., de Rose, R., Samson, F., Biaudet, V., Lecharny, A., Cruaud, C., Weissenbach, J., Caboche, M. and Lepiniec, L. (2001) Improved PCR-walking for large-scale isolation of plant T-DNA borders. *Biotechniques*, **30**, 496–498.502,504.

Christian, M., Cermak, T., Doyle, E.L., Schmidt, C., Zhang, F., Hummel, A., Bogdanove, A.J. and Voytas, D.F. (2010) Targeting DNA double-strand breaks with TAL effector nucleases. *Genetics*, **186**, 757–761.

Cotsaftis, O. and Guiderdoni, E. (2005) Enhancing gene targeting efficiency in higher plants: rice is on the move. *Transgenic Res.* **184**, 1–14.

D'Halluin, K., Vanderstraeten, C., Stals, E., Cornelissen, M. and Ruiter, R. (2008) Homologous recombination: a basis for targeted genome optimization in crop species such as maize. *Plant Biotechnol. J.*, **6**, 93–102.

Fauser, F., Roth, N., Pacher, M., Ilg, G., Sànchez-Fernàndez, R., Biesgen, C. and Puchta, H. (2012) In planta gene targeting. *Proc. Natl Acad. Sci. USA*, **109**, 7535–7540.

Gong, W.J. and Golic, K.G. (2003) Ends-out, or replacement, gene targeting in *Drosophila*. *Proc. Natl Acad. Sci. USA*, **100**, 2556–2561.

Grelon, M., Vezon, D., Gendrot, G. and Pelletier, G. (2001) AtSPO11-1 is necessary for efficient meiotic recombination in plants. *EMBO J.*, **20**, 589–600.

Hanin, M. and Paszkowski, J. (2003) Plant genome modification by homologous recombination. *Curr. Opin. Plant Biol.*, **6**, 157–162.

Hanin, M., Volrath, S., Bogucki, A., Briker, M., Ward, E. and Paszkowski, J. (2001) Gene targeting in *Arabidopsis*. *Plant J.*, **28**, 671–677.

Hoekema, A., Hirsch, P.R., Hooykaas, P.J.J. and Schilperoort, R.A. (1983) A binary plant vector strategy based on separation of vir- and T-region of the *Agrobacterium tumefaciens* Ti-plasmid. *Nature*, **303**, 179–180.

Huang, J., Zhou, W., Watson, A.M., Jan, Y.N. and Hong, Y. (2008) Efficient ends-out gene targeting in *Drosophila*. *Genetics*, **180**, 703–707.

Ishida, Y., Saito, H., Ohta, S., Hiei, Y., Komari, T. and Kumashiro, T. (1996) High efficiency transformation of maize (*Zea mays* L.) mediated by *Agrobacterium tumefaciens*. *Nat. Biotechnol.*, **14**, 745–750.

Jacquier, A. and Dujon, B. (1985) An intron-encoded protein is active in a gene conversion process that spreads an intron into a mitochondrial gene. *Cell*, **41**, 383–394.

Khanna, K.K., Lavin, M.F., Jackson, S.P. and Mulhern, T.D. (2001) ATM, a central controller of cellular responses to DNA damage. *Cell Death Differ.*, **8**, 1052–1065.

Komari, T., Hiei, Y., Saito, Y., Murai, N. and Kumashiro, T. (1996) Vectors carrying two separate T-DNAs for co-transformation of higher plants mediated by *Agrobacterium tumefaciens* and segregation of transformants free from selection markers. *Plant J.*, **10**, 165–174.

Krysan, P.J., Young, J.C. and Sussman, M.R. (1999) T-DNA as an insertional mutagen in *Arabidopsis*. *Plant Cell*, **11**, 2283–2290.

Lu, C., Vasil, V. and Vasil, I.K. (1983) Improved efficiency of somatic embryogenesis and plant regeneration in tissue cultures of maize (*Zea mays* L.). *Theor. Appl. Genet.*, **66**, 285–289.

Matzke, A.J.M. and Matzke, M.A. (1998) Position effects and epigenetic silencing of plant transgenes. *Curr. Opin. Plant Biol.*, **1**, 142–148.

McElroy, D., Blowers, A.D., Jenes, B. and Wu, R. (1991) Construction of expression vectors based on the rice actin 1 (Act1) 5' region for use in monocot transformation. *Mol. Gen. Genet.*, **231**, 150–160.

Paszkowski, J., Baur, M., Bogucki, A. and Potrykus, I. (1988) Gene targeting in plants. *EMBO J.*, **7**, 4021–4026.

De.Pater, S., Neuteboom, L.W., Pinas, J.E., Hooykaas, P.J.J. and Van Der Zaal, B.J. (2009) ZFN-induced mutagenesis and gene-targeting in *Arabidopsis* through *Agrobacterium*-mediated floral dip transformation. *Plant Biotechnol. J.*, **7**, 821–835.

Peterhans, A., Schlupmann, H., Basse, C. and Paszkowski, J. (1990) Intrachromosomal recombination in plants. *EMBO J.*, **9**, 3437–3445.

Puchta, H. (1999) Double-strand break-induced recombination between ectopic homologous sequences in somatic plant cells. *Genetics*, **152**, 1173–1181.

Puchta, H., Dujon, B. and Hohn, B. (1993) Homologous recombination in plant cells is enhanced by in vivo induction of double strand breaks into DNA by a site-specific endonuclease. *Nucleic Acids Res.*, **21**, 5034–5040.

Puchta, H., Dujon, B. and Hohn, B. (1996) Two different but related mechanisms are used in plants for the repair of genomic double-strand breaks by homologous recombination. *Proc. Natl Acad. Sci. USA*, **93**, 5055–5060.

Rong, Y.S. (2002) Gene targeting by homologous recombination: a powerful addition to the genetic arsenal for *Drosophila* geneticists. *Biochem. Biophys. Res. Commun.*, **297**, 1–5.

Sallaud, C., Meynard, D., van Boxtel, J., Gay, C., Bes, M., Brizard, J.P., Larmande, P., Ortega, D., Raynal, M., Portefaix, M., Ouwerkerk, P.B., Rueb, S., Delseny, M. and Guiderdoni, E. (2003) Highly efficient production and characterization of T-DNA plants for rice (*Oryza sativa* L.) functional genomics. *Theor. Appl. Genet.*, **106**, 1396–1408.

Sambrook, J. and Russell, D.W. (2006) *The condensed protocols from Molecular cloning: a laboratory manual*. Cold Spring Harbor, NY: Cold Spring Harbor Laboratory Press.

Schünmann, P.H.D., Llewellyn, D.J., Surin, B., Boevink, P., Feyter, R.C.D. and Waterhouse, P.M. (2003) A suite of novel promoters and terminators for plant biotechnology. *Funct. Plant Biol.*, **30**, 443–452.

Shukla, V.K., Doyon, Y., Miller, J.C., DeKelver, R.C., Moehle, E.A., Worden, S.E., Mitchell, J.C., Arnold, N.L., Gopalan, S., Meng, X., Choi, V.M., Rock, J.M., Wu, Y.Y., Katibah, G.E., Zhifang, G., McCaskill, D., Simpson, M.A., Blakeslee, B., Greenwalt, S.A., Butler, H.J., Hinkley, S.J., Zhang, L., Rebar, E.J., Gregory, P.D. and Urnov, F.D. (2009) Precise genome modification in the crop species *Zea mays* using zinc-finger nucleases. *Nature*, **459**, 437–441.

Szostak, J.W., Orr-Weaver, T.L., Rothstein, R.J. and Stahl, F.W. (1983) The double-strand-break repair model for recombination. *Cell*, **33**, 25–35.

Tuteja, N., Ahmad, P., Panda, B.B. and Tuteja, R. (2009) Genotoxic stress in plants: shedding light on DNA damage, repair and DNA repair helicases. *Mutation Res.*, **681**, 134–149.

Tzfira, T. and White, C. (2005) Towards targeted mutagenesis and gene replacement in plants. *Trends Biotechnol.*, **23**, 567–569.

Tzfira, T., Li, J., Lacroix, B.İ.t. and Citovsky, V. (2004) *Agrobacterium* T-DNA integration: molecules and models. *Trends Genet.*, **20**, 375–383.

Tzfira, T., Weinthal, D., Marton, I., Zeevi, V., Zuker, A. and Vainstein, A. (2012) Genome modifications in plant cells by custom-made restriction enzymes. *Plant Biotechnol. J.*, **10**, 373–389.

Verdaguer, B., de Kochko, A., Fux, C.I., Beachy, R.N. and Fauquet, C. (1998) Functional organization of the cassava vein mosaic virus (CsVMV) promoter. *Plant Mol. Biol.*, **37**, 1055–1067.

Vergunst, A.C., Jansen, L.E. and Hooykaas, P.J. (1998) Site-specific integration of *Agrobacterium* T-DNA in *Arabidopsis thaliana* mediated by Cre recombinase. *Nucleic Acids Res.*, **26**, 2729–2734.

Wehrkamp-Richter, S., Degroote, F., Laffaire, J.B., Paul, W., Perez, P. and Picard, G. (2009) Characterisation of a new reporter system allowing high throughput in planta screening for recombination events before and after controlled DNA double strand break induction. *Plant Physiol. Biochem.*, **47**, 248–255.

White, J., Chang, S.Y. and Bibb, M.J. (1990) A cassette containing the bar gene of *Streptomyces hygroscopicus*: a selectable marker for plant transformation. *Nucleic Acids Res.*, **18**, 1062.

Yang, M., Djukanovic, V., Stagg, J., Lenderts, B., Bidney, D., Carl Falco, S. and Alexander Lyznik, L. (2009) Targeted mutagenesis in the progeny of maize transgenic plants. *Plant Mol. Biol.*, **70**, 669–679.

Permissions

List of Contributors

Gaurav Agarwal, Vanika Garg, Himabindu Kudapa, Dadakhalandar Doddamani, Lekha T. Pazhamala, Aamir W. Khan and Mahendar Thudi
International Crops Research Institute for the Semi-Arid Tropics (ICRISAT), Hyderabad, India

Rajeev K. Varshney
International Crops Research Institute for the Semi-Arid Tropics (ICRISAT), Hyderabad, India
School of Plant Biology, Institute of Agriculture, The University of Western Australia, Crawley, WA, Australia

Suk-Ha Lee
Department of Plant Science, Research Institute for Agriculture and Life Sciences, Seoul National University, Seoul, Korea
Plant Genomics and Breeding Institute, Seoul National University, Seoul, Korea

Bilgin Candar-Cakir
Programme of Molecular Biology and Genetics, Institute of Science, Istanbul University, Vezneciler, Istanbul, Turkey
Department of Biology, East Carolina University, Greenville, NC, USA

Baohong Zhang
Department of Biology, East Carolina University, Greenville, NC, USA

Ercan Arican
Department of Molecular Biology and Genetics, Faculty of Science, Istanbul University, Vezneciler, Istanbul, Turkey

Maxime Chantreau, Simon Hawkins and Godfrey Neutelings
Umr Inra 1281 Stress Abiotiquesét Differenciation des Végétaux Cultives, Universite Lille Nord de France Lille 1, Villeneuve d'Ascq, France

Brigitte Chabbert
Inra, Umr 614 Fractionnement Des AgroRessources et Environnement, Reims, France
Umr 614 Fractionnement Des Agroressources Et Environnement, Universite de Reims Champagne-Ardenne, Reims, France

Sylvain Billiard
UMR CNRS 8198 Laboratoire de Genetique and Evolution des Populations Vegetales, Universite Lille Nord de France Lille 1, Villeneuve d'Ascq, France

Si Nian Char, Erica Unger-Wallace, Sarah A. Briggs, Martin H. Spalding, Erik Vollbrecht, Anjanasree K. Neelakandan, Philip W. Becraft and Bing Yang
Department of Genetics, Development and Cell Biology, Iowa State University, Ames, IA, USA

Bronwyn Frame, Marcy Main and Kan Wang
Department of Agronomy, Iowa State University, Ames, IA, USA

Hartinio Nahampun, Bronwyn Frame, Marcy Main and Kan Wang
Department of Agronomy, Iowa State University, Ames, IA, USA

Blake C. Meyers
Donald Danforth Plant Science Center, St. Louis, MO, USA

Virginia Walbot
Department of Biology, Stanford University, Stanford, CA, USA

Rekha Chawla, Roshani Shakya and Caius M. Rommens
Simplot Plant Sciences, J. R. Simplot Company, Boise, ID, USA

Xiaoping Chen, Haifen Li, Yanbin Hong, Fanghe Zhu, Wei Zhu, Haiyan Liu, Erhua Zhang, Ni Zhong, Shijie Wen, Xingyu Li, Guiyuan Zhou, Shaoxiong Li and Xuanqiang Liang
Crops Research Institute, Guangdong Academy of Agricultural Sciences (GAAS), South China Peanut Sub-center of National Center of Oilseed Crops Improvement, Guangdong Key Laboratory for Crops Genetic Improvement, Guangzhou, China

Rajeev Varshney
Crops Research Institute, Guangdong Academy of Agricultural Sciences (Gaas), South China Peanut Sub-center of National Center of Oilseed Crops Improvement, Guangdong Key Laboratory for Crops Genetic Improvement, Guangzhou, China
International Crops Research Institute for the Semi-Arid Tropics (Icrisat), Patancheru, India

Lijuan Pan, Na Chen, Xiaoyuan Chi, Mingna Chen, Zhen Yang, Tong Wang, Mian Wang and Shanlin Yu
Shandong Peanut Research Institute, Shandong Academy of Agricultural Sciences, Qingdao, China

Qingli Yang
Shandong Peanut Research Institute, Shandong Academy of Agricultural Sciences, Qingdao, China
College of Food Science and Engineering of Qingdao Agricultural University, Qingdao, China

Heying Li, Hong Liu and Hong Wu
South China Agricultural University, Guangzhou, China

Yingchun Zhang, Zheyong Xue, Laibao Feng and Xiaoquan Qi
Key Laboratory of Plant Molecular Physiology, Institute of Botany, Chinese Academy of Sciences, Beijing, China

Xu Chi
Graduate University of Chinese Academy of Sciences, Beijing, China

Huaqing Liu and Feng Wang
Fujian Provincial Key Laboratory of Genetic Engineering for Agriculture, Fujian Academy of Agricultural Sciences, Fuzhou, China

A. Mark Cigan, Manjit Singh, Geoffrey Benn, Lanie Feigenbutz, Manish Kumar, Sergei Svitashev and Joshua Young
Trait Technologies, DuPont Pioneer, Johnston, IA, USA

Myeong-Je Cho
Trait Technologies, DuPont Pioneer, Hayward, CA, USA

Cécile Collonnier, Aline Epert, Kostlend Mara, François Maclot, Anouchka Guyon-Debast, Florence Charlot and Fabien Nogué
Inra Centre De Versailles-Grignon, Ijpb (Umr1318), Versailles Cedex, France

Charles White
Génétique, Reproduction Et Développement, Umr Cnrs 6293, Clermont Université, Inserm U1103, Université Blaise Pascal, Clermont Ferrand, France

Didier G. Schaefer
Laboratoire de Biologie Moleculaire et Cellulaire, Institut de Biologie, Universite de Neuch^atel, Neuchâtel, Switzerland

Maryam Nasr Esfahani
Department of Biology, Faculty of Sciences, Lorestan University, Khorramabad, Iran

Lam-Son Phan Tran
Signaling Pathway Research Unit, RIKEN Center for Sustainable Resource Science (Csrs), Suehiro-cho, Tsurumi, Yokohama, Japan

Saad Sulieman
Signaling Pathway Research Unit, Riken Center For Sustainable Resource Science (Csrs), Suehiro-cho, Tsurumi, Yokohama, Japan
Department of Agronomy, Faculty of Agriculture, University of Khartoum, Shambat, Khartoum North, Sudan

Joachim Schulze
Section of Plant Nutrition, Department of Crop Sciences, Georg-August-University of Göttingen, Göttingen, Germany

Kazuko Yamaguchi-Shinozaki
Laboratory of Plant Molecular Physiology, Graduate School of Agricultural and Life Sciences, University of Tokyo, Bunkyo-ku, Tokyo, Japan

Kazuo Shinozaki
Gene Discovery Research Group, RIKEN Center for Sustainable Resource Science (CSRS), Tsurumi, Yokohama, Japan

Yucheng Wang, Liuqiang Wang, Ping Hu, Yuanyuan Jia, Chunrui Zhang, Yu Zhang, Yiming Zhang, Chao Wang and Chuanping Yang
State Key Laboratory of Tree Genetics and Breeding, Northeast Forestry University, Harbin, China

Huiyan Guo
State Key Laboratory of Tree Genetics and Breeding, Northeast Forestry University, Harbin, China
Department of Life Science and Technology, Mudanjiang Normal College, Mudanjiang, China

Yanmin Wang
State Key Laboratory of Tree Genetics and Breeding, Northeast Forestry University, Harbin, China
Key Laboratory of Fast-Growing Tree Cultivating of Heilongjiang Province, Forestry Science Research Institute of Heilongjiang Province, Harbin, China

Adam L. Heuberger, Corey D. Broeckling and Kaylyn R. Kirkpatrick
Proteomics and Metabolomics Facility, Colorado State University, Fort Collins, CO, USA

Jessica E. Prenni
Proteomics and Metabolomics Facility, Colorado State University, Fort Collins, CO, USA
Department of Biochemistry and Molecular Biology, Colorado State University, Fort Collins, CO, USA

Qing Liu, Pushkar Shrestha, James Petrie, Allan G. Green and Surinder P. Singh
Csiro Agriculture and Food, Canberra, Act, Australia

Man Wu
Csiro Agriculture and Food, Canberra, Act, Australia
State Key Laboratory of Cotton Biology, Cotton
Research Institute, Chinese Academy of Agricultural
Sciences, Anyang, China

Baolong Zhang
Csiro Agriculture and Food, Canberra, Act, Australia
Jiangsu Provincial Key Laboratory of Agrobiology,
Jiangsu Academy of Agricultural Sciences, Nanjing,
China

**Bradley C. Campbell, Edward K. Gilding and Ian
D. Godwin**
School of Agriculture and Food Sciences, The University
of Queensland, Brisbane, Qld, Australia

Emma S. Mace
Department of Agriculture and Fisheries (Daf),
Warwick, Qld, Australia

Shuaishuai Tai
BGI-Shenzhen, Shenzhen, China

Yongfu Tao and David R. Jordan
Queensland Alliance for Agriculture and Food
Innovation, The University of Queensland, Warwick,
Qld, Australia

Peter J. Prentis
Science and Engineering Faculty, Queensland
University of Technology (Qut), Brisbane, Qld,
Australia

Pauline Thomelin
Australian Centre for Plant Functional Genomics, Glen
Osmond, SA, Australia

Peter C. Bundock and Robert J. Henry
Co-operative Research Centre for Sugar Industry
Innovation through Biotechnology, Southern Cross
Plant Science, Southern Cross University, Lismore,
NSW, Australia

Rosanne E. Casu
Co-operative Research Centre for Sugar Industry
Innovation through Biotechnology, Csiro Plant
Industry, Queensland Bioscience Precinct, St Lucia,
Qld, Australia

**Jean-Baptiste Laffaire, Samuel Le Goff, Julien Levy,
Sandrine Chaignon, Hajer Salmi, Alexandra Lepicard,
Christophe Sallaud and Wyatt Paul**
Biogemma, Centre de Recherche de Chappes, Chappes,
France

Ayhan Ayar and Sophie Wehrkamp-Richter
Biogemma, Centre de Recherche de Chappes, Chappes,
France
Cnrs Umr 6293,Clermont Universite´ Inserm U1103,
Aubie`Re, France

Maria E. Gallego and Charles I. White
Cnrs Umr 6293,Clermont Universite´ Inserm U1103,
Aubie`Re, France

Index